Springer Geology

For further volumes:
http://www.springer.com/series/10172

Sreepat Jain

Fundamentals of Physical Geology

Springer

Sreepat Jain
New Delhi, Delhi
India

ISBN 978-81-322-1538-7 ISBN 978-81-322-1539-4 (eBook)
DOI 10.1007/978-81-322-1539-4
Springer New Delhi Heidelberg New York Dordrecht London

Library of Congress Control Number: 2013944780

Printed on acid-free paper

Springer is part of Springer Science+Business Media (www.springer.com)

For my wife Archna and my son Parth

Preface

We must always remember that our Garden of Eden was forged in the fires of hell.

This book does not invent the wheel but merely intends to put together sets of updated material with lots of illustrations. This book is specifically designed as a single semester introductory course book for Geology and Earth Science majors and non-majors; it aims to provide a basic understanding of physical geology, its processes, and how they work. The book stems from the idea where a diagram would encapsulate a topic in discussion. Hence, all illustrations are simplified, well-labeled, and hand-made to give a classroom-feel. All effort has been made to make chapters' self-contained.

The book is divided into four parts. Part I: *The Solar System and Cosmic Bodies* deals with elements of our Solar System and the cosmic bodies around it (like planets, meteorites, asteroids, comets, etc.). Part II: *The Earth Materials* deals with earth, its components (Atmosphere, Lithosphere, Hydrosphere, and Biosphere), hypothesis explaining its origin, its internal structure and various absolute and relative dating methods to access its age, building of the geological column, and it's details. Part III: *The Hydrologic System* deals with the hydrological system and processes of the earth including Weathering and Mass Wasting, Streams, Groundwater, Karst, Glaciers, Oceans, and Aeolian processes and landforms, and Part IV: *The Tectonic System* deals with the Tectonic System such as Plate Tectonics, Earthquakes, and Volcanoes. Each chapter ends with a chapter summary highlighting the major points discussed in the chapter. An exhaustive Glossary is also provided which explains terms used in the book. Examples are given for events and processes explained in the text, wherever possible.

The credit for this book goes to my wife Archna, as without her, this book would have never seen the light of the day. It is dedicated to her immense help, understanding and patience in dealing with my odd hour writings, and to my newborn son—Parth.

I also thank Professors Eric H. Christiansen, Phil Stoffer, Ludovic Ferrière, and James D. L. White for giving me permission to reproduce/redraw their figures. Their websites are a treasure trove of information and a must visit for all budding geoscientists. I also thank Rahul Garg for gifting me the classic book—Principles of Physical Geology by Arthur Holmes, and thus laying the foundation for this endeavor. Special thanks to Springer India, to my Commissioning Editor Aninda Bose, and to all concerned for doing a fabulous job!

Above all, this book is a humble attempt in understanding the basics of Physical Geology. Hence, as our understanding becomes refined and better, the text will also improve. Therefore, I encourage students to suggest changes and modifications that they deem fit will make this book better and more readable. Please share your suggestions at jainphysicalgeology@gmail.com.

Sreepat Jain

Contents

1 Introduction 1
- 1.1 What is Geology? 1
- 1.2 The Branches of Geology 1
  - 1.2.1 Physical Geology 2
  - 1.2.2 Geomorphology 5
  - 1.2.3 Structural Geology 6
  - 1.2.4 Sedimentology 6
  - 1.2.5 Mineralogy 6
  - 1.2.6 Economic Geology 7
  - 1.2.7 Geochemistry 7
  - 1.2.8 Isotope Geology 7
  - 1.2.9 Historical Geology 7
  - 1.2.10 Applied Geology 9
- 1.3 Summary 10
- References 10

Part I The Solar System and Cosmic Bodies

2 The Solar System: Sun and Planets 13
- 2.1 Introduction 13
- 2.2 The Solar System 14
- 2.3 Sun 16
- 2.4 The Planets 18
- 2.5 Classification of Planets 18
- 2.6 Planet Description 20
  - 2.6.1 Mercury 20
  - 2.6.2 Venus 21
  - 2.6.3 Mars 24
  - 2.6.4 Saturn 27
  - 2.6.5 Uranus 28
  - 2.6.6 Neptune 29

2.7 Pluto 30
2.8 Summary 32
References 34

3 **The Cosmic Bodies** 37
3.1 Introduction 37
3.2 Asteroids 37
3.2.1 Classification of Asteroids 40
3.3 Meteors and Meteorites 42
3.3.1 Meteorite Types 44
3.3.2 Falls and Finds 47
3.3.3 Meteor Showers 48
3.4 Comet 48
3.4.1 Parts of Comets 50
3.4.2 The Orbit of a Comet 50
3.5 Summary 50
References 53

Part II The Earth Materials

4 **Earth as a Planet** 57
4.1 Introduction 57
4.2 Earth's Components 57
4.2.1 Atmosphere 57
4.2.2 Lithosphere 60
4.2.3 Hydrosphere 60
4.2.4 Biosphere 64
4.3 Moon 64
4.4 The Origin of Earth 65
4.4.1 Uniformitarian or Uniparental Hypothesis 66
4.4.2 Cataclysmic or Bi-Parental Hypothesis 67
4.5 The Age of the Earth 69
4.5.1 Radioactive Dating 70
4.6 Summary 73
References 75

5 **Interior of the Earth** 77
5.1 Introduction 77
5.2 Structure 77
5.2.1 Continental Crust (0–75 km) 77
5.2.2 Oceanic Crust (0–10 km) 78
5.2.3 Upper Mantle (10–400 km) 78
5.2.4 Transition Region (400–650 km) 78

5.2.5 Lower Mantle (650–2890 km) 79
5.2.6 D″ Layer (2700–2890 km) 79
5.2.7 Outer Core (2890–5150 km) 79
5.2.8 Inner Core (5150–6378 km) 79
5.3 Methods to Infer Earth's Interior 81
5.3.1 Xenoliths 81
5.3.2 Ophiolites 81
5.3.3 Volcanism 83
5.3.4 Drilling 83
5.3.5 Meteorites 84
5.3.6 Seismic Waves 84
5.4 Seismology and Earth's Interior 84
5.4.1 Reflection and Refraction of Seismic Waves 86
5.4.2 Velocity of Seismic Waves 87
5.5 Layers of the Earth 87
5.5.1 Physical and Tectonic Layers 88
5.6 Internal Structure 89
5.6.1 Crust 89
5.6.2 Mantle 90
5.6.3 Core 91
5.7 Summary 93
References 94

6 The Geological Timescale 95
6.1 Introduction 95
6.2 Early Attempts to Determine Earth's Age 95
6.2.1 Early Scientific Efforts 95
6.3 The Principle of Uniformitarianism 98
6.4 Relative Dating Methods 98
6.4.1 Principle of Original Horizontality 99
6.4.2 Principle (Law) of Superposition 100
6.4.3 Principle of Lateral Continuity 100
6.4.4 Law of Cross-Cutting Relationships 100
6.4.5 Principle of Inclusion 100
6.4.6 Principle of Faunal Succession 100
6.5 Unconformities 101
6.5.1 Disconformity 101
6.5.2 Angular Unconformity 102
6.5.3 Nonconformity 102
6.6 Correlation of Sedimentary Rocks 102
6.6.1 Correlation by Physical Features 102
6.6.2 Correlation by Fossils 103
6.7 Absolute Dating Methods 103
6.7.1 Radioactivity 103

6.8 Radiometric Dating . . . 108
6.8.1 Long-Lived Radioactive Isotope Pairs . . . 109
6.8.2 Fission-Track Dating . . . 109
6.8.3 Radiocarbon Dating . . . 109
6.9 Tree Ring Dating Methods . . . 111
6.10 Varves . . . 112
6.11 The Geological Column . . . 113
6.12 Summary . . . 113
References . . . 125

Part III The Hydrologic System

7 Weathering and Mass Wasting . . . 129
7.1 Introduction . . . 129
7.2 The Rock Cycle . . . 130
7.3 Types of Weathering . . . 131
7.3.1 Physical Weathering . . . 131
7.3.2 Chemical Weathering . . . 136
7.3.3 Biological Weathering . . . 143
7.4 Weathering Products . . . 144
7.5 Weathering and Associated Rocks . . . 145
7.5.1 Exfoliation . . . 145
7.5.2 Spheroidal Weathering . . . 145
7.5.3 Weathering Rinds . . . 146
7.6 Rates of Weathering . . . 146
7.6.1 Time . . . 146
7.6.2 Slope (Topography) . . . 147
7.6.3 Climate . . . 147
7.6.4 Type of Rock (Mineral Composition) . . . 147
7.6.5 Exposure . . . 148
7.6.6 Particle Size . . . 148
7.6.7 Rock Structure . . . 148
7.6.8 Gravity . . . 149
7.6.9 Animals . . . 149
7.7 The Stability of Minerals . . . 149
7.8 Weathering of Common Rocks . . . 149
7.9 Mass Wasting . . . 152
7.10 The Mass Wasting Process . . . 152
7.10.1 Rockfall . . . 153
7.10.2 Rock Avalanche . . . 154
7.10.3 Rockslide . . . 155
7.10.4 Debris Fall . . . 155
7.10.5 Debris Avalanches . . . 155

7.10.6 Debris Slide 156
7.10.7 Mudflows 156
7.10.8 Earthflow 156
7.10.9 Slumps 157
7.10.10 Debris Flow 157
7.10.11 Solifluction 157
7.10.12 Earth Creep 157
7.10.13 Sediment Flows 158
7.10.14 Grain Flows 159
7.11 Factors Responsible for Mass Wasting 159
7.11.1 Slope 159
7.11.2 Water 159
7.11.3 Geology and Relief 159
7.11.4 Precipitation 160
7.11.5 Slope Modification 160
7.11.6 Undercutting 160
7.11.7 Shocks 160
7.11.8 Volcanic Eruptions 160
7.11.9 Submarine Slope Failures 160
7.11.10 Frost Heaving 161
7.11.11 Gelifluction 161
7.11.12 Rock Glaciers 161
7.12 Summary 161
References 163

8 Streams 165
8.1 Introduction 165
8.2 Development Stages 165
8.2.1 Initial (Youthful Stage; Upper Course) 166
8.2.2 Intermediate (Mature Stage; Middle Course) 168
8.2.3 Terminal (Old Stage; Lower Course) 171
8.2.4 Rejuvenated Rivers 172
8.3 Classification of Streams 172
8.3.1 Constancy of Flow 172
8.3.2 Groundwater Table Conditions 175
8.3.3 Stream Equilibrium Conditions 175
8.4 Genetic Classification of Streams 176
8.4.1 Consequent Streams 176
8.4.2 Subsequent Streams 176
8.4.3 Obsequent Streams 176
8.4.4 Resequent Streams 176
8.4.5 Insequent Streams 177
8.4.6 Superimposed Stream 177
8.4.7 Antecedent Stream 177

8.4.8 Incised River 177
8.5 Stream Channels 178
8.5.1 Geometry 178
8.5.2 Dynamics of Discharge 179
8.6 Channel Morphology 180
8.6.1 Straight Channels 181
8.6.2 Meandering Channels 181
8.6.3 Braided Channels 181
8.6.4 Stream Capture 183
8.7 Drainage Basin and System 184
8.7.1 Drainage Basins 184
8.7.2 Drainage Divide 184
8.8 Basin Morphometry 186
8.8.1 Stream Order 186
8.8.2 Drainage Patterns 187
8.8.3 Drainage Texture and Drainage Density 190
8.9 Geological Work of Streams 191
8.9.1 Stream Erosion 191
8.9.2 Stream Transport 193
8.9.3 Deposition 195
8.10 Fluvial Landforms 196
8.10.1 Erosional Landform 196
8.10.2 Depositional Landform 202
8.11 Factors Affecting Fluvial Systems 206
8.11.1 Structure and Tectonics 206
8.11.2 Lithology 207
8.12 Summary 208
References 210

9 Groundwater 211
9.1 Introduction 211
9.2 Vertical Distribution 211
9.2.1 Zone of Aeration 213
9.2.2 Zone of Saturation 214
9.3 Hydrological Property of Rocks 215
9.3.1 Porosity 215
9.3.2 Permeability 216
9.4 Aquifers and Aquicludes 219
9.4.1 Aquifers 219
9.4.2 Aquicludes 220
9.5 The Water Table 221

9.6 Groundwater Movement ... 221
9.6.1 Darcy's Law ... 222
9.7 Geologic Activity of Groundwater ... 222
9.7.1 Caves, and Caverns ... 222
9.7.2 Sinkholes ... 223
9.7.3 Dissolution ... 224
9.7.4 Chemical Cementation and Replacement ... 224
9.8 Karst Topography ... 224
9.9 Summary ... 225
References ... 226

10 Karst ... 227
10.1 Introduction ... 227
10.2 Factors Affecting Karst Development ... 228
10.2.1 Lithology ... 228
10.2.2 Structure ... 228
10.2.3 Topographic Relief ... 231
10.2.4 Hydrology ... 231
10.2.5 Climate ... 231
10.2.6 Vegetation ... 232
10.3 Karst Landforms ... 232
10.3.1 Sinkholes (Dolines) ... 232
10.3.2 Karren Features ... 234
10.3.3 Clints and Grikes ... 234
10.3.4 Runnels, Pits, and Pans ... 234
10.3.5 Karst Lakes ... 236
10.3.6 Karst Valleys ... 236
10.4 Major Karst Areas of the World ... 236
10.5 Summary ... 236
References ... 238

11 Glaciers ... 241
11.1 Introduction ... 241
11.2 Formation of Glaciers ... 241
11.3 Glacier Movement ... 242
11.3.1 Basal Slip (Sliding) ... 245
11.3.2 Internal Flow (Plastic Flow) ... 245
11.4 Glacier Types ... 247
11.4.1 Glacier Classification ... 247
11.5 Glacial Processes ... 251
11.5.1 Abrasion ... 251
11.5.2 Plucking ... 251
11.5.3 Freezing-On ... 251
11.5.4 Glacial Marine Drift ... 252
11.5.5 Stratified Drift ... 252

11.6 Erosional Landforms . . . 252
11.6.1 Alpine Glaciers . . . 253
11.7 Depositional Landforms . . . 256
11.7.1 Till Deposits . . . 256
11.7.2 Stratified Drift Deposits . . . 259
11.8 Summary . . . 261
References . . . 262

12 Ocean . . . 263
12.1 Introduction . . . 263
12.2 Features of an Ocean . . . 263
12.2.1 Continental Shelf . . . 264
12.2.2 Continental Slope . . . 264
12.2.3 Continental Rise . . . 264
12.2.4 Abyssal Plains . . . 265
12.2.5 Oceanic Trenches . . . 265
12.2.6 Seamounts . . . 265
12.2.7 Guyots . . . 266
12.3 The Ocean Sediments . . . 266
12.3.1 Terrigenous Sediments . . . 266
12.3.2 Biogenous Sediment . . . 266
12.3.3 Hydrogenous Sediment . . . 269
12.3.4 Cosmogenous Sediments . . . 269
12.4 Shore Processes . . . 269
12.4.1 Ocean Waves . . . 269
12.4.2 Refraction . . . 273
12.4.3 Ocean currents . . . 274
12.5 Coastal Erosion and Sediment Transport . . . 274
12.5.1 Erosional Features . . . 275
12.6 Depositional Features Formed by Waves . . . 277
12.6.1 Beach . . . 277
12.6.2 Deltas . . . 278
12.6.3 Spits . . . 279
12.6.4 Bay Barriers . . . 281
12.6.5 Tombolos . . . 281
12.6.6 Barrier Islands . . . 281
12.6.7 Reefs and Atolls . . . 281
12.7 Summary . . . 282
References . . . 283

13 Aeolian Processes . . . 285
13.1 Introduction . . . 285
13.2 Process of Wind Erosion . . . 285
13.2.1 Deflation . . . 285

13.2.2 Abrasion 286
13.2.3 Attrition 286
13.3 Sediment Transport by Wind 287
13.4 Erosional Landforms 289
13.4.1 Desert Pavement 289
13.4.2 Rock Pedestal 289
13.4.3 Zeugen 289
13.4.4 Yardang 290
13.4.5 Mesas and Buttes 290
13.4.6 Inselberg 291
13.4.7 Ventifacts 292
13.5 Depositional Landform 293
13.5.1 Sand Seas or Ergs 293
13.5.2 Sand Sheets 294
13.5.3 Ripples 294
13.5.4 Windblown Dust 294
13.5.5 Loess 294
13.5.6 Dust in Ocean Sediments 295
13.5.7 Volcanic Ash 296
13.5.8 Dunes 296
13.6 Deserts and Desert Types 299
13.6.1 Subtropical Deserts (Trade Wind Deserts) 300
13.6.2 Continental Deserts 300
13.6.3 Rainshadow Deserts 301
13.6.4 Coastal Deserts 301
13.6.5 Polar Deserts 301
13.6.6 Monsoon Desert 302
13.6.7 Paleodeserts 302
13.6.8 Extraterrestrial Deserts 302
13.7 Surface Processes and Desert Landforms 303
13.7.1 Weathering and Mass Wasting Processes 303
13.7.2 Streams and Fluvial Landforms 303
13.7.3 Playa Lake 303
13.7.4 Alluvial Fan and Bajada 304
13.7.5 Pediment 304
13.7.6 Inselberg 304
13.7.7 Desertification 304
13.8 Aridity 305
13.8.1 Causes of Aridity 305
13.9 Summary 308
References 310

Part IV The Tectonic System

14 Plate Tectonics 313
14.1 Introduction 313
14.2 Continental Drift 313
14.2.1 Paleontological Evidence 313
14.2.2 Geologic Features 316
14.2.3 Glacial Evidence 318
14.2.4 Paleoclimatic Evidence 320
14.3 Plate Tectonics 321
14.3.1 Plate Boundaries 321
14.3.2 Divergent Plate Boundary 325
14.3.3 Transform Plate Boundary 327
14.3.4 Accreted Terrane Margins 328
14.4 Continental Crust and Plate Tectonics 330
14.4.1 Cratons 330
14.4.2 Orogens and Folded Mountain Belts 333
14.5 Cause of Plate Tectonics 333
14.6 Summary 335
References 336

15 Earthquakes 337
15.1 Introduction 337
15.2 Locating an Earthquake (and Some Terminologies) 339
15.2.1 Seismology 341
15.2.2 Seismographs 341
15.2.3 Seismogram 342
15.2.4 Focus 342
15.2.5 Focal Depth 342
15.2.6 Epicenter 342
15.2.7 Seismic Waves 343
15.2.8 Shadow Zone 345
15.3 How to Measure an Earthquake 346
15.3.1 Magnitude 346
15.3.2 Intensity 347
15.4 Classification of Earthquakes 349
15.4.1 Depth of Focus 349
15.4.2 The Cause of Their Origin 349
15.4.3 Intensity and Magnitude 357
15.5 History of Earthquakes 357
15.5.1 World 357
15.6 World Distribution of Earthquakes 358

15.7 Earthquakes and Plate Boundaries. 358
15.7.1 Divergent Plate Boundaries 358
15.7.2 Convergent Plate Boundaries 359
15.7.3 Transform Fault Boundaries 360
15.8 Earthquake Hazards 360
15.8.1 Direct Effects 362
15.8.2 Secondary Effects 362
15.8.3 Transient Effects. 364
15.9 Summary 366
References 369

16 Volcanoes. 371
16.1 Introduction 371
16.2 Volcanic Features 372
16.3 Magma 373
16.3.1 Basaltic Magma 375
16.3.2 Andesitic Magma 376
16.3.3 Dacite Magma 377
16.3.4 Rhyolytic Magma 378
16.3.5 Carbonatite Magma. 378
16.4 Lava Flows 380
16.4.1 Continuous Surface 380
16.4.2 Discontinuous Surface 381
16.5 Lava Tubes or Lava Caves 381
16.6 Pressure Ridges 382
16.7 Nuées Ardentes 383
16.8 Pillow Lava 383
16.9 Global Distribution of Volcanoes 384
16.9.1 Convergent Plate Margin 384
16.9.2 Divergent Plate Margins 386
16.9.3 Hot Spots. 387
16.10 Geographic Distribution of Volcanoes 388
16.11 Classification of Volcanoes 388
16.11.1 Continuity of Eruption. 388
16.12 Nature of Eruption 391
16.13 Mode of Eruption 391
16.13.1 Icelandic 396
16.13.2 Hawaiian 396
16.13.3 Strombolian 397
16.13.4 Vulcanian. 399
16.13.5 Vesuvian 399
16.13.6 Krakatoan 400
16.13.7 Peléan 401
16.13.8 Plinian. 401
16.13.9 Other Eruptions 403

16.14 Volcanic Explosivity Index 408
16.15 Volcanic Landforms 409
16.15.1 Type of Volcanoes 409
16.15.2 Calderas. 416
16.16 Submarine Volcanoes 418
16.17 Other Types of Landforms. 419
16.17.1 Volcanic Necks and Plugs (or Lava Neck). 419
16.17.2 Maars (Tuff Cones). 420
16.17.3 Non-Volcanic Craters (Cryptovolcanic) 420
16.17.4 Geysers, Fumaroles, Hot Springs and Solfataras . . . 422
16.17.5 Geysers 422
16.17.6 Hot Springs (Thermal Springs) 422
16.17.7 Fumaroles and Solfataras 422
16.18 Summary 423
References 425

About the Author 427

Glossary 429

Index 477

Chapter 1
Introduction

1.1 What is Geology?

Geology deals with the study of earth, its materials, processes that affect them, the products formed and the earth's history since its birth, 4.54 billion years ago (±0.5 billion years; inferred from the Canyon Diablo iron meteorite that impacted at the Barringer Crater, Arizona, USA (USGS 2007; Dalrymple 2001, 2004). Geology not only includes processes that have shaped the earth's surface but also involves the study of ocean floor, and the interior of the earth (Fig. 1.1). Geologists (who study Geology) investigate the composition of earth materials and various geological processes to locate and exploit its mineral resources. They investigate geological phenomena such as earthquakes and volcanoes and attempt to predict and minimize their damaging effects. They study earth's geological history to determine the former positioning of continents and oceans, the nature of ancient climates, and the evolution of life as revealed by fossil records.

However, in order to have a better understanding of the earth and its working, it is imperative to look at processes and structures that operate today, and interpret them accordingly, as to what must have had happened in the past. This is done by studying Geology. It is thus, the science of earth, dealing with its composition, structure, and geological history; studying the origins, properties, and compositions of both rocks and minerals.

1.2 The Branches of Geology

Geology can be broadly divided into several branches like Physical Geology, Geomorphology, Structural Geology, Sedimentology, Mineralogy, Economic Geology, etc. Table 1.1 details the various branches of geology and related sciences.

S. Jain, *Fundamentals of Physical Geology*, Springer Geology,
DOI: 10.1007/978-81-322-1539-4_1,

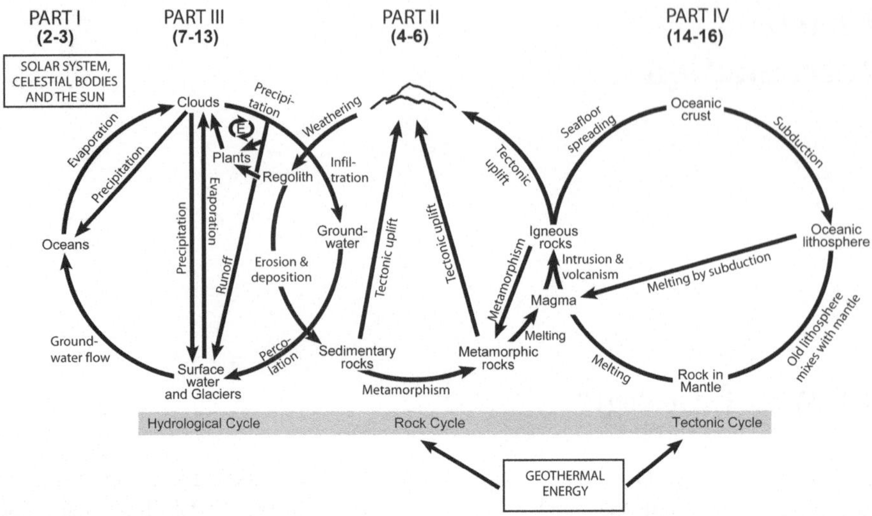

Fig. 1.1 The inter-relationship of major processes discussed in the book. Parts I–IV indicate parts of the book that discusses them. E = Evapotranspiration. Note that Igneous, Metamorphic and Sedimentary processes (Part II) are mentioned but not detailed in this book. Modified from Murck (2001)

1.2.1 Physical Geology

The term Physical Geology was coined by an English mathematician and a geologist, William Hopkins in 1883. The subject deals with the physical forces and processes that bring about changes in the earth's crust or to the surface of the earth on account of their prolonged existence and action (Fig. 1.2). Physical Geology is broadly divided into two branches—those dealing with the internal dynamics of the earth (Endogenous Geology) and those with external dynamics (Exogenous Geology).

1.2.1.1 Endogenous Geology

The movement of earth's crust, earthquakes, and volcanic eruptions, etc. (i.e., the endogenous processes) form the preview of this branch (Figs. 1.1, 1.2). It is further divided into the following sub-branches:

Geotectonics

It pertains to the study of the conditions of the occurrence of rocks, movement of earth's crust, and the deformation caused by them (refer to the Tectonic cycle of Fig. 1.1).

Table 1.1 The diverse fields of geology and their relationship with other Science subjects

Branches			Area of study	Related science
Physical Geology	Endogenous Geology			
		Geotectonics	Earth's crust (condition and deformation)	Physics
		Metamorphism	Transformation of existing rocks	
		Magmatism	Interior of earth	
		Volcanism	Volcanoes	
		Seismology	Earthquakes	Physics
	Exogenous Geology			
		Weathering and Mass Wasting	Alteration of sediments	Chemistry
		Oceanography	Oceans and seas	
		Marine Geology	Ocean margin and floor	Biology
		Hydrogeology	Water resources	
		Glaciology	Glaciers	
		Limnology	Bogs and lakes	
Geomorphology			Landforms	
Structural Geology			Rock deformation	Physics
Sedimentology			Sediment deposition	
Mineralogy			Minerals	
Economic Geology			Minerals and energy resources	Chemistry
	Petroleum Geology		Hydrocarbons	
	Coal Geology		Coal	
Geochemistry			Chemistry of earth	Chemistry
Isotope Geology			Radioactive elements	
Historical Geology			Earth's evolutionary history	
	Geochronology		Time and history of earth	Astronomy
	Stratigraphy		Layered sediments and correlation	
	Paleontology		Fossils	Biology
	Mircopaleontology		Microorganisms (largely single-celled)	
	Paleoecology		Ancient ecology and environment	Biology
	Paleoclimatology		Ancient climate	
	Paleography		Ancient geographic features and locations	
	Paleomagnetism		Earth's magnetic field	Physics
Applied Geology				
	Environmental Geology		Environment	Chemistry, Biology
	Engineering Geology		Geological hazards, bridges, dams, etc.	Engineering
	Hydrogeology		Water resources	
	Geophysics		Earth's interior	Physics

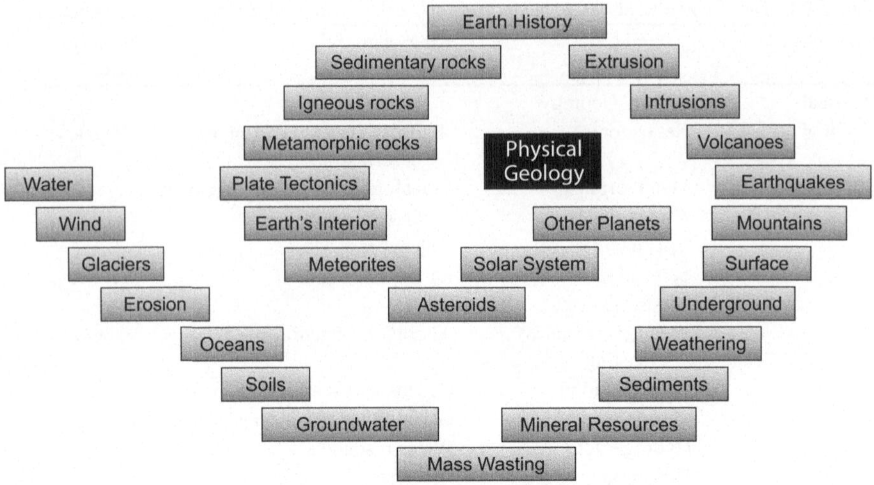

Fig. 1.2 The relationship of Physical Geology to major processes discussed in the book. Igneous, Metamorphic and Sedimentary rocks are mentioned but not detailed in the present book

Metamorphism

It studies changes in rocks in the earth's interior, under conditions of high temperature and pressure.

Magmatism

It deals with the composition of magma and the processes taking place in it.

Volcanism

It is concerned with the study of volcanoes and their activity. This sub-branch overlaps with Magmatism.

Seismology

It is the study of earthquakes and the earth's interior.

1.2.1.2 Exogenous Geology

Those processes that occur on the outer fringes of the earth come into the preview of Exogenous Geology and include the formation and development of seas, rivers, streams, landforms (underground water), erosion, weathering, and all sedimentary and allied processes. Exogenous geology can be further divided into the following sub-branches (for sake of brevity, only major ones are mentioned below).

Weathering and Mass Wasting

This branch studies the process of the alteration of rocks under the action of physical, chemical, and biological agents.

Oceanography

This branch deals with the geological activity of oceans and seas.

Marine Geology

It is the study of the ocean floor and ocean-continent margins.

Hydrogeology

It covers the geological activity of underground water.

Glaciology

The study of glaciers and their phenomena form part of this branch.

Limnology

The geological activity of bogs and lakes are studied.

1.2.2 Geomorphology

This branch deals with the study of present-day landforms, their classification, description, nature, origin, development, and relationships to the underlying structures. Geomorphology also includes the history of geologic changes as recorded by these surface features. However, Geomorphology is sometimes restricted to features produced only by erosion and deposition. Table 1.2 enumerates the hierarchy as used by most Geomorphologists in identifying landforms.

Table 1.2 Hierarchy used by Geomorphologists in identifying landforms (sorted by magnitude)

1st	Continent, ocean basin, climatic zone	~ 10,000,000 km^2
2nd	Baltic shield, mountain range	~ 1,000,000 km^2
3rd	Isolated sea, Sahel	~ 100,000 km^2
4th	Massif Central	~ 10,000 km^2
5th	River valley	~ 1,000 km^2
6th	Individual mountain or volcano, small valleys	~ 100 km^2
7th	Hill slopes, stream channels, estuary	~ 10 km^2
8th	Gully, barchan	~ 1 km^2
9th	Meter-sized features	

1.2.3 Structural Geology

This branch primarily deals with the study of rock deformation in the earth's lithosphere viewed from all scales—from the microscopic (atomic scale) to the macroscopic (continental scale). The study also includes the deformation of rocks and their structural attitudes of arrangements.

1.2.4 Sedimentology

Sedimentology deals with the study of modern sediments and understanding the processes that deposit them. The formation of a sequence of deposit and the processes that cause their formations within the uppermost part of the earth's crust form the backbone of Sedimentology.

1.2.5 Mineralogy

It involves the study of minerals, their formation, analysis, association, and classification. It also includes the study of their chemical composition, specific features of their structures, physical properties, and conditions of occurrence, their interrelationship, and their origin. Mineralogy is subdivided into several branches. The physical, chemical, and optical mineralogy are primarily concerned with the physical, chemical, and optical properties of minerals, while Crystallography deals with the structure, forms, and properties of crystals. Petrology deals with the origin, structure, occurrence, and history of rocks. Both Petrology and Petrography involve the study of different rocks, their formation, association, characteristic features, mineralogical and chemical composition, forms, structures, textures, etc. Petrology is subdivided into three branches on the basis of rock types—Igneous, Sedimentary, and Metamorphic.

1.2.6 Economic Geology

Earth materials that are used for economic and/or industrial purposes such as petroleum, coal, ores, building stones, salt, gemstones, etc., form the preview of this branch. Economic Geology is subdivided into a number of branches, but the most important of these include Petroleum and Coal Geology.

1.2.6.1 Petroleum Geology

This branch deals with the specific search for hydrocarbons (oil exploration).

1.2.6.2 Coal Geology

It is the study of coal.

1.2.7 Geochemistry

It is the study of the sources and fates of chemical species in natural environments. Geochemistry encompasses the investigation of the chemical composition of the earth, other planets, solar system, and the universe as a whole (Cosmochemistry), as well as the chemical processes that occur within them.

1.2.8 Isotope Geology

It is part of Geochemistry but has off late assumed great significance. It deals with the relative and absolute concentrations of the elements and their isotopes in the earth. By studying the isotopic compositions of various chemical elements we can infer information about their age, history, and origin of terrestrial and extraterrestrial materials.

1.2.9 Historical Geology

Historical Geology aims at depicting earth's evolutionary history in a chronological manner. It not only gives clues to the past, but also formulates the understanding about the geological makeup of the present. Hence, this branch deals with the earth's evolution and it's past. Major branches of Historical Geology are briefly mentioned below.

1.2.9.1 Geochronology

It is the study of time relative to the earth's history.

1.2.9.2 Stratigraphy

This branch deals with the succession and interrelation of the strata of the earth's crust. It is thus, concerned with unrevealing the earth's past based on geological evidences provided within the rock beds and involves the interpretation and correlation of the earth's rock strata.

1.2.9.3 Paleontology

It is the study of prehistoric plants and animals as revealed by their fossil records, relative to the earth's chronology.

1.2.9.4 Mircopaleontology

It is the study of fossil organisms that are microscopic in size.

1.2.9.5 Paleoecology

It is the study of the inter-relationship between ancient organisms and their environment.

1.2.9.6 Paleoclimatology

It is the study of climates of the geologic past.

1.2.9.7 Paleogeography

It is the study of physical geography of all or part of earth's surface in the geologic past.

1.2.9.8 Paleomagnetism

It is the study of the nature of earth's magnetic field over geologic time.

1.2.10 Applied Geology

This branch deals with the applied aspects of Geology like Engineering Geology, Geophysics, Mining Geology, Groundwater, etc. Some of the major sub-branches are mentioned below.

1.2.10.1 Environmental Geology

To develop an understanding of the environmental issues related to the science of geology including geologic hazards, waste disposal, environmental pollution of geologic environments, and the political, policy, standards, and research aspects of environmental geology, form the preview of Environmental Geology. This field also includes the study of the effects of various geologic hazards such as earthquakes, landslides, flooding, coastal erosion, mineral and energy resources, and land-use planning. It explores how humans have impacted the earth and earth processes. It is, thus, a multidisciplinary field of applied geology, closely related to engineering geology.

1.2.10.2 Engineering Geology

It deals with the application of geology to engineering practices and solving associated/allied engineering problems. Engineering geologists are concerned with the distribution and relevance of earth materials; potentially dangerous naturally occurring and human-induced geologic hazards; assessment of the risks of damage and injury associated with those hazards, their planning, location, design, construction, and maintenance of transportation systems. Engineering geology is more related to topics in Civil Engineering and studies that deal with the construction of dams, tunnels, mountain roads, rails, etc.

1.2.10.3 Hydrogeology

It is the area of geology that deals with the distribution and movement of groundwater in the soil and rocks of the earth's crust (mostly in Aquifers). Geohydrology is often used interchangeably with Hydrogeology.

1.2.10.4 Geophysics

This branch deals with the data derived from the study of the physics of earth.

There are several other allied and applied branches that are not listed here. These include limnology, pedology, speleology, metallurgy, geodesy, paleomagnetism, remote sensing, astronomy, meteorology, geological mapping, and modeling, etc. It must be noted that this chapter is only meant to give a broad idea that Geology is a multidisciplinary subject, and closely interlinked with different disciplines (Table 1.1). Physical Geology, the topic of this book, forms a major part of Geology, and is equally multidisciplinary in nature (Figs. 1.1, 1.2).

1.3 Summary

Geology deals with the study of earth, its materials, processes that affect its materials, and products formed and its history since its birth, 4.54 billion years ago (± 0.5 billion years). Geology not only includes processes, that have shaped the earth's surface, but also involves the study of ocean floor, and the interior of the earth. Geology also deals with the earth's composition, structure, and geological history; studying the origins, properties, and compositions of both rocks and minerals. Geology is a multidisciplinary subject and is broadly divided into several branches viz. Physical Geology, Geomorphology, Structural Geology, Sedimentology, Mineralogy, Economic Geology, etc. The term Physical Geology was coined by an English mathematician and a geologist, William Hopkins in 1883. Physical Geology deals with the physical forces and processes that bring about changes in the earth's crust or to the surface of the earth on account of their prolonged existence and action. Physical Geology is broadly divided into two branches—those dealing with the internal dynamics of the earth (Endogenous Geology) and those with the external dynamics (Exogenous Geology). These two branches are further divided into several disciplines.

References

Dalrymple GB (2001) The age of the earth in the twentieth century - a problem (mostly) solved. In: Lewis, CLE, and Knell SJ, (eds) The age of the Earth - from 4004 BC to AD 2002: The Geological Society, Vol 190. Special Publication, London, p. 205–221, ISBN 1-86239-093-2

Dalrymple GB (2004) Ancient Earth, ancient skies - the age of Earth and its cosmic surroundings. Stanford University Press, Stanford, California, ISBN 0-8047-4933-7

Murck BW (2001) Geology: a self-teaching guide. Wiley, Newyork, p350 (Book 154)

U.S. Geological Survey Geologic Names Committee (2007) Divisions of geologic time - Major chronostratigraphic and geochronologic units: U.S. Geological Survey Fact Sheet 2007-3015, 2 p.

Part I
The Solar System and Cosmic Bodies

Chapter 2
The Solar System: Sun and Planets

2.1 Introduction

The Universe is all the mass that exists, spanning from the largest galaxies to the tiny subatomic particle. Furthermore, as mass is exchangeable with energy, the Universe also encompasses all forces and energies that exist therein. Hence, the Universe is unlimited in time and shape and infinite in its variety of forms. Billions of cosmic bodies of differing size and structure make up the Universe; the stars, planets, and the interstellar matter are principal forms of such cosmic bodies.

Stars (massive, luminous balls of plasma held together by gravity), are large active bodies either single or united in stellar associations. They range in size from neutron stars (which can be only 20 km (~ 12 miles) wide) to supergiants roughly 1,200–1,300 times the diameter of our Sun (= ~695,500 km, or 432,450 mi). The largest star is the KY Cygni, about 1,420–2,850 times the size of the Sun (Levesque et al. 2005). Hence, our Sun is but a speck when compared to the size of the Universe. Stars are hot and have temperatures ranging from 3,600 K to 50,000 K (Kelvin: [K] = [°C] + 273.15) on their surface. The surface of our Sun, the Photosphere, has a temperature around 5,778 K (5,505 °C) with a density of tens of thousands of times greater than that of water.

Planets are comparatively smaller bodies and usually occur as a satellite of a Star. The Interstellar Medium (ISM) is a very low-density interstellar matter (of gases and dust) that fills the space between cosmic bodies. Around 99 % of the ISM is composed of interstellar gas, and of its mass, ~75 % is in the form of Hydrogen, and the remaining as Helium. ISM is of fundamental importance in understanding both the processes leading to the formation of stars (including our own solar system), and the origin of life within the Universe. The cosmic bodies are grouped into systems that are linked together by gravitational forces. A planet and its associated satellite (like Earth and Moon) are fundamental units followed by their next higher order, solar system, galaxy, archipelago, and the Universe (Table 2.1).

The whole solar system, orbits the center of our home galaxy, the Milky Way. It is a barred (central bar-shaped) spiral disk of about 400 billion stars and ~50

S. Jain, *Fundamentals of Physical Geology*, Springer Geology,
DOI: 10.1007/978-81-322-1539-4_2, © Springer India 2014

Table 2.1 The order of various systems in the Universe

Order	System	Description	Example
First	Planet	Simplest system with associated satellites	Earth and moon
Second	Solar system	Contains a star, planets, and their satellites (including other cosmic bodies)	Our solar system, consisting of a star called Sun, eight planets and their satellite, asteroids, comets, meteorites, dust and gases
Third	Galaxy	It has a more complex structure	Our Milky Way, including the solar system. It contains ~150 thousand million stars.
Fourth	Archipelago	Countless galaxies from an Archipelago	Archipelago of stellar island
Fifth	Universe	Millions of Archipelagos from a system of the Universe	Universe

billion planets. Our Sun orbits around the center of the galaxy in a galactic year, once every 225–250 million earth years. The Milky Way is part of the Local Group of galaxies and is one of 100 billion galaxies within the Universe, as known to us, so far. The diameter of the Milky Way is about 10^5 light years (a light year is a unit of length, equal to just under 10 trillion km; 10^{16} m). The nearest large galaxy is the Andromeda, also a spiral galaxy, but four times as massive and 6.8×10^5 light years away from us. The Sun's nearest known stellar neighbor is a red dwarf star called Proxima Centauri, which is at a distance of about 4.3 light years. Hence, our galaxy is just one of the billions of galaxies known within the intergalactic space.

To fully appreciate these cosmic distances, lets assume that our solar system (out to Pluto) to be the size of a US quarter (25 mm or 0.98 inches) in diameter, then, the Milky way would be a disk approximately 2,000 km (1,200 mi) in diameter, roughly one-third the area of the United States!

2.2 The Solar System

The solar system consists of an average sized star called the Sun and the cosmic bodies that are bound to it by its gravity. These bodies include the eight planets (Mercury, Venus, Earth, Mars, Jupiter, Saturn, Uranus, and Neptune), their more than 130 satellites, large number of smaller bodies (such as comets and asteroids), and the Interstellar Matter (ISM) (Miner and Wessen 2002). Sun contains 99.85 % of the solar system's known mass. The other four orbiting bodies, the gas giants (Jupiter, Saturn, Uranus, and Neptune), account for 99 % of the remaining mass, with Jupiter and Saturn together making up more than 90 %. Compositionally, the planets make up 0.135 % of all the matter within the solar system, followed by satellites of planets, comets, asteroids, meteoroids, and the interstellar medium,

Table 2.2 The mass distribution of the members of our solar system

Members of solar system	Percentage of mass (approx.)
Sun	99.85
Planets	0.135
Comets	0.01
Satellites	0.00005
Minor planets (Asteroids)	0.0000002
Meteoroids	0.0000001
Interstellar medium	0.0000001

with 0.015 % share. (Table 2.2). The diameter of the solar system is about 12,000 million km.

If the size of the solar system is reduced by a factor of a billion (10^9), the earth would be about 1.3 cm in diameter (the size of a grape), the moon orbits about a foot away, the Sun is 1.5 m in diameter (about the height of a man) and 150 m (about a city block) from the earth. Jupiter would be 15 cm in diameter (the size of a large grapefruit) and 5 blocks away from the Sun. Saturn (the size of an orange) would be 10 blocks away; Uranus and Neptune (Lemons) 20 and 30 blocks away. A human on this scale would be the size of an atom and the nearest star would be over 40,000 km away. Details of planets and their distances from the Sun are shown in Fig. 2.1 and their orbits (including that of Pluto) in Fig. 2.2.

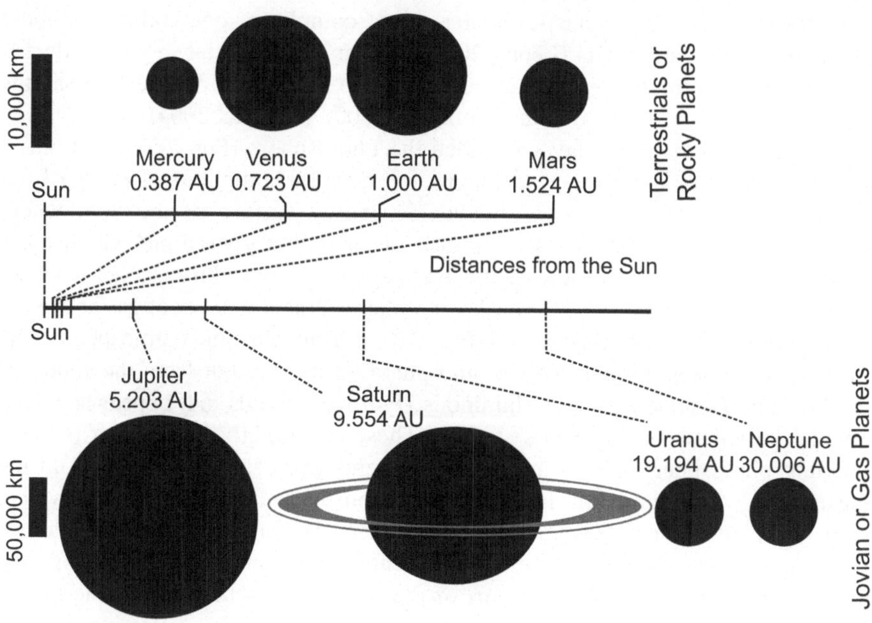

Fig. 2.1 Planets and their distances from the Sun. AU = Astronomical Unit; a distance unit based on the average distance of the Earth from the Sun (1 AU = 149,597,871 km)

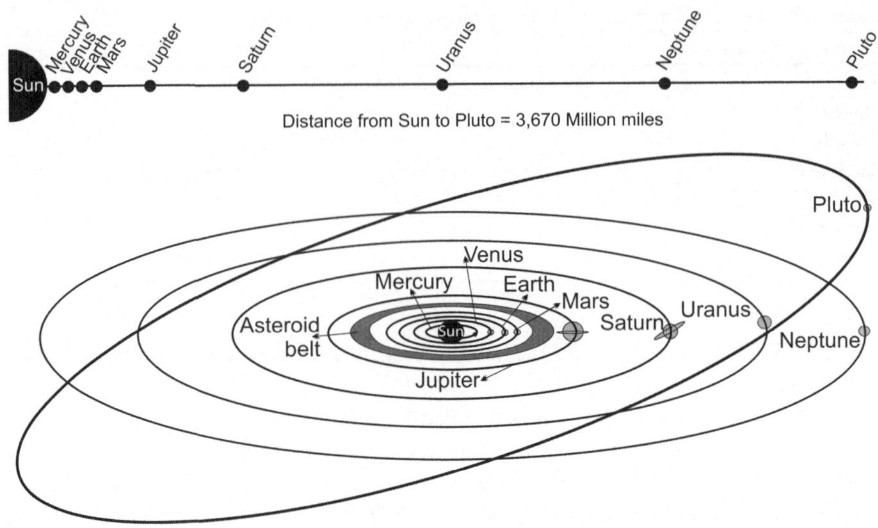

Fig. 2.2 Orbits of planets. Pluto is not considered as a planet (see text for explanation)

2.3 Sun

The Sun orbits the center of the Milky Way at a distance of about 24,000–26,000 light years from the galactic center, and completes one orbit in about 225–250 million earth years (Leong 2002). It is an ordinary star (a yellow dwarf) with an active thermonuclear process. It is also the most prominent feature and the largest object in our solar system (Woolfson 2000; Williams 2004).

The Sun's outer visible layer is called the Photosphere (Fig. 2.3) which has a spotted appearance due to the continuous and turbulent eruptions of energy from its surface. The Photosphere is the visible energy surface of the Sun, where temperatures reach 5,500 °C. This is also the part that gives light which takes 8 min to reach the earth. Above the photosphere is the chromosphere, where Faculae and Flares arise (Fig. 2.3). Chromosphere is the thin layer of gas above the Photosphere. Along with the Corona (Fig. 2.3), it forms the Sun's atmosphere. The solar energy passes through the Chromosphere on its way out from the center of the Sun. The Faculae are bright luminous Hydrogen clouds that appear as bright granular structures on the Sun's surface. These are slightly hotter than the surrounding photosphere. Flares are bright filaments of hot gas emerging from the Sunspot (Fig. 2.3) and form dark depressions on the photosphere, with a typical temperature of about 4,000 °C. The photosphere exhibits a mottled appearance (Granulation; see Fig. 2.3) formed due heat convection occurring below the photosphere. Corona is the outer part of the Sun's atmosphere and is the region where prominences appear. Prominences are immense clouds of glowing gas that erupt from the upper Chromosphere. The outer region of the Corona stretches far into space (millions of kms) and consists of particles traveling slowly away from

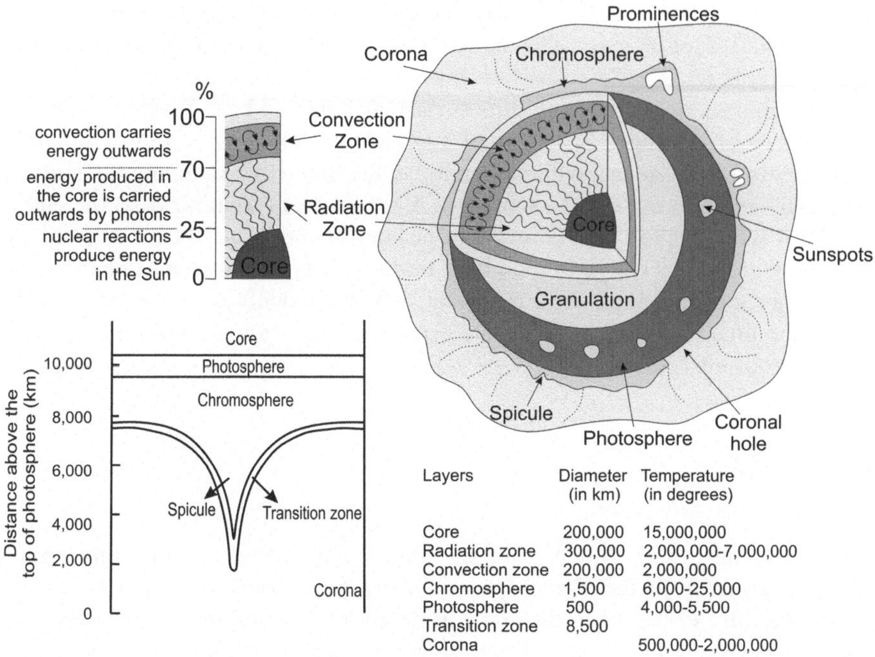

Layers	Diameter (in km)	Temperature (in degrees)
Core	200,000	15,000,000
Radiation zone	300,000	2,000,000-7,000,000
Convection zone	200,000	2,000,000
Chromosphere	1,500	6,000-25,000
Photosphere	500	4,000-5,500
Transition zone	8,500	
Corona		500,000-2,000,000

Fig. 2.3 The structure of the Sun with details of the Spicule (see text for explanation)

the Sun. Along with the Chromosphere, Corona is only visible during total solar eclipse, when the Sun's surface is completely hidden behind the Moon. The core of the Sun is considered to extend from the center to 25 % of the solar radius (Fig. 2.3). Solar energy is created deep within the core of the Sun. It is here that both temperature ($\sim$15,000,000 °C) and pressure (about 340 billion times earth's air pressure at sea level) are so massive that nuclear reactions occur, causing four protons (Hydrogen nuclei) to fuse together to form one alpha particle (or a Helium nucleus) (Woolfson 2000; Williams 2004). The alpha particle is $\sim$0.7 % less massive than the four protons. This difference in mass is expelled as energy and is carried to the surface of the Sun, through a process called Convection.

The core of the Sun contains 50 % of its mass and 2 % of its volume (García et al. 2007). It primarily contains the elements Hydrogen (74 % by mass) and Helium (25 % by mass). Surrounding the core of the Sun is a huge shell known as the Radiation zone (Fig. 2.3). This zone extends 70 % of the way to the photosphere. It constitutes 30 % of the Sun's volume and 48 % of its mass. Through the Radiation zone, energy is transferred from the Core by radiation; one photon takes over a million year to pass through the Radiation zone, primarily due to scattering. The temperature with the Radiation zone ranges from 7 million K (closest to the core) to 2 million K (at the top of the zone). Following the Radiation zone is the Convection zone, the outermost layer of the Sun's interior where energy is transferred through convection; areas of hot plasma (they rise, cool and sink), form a convection cell (Fig. 2.3). The Convection zone makes up 66 % of the Sun's volume and 2 % of its mass. The temperature at the top of the Convection zone is about 5,800 K (5,527 °C).

The Sun's energy is released at its surface as light and heat. Every second, 700 million tons of Hydrogen is converted into Helium, releasing ~5 million tons of energy (Stix 2003). Hence, with passing time, the Sun will become lighter and will eventually cease to exist, after another ~5 billion years from now. At the end of its life, it will start to fuse Helium into heavier elements and begin to swell up; ultimately growing so large (a red giant) that it will eventually consume the earth. Then, after a billion years as a red giant, it will suddenly collapse into a white dwarf, the final end product of a star, like ours. Thereafter, it will take a trillion years to cool off completely.

2.4 The Planets

Mercury, Venus, Earth, Jupiter, Saturn, Uranus, and Neptune are the eight planets arranged away from the Sun; Mercury being the closest. All planets revolve around the Sun in the same direction in nearly circular orbits and in almost the same plane (Fig. 2.2) (see also Hamblin and Christiansen 2008). Details of each planets are given in Table 2.3.

2.5 Classification of Planets

Planets are divided into two broad groups according to their mass, density, and other parameters:

(i) The inner planets (Mercury, Venus, Earth, and Mars), also called the Terrestrial planets and
(ii) The outer planets (Jupiter, Saturn, Uranus, and Neptune), also called the Giant, Jovian or Gas planets (Fig. 2.4).

Planets have been classified based on their chemical composition and/or their point of origin. However, these parameters are untenable as they result either in too many classes (groups) or have too many exceptions. The eight bodies referred to as planets are often further classified based on their composition, size, position from the Sun, position from Earth, and based on their history (Table 2.4; see also USGS Gazetteer of Planetary Nomenclature). Their vital statistics (and also of Pluto) are given in Table 2.5 and their atmospheric composition in Table 2.6.

Table 2.3 Statistical information of planets and the Sun

Planet	Distance (in AU)	Radius	Mass	Rotation	No. of satellites	Orbital inclination	Orbital eccentricity	Obliquity (in degrees)	Density (g/cm^3)
Compared to Sun's		Compared to Earth's							
Sun	0	109	332800	25–36	9	–	–	–	1.410
Mercury	0.387	0.38	0.040	58.80	0	7	0.2056	0.1	5.43
Venus	0.723	0.95	0.810	244.00	0	3.394	0.0068	177.4	5.25
Earth	1.000	1.00	1.000	1.00	1	0.000	0.0167	23.45	5.52
Mars	1.524	0.53	0.110	1.029	2	1.850	0.0934	25.19	3.95
Jupiter	5.203	11.0	316.94	0.411	16	1.308	0.0483	3.12	1.33
Saturn	9.554	9.0	94.900	0.428	18	2.488	0.0560	26.73	0.69
Uranus	19.194	4.0	14.660	0.748	15	0.774	0.0461	97.86	1.29
Neptune	30.006	4.0	17.160	0.802	8	1.774	0.0097	29.56	1.64

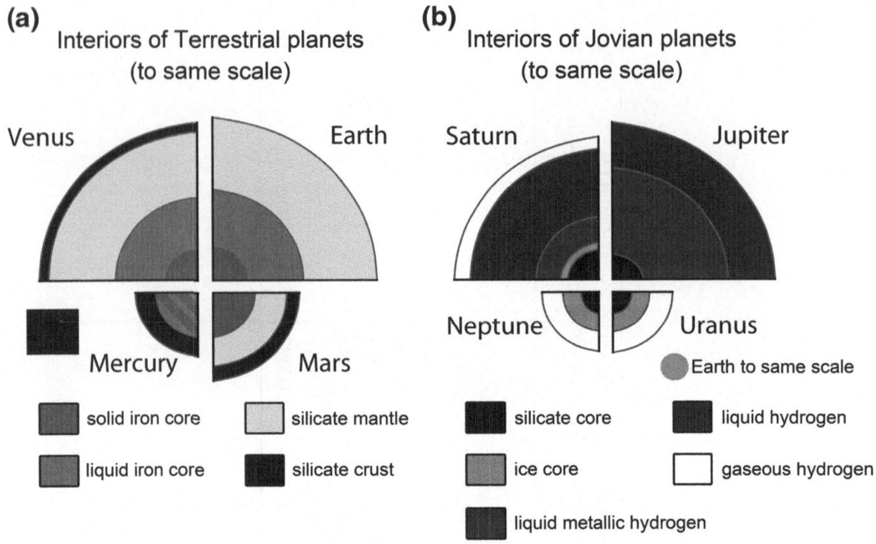

Fig. 2.4 Comparison of the internal structure of Terrestrial and Jovian planets (the earth is taken as reference; in *green*) Modified after Nick Strobel (www.astronomynotes.com)

2.6 Planet Description

2.6.1 Mercury

Mercury is the closest planet to the Sun and the second smallest planet within the solar system (for further details on Mercury see Lyttleton 1969; Davies et al. 1978; Vilas et al. 1988; Mercury Fact Sheet NASA). Its diameter is ~40 % smaller than earth's and ~40 % larger than Moon's. It is even smaller than the Jupiter's moon Ganymede and Saturn's moon Titan. Mercury's sky is always black as it has virtually no atmosphere to cause scattering of light. Intercrater and smooth plains cover most of its surface and its dust-covered hills have been eroded from the constant bombardment of meteorites. Mercury's high density (5.44 gm/cm^3; see also Hamblin and Christiansen 2008) suggests the presence of a large metallic molten core (Fig. 2.4a). Depending on the details of its composition (especially on how much Sulfur it contains), geophysicists calculated that its core makes up about 70 to 80 weight percent of the planet (Fig. 2.4). For comparison, earth's metallic core makes up only about 32 weight percent (Fig. 2.4). Geologists have established a sequence of four major events (same as in the Moon) for the formation of Mercury. These are: (a) accretion, planetary differentiation, and intense meteorite bombardment, (b) formation of multi-ring basins, (c) flooding of basins by the extrusion of basaltic lava, and (d) light meteorite bombardment. Mercury's vital

Table 2.4 Various classifications of planets

Basis of classification	Types	Characteristics	Examples
Composition	Terrestrial or rocky planets	They are composed primarily of rock and metal. They have relatively high densities, slow rotation, solid surfaces, no rings, and few satellites	Mercury, Venus, Earth, and Mars
	Jovian (Jupiter like) or Gas planets	They are composed primarily of Hydrogen and Helium. They have low densities, rapid rotation, deep atmospheres, rings, and lots of satellites	Jupiter, Saturn, Uranus, and Neptune
	Another		Pluto
Size	Small planets	Their diameter is less than 13,000 km. Mercury is referred as lesser planets (term not to be confused with asteroids)	Mercury, Venus, Earth, and Mars
	Giant planets	The giant planets have diameters >48,000 km and are sometimes referred as gas giants	Jupiter, Saturn, Uranus, and Neptune.
Position from the Sun	Inner planets		Mercury, Venus, Earth, and Mars
	Outer planets		Jupiter, Saturn, Uranus and Neptune
	Another planets	The belt between the inner solar system, and the outer solar system	Asteroid
Position from Earth	Inferior planets	Closer to the Sun than Earth. They show phases like the Moon, when viewed from the Earth	Mercury and Venus
	Superior planets	Farther from the Sun than Earth. They always appear full or nearly so	Mars thru Pluto
	Another		Earth
History	Classical planets	Known since prehistoric times and visible to the unaided eye	Mercury, Venus, Mars, Jupiter, and Saturn
	Modern planets	Discovered in modern times and visible only with telescopes	Uranus and Neptune
	Another		Earth

statistics and atmospheric composition are given in Tables 2.5 and 2.6, respectively.

2.6.2 Venus

Venus is the brightest "star" in the sky and is considered as earth's sister planet due to its similarity in size, mass, density and volume. It has a central iron core and a rocky mantle, similar to the composition of the earth (Fig. 2.4). However, Venus

Table 2.5 Vital statistics of all planets and Pluto

Parameters/Planets	Sun	Mercury	Venus	Mars	Jupiter	Saturn	Uranus	Neptune
Mass (kg)	–	3.303×10^{23}	4.87×10^{24}	6.42×10^{23}	1.900×10^{27}	5.69×10^{26}	8.69×10^{25}	1.02×10^{26}
Mass (Earth = 1)	332830	5.5271×10^{-2}	0.81476	1.07×10^{-1}	3.1794×10^{2}	9.52×10^{1}	1.45×10^{1}	1.71×10^{1}
Equatorial radius (km)	695,000	2,439.70	6,051.80	3,397.20	71,492	60,268	25,559	24,746
Equatorial radius (Earth = 1)	108.97	3.8252×10^{-01}	0.94886	5.33E − 01	1.1209×10^{1}	9.45E + 00	4.0074	3.88E + 00
Mean density (gm/cm^3)	1.41	5.42	5.25	3.94	1.33	0.69	1.29	1.64
Mean distance from the Sun (in 1000 km)	–	57,910	108,200	227,940	778,330	1,429,400	2,870,990	4,504,300
Mean distance from the Sun (Earth = 1)	–	0.3871	0.7233	1.5237	5.2028	9.5388	19.1914	30.0611
Rotational period (days)	25–36	58.6462	243.0187	1.025957	0.41354	10.233[a]	17.9[a]	16.11[a]
Orbital period (days)	–	87.969	224.701	686.98	4332.71	29.458[b]	84.01[b]	164.79[b]
Mean orbital velocity (km/sec)	–	47.88	35.02	24.13	13.07	9.67	6.81	5.45
Orbital eccentricity	–	0.2056	0.0068	0.0934	0.0483	0.056	0.0461	0.0097
Tilt of axis (degrees)	–	0	177.36	25.19	3.13	25.33	97.86	28.31
Orbital inclination (degrees)	–	7.004	3.394	1.85	1.308	2.488	0.774	1.774
Equatorial surface gravity (m/sec^2)	–	2.78	8.87	3.72	22.88	9.05	7.77	11
Equatorial escape velocity (km/sec)	618.02	4.25	10.36	5.02	59.56	35.49	21.3	23.5
Visual geometric albedo	–	0.1	0.65	0.15	0.52	0.47	0.51	0.41
Magnitude (Vo)	–26.8	–1.9	–4.4	–2.01	–2.7	0.67	5.52	7.84
Mean surface temperature (in °C)	6,000	179	482	–140	–121	–125	–193	–193 to –153
Maximum surface temperature (in °C)	–	427	420	−63	–	–	–	–
Minimum surface temperature (in °C)	–	–173	–220	20	–	–	–	–
Atmospheric pressure (bars)	–	–	92	0.007	0.7	1.4	1.2	1–3

[a] Data in hours

[b] Data in years. The Luminosity of Sun is 3.827×10^{23} ergs/s

Table 2.6 Atmospheric composition of planets

Atmospheric composition (in percent)	Planets							
	Sun	Mercury	Venus[a]	Mars	Jupiter	Saturn	Uranus	Neptune
Hydrogen	92.1		–	–	90	97	83	85
Helium	7.8	42	–	–	10	3	15	13
Methane	–	–	–	–	–	–	2	2
Nitrogen (N_2)	0.0084	–	>3	2.7	–	–	–	–
Carbon dioxide (CO_2)	–	–	96	95.32	–	–	–	–
Sodium	–	42	–	–	–	–	–	–
Oxygen	0.061	15	–	–	–	–	–	–
Other	–	1	–	–	–	–	–	–
Argon (Ar)	–	–	–	1.6	–	–	–	–
Oxygen (O_2)	–	–	–	0.13	–	–	–	–
Carbon monoxide (CO)	–	–	–	0.07	–	–	–	–
Water (H_2O)	–	–	–	0.03	–	–	–	–
Neon (Ne)	0.0076	–	–	0.00025	–	–	–	–
Krypton (Kr)	–	–	–	0.00003	–	–	–	–
Xenon (Xe)	–	–	–	8 ×				10^{-6}
–	–	–	–					
Ozone (O_3)	–	–	–	3 ×				10^{-6}
–	–	–	–					
Carbon	0.03	–	–	–	–	–	–	–
Iron	0.0037	–	–	–	–	–	–	–
Silicon	0.0031	–	–	–	–	–	–	–
Magnesium	0.0024	–	–	–	–	–	–	–
Sulfur	0.0015	–	–	–	–	–	–	–

[a] Venus has trace amounts of sulfur dioxide (SO_2), water vapor, carbon monoxide, argon, helium, neon, hydrogen chloride, and hydrogen fluoride (all = and <1 %)

is very different as it has no oceans, is surrounded by a heavy atmosphere composed primarily of carbon dioxide (CO_2) with virtually no water vapor and its clouds are composed of droplets of sulfuric acid (Hunten et al. 1983; Barsukov et al. 1992; Robinson 1995). The surface atmospheric pressure is 92 times that of earth's with a surface temperature of ~482 °C (Table 2.5). This high temperature is primarily due to the runaway greenhouse effect caused by CO_2. As sunlight passes through its atmosphere to heat the surface of the planet, it is not radiated back, but is trapped by the dense atmosphere, thereby, making Venus hotter than Mercury. The crust of Venus is dark red, mantle lighter orange-red, and the core yellow. The topographic features of Venus include vast plains covered by lava flows and mountain or highland regions deformed by geological activity; at least 85 % of the surface is covered with volcanic rocks. Giant calderas of more than 100 km in diameter have also been noted on Venus (Hunten et al. 1983; Barsukov et al. 1992). The planet rotates from East to West. Its vital statistics and atmospheric composition are given in Tables 2.5 and 2.6, respectively.

2.6.3 *Mars*

There is evidence on Mars that in the past a denser Martian atmosphere may have allowed water to flow on the planet. Physical features closely resembling shorelines, gorges, riverbeds and islands indicate that rivers once marked the planet (Mutch et al. 1976; Greeley et al. 1978). The rocks, soil and sky have a red or pink shade; hence, it is also called the Red Planet. Straight lines, like irrigation canals, crisscross its surface (Carr 1981; Sheehan 1996; Raeburn 1998). Seasonal color changes are also noted. The surface of Mars contains several craters and naturally occurring channels but, so far, there is no hard evidence of artificial canals or flowing water. Atmosphere of Mars is composed primarily of CO_2 with small amounts of other gases (Table 2.6). Martian air contains only about 1/1,000 as much water as the earth's air, but even this small amount can condense out, forming clouds that ride high in the atmosphere or swirl around the slopes of towering volcanoes. Interestingly, Mars has the same basic internal structure as earth's and other terrestrial (rocky) planets (Fig. 2.4). The largest volcano in the solar system is found in its northern hemisphere. If present on earth, it would stretch from San Francisco to New York (around 4,680 km; 2,908 mi)! The vital statistics and atmospheric composition of Mars are given in Tables 2.5 and 2.6, respectively.

2.6.3.1 Moons of Mars

Mars has two moons, Phobos and Deimos (Blunck 1982). The details of moons of all planets are given in Table 2.7.

2.6.3.2 Jupiter

It is a huge gas planet that can house more than a thousand earths. It probably has a solid, rocky core (Fig. 2.4), about 10 to 15 times the mass of the earth (Beebe 1994). It has more matter than all of the other planets combined. Jupiter's atmosphere is very deep and is composed primarily of Hydrogen and Helium, with small amounts of Methane, Ammonia, Water vapor and other compounds (Table 2.5). At depths, pressure is very high, so much so that the Hydrogen atoms are broken up and electrons are freed. These bare protons, thus, produce a state in which Hydrogen becomes metallic (Yeates et al. 1985). Colorful latitudinal bands, atmospheric clouds, and storms illustrate Jupiter's dynamic weather systems that changes in a matter of hours or days. The Great Red Spot is a complex storm moving in a counter-clockwise direction. At the outer edge, materials appear to rotate within four to six days; although, near the center, motions are small and nearly random in direction. An array of other smaller storms and eddies are noted throughout the banded clouds. As such, Jupiter is a very windy planet where wind

Table 2.7 Detailed statistics (radius, mass, distance from the planet center, discoverer and the date of discovery) of the moons of all planets discussed in the text

Nos.	Moon	Radius (km)	Mass (kg)	Distance (km)	Discoverer	Years
Mars						
1	Phobos	13.5 × 10.8 × 9.4	1.08×10^{16}	9,380	A. Hall	1877
2	Deimos	7.5 × 6.1 × 5.5	1.80×10^{15}	23,460	A. Hall	1877
Jupiter						
1	Metis	20	9.56E + 16	127,969	S. Synnott	1979
2	Adrastea	12.5 × 10 × 7.5	1.91E + 16	128,971	Jewitt-Danielson	1979
3	Thebe	55 × 45	7.77E + 17	221,895	S. Synnott	1979
4	Leda	8	5.68E + 15	11,094,000	C. Kowal	1974
5	Ananke	15	3.82E + 16	21,200,000	S. Nicholson	1951
6	Lysithea	18	7.77E + 16	11,720,000		1938
7	Carme	20	9.56E + 16	22,600,000		1938
8	Sinope	18	7.77E + 16	23,700,000		1914
9	Pasiphae	25	1.91E + 17	23,500,000	P. Melotte	1908
10	Elara	38	7.77E + 17	11,737,000	C. Perrine	1905
11	Himalia	93	9.56E + 18	11,480,000		1904
12	Am althea	135 × 84 × 75	7.17E + 18	181,300	E. Barnard	1892
13	Io	1,815	8.94E + 22	421,600	Marius-Galileo	1610
14	Europa	1,569	4.80E + 22	670,900		1610
15	Ganymede	2,631	1.48E + 23	1,070,000		1610
16	Callisto	2,400	1.08E + 23	1,883,000		1610
Saturn						
1	Metis	20	9.56E + 16	127,969	S. Synnott	1979
2	Adrastea	12.5 × 10 × 7.5	1.91E + 16	128,971	Jewitt-Danielson	1979
3	Thebe	55 × 45	7.77E + 17	221,895	S. Synnott	1979
4	Leda	8	5.68E + 15	11,094,000	C. Kowal	1974
5	Ananke	15	3.82E + 16	21,200,000	S. Nicholson	1951
6	Lysithea	18	7.77E + 16	11,720,000		1938
7	Carme	20	9.56E + 16	22,600,000		1938
8	Sinope	18	7.77E + 16	23,700,000		1914
9	Pasiphae	25	1.91E + 17	23,500,000	P. Melotte	1908
10	Elara	38	7.77E + 17	11,737,000	C. Perrine	1905
11	Himalia	93	9.56E + 18	11,480,000		1904
12	Am althea	135 × 84 × 75	7.17E + 18	181,300	E. Barnard	1892
13	Io	1,815	8.94E + 22	421,600	Marius-Galileo	1610
14	Europa	1,569	4.80E + 22	670,900		1610
15	Ganymede	2,631	1.48E + 23	1,070,000		1610
16	Callisto	2,400	1.08E + 23	1,883,000		1610
Uranus						
1	1986U10	40	–	75,000	Karkoschka	1999
2	Stephano	10	–	7,948,000	Gladman	1999

(continued)

Table 2.7 (continued)

Nos.	Moon	Radius (km)	Mass (kg)	Distance (km)	Discoverer	Years
3	Prospero	15	–	16,568,000	Holman	1999
4	Setebos	15	–	17,681,000	Kavelaars	1999
5	Caliban	49	–	7,169,000	Gladman	1997
6	Sycorax	95	–	12,213,000	Nicholson	1997
7	Cordelia	13	–	49,750	Voyager 2 space craft	1986
8	Ophelia	16	–	53,760		1986
9	Bianca	22	–	59,160		1986
10	Cressida	33	–	61,770		1986
11	Desdemona	29	–	62,660		1986
12	Juliet	42	–	64,360		1986
13	Portia	55	–	66,100		1986
14	Rosalind	27	–	69,930		1986
15	Belinda	34	–	75,260		1986
16	Puck	77	–	86,010		1985
17	Miranda	235.8	6.33E + 19	129,780	G. Kuiper	1948
18	Ariel	578.9	1.27E + 21	191,240	W. Lassell	1851
19	Umbriel	584.7	1.27E + 21	265,970		1851
20	Titania	788.9	3.49E + 21	435,840	W. Herschel	1787
21	Oberon	761.4	3.03E + 21	582,600		1787
Neptune						
1	Naiad	29		48,000	Voyager 2 space craft	1989
2	Thalassa	40		50,000		1989
3	Despina	74		52,500		1989
4	Galatea	79		62,000		1989
5	Larissa	104 × 89		73,600		1989
6	Proteus	200		117,600		1989
7	Nereid	170		5,513,400	G. Kuiper	1949
8	Triton	1,350		354,800	W. Lassell	1846

speed ranges from 192 mph to more than 400 mph (308 kmph to >644 kmph). Auroral emissions, like the earth's northern and southern lights, are observed in the polar regions of Jupiter. Cloud-top lightning bolts, like the superbolts in the earth's high atmosphere, have also been noticed. The vital statistics and atmospheric composition of Jupiter are given in Tables 2.5 and 2.6, respectively.

2.6.3.3 Rings of Jupiter

Jupiter has four rings and the brightest is called the Main ring. Others include Halo and the rest two are called Gossamer rings (Table 2.8). These are much less visible than the other rings and are primarily composed of dust particles. So are the more visible rings but the other two (Halo and Main) are made up of denser collections

Table 2.8 Details of Jupiter's rings

Name	Distance[a] (in Km)	Width (in Km)	Thickness (in Km)	Mass (in Kg[b])
Halo	92,000	30,500	20,000	–
Main	122,500	6,440	<30	1×10^{13}
Inner gossamer	128,940	52,060	–	–
Outer gossamer	181,000	40,000	–	–

[a] The distance is measured from the planet center to the start of the ring
[b] Kilogram

of largely microscopic dust particle that are kicked up when interplanetary meteoroids smash into Jupiter's four small inner moons (Metis, Adrastea, Thebe, and Amalthea; Table 2.7). The details of the rings of Jupiter are given in Table 2.8.

2.6.3.4 Moons of Jupiter

Galileo Galilei, an Italian physicist, mathematician, astronomer, and philosopher, in 1610, discovered four of the 28 moons of Jupiter. Twelve moons are relatively small and may have been captured rather than formed. The four large ones—Lo, Europa, Ganymede and Callisto, are believed to have been accreted as part of the process by which Jupiter itself was formed. The details of sixteen major moons are given in Table 2.7.

2.6.4 Saturn

Saturn is flattened at the poles, due to its very fast rotation on its axis (10.23 earth hours). Its atmosphere is mainly composed of Hydrogen with small amounts of Helium and Methane (Table 2.7); its outer layer is composed of molecular Hydrogen. It is believed that Saturn has a rocky or rocky-ice core with iron and nickel (Hunt and Moore 1982) (Fig. 2.4). It is the only planet that is less dense than water ($\sim$30 % less). Winds, near the equator, mostly blow in an easterly direction, and reach velocities of 1,770 km/h. On earth, the fastest recorded winds generally do not exceed more than 240 km/h.

2.6.4.1 Rings of Saturn

Saturn's ring system is split into a number of different parts, which include the bright A and B rings and a fainter C ring. Saturn's ring system has various gaps. The most notable is the Cassini division, named after Giovanni Cassini (an Italian/French mathematician, astronomer, engineer, and astrologer), who discovered it in

1675. This division separates A and B rings. Another gap is the Encke division, which splits the A ring, and is named after Johann Encke (a German astronomer), who discovered it in 1837. The main rings are made up of a large number of narrow ringlets. It is thought that the rings may have been formed from larger moons that were shattered by impacts of comets and meteoroids. The ring composition is not known, but they do show a significant amount of water. The rings may be composed of icebergs and/or snowballs from a few cm to a few meters in size. If measured, the width of Saturn's rings would reach from Earth to Moon.

2.6.4.2 Moons of Saturn

Saturn has eighteen moons and more than a dozen satellites that have a nearly circular orbit and lie in the equatorial plane. The two exceptions are Iapetus and Phoebe. All Saturn satellites have a density of <2 gm/cm^3 suggesting that they are composed of 30–40 % rock and the rest, water ice. Most of the satellites reflect 60–90 % of the light that strikes them; Phoebe reflects only 2 % (Soderblom and Torrence 1982). Details of sixteen major moons of Saturn are given in Table 2.7.

2.6.5 Uranus

The atmosphere of Uranus is composed of Hydrogen, Helium, Methane and small amounts of Acetylene and other hydrocarbons (Table 2.6). The blue-green color of Uranus is due to the presence of Methane in its upper atmosphere that absorbs red light (Miner 1998; Bergstralh et al. 1991). The atmosphere of Uranus is arranged into clouds running at constant latitudes, similar to the orientation of the more vivid latitudinal bands seen on Jupiter and Saturn. Winds at mid-latitudes blow in the direction of the planet's rotation at velocities of 40–160 m/s. The internal structure of Uranus is similar to that of Neptune (Fig. 2.4) except for the fact that it is less active in terms of atmospheric dynamics and interior heat flow. Uranus has at least twenty moons; Titania and Oberon are the largest ones (Table 2.7).

2.6.5.1 Rings of Uranus

There are eleven rings (Table 2.9) and these are distinctly different from those of Jupiter and Saturn. The outermost epsilon ring is composed mostly of ice boulders several feet across. A very tenuous distribution of fine dust is noted throughout the Uranian ring system.

Table 2.9 Details of the rings of Uranus

S. No.	Name	Distance[a] (km)	Width (km)	Thickness (km)
1	1986U2R	38,000	2,500	0.1
2	6	41,840	1–3	0.1
3	5	42,230	2–3	0.1
4	4	42,580	2–3	0.1
5	Alpha	44,720	7–12	0.1
6	Beta	45,670	7–12	0.1
7	Eta	47,190	0–2	0.1
8	Gamma	47,630	1–4	0.1
9	Delta	48,290	3–9	0.1
10	1986U1R	50,020	1–2	0.1
11	Epsilon	51,140	20–100	< 0.15

[a] The distance is measured from the center of planet to the start of ring

2.6.5.2 Moons of Uranus

Uranus has twenty one moons (Table 2.7). Analysis shows that the five large ones (Miranda, Ariel, Umbriel, Titania, and Oberon) are ice-rock conglomerates similar to the satellites of Saturn. Titania and Oberon are about 1,600 km in diameter, roughly half the size of the Earth's Moon (Bergstralh et al. 1991). Larger Uranian moons contain ~50 % water ice, 20 % carbon- and nitrogen-based materials, and 30 % rock. Their surface, almost uniformly dark gray in color, displays varying degrees of geologic history. Very ancient, heavily cratered surfaces are apparent on some of the moons, while others show strong evidence of internal geologic activity.

2.6.6 Neptune

The blue colored Neptune is the eighth planet of the solar system. It is about 39 times bigger and 17 times more massive than the earth. Neptune is a dynamic planet with several large, dark spots reminiscent of Jupiter's hurricane-like storms. The largest spot, known as the Great Dark Spot, is about the size of the earth and is similar to the Great Red Spot on Jupiter (Hunt and Moore 1994). Two thirds of Neptune is composed of a mixture of molten rock, water, liquid Ammonia and Methane (Fig. 2.4; Table 2.7). One third is a mixture of heated gases consisting of Hydrogen, Helium, water and Methane. Methane gives Neptune its blue cloud color. Long bright clouds, similar to cirrus clouds on earth are noted high in Neptune's atmosphere. The strongest winds on any planet are measured on Neptune. Most of the winds blow westward, opposite to the rotation of the planet. Near the Great Dark Spot, winds blow up to 2,000 km/h. Neptune's atmosphere is dominantly hydrogen (80 %), with helium (19 %) and methane (1 %), and a very small mixture of other compounds. The average temperature is –235 °C.

Table 2.10 Detailed statistics of the rings of Neptune

Name	Distance[a] (km)	Width (km)
1989N3R	41,900	15
1989N2R	53,200	15
1989N4R	53,200	5,800
1989N1R	62,930	<50

[a] Distance is measured from the planet center to the start of the ring

2.6.6.1 Rings of Neptune

Neptune has four rings, which are narrow and very faint and made up of dust particles. Table 2.10 gives a summary of the rings of Neptune.

2.6.6.2 Moons of Neptune

Neptune has eight moons (Table 2.7). Triton is the largest and the only really planet-sized moon and makes up nearly all of the mass of the Neptunian system; all other moons together make up only one-third of 1 %. Triton is also one of the coldest and darkest spots within the solar system as its icy, spherical surface reflects almost all of the sunlight that it receives. Triton is the only moon in the solar system that circles its planet in a direction opposite to the planet's rotation (i.e. a retrograde orbit). This suggests that Triton may once have been an independent object and that Neptune captured it, later.

2.7 Pluto

Pluto, the farthest from the Sun, was once considered as the ninth planet. Its orbit, rotational relationship with its satellite, spin axis, and light variations are unique (White 1980). However, due to its eccentric orbit, it is closer than Neptune for 20 years out of its 249 years orbit. As Pluto approaches the perihelion, it reaches its maximum distance from the ecliptic due to its 17° inclination. Pluto is also the only object to rotate synchronously with the orbit of its satellite. Thus, being tidally locked, Pluto and Charon (its satellite) continuously face each other as they travel through space (Stern and Tholen 1998). Unlike most other planets, but similar to Uranus, Pluto rotates with its poles almost in its orbital plane. Pluto's rotational axis is tipped at 122°. Pluto has a highly reflective south polar cap, a dimmer north polar cap, and both bright and dark features at the equatorial region. Pluto's icy surface is mostly composed of Nitrogen (N_2), Methane (CH_4) and traces of Carbon Monoxide (CO) (Table 2.6). Solid Methane indicates that Pluto is colder than 70 °K (–203 °C). Pluto's temperature varies widely during the course of its orbit as Pluto can be as close to the Sun as 30 AU and as far away as 50 AU.

Table 2.11 Characteristics of Pluto

Parameters/Planets	Pluto
Mass (kg)	1.27×10^{22}
Mass (Earth = 1)	2.125×10^{-03}
Equatorial radius (km)	1,137
Equatorial radius (Earth = 1)	0.1783
Mean density (gm/cm^3)	2.05
Mean distance from the Sun (km)	5,913,520,000
Mean distance from the Sun (Earth = 1)	39.5294
Rotational period (days)	6.3872
Orbital period (days)	248.54
Mean orbital velocity (km/s)	4.74
Orbital eccentricity	0.2482
Tilt of axis (degrees)	122.52
Orbital inclination (degrees)	17.148
Equatorial surface gravity (m/s^2)	0.4
Equatorial escape velocity (km/s)	1.22
Visual geometric albedo	0.3
Magnitude (Vo)	15.12
Mean surface temperature (in °C)	−228
Maximum surface temperature (in °C)	−223
Minimum surface temperature (in °C)	−233

Due to its great distance from the Sun, Pluto's surface is believed to reach temperatures as low as –240 °C. There is a thin atmosphere that freezes and falls to the surface as Pluto moves away from the Sun. Characteristics of Pluto is given in Table 2.11.

So, why is Pluto now not considered a planet? According to the IAU (International Astronomical Union; IAU 2006; Brit 2006), for an object to be a planet, it must meet three requirements: (a) it needs to be in orbit around the Sun, (b) it needs to have enough gravity to pull itself into a spherical shape and (c) it needs to have "cleared the neighborhood" of its orbit. However, according to the third point, Pluto is not a planet. The "cleared its neighborhood" means that as planets form, they become the dominant gravitational body in their orbit within the solar system (IAU 2006; Brit 2006). As they interact with other, smaller objects, they either consume them, or sling them away with their gravity. But, Pluto is only 0.07 times the mass of other objects in its orbit. Even, the earth, in comparison, has 1.7 million times the mass of the other objects in its orbit. Thus, any object that doesn't meet the 3rd criteria is considered a Dwarf planet. And so, Pluto is one such dwarf planet. Until Pluto gains mass (by crashing into objects in its orbit) it will continue to be considered as a dwarf planet.

A summary of physical characteristics of planets and their selected moons are given in Table 2.12.

Table 2.12 Physical characteristics of the planets and selected Moons

Planetary body	Density (g/cm^3)	Diameter (km)	Surface composition	Atmospheric composition	Known Moons
Mercury	5.43	4880	Silicate	–	0
Venus	5.24	12,104	Silicate	CO_2	–
Earth	5.52	12,756	Silicate and water	N_2 and O_2	1
Moon	3.34	3,476	Silicate	–	–
Mars	3.93	6,787	Silicate	CO_2	2
Asteroids					
Eros	2.7	33	Silicate	–	–
Vesta	3.3–3.9	549	Silicate	–	–
Ida	2.2–2.9	56	Silicate and iron	–	1
Ceres	2.0–2.7	1,020	Silicate and carbon	–	–
Jupiter	1.33	143,800	–	H_2 and He	60
Io	3.53	3,640	Silicates and sulfur	SO_2 (thin)	–
Europa	2.99	3,130	Water ice	–	–
Ganymede	1.94	5,280	Water ice	–	–
Callisto	1.85	4,840	Water ice	–	–
Saturn	0.69	120,660	–	H_2 and He	31
Mimas	1.14	392	Water ice	–	–
Enceladus	1.12	500	Water ice	–	–
Tethys	1	1,060	Water ice	–	–
Dione	1.44	1,120	Water ice	–	–
Rhea	1.24	1,530	Water ice	–	–
Titan	1.88	5,150	Water ice	N_2	–
Uranus	1.32	51,120	–	H_2 and He	21
Miranda	1.2	470	Water ice	–	–
Ariel	1.67	1,150	Water ice	–	–
Umbriel	1.4	1,170	Water ice	–	–
Titania	1.71	1,580	Water ice	–	–
Oberon	1.63	1,520	Water ice	–	–
Neptune	1.64	49,560	–	H_2 and He	11
Triton	2.05	2,700	N_2 and CH_4 ice	N_2, CH_4 (thin)	–
Pluto	2.06	2,284	Nitrogen ice	N_2	1
Charon	2.24	1,170	Nitrogen ice	–	–

2.8 Summary

The Universe is unlimited in time and shape and infinite in its variety of forms. Billions of cosmic bodies of differing size and structure make up the Universe; the stars, planets, and the interstellar matter are principal forms. Stars are large active bodies either single or united in stellar associations. Planets are comparatively smaller bodies and usually occur as a satellite of a Star. The Interstellar Medium (ISM) is a very low-density interstellar matter (of gases and dust) that fills the space between cosmic bodies. ISM is of fundamental importance in understanding both the processes leading to the formation of stars (including our own solar system), and the origin of life in the Universe.

The solar system orbits the center of our home galaxy, the Milky Way which is a bar-shaped spiral disk of about 400 billion stars and ~50 billion planets. Our Sun orbits around the center of the galaxy in a galactic year—once every 225–250 million earth years. The Milky Way is part of the Local Group of galaxies and is one of 100 billion galaxies within the Universe, as known to us, so far. Our galaxy is just one of the billions of galaxies known within the intergalactic space.

The solar system consists of an average sized star called the Sun and the cosmic bodies that are bound to it by its gravity. These bodies include the eight planets (Mercury, Venus, Earth, Mars, Jupiter, Saturn, Uranus, and Neptune), their ~130 satellites, smaller bodies like comets and asteroids, and the Interstellar Matter. Sun contains 99.85 % of the solar system's known mass; the gas giants (Jupiter, Saturn, Uranus, and Neptune) account for 99 % of the remaining mass, with Jupiter and Saturn together making up more than 90 %. Compositionally, the planets make up 0.135 % of all the matter within the solar system, followed by satellites of planets, comets, asteroids, meteoroids, and the interstellar medium, with 0.015 % share.

The Sun, an ordinary star with an active thermonuclear process is also the most prominent, brightest and the largest object of our solar system. All planets revolve around the Sun in the same direction in nearly circular orbits and in almost the same plane. Planets, based on their mass, density and other parameters are divided into two groups: (i) The inner planets (Mercury, Venus, Earth and Mars), also called the Terrestrial planets and (ii) The outer planets (Jupiter, Saturn, Uranus and Neptune), also called the Giant, Jovian or Gas planets.

Mercury is the closest planet to the Sun and the second smallest planet within the solar system. Its diameter is ~40 % smaller than Earth's and ~ 40% larger than Moon's. Geologists have established a sequence of four major events (same as in the Moon) for Mercury's formation: (a) accretion, planetary differentiation, and intense meteorite bombardment, (b) formation of multi-ring basins, (c) flooding of basins by the extrusion of basaltic lava and (d) light meteorite bombardment.

Venus is the brightest "star" in the sky and is considered as earth's sister planet due to its similarity in size, mass, density and volume. It has a central iron core and a rocky mantle, similar to the earth.

Mars, in the past possessed a denser atmosphere that might have allowed water to flow on the planet as evidenced by the presence of feathures resembling shorelines, gorges, riverbeds and islands. Straight lines, like irrigation canals, crisscross its surface, but so far, there is no hard evidence of artificial canals or flowing water. The rocks, soil and sky have a red or pink shade; hence, it is also called the Red Planet. The largest volcano in the solar system is found in its northern hemisphere. If present on earth, it would stretch from San Francisco to New York (around 4,680 km)!

Jupiter is a huge gas planet that can house more than a thousand earths and has more matter than all of the other planets combined. Jupiter's atmosphere is very deep and is composed primarily of Hydrogen and Helium, with small amounts of Methane, Ammonia, water vapor and few other compounds. The Great Red Spot is a complex storm moving in a counter-clockwise direction. Jupiter is a very windy planet where wind speed ranges from 300 kph to more than 600 kph. Auroral emissions and cloud-top lightning bolts, as in earth, are common. Jupiter has four rings and twenty eight moons.

Saturn is flattened at the poles, due to its very fast rotation on its axis (10.23 earth hours) and is the only planet that is less dense than water ($\sim$30 % less). Winds, near its equator reach velocities of 500 m/s (1,100 miles/hour). On earth, the fastest recorded winds generally do not exceed more than 150 mi/hr (240 km/h). Saturn's ring system is split into a number of different parts, which include the bright A and B rings and a fainter C ring. The ring system has various gaps; notable among them are the Cassini and Encke divisions. Saturn has eighteen moons and more than a dozen satellites.

Uranus has a similar internal structure as Neptune, except for the fact that it is less active in terms of atmospheric dynamics and interior heat flow. Uranus has at least twenty moons and eleven rings.

Neptune, the blue colored eighth planet of the solar system, is a dynamic planet with several large, dark spots reminiscent of Jupiter's hurricane-like storms. The largest spot, known as the Great Dark Spot, is about the size of the earth and is similar to the Great Red Spot on Jupiter. Methane gives Neptune its blue cloud color. The strongest winds on any planet are measured on Neptune. Neptune has four rings and eight moons.

Pluto, the farthest from the Sun, was once considered as the ninth planet. So, why is it now not considered a planet? According to the IAU, for an object to be a planet, it must meet three requirements: (a) it needs to be in orbit around the Sun, (b) it needs to have enough gravity to pull itself into a spherical shape and (c) it needs to have "cleared the neighborhood" of its orbit. Pluto does not meet the third criteria and any object that doesn't meet this is considered a Dwarf planet. Due to its great distance from the Sun, Pluto's surface is believed to reach temperatures as low as –240 °C. There is a thin atmosphere that freezes and falls to the surface as Pluto moves away from the Sun.

References

Barsukov V, Basilevsky A, Volkov V, Zharkov V (1992) Venus Geology. Research Results from the USSR, University of Arizona Press, Tucson, Geochemistry and Geophysics

Beebe R (1994) Jupiter, the Giant Planet. Smithsonian Institution Press, Washington, D.C.

Bergstralh J, Miner E, Matthews M (1991) Uranus. University of Arizona Press, Tucson

Blunck J (1982) Mars and Its Satellites—A Detailed Commentary on the Nomenclature. Exposition Press, Smithtown, New York

Britt RR (2006). Pluto demoted in highly controversial definition. Space.com. http://www.space.com/scienceastronomy/060824_planet_definition.html. Retrieved 2013-03-24

Carr MH (1981) The surface of Mars. Yale University Press, Newhaven, p 232

Davies M, Dwornik S, Gault D, Strom R (1978) Atlas of Mercury, NASA SP-423. Washington, D.C.

García R et al (2007) Tracking solar gravity modes: the dynamics of the solar core. Science 316(5831):1591–1593

Greeley R, Papson R, Veverka J (1978) Crater streaks in the Chryse Planitia Region of Mars, early Viking results. Icarus 34(3):556–567

Hamblin, W. K. and Christiansen, E. H. 2008. Earth's Dynamic System, Prentice Hall, USA (Web edition 10)

Hunt GE, Moore P (1982) Saturn. Rand McNally, New York

Hunt, G. E., and Moore, P. 1994. Atlas of Neptune. Cambridge University Press, 86p

Hunten, D. Colin, L., Donahue, T. and Moroz, V. 1983. Venus, University of Arizona Press, Tucson

IAU 2006 General Assembly: Resolutions 5 and 6. IAU. 2006-08-24. http://www.iau.org/static/resolutions/Resolution_GA26-5-6.pdf. Retrieved 2013-02-13

Leong S (2002) Period of the Sun's Orbit around the Galaxy (Cosmic Year). The Physics Factbook. http://hypertextbook.com/facts/2002/StacyLeong.shtml. Retrieved 2013-03-24.

Levesque EM, Massey P, Olsen KAG, Plez B, Josselin E, Maeder A, Meynet G (2005) The Effective Temperature Scale of Galactic Red Supergiants: Cool, but Not As Cool As We Thought. Astrophys J 628(2):973–985

Lyttleton RA (1969) On the Internal Structures of Mercury and Venus. Astrophys Space Sci 5(1):18

Mercury Fact Sheet. NASA Goddard Space Flight Center. November 30, 2007. http://nssdc.gsfc.nasa.gov/planetary/factsheet/mercuryfact.html. Retrieved 2013-03-20

Miner ED (1998) Uranus - The Planet, Rings, and Satellites. Wiley, New York

Miner ED, Wessen RR (2002) Neptune: the Planet, Rings, and Satellites, Springer/Praxis 2002, pp 111–114

Mutch T, Arvidson R, Head J, Jones K (1976) and Saunders, R. The Geology of Mars Princeton University Press, Princeton

Raeburn, P. 1998. Uncovering the Secrets of the Red Planet, Mars National Geographic Society, Washington, D.C. Pioneer Venus Special Issue. 1980. Journal of Geophysical Research, Vol. 85

Robinson, C., 1995. Magellan Reveals Venus. Astronomy, 32-41

Sheehan, W. 1996. The Planet Mars, a history of observation and discovery, University of Arizona Press

Soderblom, L. A. and Torrence V. J., 1982. The Moons of Saturn. Scientific American

Stern, S. A., and Tholen, D. J. 1998. Pluto and Charon. University of Arizona Press, Tucson, USA. 728 pages

Stix M (2003) On the time scale of energy transport in the sun. Sol Phys 212(1):3–6

Strobel, N. www.astronomynotes.com. Retrieved 2013-03-24

USGS Gazetteer of Planetary Nomenclature. http://planetarynames.wr.usgs.gov/append7.html. Retrieved 2013-01-17

Vilas F, Chapman C, Matthews M (1988) Mercury. University of Arizona Press, Tucson

White A (1980) The Planet Pluto. Pergamon, New York

Williams, D. R. 2004. Sun Fact Sheet. NASA. http://nssdc.gsfc.nasa.gov/planetary/factsheet/sunfact.html. Retrieved 2013-03-20

Woolfson, M. 2000. The origin and evolution of the solar system. Astronomy and Geophysics, 41(1): 1.12. doi:10.1046/j.1468-4004.2000.00012.x

Yeates et al., 1985. Galileo, Exploration of Jupiter's System NASA SP-479, Washington, D.C

Chapter 3
The Cosmic Bodies

3.1 Introduction

Our Solar system is home to all cosmic bodies—from Asteroids and Comets to Satellites and Planets. Asteroids include all Meteoroids, Meteors, and Meteorites (see also Seymour 1994; Lodders and Fegley 1998). To avoid further confusion of their term usage, a brief nomenclature of cosmic bodies used in this book is given in Table 3.1.

3.2 Asteroids

There are more than 90,000 known Asteroids but many others are far too small to be seen even through the best telescopes (Peebles 2000). Some estimates have put the figure as high as 750,000 for all objects in the size range of 1 km. The Asteroids are found between Earth's and Saturn's orbit and few, beyond. However, most are concentrated within and between the orbits of Mars and Jupiter (the Main asteroid belt; Fig. 3.1). The gravitational force of Jupiter prevents them from accreting into a single larger planet. Asteroids range in size from pebble size to 1,000 km across. Ceres is the biggest with a diameter of 966 km (Table 3.2), containing about 25 % of the mass of all asteroids (Cunningham 2001; Chapman et al. 1975). Ceres is also the first asteroid to be discovered. Giuseppe Piazzi discovered it in 1801 orbiting between Mars and Jupiter. Although, Ceres is now considered as a dwarf planet as (a): it orbits the Sun, (b): has enough mass to form in a sphere, (c): has not cleared the area around the orbit and (d): is not a satellite. There are 26 known asteroids that are larger than 200 km in diameter and around 200 that are larger than 100 km. About 99 % of the asteroids are larger than 100 km and in the 10–100 km range, about half of them have been cataloged. But, very few of the smaller ones have been cataloged and there are probably more than a million in the 1 km range. Table 3.2 lists some of the well-known asteroids.

S. Jain, *Fundamentals of Physical Geology*, Springer Geology,
DOI: 10.1007/978-81-322-1539-4_3, © Springer India 2014

Table 3.1 Terminology used and basic differences among cosmic bodies used in this book

Cosmic bodies			
Asteroids			Comets
Asteroid are small or minor planets orbiting the Sun. They are made of rock and metal and may also contain organic compounds. They are found inside the Earth's orbit and extend to Saturn's orbit and beyond. The largest asteroid is over 966 km in diameter. All Meteoroids, Meteors, and Meteorites are Asteroids			These are small, irregularly shaped bodies that orbit the Sun. They are most remote from the Sun and are composed of a mixture of non-volatile grains and frozen gases. When close to the Sun they display a visible coma (a fuzzy outline generated by solar radiation) and sometimes a tail
Meteoroids	*Meteors*	*Meteorites*	
These are cometary rocks or debris in our Solar System. They are on a collision course with Earth. They range in size from dust to ~ 10 m in diameter. The larger ones are called Asteroids	These are Meteoroids that have burned up as they pass through the Earth's atmosphere (largely due to friction). A streak of light, a shooting star as we call it, is actually seeing a Meteor	These are Meteoroid that have survived falling through the Earth's atmosphere and have collided with the Earth's surface. They do not burn up completely	

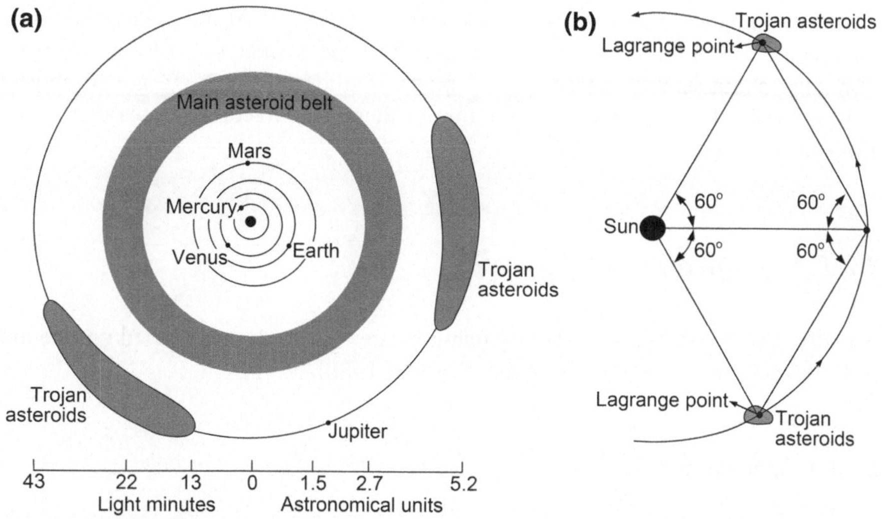

Fig. 3.1 **a** The asteroid belt, **b** position of trojan asteroids and lagrange points

Table 3.2 Brief list of some famous Asteroids

Name	Radius (km)	Distance[a] (10^6 km)	Discoverer	Years
Agamemnon	88	778.1	Reinmuth	1919
Amun	Not known	145.71	Shoemaker	1986
Apollo	0.7	220.061	Reinmuth	1932
Aten	0.5	144.514	Helin	1976
Ceres	483	413.9	G. Piazzi	1801
Chiron	85	2051.9	Kowal	1977
Davida	168	475.4	R. Dugan	1903
Eros	33 × 13 × 13	172.8	Witt	1989
Eros	17.5 × 6.5	218	G. Witt, A. Charlois	1893
Eunomia	136	395.5	De Gasparis	1851
Europa	156	463.3	Goldschmidt	1858
Gaspra	8 × 17 × 10	330	Neujmin	1916
Hephaistos	4.4	323.884	Chernykh	1978
Hygiea	215	470.3	De Gasparis	1849
Icarus	0.7	161.269	Baade	1949
Ida	35	428	J. Palisa	1884
Interamnia	167	458.1	V. Cerulli	1910
Juno	123	399.4	Harding	1804
Mathilde	28.5 × 25	396	J. Palisa	1885
Pallas	261	414.5	H. Olbers	1802
Psyche	132	437.1	De Gasparis	1852
Sylvia	136	521.5	N. Pogson	1866
Vesta	262.5	353.4	H. Olbers	1807

[a] Mean distance from the Sun

If the estimated total mass of all asteroids are put together, the object would be less than ~1,500 km across (i.e., less than half the diameter of Moon). Compositionally, about 92.8 % meteorites are made of silicates and 5.7 % are composed of Iron and Nickel. The rest are a mixture of these. The average temperature of an asteroid is −73 °C.

3.2.1 *Classification of Asteroids*

Broadly, two types of asteroid classifications are followed, those based on Spectra and Albedo and on Position (see also Tholen 1988).

3.2.1.1 Spectra and Albedo

Asteroids are classified according to their Spectra (and hence, their chemical composition) and Albedo (Table 3.3). Albedo is the measure of the reflecting power of a non-luminous object, such as a planet, moon, or an asteroid. Its value ranges from 0 (for a perfectly black surface) to 1 (for a totally reflective surface).

Table 3.3 Classification of asteroids based on their chemical composition and albedo

Type	Percentage of all asteroids	Albedo[a]	Composition/Content
C-type	>75	Extremely dark (0.03)	They are similar to the Carbonaceous chondrites, and have approximately the same chemical composition as the Sun (minus Hydrogen, Helium and other volatiles). They are the most common types and are also the most ancient objects in our Solar system.
S-type	~15	Relatively bright (0.10–0.22)	Metallic nickel–iron and mixed with iron- and magnesium-silicates (Stony)
M-type	~5	Bright (0.10–0.18)	Pure nickel–iron
X-type	Similar to M-type	–	Mostly metallic asteroids

[a] Albedo is the measure of the reflecting power of a non-luminous object

Table 3.4 Classification of asteroids based on their position within our Solar System

Main Belt	Near-Earth asteroids (NEAs)			Trojans[d]	Kirkwood gaps	Centaurs[d]
It is located at a distance of 2–4 AU[a] from the Sun, between Mars and Jupiter	These are close to the earth. They travel within 1.3 AU (195 million km) of the Sun. They number more than 250. They are considered primitive leftover building blocks of the Solar System formation process and hence, offer clues to the chemical mixture from which the planets were formed some ~4.54 billion years ago. Thus, they are the DNA of the Solar System that gives us some clue on the origin of planets and life on earth. They are grouped into three subgroups (Fig. **3.2**):			They are located near Jupiter's Lagrange points[c] (60° ahead and behind Jupiter in its orbit; Fig. 3.1b). They number >1000 within the lagrange points of venus and earth	These are specific regions within the asteroid Main belt where few asteroids are found due to jupiter's gravitational influence	They have planet-crossing orbits and are found in the outer Solar System. They are inherently unstable and are likely to be perturbed in future. The composition of these objects is more like that of comets or the Kuiper Belt objects. Chiron (once thought to be the most distant known asteroid) is now classified as a comet. It has an albedo of 0.15 and a diameter of ~180 km
	Atens	*Apollos*	*Amors*		These are regions where an object's orbital period is a simple fraction of that of jupiter (i.e., an orbital period close to 1/2, 1/3, or 2/5 of jupiter)	
	Its semi-major axis is <1.0 AU	>1.0 AU	between 1 and 1.523 AU			
	They have a perihelion[b] distance of 0.092 AU	<1.017 AU	1.017–1.3 AU.			
	They cross earth's orbit and their orbital period is <1 year	They cross earth's orbit and their orbital period is >1 year	They cross the orbit of Mars but do not cross the orbit of earth			
	Examples					
	2340 Hathor	1620 Geographos	433 Eros			

[a] AU is an astronomical unit, a unit of length equal to the average or mean distance between the Earth and Sun. [b] Perihelion is that point of the orbit that has the nearest approach to Sun.[c] Lagrange points (L-points) are those where an asteroid will remain stationary with respect to the earth. There are 5 such points on or near the Earth's orbit. [d] Both Trojansand Centaurs characteristically have low albedo (0.03–0.013) and red color. The former are formed near the Jupiter's orbit, while the latter are formed far beyond its orbit. However, both are formed at low temperatures at which water exists as solid ice

3.2.1.2 Position of Asteroids

This classification uses the position of asteroids within our Solar System. Based on this, there are five categories—Main Belt, Near-Earth Asteroids (NEAs), Trojans, Kirkwood gaps, and Centaurs (Table 3.4).

3.3 Meteors and Meteorites

A meteor is the visible path of a meteoroid that enters the earth's atmosphere. They are commonly called Shooting stars. Very bright ones are called Fireballs or Bolides. A Meteor describes a streak of light, temporary incandescence, as matter within the Solar System falls into the earth's atmosphere. This incandescence is caused by atmospheric friction and typically occurs at heights from 80 to 110 km above the earth's surface (Fig. 3.2).

The mineral composition of a meteorite is variable as similar elements often substitute for each other within the crystal structure (for example Mg, Fe, Ca in silicates or Mg, Fe, Cr, Al in oxides) (see also Meteoritical Society Guidelines for Meteorite Nomenclature 2013). These variations are important for meteorite classification. Table 3.5 lists minerals by group such as silicates, metal, sulfides, oxides, phosphates, and carbon compounds.

Most meteorites are quite small. The largest is the metallic Hoba meteorite (also called Hoba West) found in Northern Namibia (SW Africa) discovered in 1920 (Table 3.6). It measures 2.7 × 2.7 × 1 m and weighs about 60 tons (1 ton = 1000 kilograms). Its chemical composition is 82.4 % iron, 16.4 % nickel,

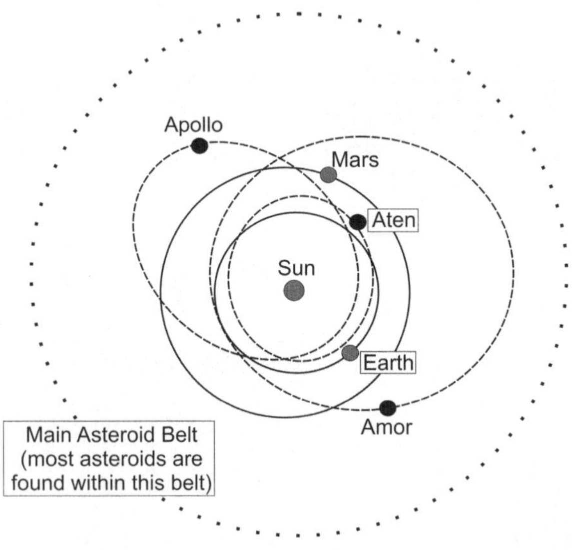

Fig. 3.2 The position of the NEAs, Aten, Apollo, and Amors

Table 3.5 List of major minerals in meteorites, their composition and occurrence

Mineral constituent	Composition	Occurrence in meteorite
Silicates olivine pyroxene feldspar clay minerals	$(Mg, Fe)Si_2O_4$ $(Mg, Fe, Ca)SiO_3$ $CaAl_2Si_2O_8$-$NaAlSi_3O_8$ $(Mg, Fe, Ca)_{3–6}$ Si_4O_{10} $(OH)_2$ H_2O	Stony and stony-iron chondrites, Stony-iron Chondrites, Achondrites, Stony-irons most abundant in achondrites mostly in carbonaceous chondrites
Metal Kamacite, Taenite	Fe (low and high Ni)	Abundant in irons, stony irons common in most chondrites
Sulfides troilite, Pyrrotite	FeS, Fe_7S_8	Abundant in irons, stony irons minor in stony meteorites
Oxides		Minor in most meteorites
Spinel, magnetite, Chromite	$(Mg, Fe, Cr, Al)_3O_4$	Composition depends on type
Phosphates apatite, Whitlockite	$Ca_5(F, Cl, OH)(PO_4)_3$, Ca_2PO_4	Minor in stony meteorites
Carbon compounds diamond, Graphite organic molecules, Amino acids	C (elemental carbon) C, H, O, N compounds	Carbonaceous chondrites

Many minerals occur in small abundances and hence are not listed here

Table 3.6 Ten largest known iron meteorites on earth

Meteorite	Locality/year	Weight in tons	Location
Hoba	Hoba West, Namibia/1920	60	in situ
Campo de Cielo/El Chaco	Chaco, Argentina/1969	37	in situ
Ahnigito/Cape York	West Greenland/1894	31	Museum of Natural History, New York
Armanty	Xinjiang, China/1898	28	Urumqui Museum, Xinjian, China
Bacubirito	Sinaloa, Mexico/1806	22	Centro de Ciencias, Sinaloa, Mexico
Agpalilik	East Greenland/1963	20	Copenhagen, Denmark
Mbosi	Rungwe, Tanzania/1930	16	in situ
Willamete	Clackamas, Oregon/1902	14	Museum of Natural History, New York
Chupaderas I	Chihuahua, Mexico/1852	14	Palacio de Minerais, Mexico City
Mundrabilla	West Australia/1966	11.5	Museum of West Australia, Perth

0.8 % cobalt and traces of other metals. Antarctica is an important site for locating meteorites as chunks of rock are easily spotted on ice (Harvey 2003). Interestingly, several Antarctic meteorites compositionally come either from Moon or Mars (see also The NHM Catalog of Meteorites 2013).

3.3.1 Meteorite Types

Meteorites are considered pristine samples of the early Solar System. But, in some cases their properties are altered by thermal metamorphism or icy alteration during their long space journey. Compositionally, they are close to the terrestrial planets (like Mercury, Venus, Mars, etc.). Their radiometric dating places them to be 4.55 billion years old, also the approximate age of our solar system. Three broad groups of meteorites are recognized: Iron, Stony iron and Stony; the latter type is the most common (see also Norton 2002; Krot et al. 2003; Binze et al. 2006) (Table 3.7). Each of these meteorite classes can be further subdivided into smaller groups with distinct properties (Table 3.7) that enable us to indicate the probable location from where they were formed. The Enstatite chondrite (a sub-type; see Table 3.7) contains the most refractory elements and hence, are believed to have been formed in the inner reaches of our Solar System. Ordinary chondrites (the most common sub-type), contain both volatile and oxidized elements, and are thus, thought to have been formed within the inner asteroid belt. Carbonaceous chondrites, which have the highest proportion of volatile elements and are the most oxidized, probably originated at even greater solar distances.

Other meteorite types include Achondrites, Pallasites and Irons. Achondrites are Stony meteorites (Table 3.7) possess distinct texture and mineralogy indicative of the igneous process noted on the earth. Pallasites are Stony−iron meteorites. They are very rare and are believed to have been formed on differentiated bodies in the transition area of the asteroid, between the metal-rich core and the olivine-rich mantle, where olivine cooled slowly enough to form relatively large crystals. Iron meteorites are classified into 13 major groups and consist primarily of iron-nickel alloys with minor amounts of carbon, sulfur, and phosphorus. Iron meteorites formed when molten metal segregated from the less dense silicate material; hence, they contain evidence of changes that occurred on the parent body from which they were removed, probably by an impact.

Meteoroids that are larger than a few hundred tons make craters (Table 3.8). A good example is the Barringer Crater (also called the Meteor Crater) near Winslow, Arizona (USA). It was formed about 49,000 years ago by a nickel/iron meteor of about 30–50 m in diameter, traveling at a speed of about 11 km per sec. This meteor made a crater of 1,200 m in diameter and 200 m in depth. Another significant and more recent impact occurred in 1908 in the remote uninhabited region of western Siberia known as Tunguska. The impact was about 60 m in diameter and probably consisted of many loosely bound pieces. In contrast to the Barringer Crater event, the Tunguska object completely disintegrated before hitting the ground and so no crater was formed. Nevertheless, all the trees were flattened in an area 50 km across. About 182 impact craters have since been identified on our earth (Fig. 3.3; see also Earth Impact Database 2010). Table 3.9 lists the top 20 largest terrestrial impact craters so far identified.

Table 3.7 Classification of meteorites

Types	Iron	Stony Iron	Stony	
			Stony meteorites are divided into 2 main groups	
			Chondrites	Achondrites
Major characteristics	They are also called Siderites and are similar to the M-type asteroids	They are rare and contain iron and stony materials. They are similar in composition to the S-type asteroids	They contain tiny, whitish, glassy spheres (droplets of olivine and pyroxene) called chondrules. These occur scattered throughout the dark-gray colored rock matrix. In composition, the chondrites have remained the same since their inception, some 4.54 billion years ago. The Carbonaceous chondrites, a type of a chondrite, contains amino acids, the essential ingredient of life on earth	Although the majority of achondrites are of asteroidal origin, some are known to have come from the highland regions of the moon's far side and from mars. A meteorite found in the sahara desert in 1999, is suspected to have originated on mercury
Composition	Siderites are mainly composed of Nickel with minor Cobalt. The metallic Iron content is very high (>90 %)	They are also called Breccias, and are made of rock fragments welded together by high heat and pressure, possibly during the high-speed impact of a meteoroid and an asteroid	Chondrites mostly contain iron and magnesium bearing silicate minerals and are similar in composition to the crust and mantle of terrestrial planets. They are very similar in composition to the Sun and C-type asteroids. The carbonaceous chondrites have less amount of volatile content and are considered the most primitive and unaltered type of meteorites known. They have an elemental composition probably similar to that of the nebula from which our solar System formed	Most achondrites are chemically similar to basalts and are thought to be the product of melting on large asteroids, moons, and planets. They contain plagioclase, pyroxene, and olivine. They generally, but not always, lack small rounded inclusions (glassy beads; chondrules) noted in carbonaceous chondrites
Percentage	~5 % of all the falls[a]	<1 % of all the falls	~5 % of all Chondritic falls and 94 % of all falls	~9 % of all meteorite falls

[a] Meteorites that arrive on Earth are "falls" and those discovered are called "finds" (for further discussion see Sect. 3.3.2

Table 3.8 Supposition about the consequences of impacts of various meteor sizes

Impactor diameter (in m)	Yield (in megatons)	Interval (in years)	Consequences
<50	<10	<1	Meteors in the upper atmosphere but most do not reach the surface of the earth
75	10–100	1,000	Irons make craters like the meteor crater; stones produce airbursts like the Tunguska; land impact destroys an areas equal to the size of a city
160	100–1,000	5,000	Irons, stones hit ground; comets produce airbursts; land impact destroys an area equal to the size of a large urban area (like New York and Tokyo)
350	1,000–10,000	15,000	Land impact destroys an area equal to the size of a small state; ocean impact produces mild Tsunamis
700	10,000–100,000	63,000	Land impact destroys an area equal to the size of a moderate state (like Virginia, USA); ocean impact makes big Tsunamis
1,700	100,000–1,000,000	250,000	Land impact raises dust with global implication; destroys an area equal to the size of large country (like France)

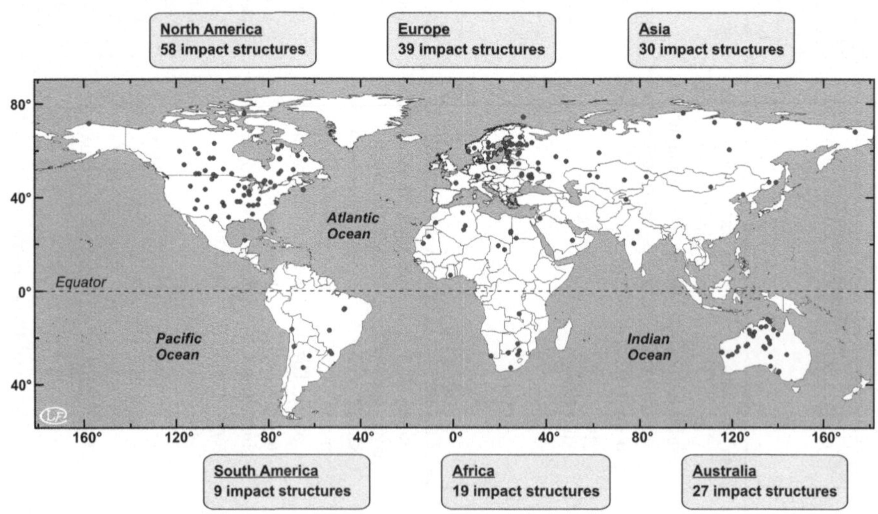

Fig. 3.3 The distribution map of the 182 confirmed meteorite impact structures on earth. A more recent "small impact crater" was recorded on the 15th of September 2007, the Carancas crater (Peru), about 13.5 m in diameter (see Kenkmann et al. 2009). However, no large impact crater was formed during the last thousands years on earth. Reproduced with permission from Dr Ludovic Ferrière's website; www.MeteorImpactOnEarth.com

Table 3.9 List of top 20 (largest in size) terrestrial meteorite impact craters so far identified on earth (see also Earth Impact Database, 2010)

Nos.	Diameter (Km)	Impact site	Latitude	Longitude	Age (Ma)
1	200	Sudbury, Ontario, Canada	46°36′N	81°11′W	1850.00 ± 3
2	170	Chicxulub, Mexico	21°20′N	89°30′W	64.98 ± 0.05
3	160	Acraman, Australia	32°1′S	135°27′E	570
4	140	Vredefort, South Africa	27°0′S	27°30′E	1970 ± 100
5	100	Manicouagan, Quebec, Canada	51°23′N	68°42′W	212 ± 1
6	100	Popigai, Russia	71°30′N	111°0′E	35 ± 5
7	85	Chesapeake Bay, Virginia, USA	37°15′N	76°5′W	35.5 ± 0.6
8	80	Puchezh-Katunki, Russia	57°6′N	43°35′E	220 ± 10
9	65	Kara, Russia	69°5′N	64°18′E	73 ± 3
10	60	Beaverhead, Montana, USA	44°36′N	113°0′W	600
11	55	Tookoonooka, Queensland, Australia	27°0′S	143°0′E	128 ± 5
12	55	Siljan, Sweden	61°2′N	14°52′E	368 ± 1.1
13	54	Charlevoix, Canada	47°32′N	70°18′W	357 ± 15
14	52	Kara-Kul, Tajikistan	39°1′N	73°27′E	25
15	45	Montagnais, Nova Scotia, Canada	42°53′N	64°13′W	50.50 ± 0.76
16	40	Araguainha Dome, Brazil	16°46′S	52°59′W	249 ± 19
17	40	Saint Martin, Canada	51°47′N	98°32′W	220 ± 32
18	40	Mjolnir, Norway (underwater)	73°48′N	29°40′E	143 ± 20
19	39	Carswell, Saskatchewan, Canada	58°27′N	109°30′W	115 ± 10
20	35	Manson, Iowa	42°35′N	94°31′W	65.70 ± 1

The Vredefort in South Africa is the oldest known crater

Table 3.10 Table lists meteorite groups, falls, and finds during the period between 1740 and 1990 (meteorites from Antarctica are excluded)

Type	Fall (%)	Find (%)	Fall weight (in kg)	Find weight (in kg)
Stony	95.0	79.8	15,200	8,300
Chondrites	88.0	–	–	–
Achondrites	7.0	–	–	–
Stony Iron	1.0	1.6	525	8,600
Iron	4.0	18.6	27,000	435,000

3.3.2 *Falls and Finds*

As much as one million kilogram (1,000 tons) of meteoritic material rains down on earth each day. Meteorites that are seen to arrive are referred as "Falls" and those that are discovered are called "Finds" (Table 3.10). Interestingly, a region's terrain largely determines what sort of, and how many meteorites will be recovered. For example, meteorites found in regions with similar colored rocks are difficult to distinguish from the rocks or soil. However, in contrast, the barren wastelands of Antarctica are an ideal place to look for these "stones from the sky."

They stand out in the white background. Here, about 33 % of the meteorites are witnessed as Falls.

3.3.3 Meteor Showers

These are spawned by comets as they travel through our solar system. The dust spreads out along the comet's orbit and forms an elliptical trail of debris that passes around the Sun and crosses the orbits of the planets. The shower occurs when earth passes through this trail of debris during its yearly orbit around the Sun. The following year, earth passes through that same debris trail again, on about the same date. This is why meteor showers are predictable annual events. Meteor showers are often referred to as a "shooting star" or "falling star." Very bright meteors are called Fireballs. However, these are quite rare. A shower usually lasts only for few days. On an ordinary moonless night, three or four meteors can be seen each hour. This number goes up to several dozen during meteor showers. Table 3.11 lists of some of the famous meteor showers.

3.4 Comet

They are leftover debris from the formation of the Solar System (Sagan and Druyan 1986). Comets are small, icy cosmic bodies that orbit around the Sun. They are irregularly shaped and composed of compounds of carbon, oxygen, nitrogen, and hydrogen. They also contain water, ammonia, methane, carbon monoxide with smaller amounts of other more complex compounds (for details see also Burnham 2001). Comets have very elongated and elliptical orbits that go beyond planetary orbits. As comets approach the Sun they develop enormous tails of luminous material that extend for millions of kilometers, away from the Sun. When they are further from the Sun, their nucleus is very cold and its material is frozen solid. They are now referred to as "dirty iceberg" or "dirty snowball," as over half of their material is ice. When a comet approaches within a few Astronomical Unit (AU) of the Sun, the surface of the nucleus begins to warm, and volatiles start evaporating. The evaporated molecules boil off and carry small solid particles with them, forming the comet's Coma (of gas and dust) (see also Kronk 1984). The dust is composed of silicates and some metals. Hence, each time a comet visits the Sun; it loses some of its volatiles. Eventually, with time, this becomes just another rocky mass in the Solar System. For this reason, on a cosmological timescale, the existence of comets is short-lived. Many scientists believe that some asteroids are extinct comet nuclei; those comets that have lost all of their volatiles.

Table 3.11 List of some meteor showers

Shower name	Date of maximum	Normal limits	Best time to look	Rate/hour	Description
Quadrantids	4th January	1st–6th January	Before dawn	60	Blue meteors with fine trains
Eta Aquarids	5th May	24th Apr–20th May	Just before dawn	35	Low in sky. Associated with comet halley
Capricornids	8 –26th July	July–August	Midnight onwards	5	Bright meteors
Alpha Capricornids	2nd August	15th July–25th August	All night	5	Yellow slow fireballs
Perseids	12–13th August	23rd July–20th August	Best before dawn	75	Many bright fast meteors with trains. Associated with comet swift-tuttle (in 1737, 1862, 1992)
Orionids	22nd October	16–27th October	After midnight	25	Fast with fine trains. Associated with comet halley
Taurids	4th November	20th Oct–30th November	All night	10	Very slow meteors
Leonids	17–18th November	15–20th November	After midnight	10	Fast bright meteors with fine trains
Geminids	14th December	7–16th December	After midnight	75	Plenty of bright meteors, few trains

3.4.1 *Parts of Comets*

The structure of a comet is very diverse and dynamic. However, they all develop a surrounding cloud of diffuse material, called Coma (see also Kronk 1984). The Coma grows in size and brightness as the comet approaches the Sun. Often a small, bright nucleus (<10 km in diameter) is visible in the middle of the Coma. Coma and Nucleus constitute the Head of the comet (Table 3.12). The Head consists of an aggregation of solid masses like small pieces of rocks, ice and solid gas varying in size that are held together by gravity. The particles of the Head emit luminous gases form the Coma. These gases become more and more rarefied with increasing distance from the Sun and this is how a comet's tail is formed. The tail always points away from the Sun and can be thousands of millions of km long, even for comets that have a comparatively small mass.

3.4.2 *The Orbit of a Comet*

Most comets travel in highly elliptical orbits around the Sun with orbital periods ranging from three to millions of years. Their velocity increases greatly when they are near the Sun and slows down at the far reaches of their orbit. Comet is visible only when it is near the Sun (as it is vaporizing), and are dark (and virtually invisible) throughout most of their orbit, hence, they are visible only at sunrise or sunset. Some comets, called "periodic comets'" return every few years, and travel no further than the orbit of Jupiter. Others have periods of several millions of years and their orbit takes them far beyond the orbit of Pluto. Most comets have highly eccentric orbits and some are seen once and then disappear for millennia. Only the short and intermediate-period comets (like Halley's Comet; Fig. 3.4), stay within the orbit of Pluto for a significant fraction of their orbit (see also Schneider and Etter 1985; Schaaf 1997).

3.5 Summary

Our Solar system is home to all cosmic bodies—from Asteroids and Comets to Satellites and Planets. Asteroids include all Meteoroids, Meteors, and Meteorites.

There are more than 90,000 known asteroids; most are concentrated within and between the orbits of Mars and Jupiter. About 99 % of the asteroids are larger than 100 km and there are probably more than a million in the 1 km range. If the estimated total mass of all asteroids put together will make an object ~1,500 km across. Compositionally, about 92.8 % meteorites are made of silicates and 5.7 % are composed of iron and nickel; the rest are a mixture of these. Two types of asteroid classifications are used: based on Spectra and Albedo and on Position.

Table 3.12 Details of the parts of a comet

Parts of a comet				
Head		Hydrogen envelope	Tail	
Nucleus	Coma			
It is a dusty snowball consisting of tiny pieces of rocky materials embedded in a mass of ice and frozen gases. The nuclei of most comets range from 1 to 10 km in diameter	This is a dense cloud of water vapor, carbon dioxide gas, ammonia, dust, and neutral gases that have sublimed from the solid nucleus. It is about a million km across	In some, a cloud of Hydrogen gas surrounds the Coma. This envelope is about 10 million km across at the nucleus of the comet and about 100 million km long. It is bigger when the comet is near the Sun	When comets come near the Sun, their tail becomes visible due to heat. Sun's heat vaporizes some of the icy nucleus (or Head) and sunlight is reflected from the vapor. The solar wind pushes the vapor in a direction away from the Sun, forming the comet's tail. Hence, the comet's tails generally points away from the Sun. The comet's tail is often very long. When the great comet of 1843 was near the sun, its tail was as long as 475 million km, i.e., extending beyond the orbit of mars. There are two types of tails: ion and dust	
			Ion	*Dust*
			The tail is long and composed of plasma, laced with rays and streamers caused by the interactions with the solar wind.	The tail is composed of microscopic dust particles that are buffeted by photons emitted from the Sun. This is the most prominent part of a comet.
			>100 million km long	~10 million km long

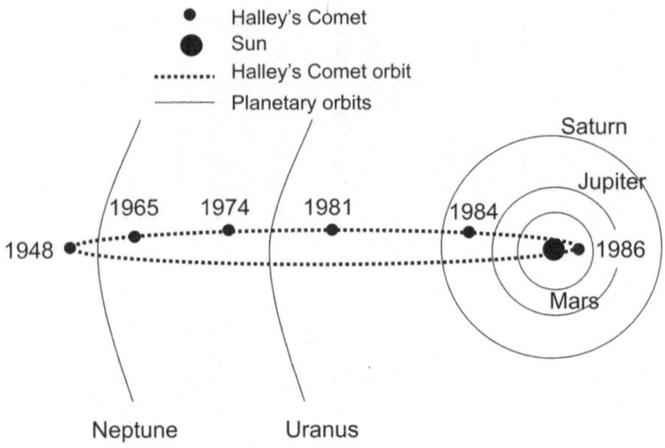

Fig. 3.4 The orbit of Halley's Comet

A Meteor is the visible path of a meteoroid that enters the earth's atmosphere. These are also called Shooting stars; very bright ones are called Fireballs or Bolides. A meteor describes a streak of light, temporary incandescence, as matter within the Solar System falls into the earth's atmosphere. Most meteorites are quite small. The largest known is the metallic Hoba meteorite found in Namibia (SW Africa).

Meteorites are considered pristine samples of the early Solar System. Three broad groups of Meteorites are recognized: Iron, Stony iron, and Stony; the last type is the most common. Other meteorite types include Achondrites and Pallasites. Achondrites are Stony meteorites and possess distinct texture and mineralogy indicative of the igneous process noted on the earth. Pallasites are Stony–iron meteorites. The radiometric dating of Stony meteorites places them to be 4.55 billion years old, the approximate age of our Solar System. Compositionally, they are close to Terrestrial planet characteristics (like Mercury, Venus, Mars, etc.). As much as one million kilogram of meteoritic material rains down on earth each day. Meteorites that are seen to arrive are called "Falls" and those that are discovered as "Finds." Antarctica is an important site for locating meteorites.

The meteor showers are spawned by comets as they travel through our Solar System. The dust spreads out along the comet's orbit and forms an elliptical trail of debris that passes around the Sun and crosses the orbits of planets. The shower occurs when earth passes through this trail of debris during its yearly orbit around the Sun.

Comets are leftover debris from the formation of the Solar System. They are small, icy cosmic bodies that orbit around the Sun. Their structure is very diverse and dynamic. Most comets travel in highly elliptical orbits around the Sun with orbital periods ranging from three to millions of years. Their velocity increases greatly when they are near the Sun and slows down at the far reaches of their orbit. Comet is visible only when it is near the Sun (as it is vaporizing), and are dark (and virtually invisible) throughout most of their orbit, hence, they are visible only at sunrise or sunset.

References

Binze DS, Lauretta HY, McSween Jr (eds) (2006) Systematics and evaluation of meteorite classification. Meteorites and the early solar system II. University of Arizona Press, Tucson, pp 19–52. ISBN 978-0816525621

Burnham R (2001) Great comets. Cambridge University Press, England

Chapman CR, Morrison D, Zellner B (1975) Surface properties of asteroids: a synthesis of polarimetry, radiometry, and spectrophotometry. Icarus 25(1):104–130

Cunningham C (2001) The First asteroid, ceres, 1801–2001. Star Lab Press, Surfside

Earth Impact Database (2010). http://www.unb.ca/fredericton/science/research/passc/ Accessed 2 April 2013

Harvey R (2003) The origin and significance of Antarctic meteorites. Chem Erde 63(2):93–147

Kenkmann T, Artemieva NA, Wünnemann K, Poelchau MH, Elbeshausen D, Núñez Del Prado H (2009) The Carancas meteorite impact crater, Peru: geologic surveying and modeling of crater formation and atmospheric passage. Meteor Planet Sci 44:985–1000

Kronk G (1984) Comets—a descriptive catalog. Enslow Publishers, Hillside

Krot AN, Keil K, Scott ERD, Goodrich CA, Weisberg MK (2003) Classification of meteorites. In: Holland HD, Turekian KK (eds) Treatise on geochemistry, 1st edn. Elsevier, Oxford, pp 83–128. doi:10.1016/B0-08-043751-6/01062-8. ISBN 978-0-08-043751-4

Lodders K, Fegley B (1998) The planetary scientist's companion. Oxford University Press, New York

Meteoritical Society Guidelines for Meteorite Nomenclature (2013) http://meteoriticalsociety.org/?page_id=59. Retrieved 2013-03-24

Norton OR (2002) The Cambridge encyclopedia of meteorites. Cambridge University Press, Cambridge

Peebles C (2000) Asteroids, a history. Smithsonian Institution Press, Washington

Sagan C, Druyan A (1986) Comets. Random House, New York

Schaaf F (1997) Comet of the Century—from Halley to Hale-Bopp. Copernicus, New York

Schneider S, Etter R (1985) Halley's comet-memories of 1910. Abbeville Press, New York

Seymour S (1994) Comets, meteors, and asteroids. Morrow Jr Books, New York

NHM Catalogue of Meteorites (2013). The Meteorite Catalogue Database at the Natural History Museum. http://www.nhm.ac.uk/research-curation/research/projects/metcat/. Retrieved 2013-03-24.

Tholen DJ (1988) Asteroid taxonomic classifications. Asteroids II. In: Proceedings of the Conference, University of Arizona Press, Tucson, pp 1139–1150

Part II
The Earth Materials

Chapter 4
Earth as a Planet

4.1 Introduction

Earth has been referred by several names as *Tellus* (the Roman goddess of Earth and a symbol of fertility), *Gaia* (ancient Greek Earth goddess), and *Terra*, from the Latin meaning Earth. Earth is the third planet of our Solar System and is placed at a distance of about 150 million km from the Sun. The mean distance between Earth and Sun is used to measure distances within our Solar System. This is called an Astronomical Unit (AU). Detailed statistics of the earth are given in Tables 2.3, 2.5, and 2.6 (see Chap. 2).

Earth's atmosphere is mainly composed of Nitrogen and Oxygen. Its extensive magnetic field, generated by its rapid spin and molten nickel-iron core, gives its atmosphere the shield, from nearly all harmful radiations coming from the Sun and other stars. This atmosphere also protects the earth from meteors, most of which burn before reaching it (for details on Meteors see Chap. 3).

4.2 Earth's Components

Earth, for now, is the only planet to sustain life and Mars is the only other planet to remotely have a possibility of sustaining life. Earth has four layers: Atmosphere, Lithosphere, Hydrosphere, and Biosphere (Fig. 4.1). Each layer is diagrammatically illustrated in Fig. 4.1 and briefly described below.

4.2.1 Atmosphere

It extends as far as 700 km above and beyond and is made up of a mixture of gases dominated by Nitrogen and Oxygen (Table 4.1).

S. Jain, *Fundamentals of Physical Geology*, Springer Geology,
DOI: 10.1007/978-81-322-1539-4_4, © Springer India 2014

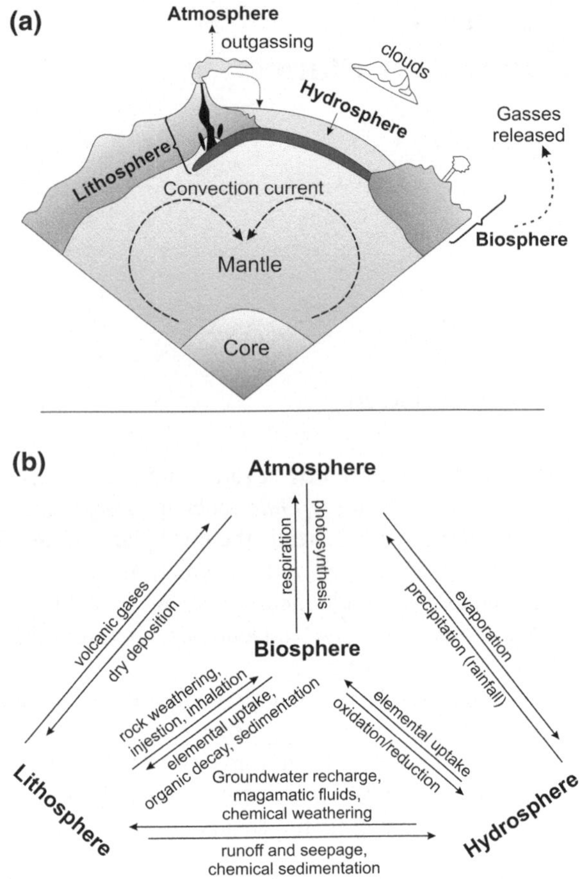

Fig. 4.1 The four components of the earth (Atmosphere, Lithosphere, Hydrosphere, and Biosphere). **a** Illustrated and **b** Their interrelationship

Table 4.1 Chemical composition of the Atmosphere

Component	Volume (in %)
Nitrogen (N_2)	78.084
Oxygen (O_2)	20.948
Argon (Ar)	0.934
Carbon Dioxide (CO_2)	0.033
Others (Ne, H, CH_4, Kr, H_2, Xe, etc.)	0.002

The temperature, within the Atmosphere, varies noticeably. At some levels, the temperature falls with increasing height, while at others, it remains constant and in still others, it rises with increasing height (Fig. 4.2). The Atmosphere is divided into four zones: Troposphere, Stratosphere, Mesosphere, and Thermosphere (Fig. 4.2).

The layer of air immediately above the earth's surface is called the Troposphere (Fig. 4.2). Here, the air circulates in vertical and horizontal convection currents,

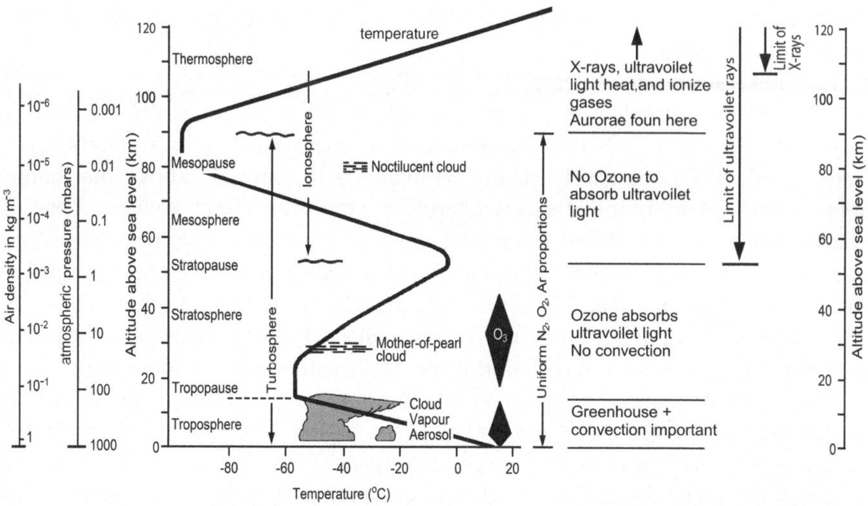

Fig. 4.2 The four zones of atmosphere (Troposphere, Stratosphere, Mesosphere, and Thermosphere) and the observed variations (in Temperature, N_2, O_2, and Argon (Ar) gas proportions, Atmospheric pressure, and cloud distribution) in them

thus, redistributing heat and moisture across the globe. It is an important component of the atmosphere as it maintains the earth's natural thermostat, and thus, enables life to exist. It also hosts the water cycle and the pressure systems that help purify and deliver water to most regions of the globe. The majority of clouds and water vapor in the atmosphere occur in the Troposphere (Fig. 4.2), primarily due to the presence of dust. The Troposphere extends an average of 11 km; ~12 km over Equator, and ~7 km over Poles. The Troposphere contains ~80 % of the total mass of the atmosphere and is denser than any other layer. A sudden reversal of temperature gradient at the top of the Troposphere creates a sharp boundary called the Tropopause, which limits the mixing between the Troposphere and layer above, called the Stratosphere (Fig. 4.2).

The Stratosphere is the next layer that extends from the Tropopause up to ~50 km above. The Ozone layer is located here (Fig. 4.2). The Stratosphere is vastly more dilute than the Troposphere and has almost no water vapor with nearly 1,000 times more Ozone. The height of the bottom of the Stratosphere varies both with latitude and seasons, occurring between ~8 and 16 km; ~16 km near the Equator ~10 km at mid-latitudes; and ~8 km near the Poles. It is lower in winter at mid- and high-latitudes and higher in summer. Stratopause is the boundary between the Stratosphere and the overlying Mesosphere.

The Mesosphere is the third thermal zone of the atmosphere (Fig. 4.2). Here, temperature decreases with increasing height. It is the coldest naturally occurring place on earth; a temperature minimum of ~−90 °C or lower, is reached at ~85 km (Fig. 4.2). Mesosphere begins at ~50 km and extends up to 85 km. Its upper boundary is called the Mesopause which is usually at heights near

100 km, except at middle and high latitudes in summer where it descends down to heights of ~85 km.

The next layer, the Thermosphere (also called the heated layer) begins at ~90 km and extends between 500 and 1,000 km above the earth (Fig. 4.2). In its lower part (until 200–300 km altitude), the temperature climbs sharply, then plateaus off, and holds fairly steady with increasing altitude above that height. Solar activity strongly influences temperature within the Thermosphere. Although Thermosphere is considered as part of earth's atmosphere, here, air density is extremely low, so it is generally considered as outer space. In fact, the general understanding is that space begins at an altitude of 100 km, slightly above the Mesopause, and at the bottom of the Thermosphere. Both the Space shuttle and the International Space Station (ISS) orbit the earth within the thermosphere!

The region of atmosphere beyond 700 km is termed as Exosphere (some even consider 600 km as the upper limit). This is the region where atoms and molecules escape into space. It is an extremely rarefied space, with very low density and high temperatures and with minimum atomic collusions. The boundary between the Thermosphere and the Exosphere above is called the Thermopause. A summary of the structure and composition of all the atmospheric layers is given in Table 4.2.

Earth's other three components are Lithosphere, Hydrosphere, and Biosphere.

4.2.2 Lithosphere

It is the land part of the earth (litho = stone) and includes all the solid materials constituting earth, from surface downward (Fig. 4.3). Both crust and upper mantle make up the Lithosphere which extends ~80–100 km below. Under the continents, the lithosphere is thickest, at about 120 km or so. Under middle of the oceans, the lithosphere is only a few kilometers thick. Here, just below the surface, the mantle's temperature is about 1,000 °C. As such, as one moves down, temperatures in the lithosphere increases by 25 °C per 1,000 m and reaches 3,700 °C at the core-mantle boundary. Lithosphere is fragmented into massive plates that fit around the globe like pieces in a jig saw puzzle (Fig. 4.3b). These plates move independently and slowly, relative to one another and slide on top of a somewhat fluid (plastic) part of the mantle called the Asthenosphere (Fig. 4.3a). Movement of plates causes earthquakes. Details about these plates are discussed in more detail in the next chapter—The Interior of the Earth.

4.2.3 Hydrosphere

Hydrosphere includes all naturally occurring water on or below the surface. About 97.25 % of the Hydrosphere is made up of large surface bodies of saline water called Ocean. Around 72.5 % of the surface area of the earth (~361 million km^2)

Table 4.2 Structure and composition of atmospheric layers

Avg. height (in km)	Percent mass[a]	Layers	Sub-layers	Water vapor	Temperature	Avg. temp. (in °C)	Gases	Other major characteristics
500	< 0.001	Thermosphere			Slowly increases with height	1200	Rarefied	Temperatures in the upper thermosphere can range from about 500 to 2,000 °C or even higher
			Ionosphere	None	Rapidly increases with height		Atomic H, O, N, He	
80	0.099	Mesosphere		None	Decreases with height (from −15 °C to −120 °C)	−90	Ozone (O_3) layer	Temperature drops again, reaching about −100 °C at the top
50	9.9	Stratosphere		Little	Increases with height (from −60 °C to −15 °C)	0	Diatomic O_2	Ozone absorbs ultra-violet (UV) solar radiations, thereby, producing warmer temperatures toward the top
11	90	Troposphere		Maximum with vapor Clouds	Decreases with height (from 17 °C to −51 °C)	15	N_2, O_2, CO_2, etc.	Air temperature drops rapidly with increasing altitude (6.3–6.5 °C per km), reaching a temperature of ~−60 °C at the top

The average temperature of the atmosphere at the surface of earth is 14 °C (57 °F)

[a] Percent mass of the total atmosphere. 50 % of the mass of the atmosphere lies below 5.5 km elevation; 75 % lies below 10.7 km; 80 % lies below 11.9 km (the common cruising altitude of Jetliners); 90 % lies below 16 km; and 99.9 % lies below 50 km

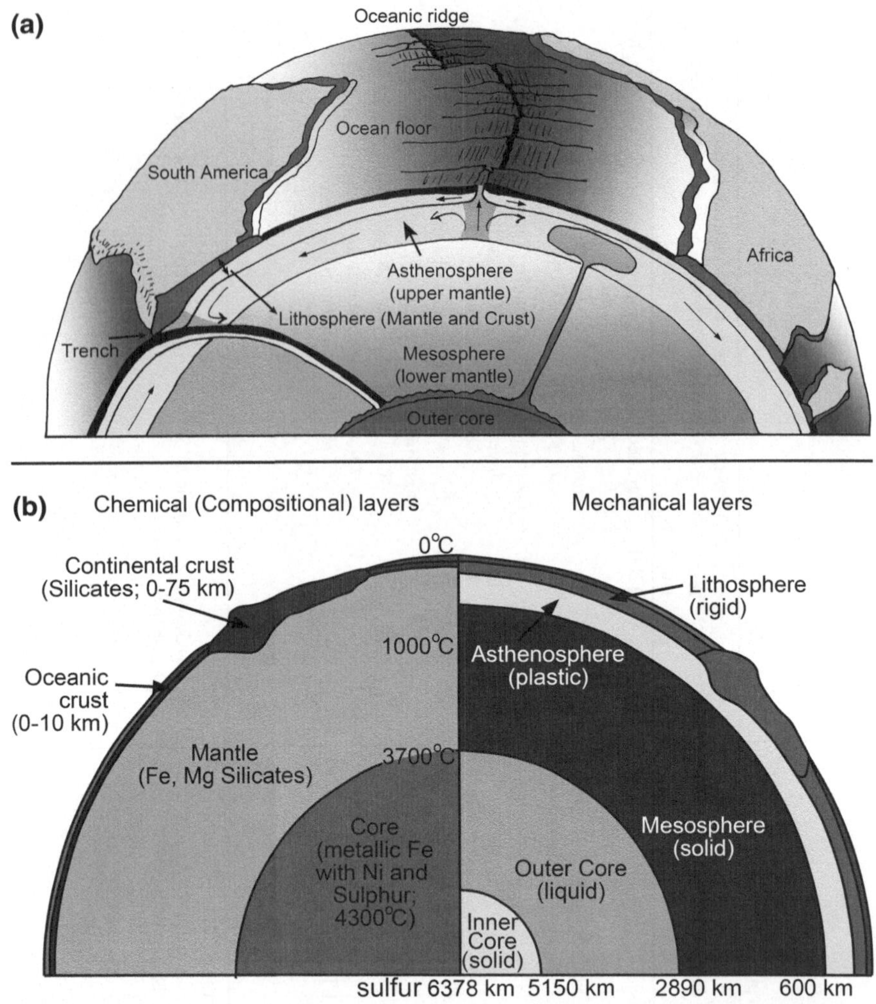

Fig. 4.3 **a** Tectonic. **b** Chemical (Compositional) and Mechanical layers of the earth and their major characteristics. (a): Redrawn from Earth's Dynamic Systems with permission of Eric H. Christiansen

is covered by five oceans namely the Pacific, Atlantic, Indian, Arctic, and the Antarctic. These and their associated extensions are called Seas and Bays. Rivers, lakes, bodies of frozen water (ice and snow), and groundwater form parts of the Hydrosphere (their detailed inventory is given in Table 4.3). Interestingly, the Hydrosphere makes only about 0.03 % of the total mass of earth.

Table 4.3 Hydrosphere inventory

Earth's surface	Percent of total surface	Area in km^2	Reservoir	Volume ($km^3 \times 10^6$)	Percent of total	Oceans	Percent of total surface	Area in km^2
Area covered by land	29.2	148,940,000						
Area covered by water	70.8	361,132,000	Oceans	1370	97.25	Pacific	30.5	155,557,000
						Atlantic	20.8	76,762,000
						Indian	14.4	68,556,000
						Southern	4	20,327,000
						Arctic	2.8	14,056,000
			Ice caps and glaciers	29	2.05			
			Groundwater	9.5	0.68			
			Lakes	0.125	0.01			
			Soil moisture	0.065	0.005			
			Atmosphere	0.013	0.001			
			Streams and Rivers	0.0017	0.0001			
			Biosphere	0.0006	0.00004			

Table 4.4 The hierarchical nature of the biosphere

Ecological	Taxonomical
Biosphere	Kingdom
Biome	Phylum
Landscape	Class
Ecosystem	Order
Biotic community	Family
Population (species)	Genus
Individual organisms	Species

4.2.4 Biosphere

The biosphere encompasses all zones (Lithosphere, Hydrosphere, and Atmosphere) on the earth in which life is present. Life on earth requires water, a source of energy (Sun light), and various nutrients found in soil, water, and air. The suitable combinations of these exist only in a narrow layer near the surface of the earth, i.e., within the upper layer of the earth's crust. This Biosphere is hierarchical in nature (Table 4.4).

4.3 Moon

Moon is earth's only satellite, formed 4.527 ± 0.01 billion years ago. Moon is a differentiated body (by igneous processes) with a geochemically distinct crust, mantle, and core. The crust is 50 km thick. Assuming that it is uniform, it would make up ~10 % of the Moon's volume as compared to <1 % for earth. Recent seismic data suggest that there are no evidence that the Moon has an iron-rich core, unless if it is very small. Rocks more than 4 billion years old exist on Moon, yielding information about the early history of the Solar System, so far unavailable on earth. The geological activity on Moon includes occasional large impacts and the continued formation of regolith. Moon is considered as geologically dead and fossilized in time. The most acceptable hypothesis for its formation is that the Earth-Moon system formed as a result of a massive impact. A Mars sized cosmic body hits the nearly formed proto-earth, thereby, blasting material into the orbit around the proto-earth. These, then, accreted to form the Moon. The near-identical isotopic compositions of the Earth and Moon reaffirm this theory.

Moon has no atmosphere. Its relief is made up of wide plains and mountainous regions. The wide plains are called Lunar Maria (Latin for "Seas"; singular Mare). These are vast solidified pools of ancient basaltic lava, similar to the terrestrial basalts on the earth. However, these Mare basalts have much higher abundances of iron and completely lack minerals altered by water.

Table 4.5 The Moon statistics

Moon statistics mass (kg)	7.35×10^{22}
Mass (Earth = 1)	1.23×10^{-2}
Equatorial radius (km)	1,737.40
Equatorial radius (Earth = 1)	2.72×10^{-1}
Mean density (gm/cm^3)	3.34
Mean distance from Earth (km)	3,84,400
Rotational period (days)	27.32166
Orbital period (days)	27.32166
Average length of lunar day (days)	29.53059
Mean orbital velocity (km/s)	1.03
Orbital eccentricity	0.0549
Tilt of axis (degrees)	1.5424
Orbital inclination (degrees)	5.1454
Equatorial surface gravity (m/s^2)	1.62
Equatorial escape velocity (km/s)	2.38
Visual geometric albedo	0.12
Magnitude (Vo)	−12.74
Mean surface temperature (day) (in °C)	107
Mean surface temperature (night) (in °C)	−153
Maximum surface temperature (in °C)	123
Minimum surface temperature (in °C)	−233

The first human to step on the Moon was Neil Armstrong on 21st July, 1969 of the Apollo 11 mission. The Moon affects earth in two ways: First, when it comes between the Earth and the Sun (the Eclipse) and second, through the phenomenon of Tides. The Moon statistics are given in Table 4.5.

4.4 The Origin of Earth

The origin of both the Solar System and Earth is closely interlinked. French naturalist Georges-Louis Leclerc, Comte de Buffon (1707–1788) in 1745 was one of the first scientist to put forward a hypothesis for the origin of earth. According to him, our planet was formed through the cooling of a blob of solar matter ejected from the Sun during a cataclysmic collision with another large comet. Buffon estimated that this entire process took over 70,000 years. After Buffon, many hypotheses were put forward for the origin of our planet and these can be broadly grouped into two: Uniformitarian or Uniparental hypothesis and Cataclysmic or Bi-parental hypothesis (Table 4.6).

Table 4.6 The two groups of hypothesis put forward to explain the origin of earth

Uniformitarian or Uniparental Hypothesis	Cataclysmic or Bi-parental Hypothesis
Nebular hypothesis by Kant 1755	Buffon's collision hypothesis 1749
Nebular hypothesis by Laplace, 1796	Chamberlin Moulton's Planetesimal hypothesis 1905
Nebular cloud hypothesis by Keupers 1957	Tidal hypothesis of Jeans and Jeffreys 1918
Electromagnetic theory by Hans Alfvens 1942	Binary star hypothesis of Russel and Lyttleton 1936
Meteoritic hypothesis by Schmidt 1943	Fission hypothesis of Ross Gunn 1941
Nebular hypothesis by von Weizsacker 1944	Cepheid hypothesis of Bannerji 1942
Magnetic theory by Hoyles 1958	Nova hypothesis of Fred Hoyle 1945

4.4.1 Uniformitarian or Uniparental Hypothesis

4.4.1.1 Nebular Hypothesis

The German philosopher Kant (in 1755) and the French mathematician Laplace (in 1796) independently developed the Nebular hypothesis based on the hot origin of our planets (Fig. 4.4). The hypothesis suggests that a single incandescent gaseous nebula formed the Solar System (Nebula is a Latin word meaning *mist*). Due to the rotation around its axis, this nebula assumed the shape of a disk. Gradually, as it cooled, it contracted in size, resulting in an increase in its rate of rotation and centrifugal force. When the centrifugal force exceeded the force of gravity in the equatorial part of the nebula, gaseous rings began to spin off along the whole periphery of the disk. Further, cooling of the rings led to the formation of planets and their satellites. The Sun was formed at the nucleus of the nebula (Fig. 4.4).

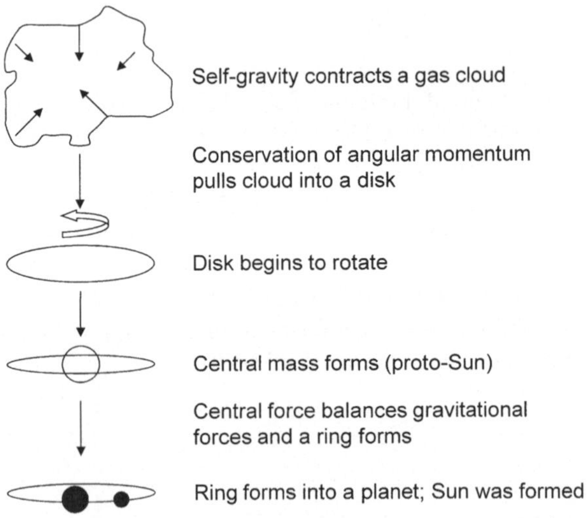

Fig. 4.4 The Nebular hypothesis. Redrawn with permission from Professor James Schombert

However, the hypothesis failed to explain the regularities of planetary bodies. Other major objections were: (a) the Sun should have the greatest angular momentum because of its mass and due to its position in the center, however, it has only 2 % of the momentum of the Solar System, and (b) no mechanism is proposed to explain how the hot gaseous material condensed into rings.

4.4.1.2 Meteorite Hypothesis

Later, a new idea was put forward for the cold origin of the earth. Among them, noteworthy is the Meteorite hypothesis suggested by the Soviet scientist Otto Schmidt in 1944. A similar hypothesis was put forward by the German physicist C. F. Von Weizsacker (in 1944) and the American scientist G. P. Kuiper (in 1951). These hypothesis assume the formation of the Sun and planets from various sources and at some stage in the Sun's evolution, it captured a cold cloud of gas and dust in the galaxy with its own angular momentum. Rotation in the Sun's strong gravitational field led to a complicated redistribution, with the result that some of them collapsed into the Sun. As particles collided, they began to merge forming larger aggregates in mass and thereby accreting smaller particles within their gravitational field. In this way there came about a process of the formation of planets and their satellite from the original meteorite.

This hypothesis provided a solid physico-mathematical substantiation of the meteorite model and was also able to explain the feature of the motion of the planet orbits, the various directions of rotation, distribution of mass and densities, etc. The drawbacks of the hypothesis were the (a) improbability of the Sun capturing a cold cloud of gas and dust (meteorites) and (b) difficulty in explaining the concentric inner surface of the earth.

4.4.1.3 Protoplanetary Hypothesis

The Protoplanetary hypothesis is concerned with the origin of the protoplanetary cloud to explain the origin of planets (Fig. 4.5). This was put forward by Professor F. Hoyle in 1958. According to him, earth was created in the process of differentiation of the Sun from an original nebular matter that had been undergoing contraction (Fig. 4.5). This was not a widely acceptable hypothesis.

4.4.2 Cataclysmic or Bi-parental Hypothesis

4.4.2.1 Planetesimal Hypothesis

This hypothesis is about the bi-parental origin of the Solar System. American scientists, Thomas Chamberlin and Forest Moulton, in 1905 suggested that the

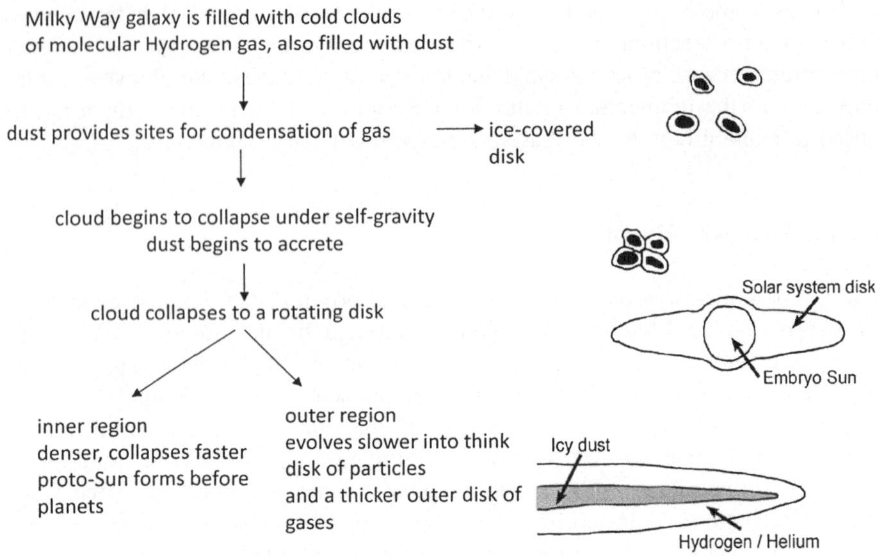

Fig. 4.5 The Protoplanetary hypothesis. Redrawn with permission from Professor James Schombert

planets were formed from the collusions and union of numerous small planetary fragments called Planetesimals. The hypothesis assumes that Sun existed even before the formation of planets. The near approach of a larger star caused tidal distortions upon the surface of the Sun. These distortions together with the eruptive forces present in the Sun, led to the distortion of the Sun's mass. This caused a number of gaseous bolts to be shot into space for great distances (due of the gravitation pull of the star, small gaseous bodies were separated from the Sun). This gaseous solar material cooled down and assumed the shape of a number of solid particles called Planetesimals. These Planetesimals rotated around the Sun in highly elliptical orbits. They intersected and collided with each other. This led to the coalescence of several large nuclei, eventually forming planets.

There are two major objections to this theory: (a) such large angular momentum cannot be produced by the passing star and (b) the theory completely fails to explain how the Planetesimals coalesced into a planet.

4.4.2.2 Gaseous Tidal Theory

In the first half of the twentieth century, James Jeans, a British astronomer in 1916 explained the origin of the planetary system. According to him, a large star came near the Sun. Due to its gravitational pull; a gaseous tide was raised on the surface of the Sun. As the star came closer, the tide increased in size and intensity. The gaseous tide detached when the star moved away. The shape of the tide was like a spindle. It broke into pieces, thus, forming the planets of the Solar System.

As briefly discussed above, there are several hypotheses explaining the origin of our planet but none have universal acceptance. Most of them explain only a few of the fundamental truths but fail, partially or completely, to explain other regularities.

4.5 The Age of the Earth

Based on available data, the planets completed their formation process very soon after the oldest components of primitive meteorites condensed out of the solar nebula, some 4.54 billion years ago (±0.5 billion years; Taylor and McLennan 1985; Windley 1995; Condie 1997). Earth grew in size as more and more Planetesimals joined together. Through this process, intense heat was generated inside the planet; short-lived radioactive elements created during the earth's formation were decaying, and giving off more energy. It is believed that at one point of time, our earth was so hot that it may have been entirely molten. These hot conditions allowed the heavier elements such as iron and nickel to sink into the earth's center and eventually formed its Core. The lighter elements remained in the Mantle and some of them formed the earth's rigid Crust (Clarke and Washington, 1924). As the planet was cooling off, various gases were released from the earth's interior, including water vapor. They enshrouded the planet in a dense cocoon, thus, forming the primordial atmosphere. It contained little, if any, Oxygen. When temperatures cooled enough, water vapor rained out of the atmosphere and the oceans were formed.

Unfortunately, so far, scientists have not been able to find a way to determine the exact age of the earth directly from its rocks as oldest rocks have been recycled and destroyed by the process of Plate Tectonics (see Chap. 14 for details). If there are any of these primordial rocks left in their original state, they have not yet been found. Nevertheless, scientists have been able to determine the probable age of the Solar System and are also able to calculate the age for earth by assuming that earth and the rest of the solid bodies within the Solar System formed at the same time and are, therefore, of the same age. The generally accepted age for the earth and the rest of the Solar System is about 4.54 billion years (±0.5 billion years) (Wilde et al. 2001).

Thousands of meteorites, which are fragments of asteroids that fall to earth, perhaps provide the best estimate of ages for the time of formation of our Solar System. The distribution of impact craters that were caused by the fall of meteorites across the world, number around 140 (these range in size from 1 to 200 km across). Of these, ages of more than 70 meteorites have been measured using radiometric-dating techniques and results show that the meteorites, and by extension, the Solar System, formed between 4.53 and 4.58 billion years ago. The best age for the earth comes not from dating individual rocks but by considering that both earth and meteorites are a part of the same evolving system in which the isotopic composition of Lead, specifically the ratio of Lead-207 to Lead-206 changes over time owing to the decay of radioactive Uranium-235 and Uranium-

Table 4.7 Major radioactive elements used in radioactive dating

Isotopes		Half-life of parent (years)	Effective dating range (years)	Minerals and materials that can be dated
Parent	Daughter			
^{238}Uranium	^{206}Lead	4.4 billion	10 million–4.6 billion	Zircon, Apatite
^{235}Uranium	^{207}Lead	0.7 billion	10 million–4.6 billion	Zircon, Apatite
^{40}Potassium	^{40}Argon	1.3 billion	50,000–4.6 billion	Muscovite, Biotite, Hornblende
^{87}Rubidium	^{87}Strontium	47 billion	10 million–4.6 billion	Muscovite, Biotite, Potassium feldspar
^{14}Carbon	^{14}Nitrogen	5730	100–70,000	Wood, charcoal, peat, bone and tissue, shells, and other calcium carbonate materials

238, respectively (Table 4.7). Scientists have used this approach to determine the time required for the isotopes in the earth's oldest Lead ores, to evolve from its primordial composition, as measured in Uranium-free phases of Iron meteorites, to its compositions at the time these lead ores separated from their mantle reservoirs. These calculations result in an age for the earth and meteorites, and hence, of the Solar System to be 4.54 billion years with an uncertainty of <1 %. This age, thus, represents the last time that the Lead isotopes were homogeneous throughout the inner Solar System and the time that Lead and Uranium were incorporated into solid bodies of the Solar System. Thus, as of today, the age of the earth is 4.54 billion years (4.54×10^9 years ± 0.05 billion years) and is also consistent with the ages of the oldest known terrestrial and lunar samples.

On Earth, the old rocks, exceeding 3.5 billion years, are found on all continents (Bowring and Williams 1999). However, a recent study revealed that the tiny crystals of mineral zircon (detrital) from the Yilgarn Block, Western Australia (Wilde et al. 2001) gave an age upwards of 4.4 billion years (the decay of radioactive Uranium was used for calculating this age). Previous oldest records were dated at 4.03 billion years from the Acasta Gneisses in NW Canada near the Great Slave Lake (Bowring and Williams 1999). An interesting feature of these ancient lake rocks is that they are not from any sort of "primordial crust" but are lava flows and sediments that were deposited in shallow waters (such as detrital zircon), an indication that earth's history began well before these rocks were actually deposited. Hence, the existence of liquid water at 4.4 billion years ago has fundamental implications for the evolution of life on earth. Contextually, microfossils records do also date back to 3.5 billion years (Schopf 1993).

4.5.1 Radioactive Dating

The most direct means for calculating earth's age is by the Lead/Lead (Pb/Pb) isochron age method. Three isotopes of Lead (^{206}Pb, ^{207}Pb, and either ^{208}Pb or ^{204}Pb) are measured and a plot is constructed of ^{206}Pb/^{204}Pb versus ^{207}Pb/^{204}Pb. If the Solar System formed from a common pool of matter, which was uniformly distributed in terms of the Pb isotope ratios, then the initial plots for all objects from that pool of matter would fall on a single point. Over time, the amounts of ^{206}Pb and ^{207}Pb will change in some samples, as these isotopes are decay end-products of the Uranium decay (^{238}U decays to ^{206}Pb, and ^{235}U decays to ^{207}Pb). This causes the data points to separate from each other. The higher the Uranium-to-Lead ratio of a rock, the more the $^{206}Pb/^{204}Pb$ and $^{207}Pb/^{204}Pb$ values will change with time. If the source of the Solar System was also uniformly distributed with respect to uranium isotope ratios, then the data points will always fall on a single line. We can compute the amount of time that has passed since the pool of matter became separated into individual objects from the slope of the line.

Most other measurements for the age of the earth rest upon calculating an age for the Solar System by dating objects that are expected to have formed with the planets but are not geologically active, such as meteorites. As shown in Table 4.8, there is excellent agreement on about 4.5 billion years, between several meteorites and by several different dating methods. Further, the oldest age determinations of individual meteorites generally give concordant ages by multiple radiometric means, or multiple tests across different samples (Table 4.6). Table 4.9 gives radiometric ages of earth derived from individuals of meteorites.

(1) **Ordinary Chondrites** are common in the inner asteroid belt, and comprise about 80 % of all meteorites that fall to earth, although they are only 10–15 % of all asteroids. These include: (a) 40 % of ordinary chondrites are H chondrites which have High-iron content; (b) 50 % of them are L chondrites, with a low-iron content, and (c) 10 % of them are LL chondrites, with a Low iron and overall Low metal content.

(2) **Carbonaceous Chondrites** contain a significant percentage of carbohydrates, and are typically dark. They are common in the outer Asteroid Belt, representing about 75 % of all asteroids. Their subgroups are usually identified by a "type meteorite" for which they are named. Each group is thought to consist of meteorites sharing a common parent asteroid. Those mentioned in the table are: (a) CM group: these contain 30 % small chondrules and CAIs (Calcium-Aluminum-rich inclusions) and 70 % matrix (dust), (b) CV group: these contain abundant CAIs in a roughly equal mix of millimeter-sized chondrites and matrix. The best-known members of this group are the Allende meteorites.

(3) **E-type or Enstatite chondrites** (named for their high enstatite ($MgSiO_3$) content) are extremely reduced; their iron is elemental or sulfides. They comprise about 2 % of meteorite falls on earth. The extremely reducing environment of formation for Enstatite chondrites has created a variety of unusual minerals, including Oldhamite (CaS), Niningerite (MgS), and Perryite (Fe-Ni silicide).

Table 4.8 Radiometric ages of earth derived from meteorites

Type	Number of samples	Method	Age (in billions of years)
Chondrites (CM, CV, H, L, LL, E)	13	Sm-Nd	4.21 ± 0.76
Carbonaceous Chondrites	4	Rb-Sr	4.37 ± 0.34
Chondrites (undisturbed H, LL, E)	38	Rb-Sr	4.50 ± 0.02
Chondrites (H, L, LL, E)	50	Rb-Sr	4.43 ± 0.04
H Chondrites (undisturbed)	17	Rb-Sr	4.52 ± 0.04
H Chondrites	15	Rb-Sr	4.59 ± 0.06
L Chondrites (relatively undisturbed)	6	Rb-Sr	4.44 ± 0.12
L Chondrites	5	Rb-Sr	4.38 ± 0.12
LL Chondrites (undisturbed)	13	Rb-Sr	4.49 ± 0.02
LL Chondrites	10	Rb-Sr	4.46 ± 0.06
E Chondrites (undisturbed)	8	Rb-Sr	4.51 ± 0.04
E Chondrites	8	Rb-Sr	4.44 ± 0.13
Eucrites (polymict)	23	Rb-Sr	4.53 ± 0.19
Eucrites	11	Rb-Sr	4.44 ± 0.30
Eucrites	13	Lu-Hf	4.57 ± 0.19
Diogenites	5	Rb-Sr	4.45 ± 0.18
Iron (plus iron from St. Severin)	8	Re-Os	4.57 ± 0.21

Sm-Nd: Samarium-Neodymium; Rb-Sr: Rubidium-Strontium; Lu-Hf: Lutetium-Hafnium; Re-Os: Rhenium-Osmium. Chondrites include the following three types

Table 4.9 Radiometric ages of earth derived from individuals of meteorites

Meteorite	Dated	Method	Age (billions of years)
Allende	Whole rock	Ar-Ar[a]	4.52 ± 0.02
	Whole rock	Ar-Ar	4.53 ± 0.02
	Whole rock	Ar-Ar	4.48 ± 0.02
	Whole rock	Ar-Ar	4.55 ± 0.03
	Whole rock	Ar-Ar	4.55 ± 0.03
	Whole rock	Ar-Ar	4.57 ± 0.03
	Whole rock	Ar-Ar	4.50 ± 0.02
	Whole rock	Ar-Ar	4.56 ± 0.05
Guarena	Whole rock	Ar-Ar	4.44 ± 0.06
	13 samples	Rb-Sr[b]	4.46 ± 0.08
Shaw	Whole rock	Ar-Ar	4.43 ± 0.06
	Whole rock	Ar-Ar	4.40 ± 0.06
	Whole rock	Ar-Ar	4.29 ± 0.06
Olivenza	18 samples	Rb-Sr	4.53 ± 0.16
	Whole rock	Ar-Ar	4.49 ± 0.06
Saint Severin	4 samples	Sm-Nd[c]	4.55 ± 0.33
	10 samples	Rb-Sr	4.51 ± 0.15
	Whole rock	Ar-Ar	4.43 ± 0.04
	Whole rock	Ar-Ar	4.38 ± 0.04
	Whole rock	Ar-Ar	4.42 ± 0.04
Indarch	9 samples	Rb-Sr	4.46 ± 0.08
	12 samples	Rb-Sr	4.39 ± 0.04

(continued)

Table 4.9 (continued)

Meteorite	Dated	Method	Age (billions of years)
Juvinas	5 samples	Sm-Nd	4.56 ± 0.08
	5 samples	Rb-Sr	4.50 ± 0.07
Moama	3 samples	Sm-Nd	4.46 ± 0.03
	4 samples	Sm-Nd	4.52 ± 0.05
Y-75011	9 samples	Rb-Sr	4.50 ± 0.05
	7 samples	Sm-Nd	4.52 ± 0.16
	5 samples	Rb-Sr	4.46 ± 0.06
	4 samples	Sm-Nd	4.52 ± 0.33
Angra dos Reis	7 samples	Sm-Nd	4.55 ± 0.04
	3 samples	Sm-Nd	4.56 ± 0.04
Mundrabrilla	Silicates	Ar-Ar	4.50 ± 0.06
	Silicates	Ar-Ar	4.57 ± 0.06
	Olivine	Ar-Ar	4.54 ± 0.04
	Plagioclase	Ar-Ar	4.50 ± 0.04
Weekeroo Station	4 samples	Rb-Sr	4.39 ± 0.07
	Silicates	Ar-Ar	4.54 ± 0.03

[a] Ar-Ar: Argon-Argon or $^{40}Ar/^{39}Ar$ dating method
[b] Rb-Sr: Rubidium-Strontium
[c] Sm-Nd: Samarium-Neodymium

4.6 Summary

Earth has been referred by several names such as *Tellus*, *Gaia,* and *Terra*. It has four components: Atmosphere, Lithosphere, Hydrosphere, and Biosphere.

The Atmosphere extends as far as 700 km above and beyond and is made up of a mixture of gases dominated by Nitrogen and Oxygen. Atmosphere is divided into four zones: Troposphere, Stratosphere, Mesosphere, and Thermosphere.

In the Troposphere, air circulates in vertical and horizontal convection currents, redistributing heat and moisture across the globe. Hence, it is an important component of the atmosphere as it maintains the earth's natural thermostat, and thus, enables life to exist. A sudden reversal of temperature gradient at the top of the Troposphere creates a sharp boundary called the Tropopause, which limits the mixing between the Troposphere and layer above, the Stratosphere.

The Stratosphere extends from the Tropopause up to ~50 km above. The Ozone layer is located here. The Stratosphere is vastly more dilute than the Troposphere and has almost no water vapor with nearly 1,000 times more Ozone. Stratopause is the boundary between the Stratosphere and the overlying Mesosphere.

The Mesosphere is the coldest naturally occurring place on earth; a temperature minimum of −90 °C or lower is reached at ~85 km. Mesosphere begins at ~50 km and extends up to 85 km. Its upper boundary is called the Mesopause.

The Thermosphere (also called the heated layer) begins at ~90 km and extends between 500 and 1,000 km above the earth. Although Thermosphere is considered part of earth's atmosphere, here, air density is extremely low, so it is generally considered as outer space. In fact, the general understanding is that space begins at an altitude of 100 km, slightly above the Mesopause, and at the bottom of the Thermosphere. Both the Space shuttle and the International Space Station (ISS) orbit the earth within the thermosphere!

The Exosphere is the region of atmosphere beyond 700 km where atoms and molecules escape into space. It is an extremely rarefied space, with very low density and high temperatures and with minimum atomic collusions. The boundary between the Thermosphere and the Exosphere above is called the Thermopause.

The next component is Lithosphere. It is the land part and includes all the solid materials constituting earth, from surface downward. Both crust and upper mantle make up the Lithosphere. Under continents, it is thickest, at about 120 km or so, whereas under the middle of oceans it is only a few kilometers thick. It is fragmented into massive plates that fit around the globe like pieces in a jig saw puzzle. These plates move independently and slowly, relative to one another and slide on top of a somewhat fluid (plastic) part of the mantle called the Asthenosphere.

The Hydrosphere includes all naturally occurring water on or below the surface. About 95.96 % of the Hydrosphere is made up of large surface bodies of saline water called Ocean. Around 72.5 % of the surface area of the earth ($\sim$361 million km^2) is covered by the five oceans—the Pacific, Atlantic, Indian, Arctic, and Antarctic. These and their associated extensions are called Seas and Bays. Rivers, lakes, bodies of frozen water (ice and snow), and groundwater form parts of the Hydrosphere. The Hydrosphere makes only about 0.03 % of the total mass of the earth.

The Biosphere is the fourth component and encompasses all zones (Lithosphere, Hydrosphere, and Atmosphere) on the earth in which life is present. Life on earth requires water, a source of energy (Sun light), and various nutrients found in soil, water, and air. The suitable combinations of these exist only in a narrow layer near the surface of the earth, within the upper layer of the earth's crust. This Biosphere is hierarchical in nature.

Earth's only satellite, the Moon, formed 4.527 ±0.01 billion years ago. It is a differentiated body (by igneous processes) with a geochemically distinct crust, mantle, and core. It is considered geologically dead and fossilized in time. The most acceptable hypothesis for its formation is that the Earth-Moon system formed as a result of a massive impact. A Mars sized cosmic body hits the nearly formed proto-earth, thereby, blasting material into the orbit around the proto-earth. These, then, accreted to form the Moon. The near-identical isotopic compositions of both Earth and Moon reaffirm this theory.

The origin of both the Solar System and earth are closely interlinked. French naturalist Buffon in 1745 was the first to put forward a hypothesis for the origin of the earth. According to him, our planet was formed through the cooling of a blob of solar matter ejected from the Sun during a cataclysmic collision with another large comet. After Buffon, many hypotheses were put forward for the origin of our planet that can be broadly grouped into two categories: Uniformitarian or Uniparental hypothesis and Cataclysmic or Bi-parental hypothesis. For the former, the Nebular, Meteorite, and Protoplanetary hypothesis are well-known and for the latte, the Planetesimal Hypothesis, and the Gaseous Tidal Theory. However, most of these explain only a few of the fundamental facts but fail, partially or completely, to explain other regularities.

So far, determining the exact age of earth directly from its rocks has failed as the oldest rocks have been recycled and destroyed by the process of Plate Tectonics. Nevertheless, scientists have been able to determine the probable age of the Solar System and are also able to calculate the age for earth by assuming that earth and the rest of the solid bodies within the Solar System formed at the same time and are, therefore, of the same age. The generally accepted age for the earth and the rest of the Solar System is 4.54 billion years (±0.5 billion years).

On earth, the oldest rocks, exceeding 3.5 billion years, are found on all continents. However, a recent study revealed that tiny crystals of mineral Zircon (detrital) from the Yilgarn Block, Western Australia gave an age upward of 4.4 billion years. An interesting feature of these ancient lake rocks is that they are not from any sort of "primordial crust" but are lava flows and sediments that were deposited in shallow waters, an indication that earth's history began well before these rocks were actually deposited. Hence, the existence of liquid water at 4.4 billion years ago has fundamental implications for the evolution of life on earth. Contextually, the microfossils records do date back to 3.5 billion years. For the age of the earth, there is excellent agreement on about 4.5 billion years, between several meteorite dates and of results by several different dating methods.

References

Bowring SA, Williams IS (1999) Priscoan (4.00–4.03 Ga) orthogneisses from northwestern Canada. Contrib. Mineral. Petrol. 134(1):3–16

Clarke FW, Washington HS (1924) The composition of the Earth's crust. US Geol Surv Prof Pap 127:1–117

Condie KC (1997) Plate Tectonics and Crustal Evolution. Butterworth-Heinemann, Oxford

Schopf JW (1993) Microfossils in the early archean apex chert: new evidence for the antiquity of life. Science 260:640–646

Taylor SR, McLennan SM (1985) The Continental crust: its composition and evolution. Blackwell, Oxford

Wilde SA, Valley JW, Peck WH, Graham CM (2001) Evidence from detrital zircons for the existence of continental crust and oceans on the earth 4.4 Gyr ago. Nature 409(6817):175–178

Windley BF (1995) The evolving continents, 3rd edn. John Wiley and Sons, New York

Chapter 5
Interior of the Earth

5.1 Introduction

The Core, Mantle, and Crust constitute the three main layers of earth (Fig. 5.1). Compositionally (chemically), the outer thin crust is mostly silicate (SiO_2-based) and mantle, the layer below is accompanied by metal oxides (such as MgO, FeO, Al_2O_3, CaO, and Na_2O) in mineral composition (Taylor and McLennan 1985) (see also Fig. 4.3 in the previous chapter). Mantle is the largest by volume, making up almost 87 % of the earth. The base of the mantle to the center of the earth (the last $^1/_8$ volume), is made up of iron (90 %), nickel (5 %) with a possible admixture of carbon, silicon, oxygen, sulfur, and hydrogen (comprising around 5 % by mass). The core makes up 35 % of the total mass of the earth and is probably made of almost pure iron, perhaps even in a single crystal form.

5.2 Structure

The earth is divided into layers (Fig. 5.1) whose salient characteristics are briefly mentioned below. The discussion follows from the uttermost to the inner most layer of the earth, i.e., from top to bottom. A more detailed discussion is given later in the chapter.

5.2.1 Continental Crust (0–75 km)

This is the outer most layers and forms the surface of the earth. It is primarily composed of crystalline rocks with low-density buoyant minerals that are largely dominated by silicates (Quartz; SiO_2) and feldspars (metal-poor silicates). As cold rock deforms slowly, this rigid and brittle outer layer is also called the Lithosphere (lithos meaning rocky or strong layer) (Fig. 5.1).

S. Jain, *Fundamentals of Physical Geology*, Springer Geology,
DOI: 10.1007/978-81-322-1539-4_5, © Springer India 2014

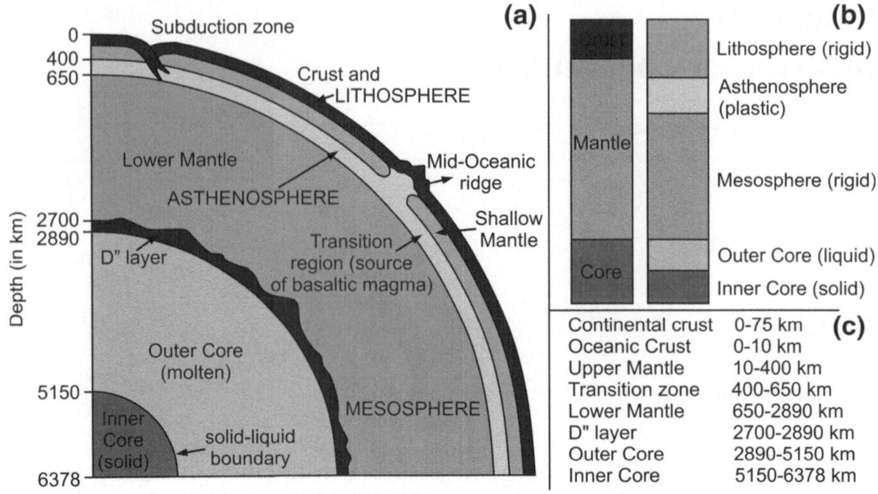

Fig. 5.1 **a** The interior structure of the earth. **b** Subdivisions of the three layers of the earth. **c** Respective thickness of these subdivisions

5.2.2 Oceanic Crust (0–10 km)

The majority of the earth's crust was made through volcanic activity. The oceanic ridge system, a 40,000 km long network of volcanoes, generates new oceanic crust at the rate of 17 km^3 per year, and covers the ocean floor with Basalt, an igneous rock. Hawaii and Iceland are two classic examples of such accumulations.

5.2.3 Upper Mantle (10–400 km)

Solid fragments of the upper mantle have been found in eroded mountain belts and volcanic eruptions. These include such minerals as Olivine [(Mg, Fe)$_2$SiO$_4$], Pyroxene [(Mg, Fe)SiO$_3$], and others that crystallize at high temperatures. The asthenosphere, part of the upper mantle (Fig. 5.1), might well be partially molten.

5.2.4 Transition Region (400–650 km)

The transition region or Mesosphere (for middle mantle) is sometimes also called the Fertile layer (Fig. 5.1). It is the source of basaltic magma and complex aluminum-bearing silicate minerals containing calcium, aluminum, and garnet. When cold, this layer is dense due to the presence of garnet. It is buoyant when hot as these minerals melt easily to form basalt which rises through the upper layers as Magma.

5.2.5 *Lower Mantle (650–2890 km)*

The lower mantle (Fig. 5.1) is probably composed of silicon, magnesium, and oxygen with some amounts of iron, calcium, and aluminum.

5.2.6 *D″ Layer (2700–2890 km)*

It is also called the D prime (D″) and is 200–300 km thick. Although it is often identified as part of the lower mantle (Fig. 5.1), seismic data suggest that this layer might differ chemically from the lower mantle.

5.2.7 *Outer Core (2890–5150 km)*

The outer core is hot and composed of electrically conducting liquid made mainly of iron and nickel. This conductive layer (Fig. 5.1) combines with earth's rotation to create a dynamo effect that maintains a system of electrical currents, thereby, creating the earth's magnetic field. It is also responsible for the subtle jerking of the earth's rotation. This layer is not as dense as pure molten iron, thus, suggesting the presence of lighter elements also. It is suspected that about 10 % of the layer is composed of sulfur and oxygen as these elements are abundant in the cosmos and also dissolve readily in molten iron.

5.2.8 *Inner Core (5150–6378 km)*

The inner core is made of solid iron and nickel and is suspended in the molten outer core, unattached to the mantle (Fig. 5.1). It is believed to have solidified as a result of pressure-freezing which occurs to most liquids under extreme pressure.

Table 5.1 enumerates salient characterisitics of the discussed layers.

Table 5.1 Detailed characteristics of earth's layers

Earth's layers	Depth (km)	Density (gm/cm^3)	Temperature (°C)	Percent of Earth's mass	Percent of mantle-crust mass	Mass fraction of mantle	Mass fraction of Earth		Volume fraction of Earth		P-wave velocity (km/sec)	Composition
Continental crust	0–55 (average = 40)	2.6–2.8	<1000	0.371	0.554	–	–	–	–	–	6	Granitic (felsic) rocks such as Granodiorite and Granite enriched in K, Na, Al, and Si
Oceanic crust	0–10 (average = 7)	3.5		0.099	0.146	0.006	0.004	–	0.008	–	7	Mafic rocks such as Basalt and Gabbro enriched in Mg and Fe
Mohorovičić discontinuity (Moho) (Crust-Mantle boundary)												
Upper mantle	10–400	3.4–4.4	1000–3700	10.21	15.3	0.154	0.1	0.68	0.17	0.84	8–12	Peridotite, Eclogite, Olivine, Spinel, Garnet, Pyroxene, and Perovskite
Transition region	400–650	–	–	7.5	11.1	0.24	0.17		0.22			Primarily Peridotite
Lower mantle	650–2890	4.4–5.6		48.01	72.9	0.6	0.41		0.44			Magnesium and Silicon Oxides
D′′ layer	(Gutenberg discontinuity)		2700–2890	–	–	3	~4	–	–	–	–	–
–	Post-Perovskite											
Outer core	2890–5150	9.9–12.2	3700–4300	29.21	–	–	–	0.32	0.154	0.16	8–10	Liquid Iron and Nickel and a bit of Sulfur
Inner core	5150–6378	12.8–13.1 ?13.5	>4300	1.6	–	–	–		0.008		11–12	Solid Iron and Nickel and a bit of Sulfur

Table 5.2 Various methods used to infer earth's internal structure and composition

Methods	Results
Ophiolites	They represent oceanic lithosphere
Xenoliths in volcanic rocks	They represent the upper Mantle
Seismic reflection	This method identifies changes in lithology
Seismic refraction	This defines velocities of seismic waves at depth
Electrical conductivity	This identifies partial melts
Geochemistry and elemental abundances	This method tells the range of earth's composition
Gravity anomalies	This identifies density differences in the earth's interior
Lithospheric flexure	This constrains Rheology
Magnetic anomalies	This method shows distribution of subsurface rocks
Mineral physics	This measures seismic velocities in rock samples
Seismic tomography	This method permits the 3D visualization of the earth's interior

5.3 Methods to Infer Earth's Interior

There are several methods that enable us to infer about the earth's interior. However, here, only major ones are briefly discussed (Table 5.2).

5.3.1 Xenoliths

Xenoliths are like chocolate chunks in a cookie dough; pieces of mantle within the lava (foreign rock inclusions in an igneous rock). Xenoliths provide information about the lower crust and the upper mantle (see Nixon 1987). They are a storehouse of valuable information about the composition of continental and oceanic crusts that the otherwise inaccessible Mantle, does not provide. The presence of coarse-grained Olivine (Peridotite) in basaltic lavas is one such example which has been brought up from far below. However, such volcanic information is only useful up to a depth of about 200 km, the depth from which Xenoliths are brought to the surface. Xenoliths range in size from a grain of sand to that of a boulder, being as big as a foot.

5.3.2 Ophiolites

Ophiolites are rich in iron-magnesium silicate minerals that once originated deep within the earth's interior. Now lying on the surface, they are unstable and hence, convert rapidly into hydrated magnesium silicate minerals, forming serpent-like bands with vivid green/brown colors in the rock. Hence, the name Ophiolites (in

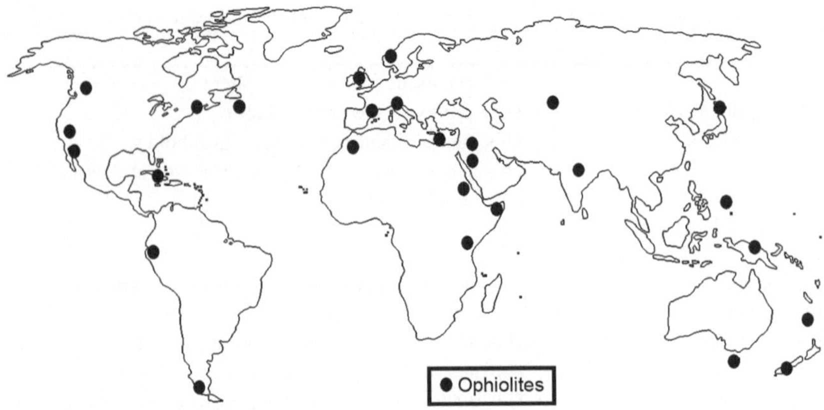

Fig. 5.2 Global distribution of ophiolites

Greek 'ophis', means snake, and 'lithos', means rock) (see also Shervais, 2001; Moores, 2003). They were first described from the Alps in the early twentieth century, and later their presence was noted from almost every continent on the earth (Fig. 5.2).

The Ophiolites provide another line of direct evidence of the earth's interior. They are sections of the earth's oceanic crust and the underlying upper mantle that has been uplifted or emplaced and exposed within the continental crustal rocks and characterize an assemblage of rocks that are formed at spreading ridges (Fig. 5.3a). Ophiolites are interpreted to be thrust sheet of ancient oceanic lithosphere that has been obducted over the continental crust in the course of orogeny. Some large ophiolitic complexes are more than 10 km thick, 100 km wide, and 500 km long.

Ophiolites show three distinct peaks of production in the earth's history, at about 750, 450, and 150 Ma, respectively (Fig. 5.3b) (see also Dilek and Robinson 2003). These time periods are referred to as Ophiolite pulses. Each pulse corresponds to a period of worldwide magmatic events represented by the intrusion of voluminous granitic rocks. Generally, Ophiolites issued by each pulse tends to form a particular ophiolite belt. For example, the Late Proterozoic ($\sim$750 Ma) ophiolites are distributed in the Pan-African orogenic belt, the Early Paleozoic ($\sim$450 Ma) ones appear in the Appalachian-Caledonian-Uralian belt, and the Mesozoic ($\sim$150 Ma) ophiolites dominate the Alpine-Himalayan belt.

Famous Ophiolite complexes include the Semail ophiolite in Oman (Mesozoic), the Troodos ophiolite in Cyprus (Mesozoic), the Papua ophiolite in Papua-New Guinea (Mesozoic), and the Bay of Islands ophiolite in Newfoundland (Paleozoic).

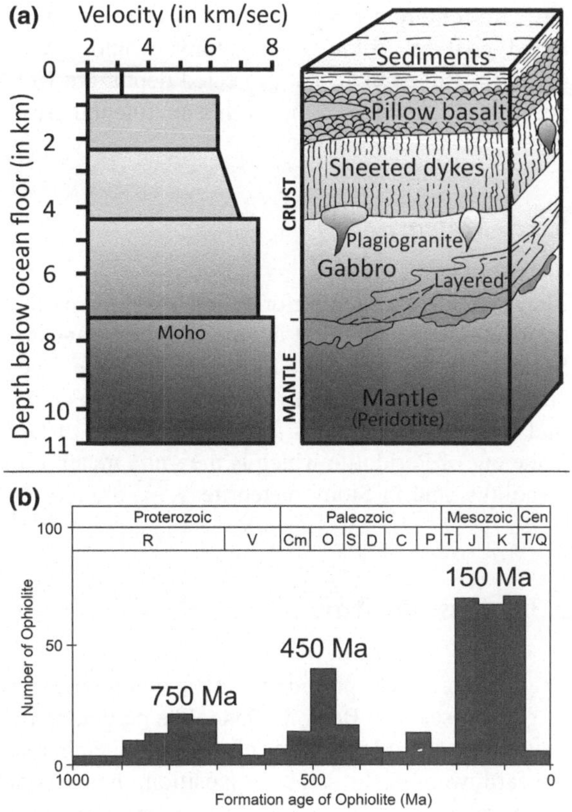

Fig. 5.3 Ophiolites are sections of the earth's oceanic crust and the underlying upper mantle that have been uplifted or emplaced and exposed within the continental crustal rocks. These assemblages of rocks are used to recognize ancient convergent plate boundaries. **a** Cross-section of an ideal Ophiolite succession. **b** Ophiolitic episodes in time

5.3.3 Volcanism

Volcanism (both Recent and Ancient i.e., Paleovolcanism) provides another evidence of information about the upper mantle. Paleovolcanologic studies reveal the geodynamical conditions that existed at the time of eruption. Such a study helps to identify belts of paleo-ridges, zones of paleo-spreading and paleo-transform faults.

5.3.4 Drilling

Drilling, as a tool to understand the earth's interior, is limited to few kilometers below due to increased geothermal gradient within the earth's crust (with each km depth the temperature increases by ~25 °C, the temperature gradient). Most drilled holes are in the upper 7 km of the crust and the deepest one (the Kola Superdeep Borehole drilled from 1989 to 1994) is about 12 km deep in the northern Kola Peninsula, NW Soviet Union, Russia. The rocks encountered were 2.7 billion years old with a bottom temperature of ~180° C (365 °F). If the hole had reached its initial goal of 15 km, temperatures would have reached an estimated 300 °C (572 °F). But even at that projected depth, the Kola project would have had only penetrated a fraction of the earth's continental crust, whose thickness is ~75 km.

5.3.5 *Meteorites*

They provide excellent information about the earth's interior and are thought to be remnants of the core and mantle of other planetary bodies from within the Solar System; all of which were formed at the same time and from the same material as our earth. Stony meteorites are very similar in composition to the materials that we find within the Xenoliths and at the bottom of Ophiolites. The earth's mantle is made out of Peridotite which is the same material that is also found in Ophiolites, Xenoliths, and in Stony meteorites.

5.3.6 *Seismic Waves*

The seismic waves provide the most comprehensive picture of the earth's interior (Fig. 5.4) (see also Benioff 1954). The particular velocity at which a seismic wave travels through a layer gives clues about the chemical composition of the layer. If the earth were of the same composition, then seismic waves would, like any other wave, take longer to travel further and die out in velocity and strength with increasing distance (this decrease is called Attenuation). However, down ~ 200 km, the seismic waves arrive with higher velocities than those within a 200 km radius (Fig. 5.4a) indicating the presence of a denser layer below; the seismic waves travel faster in denser material. Based on this fact, scientists detected a boundary within the earth's interior, a boundary between the crust and a denser layer below, the mantle. This crust-mantle boundary is also called the Mohorovičić discontinuity (better known as Moho) (Fig. 5.4b), in honor of its Croatian discoverer, Andrija Mohorovičić. Such sudden jumps in seismic velocities across a boundary are called Seismic discontinuities. Hence, a systematic study of the waves and their propagation gives a robust idea about the earth's interior, its structure and composition.

5.4 Seismology and Earth's Interior

Earthquakes produce waves that probe deep into the earth (see Dziewonski and Anderson 1984). There are three types of seismic waves: Primary (P-waves), Secondary (S-waves), and Surface (L-wave) (Fig. 5.5). L-waves travel only on the

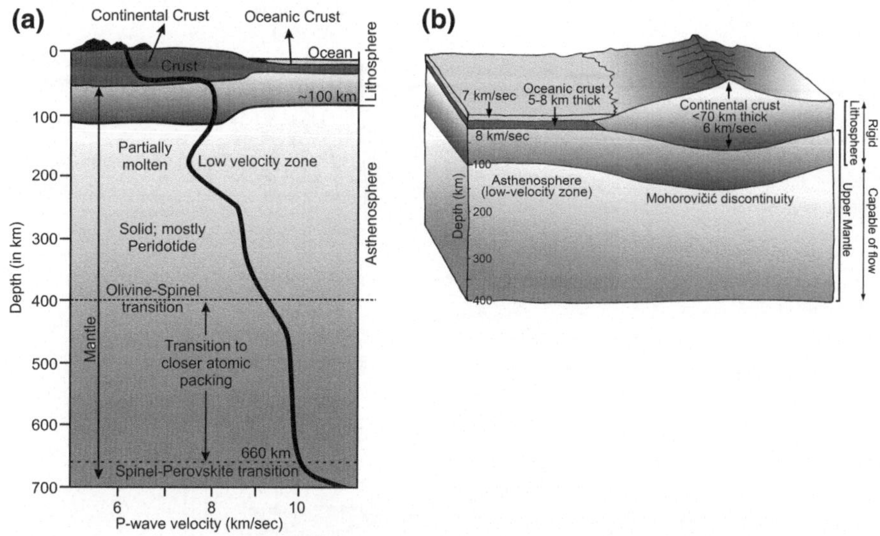

Fig. 5.4 **a** Changes in P-wave velocity and composition in earth's crust and upper mantle. Generally, the velocity of P-wave increases with depth. However, due to increase in temperature within the asthenosphere, seismic waves slow down as the rocks become plastic. This is called the Low velocity zone. At around the 660 km discontinuity, wave velocity increases rapidly marking the boundary between the upper and the lower mantle, probably because of a change in mineral content due to increasing pressure (Olivine to Perovskite). **b** The position of the Mohorovičić discontinuity (better known as Moho)

surface, and therefore are not helpful in studying the earth's interior. P-waves are the fastest and travel away from a seismic event. They are longitudinal in propagation and can travel through solids, liquids, and gases. S-waves travel slower than the P-waves and travel through solids only. Measuring of these waves is called Seismology and the instrument that does it, is called a Seismograph (Fig. 5.5b). On a Seismograph, we see the arrival of a sequence of waves; typically P-waves arrive first and L-waves, the last (Fig. 5.5a).

The properties of a material (such as composition, mineral phase, packing structure, temperature, and pressure) have a strong bearing on the velocities of seismic waves. They travel more quickly through denser material and therefore travel more quickly with depth (both pressure and density increase downwards). Hot (molten) areas slow down seismic waves as they move slowly through a liquid than a solid. Hence, molten areas slow down P-waves and stop S-waves as their shearing motion cannot be transmitted through a liquid, whereas partially molten areas slow down P-waves and weaken S-waves. Thus, the study of seismic waves gives a good idea about the earth's interior, its structure, and composition.

When seismic waves pass between geological layers with contrasting seismic velocities—Reflection and Refraction (bending) occurs (Fig. 5.6).

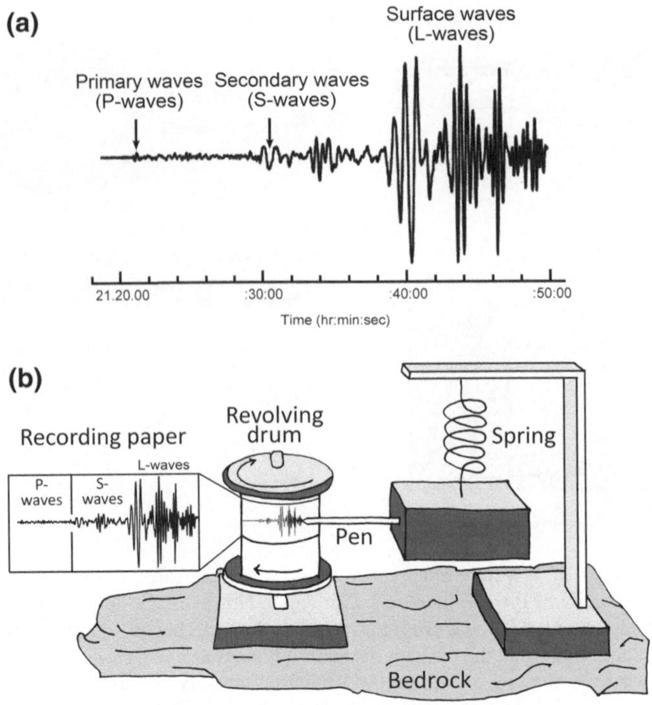

Fig. 5.5 The Seismogram and Siesmograph. **a** A schematic seismogram shows the arrival order and pattern produced by P-, S-, and L-waves. When an earthquake occurs, body and surface waves radiate out from the focus of the earthquake at the same time. Because P-waves are the fastest, they arrive at the seismograph first, followed by S-waves and then by surface waves, the slowest of them all. The difference between the arrival times of the P- and S-waves is the P–S time interval; it is a function of the distance where the seismograph station is from the focus. **b** The Siesmograph, the instrument for measuring seismic waves

5.4.1 Reflection and Refraction of Seismic Waves

The Reflection and Refraction of seismic waves at density contrasts follows exactly the same laws that govern the reflection and refraction of light through a prism. If composition (or physical properties) of a layer changes abruptly at a boundary, then seismic wave will reflect off the boundary (Fig. 5.6a) and refract (or bend) as they pass through it (Fig. 5.6b, c). If the seismic wave velocity in the rock above a boundary is less as compared to the one below, the waves will be refracted or bent upward relative to their original path (Fig. 5.6b). If the seismic wave velocity decreases when passing into the rock below the boundary, the waves will be refracted down relative to their original path (Fig. 5.6c). Refraction has an important effect on waves that travel through earth. In general, the seismic velocity in earth increases with depth and refraction of waves causes the path followed by

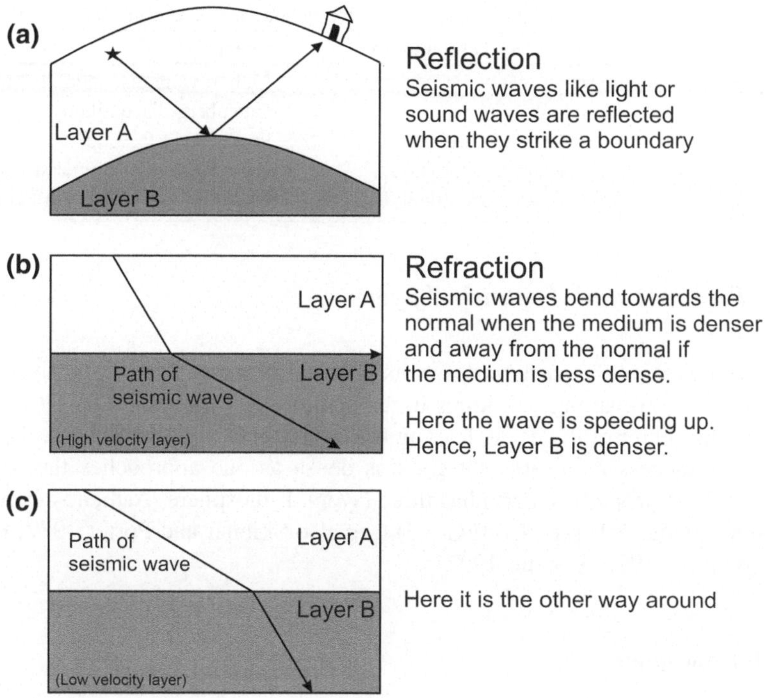

Fig. 5.6 Reflection and Refraction of seismic waves

body waves to curve upward. This shift in wave direction gives a good idea about the structure and density of the layers present in the earth's interior.

5.4.2 Velocity of Seismic Waves

The velocity of seismic wave varies with depth. Distinct boundaries (also called seismic discontinuities) are observed when there is a sudden change in their physical properties or chemical composition. From these discontinuities, the nature of various layers within the earth is revealed.

5.5 Layers of the Earth

The earth's interior can be classified into various layers on the basis of their physical and tectonic properties (Table 5.3; see also Fig. 4.3).

Table 5.3 Compositional and Phase change mark the boundaries of various layers of the earth

Type of change	Boundary/Discontinuity	Characteristics
Compositional	crust/mantle	Mohorovičić discontinuity (Moho)
	mantle/core	Gutenberg discontinuity
Phase change	inner/outer core	From liquid to solid
	400 km discontinuity	From Olivine to Spinel structure
	660 km discontinuity	From Spinel structure to Perovskite

5.5.1 Physical and Tectonic Layers

The material inside the earth varies in its physical property. Physical property is a function of the prevailing variations in temperature and pressure. Denser material settles to the center of the earth, leaving the lighter ones at the top. Thus, the earth consists of successive layers that get less dense as one approaches the surface. Based on these properties, earth has three layers: Lithosphere, Asthenosphere, and Mesosphere (Fig. 5.1; see also Fig. 4.3) (see also Skinner and Porter 1987; Fifield 1988; Windley 1995; Condie 1997).

5.5.1.1 Lithosphere

The uppermost mantle and crust together constitute the rigid layer of rock called the Lithosphere (*lithos* is the Greek word for stone). Crust is the upper part of the Lithosphere, and upper mantle is the lower part of the Lithosphere (Fig. 5.1). Within the mantle, the upper part is cooler and more rigid than the deeper. Hence, the upper part behaves more like the overlying crust. Lithosphere is thinnest under oceans and thickest under continents. The Lithosphere is broken into moving plates that contain the world's continents and oceans. The base of the Lithosphere and the top of the next layer, the asthenosphere, is marked by a sudden decrease in velocities of both P- and S-waves, at around 100 km (Fig. 5.4).

5.5.1.2 Asthenosphere

A relatively narrow, mobile zone exists below the Lithosphere and within the Mantle. This is called the asthenosphere (*Asthenes* is a Greek word for weak or plastic) (Fig. 5.1). Nearer to the surface of the earth, temperature is relatively high but the pressure is greatly reduced, thus, mantle is partially molten and accordingly, this zone is composed of hot, semi-solid material, that can soften and flow. The rigid lithosphere is thought to "float" or move about on this slowly flowing asthenosphere. The asthenosphere is the likely source of much of the basaltic magmas.

5.5.1.3 Mesosphere

Mesosphere is the layer below the Asthenosphere and includes the majority of the mantle and core. This layer is more of physical significance rather than tectonic. The boundary between Mesosphere and Asthenosphere is marked by an abrupt increase in the velocities of seismic waves at a depth of about 400 km (Fig. 5.4), representing a temperature and pressure change. This change is marked by a polymorphic phase transition, a change in the crystal structure of olivine, one of the most abundant mineral within the mantle.

5.6 Internal Structure

Based on density, composition and physical properties, earth's interior is broadly divided into Crust, Mantle, and Core.

5.6.1 Crust

It is the earth's outmost layer and is also the only layer that can be sampled directly. It is much thinner than any of the other layers. Almost 31 % of continental area is submerged beneath the oceans (Cogley 1984), and is thus, less accessible to geological sampling. For this reason, most estimates of continental crust composition come from the exposed regions of continents. Crust is quite homogenous in composition, and is primarily composed of the least dense minerals such as calcium (Ca), sodium (Na), and some aluminum-silicates. Being relatively cold, it is also rocky and brittle; hence, it fractures during earthquakes. The crust is of two types: Oceanic and Continental (Fig. 5.4). Both are less dense and contain more silica than the underlying mantle. Oceanic crust is thinner, denser, and contains less silica and aluminum and more magnesium and iron than the continental crust. It underlies the oceans. It is on an average 0–10 km thick and is made by the volcanic activity at oceanic ridges. Lack of silica makes it darker than the continental crust. Oceanic ridges generate new crust at the rate of 17 km^3 per year, whereas at continental boundaries, a similar amount of material is subducted and recycled (Fig. 5.1).

The Continental crust is composed of crystalline rocks (mainly Quartz and Feldspars) and is between 0 and 75 km in thickness. It has an average composition approximating to that of an Andesite (Clarke 1889, Clarke and Washington 1924). As the continental crust is thicker and made of less dense material than the oceanic crust, it floats on it. It sits in isostatic equilibrium on the magma and is divided into plates that move relative to each other, thereby, gradually changing the alignment of continents, over time.

5.6.1.1 The Mohorovičić Discontinuity

It marks the lower limit of the earth's crust; a discontinuous layer separating crust and mantle (Fig. 5.4b). It was discovered in 1909 by Andrija Mohorovičić, a Croatian seismologist. He noted that the velocity of a seismic wave is related to the density of the material that it is moving through. He interpreted the acceleration of seismic waves within earth's outer shell as representing a compositional change within the earth. The acceleration was caused by the presence of a higher density material at depth. This higher density material is the mantle. Hence, the Mohorovičić discontinuity is a transition from lower to higher density silicates; the transition layer being 0.1–0.5 km thick, 5–10 km below the ocean floor or 20–90 km beneath the continents. Hence, the crust is thicker under land than under oceans (Fig. 5.4b). The layer has a temperature of 500–600 °C at the base of the continental crust and 150–200 °C in the sub-oceanic discontinuity with an average density of about 4 gm/cm^3. The discontinuity is also the boundary between the felsic/mafic crust (granitic rocks in continents and basaltic rocks in ocean basins) with seismic velocity (P-waves) around 6 km/sec and the denser ultramafic mantle (ultramafic rock—pyrolite or something similar) with velocity around 8 km/sec (Fig. 5.4b).

5.6.2 *Mantle*

A 2,890 km thick, hotter, and denser layer below the crust is the Mantle (Fig. 5.1). It is ultramafic in composition (made of Fe, Mg silicates) and contains 70 % of the earth's mass. The mantle makes up the bulk of the earth's volume and is the primary source of all basaltic igneous rocks found at the surface. Here, temperature increases with depth with an average gradient of $\sim$25 °C per km. This increase in temperature is primarily caused by the radioactive decay and release of energy by the disintegration of Thorium, Uranium, and Potassium isotopes. Mantle is divisible into three layers: uppermost mantle, asthenosphere, and lower mantle.

5.6.2.1 Uppermost Mantle

It is rigid and fused to the crust and forms part of the Lithosphere with a depth of about 100 km. Parts of this upper mantle has been found in eroded mountain belts and contains Olivine and Pyroxenes. These minerals crystallize at very high temperatures, and therefore it is inferred that part of the upper mantle may be partially molten.

5.6.2.2 Asthenosphere

The asthenosphere (Fig. 5.1), at low temperatures, is dense due to the presence of garnet. However, at high temperatures, it is buoyant as the mineral (and some others such as calcium, aluminum) easily melts to form basalt, that rises through the upper layers as magma. Asthenosphere is plastic and partially molten (containing around 3 % melt).

5.6.2.3 Lower Mantle

The lower mantle accounts for nearly half of the earth's mass. Mineralogically, the lower mantle mostly consists of magnesium silicate Perovskite (Mg, Fe, A1)(Si, A1)O_3, Magnesiowustite (Mg, Fe)O, and calcium silicate Perovskite ($CaSiO_3$). Additionally, the lower mantle is composed of oxides of silicon and magnesium and probably also iron, calcium, and aluminum. Some scientists have suggested that compositionally, the lower mantle is homogeneous, however, others have found compositional discrepancies. Within the mantle, the seismic velocities gradually increase with depth, except for a low velocity zone at around 660 km (Fig. 5.4a). This discontinuity results from the change from Olivine to a denser Perovskite crystalline structure (pressure collapses the internal structure of some minerals into denser packings), which remains stable to the base of the mantle and is thought to represent a major boundary separating the less dense upper mantle from the denser lower mantle. However, others suggest that the absence of earthquakes below 660 km is an indirect evidence, although inconclusive, of a chemical boundary (that prevents penetrative convection) instead of a phase change boundary (from Ringwoodite to Perovskite and Magnesiowustite).

5.6.2.4 Gutenberg Discontinuity

The German geophysicist Beno Gutenberg in 1913 discovered a transition zone between the lower mantle and the outer core boundary (Fig. 5.1). This is called the Gutenberg discontinuity, a boundary at 2,890 km, marked by an abrupt change in seismic waves. Here, the P-waves decrease in velocity while the S-waves disappear completely. It must be noted that the S-waves cannot transmit through liquids; hence, it is believed that the unit above the discontinuity is solid, whereas the unit below is liquid or molten in nature.

5.6.3 Core

The diameter of the core is about the size of planet Mars, extending from 2,890 to 6,378 km (around 3,488 km thick). The core is nearly twice as dense as the mantle and is composed of iron with minor amounts of nickel and sulfur. It has two distinct layers, a 2,260 km-thick liquid outer core and a 1,228 km-thick solid inner core (Fig. 5.1). As earth rotates, the outer liquid core spins, and thus creates the earth's characteristic magnetic field. Hence, the outer core is a hot, electrically conducting liquid within which convective motions occur. This conductive layer combines with earth's rotation to create a dynamo effect that maintains a system of electrical currents known as the earth's magnetic field.

Seismologists, after discovering the presence of two shadow zones, P- and S-wave (Fig. 5.7). The S-wave shadow zone occurs where S-waves do not reach the area on the opposite side of the earth from the epicenter of the earthquake. Their absence (of direct S-waves) suggests that they do not pass through the core (Fig. 5.7). At larger distances, some P-waves do arrive, but no S-waves. This means that the core or at least part of the core is in a liquid state, as S-waves are not transmitted through liquids. Hence, a liquid outer core best explains the S-wave shadow zone. Between 103° (11,500 km from the epicenter) and 142° (15,500 km from the epicenter) from the epicenter (earthquake), another refraction is recognized resulting from a sudden increase in P-wave velocities (P-wave shadow zone) at a depth of 5,150 km (Fig. 5.1). This velocity increase is consistent with a change from a molten outer core to a solid inner core. Thus, earthquake waves suggest that the earth has a dense liquid core, of about one-half the radius of the earth and inside that, a solid inner core.

The inner core is a solid iron sphere with some nickel. Here, the iron is in a solid state due to immense higher pressure with only slightly higher temperature as compared to the outer core. While higher temperature melts materials, but higher

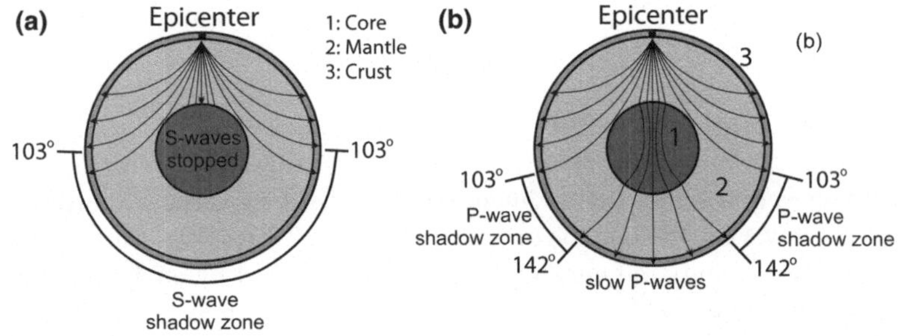

Fig. 5.7 The cross-section of the earth showing the paths of P- and S-waves. **a** The transverse S-waves cannot travel through the liquid outer core. They can travel through the mantle because the mantle behaves more like a solid than a liquid. The S-waves curve as they move through the mantle due to refraction as the density of the mantle changes. There is a large part where no S-waves are detected indicating that the outer core is liquid as it blocks S-waves (the S-wave shadow zone). **b** The longitudinal P-waves can travel through the whole planet. They also curve with the changing density of both mantle and core. The P-waves change direction suddenly at the boundary between the different layers of the earth. This is due to refraction caused by the different densities of the layers. P-waves do reach the side of earth opposite the earthquake, but their interaction with earth's core produces another shadow zone, where no P-waves are seen, called to P-waves shadow zone

pressures create solids. Hence, the outer core is liquid as the pressure is lower and the remainder of the iron-nickel inner core is solid. Another possible explanation for a solid core is that as the earth cools, and the heat flows out of the core; iron from the molten outer core solidifies onto inner core. This increases the concentration of lighter elements in outer h, and if these elements are near the saturation point they will also solidify out. However, as they are lighter than iron, they float to the top of the core and collect at the core-mantle boundary.

5.7 Summary

The Crust, Mantle, and Core constitute the three main layers of the earth. Compositionally, the outer thin crust is mostly silicate (SiO_2-based) and mantle, the layer below, is largely made up of silicates along with metal oxides in mineral compositions and the Core is probably made of almost pure iron, perhaps even in a single crystal form. Within the crust, two layers are noted: Continental crust (0–75 km), the outer most layer which is primarily composed of crystalline rocks with low-density buoyant minerals that are largely dominated by silicates (SiO_2) and feldspars (metal-poor silicates) and the oceanic crust (0–10 km) is made through volcanic activity. The mantle is divisible into three layers—upper mantle (10–400 km), transition region, middle mantle (400–650 km), and lower mantle (650–2,890 km). The asthenosphere, part of the upper mantle, might well be partially molten. The transition region or Mesosphere (for middle mantle) is sometimes also called the Fertile layer. It is the source of basaltic magma and complex aluminum-bearing silicate minerals. The lower mantle is probably composed of silicon, magnesium, and oxygen with some amounts of iron, calcium, and aluminum. The succeeding D" layer (2,700–2,890 km), also called the D prime (D'') is 200–300 km thick and is often identified as part of the lower mantle. A transition zone exists between the lower mantle and the outer core boundary. This is called the Gutenberg discontinuity, a boundary at 2,890 km, marked by an abrupt change in seismic waves. Here, the P-waves decrease in velocity while the S-waves disappear completely. The outer core (2,890–5,150 km) is hot and composed of electrically conducting liquid made mainly of iron and nickel. The inner core (5,150–6,378 km) is made of solid iron and nickel and is suspended in the molten outer core, unattached to the mantle.

There are several indirect evidences that enable us to infer about the earth's interior such as Xenoliths and Ophiolites. Xenoliths provide a glimpse of the composition of continental and oceanic crusts, that the otherwise inaccessible Mantle, does not provide. However, they provide useful information about the earth's interior up to a depth of about 200 km only. Ophiolites are rich in iron-magnesium silicate minerals that originated deep within the earth's interior. Now lying on the surface, they provide another line of direct evidence of the earth's interior. They are sections of the earth's oceanic crust and the underlying upper mantle that has been uplifted or emplaced and exposed within the continental crustal rocks and characterize an assemblage of rocks that are formed at spreading ridges. Besides these, other evidences include: Volcanism, Drilling, Meteorites and Seismic waves. There are three types of seismic waves: Surface (L-wave), Primary (P-waves), and Secondary (S-waves). L-waves travel only on the surface, and therefore are not helpful in studying the earth's interior. P-waves are the fastest and travel away from a seismic event. They are longitudinal in propagation and can travel through solids, liquids, and gases. S-waves travel slower than the P-waves and travel through solids only. Measuring of these waves is called Seismology and the instrument that does it is called a Seismograph. Hence, of all the evidences, the seismic waves provide the most comprehensive picture of the earth's interior and have enabled to detect a crust-mantle boundary within the earth's interior. It is called the Mohorovičić discontinuity (better known as Moho).

References

Benioff H (1954) Orogenesis and deep crustal structure - additional evidence from seismology. Geol Soc Am Bull 65:385–400

Clarke FW (1889) The relative abundance of the chemical elements. Phil. Soc. Washington Bull XI:131–142

Clarke FW, Washington HS (1924) The composition of the earth's crust. USGS Prof Pap 127:1–117

Cogley JG (1984) Continental margins and the extent and number of continents. Rev Geophys 22(2):101–122

Condie KC (1997) Plate tectonics and crustal evolution. Butterworth-Heinemann, Oxford

Dilek Y, Robinson PT (eds.) (2003) Ophiolites in earth history. Geological Society, London (special publications 218)

Dziewonski AM, Anderson DL (1984) Seismic tomography of the earth's interior. Am Sci 72:483–494

Fifield R (1988) Structure of the earth. New Scientist, Inside Science 1–4

Moores EM (2003) A personal history of the ophiolite concept. In Dilek and Newcomb (eds) Ophiolite concept and the evolution of geologic thought, vol 373. Geological Society of America, Special Publication, pp 17–29

Nixon PH (1987) Mantle Xenoliths. John Wiley & Sons, Chichester p 844. ISBN 0-471-91209-3

Shervais JW (2001) Birth, death, and resurrection: the life cycle of suprasubduction zone ophiolites geochemistry, Geophysics, Geosystems, vol 2. doi:10.1029/2000GC0000802000GC000080

Skinner BJ, Porter SC (1987) Physical geology the earth: inside and out. John Wiley and Sons, New York, ISBN 0-471-05668-5

Taylor SR, McLennan SM (1985) The continental crust: Its composition and evolution. Blackwell, Oxford

Windley BF (1995) The evolving continents. John Wiley and Sons, New York

Chapter 6
The Geological Timescale

6.1 Introduction

Scientists are continuing to unravel the events that might have occurred millions and billions of years ago in the earth's history. However, from our human perspective, it is very difficult to appreciate the immensity of geologic time, considering the fact that the first modern man (*Homo sapiens*), appeared only a mere 0.1 second ago in the earth's $\sim$4.54 billion year long history. Figure 6.1 gives a brief glimpse of the expanse of time for processes and events that commonly occur around us. There are two basic ways we can try to make sense of this vast geologic time, i.e., through relative and absolute dating.

6.2 Early Attempts to Determine Earth's Age

Through genealogies and the history recorded in the Bible, Archbishop James Ussher (1664) determined the date of creation to be 4004 BC, which required the earth and all its features to be no more than about 6,000 years old. This and similar ideas about earth's history dominated Western thinking until the eighteenth century (Brush 1982; Berry 1988) (Fig. 6.2).

6.2.1 Early Scientific Efforts

During the eighteenth and nineteenth centuries, scientific attempts were made to determine the earth's age (Albritton 1980; Burchfiel 1990; Dalrymple 1991) (Fig. 6.2). Georges Louis de Buffon (1707–1788; French naturalist, mathematician, and cosmologist) assumed that earth was originally molten and calculated its age to be at least 75,000 years from the earth's present temperature. He assumed a rapid rate of cooling. Later, the rate of sedimentation was used to calculate the age of the

S. Jain, *Fundamentals of Physical Geology*, Springer Geology,
DOI: 10.1007/978-81-322-1539-4_6, © Springer India 2014

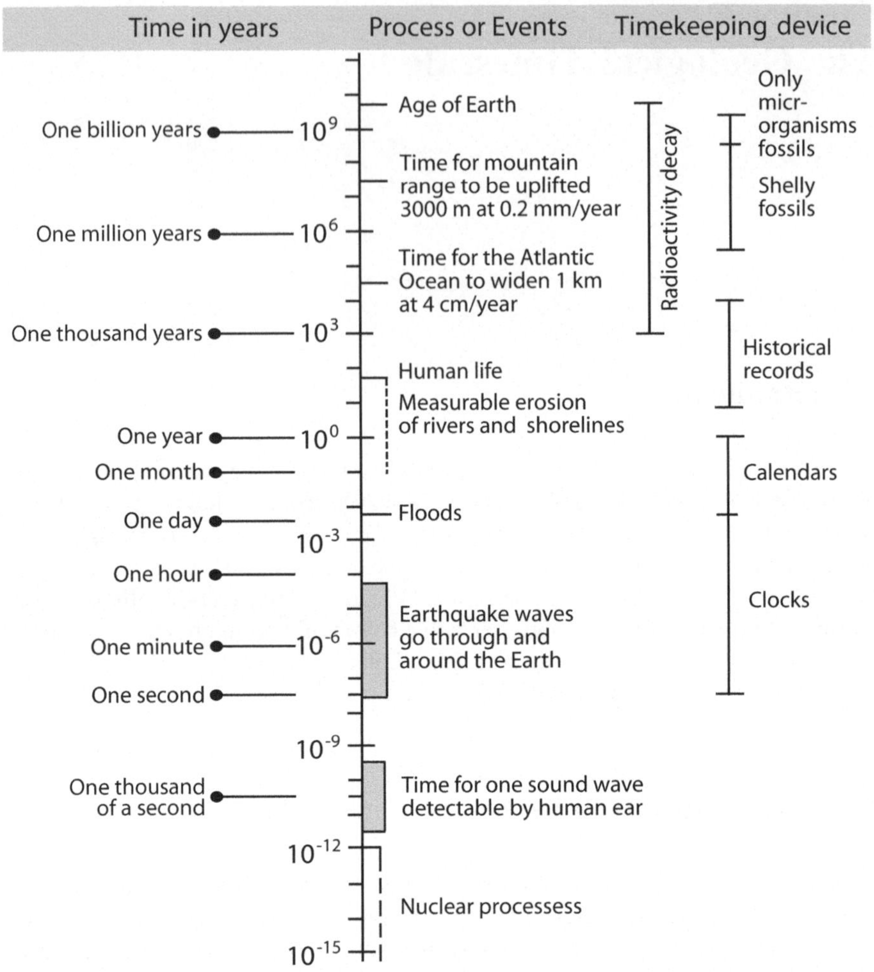

Fig. 6.1 A brief glimpse to appreciate the expanse of time for processes or events that commonly operate around us

earth. The age of the earth was also calculated from deposition rate of various sediments and the thickness of sediments in the earth's crust. This resulted in ages ranging from 3 million to 1.5 billion years for the earth's origin. Problem with these rate-based approaches is that they are never uniform for particular sediment; the sediment is removed by erosion, modified by compaction, or sometimes not even deposited at all. Hence, a complete record of sedimentation does not exist and the rates are never uniform, either. John Joly (1857–1933; an Irish physicist) used Ocean's salinity to calculate the age of the earth. He used the current salinity of the

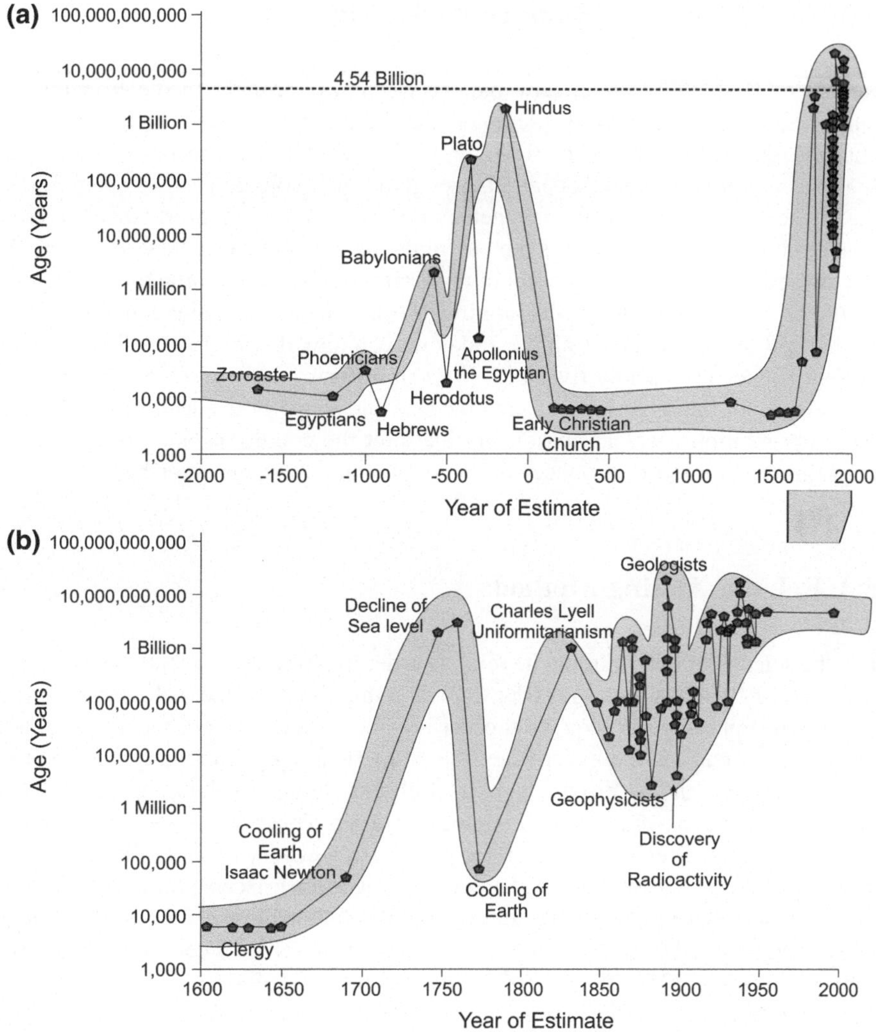

Fig. 6.2 The age of the earth. **a** Early scientific efforts. **b** Post-sixteenth century estimates heralding the advent of radioactivity and the beginning of more precise scientific dating

ocean and assumed that the water was originally pure and that the salt in it was derived from the erosion of continents. The age of the earth was estimated at about 90 Ma. However, the problem with this salt-based approach was that the rate of erosion is not constant, and hence, the loss and recycling of salt was never taken into account! Additionally, salt does not exclusively come from continents. Thus, most early attempts in deciphering the age of the earth remained tentative.

6.3 The Principle of Uniformitarianism

Recent understanding about how old the earth was brought in the Principle of Uniformitarianism. This principle holds that there are laws of nature that have not changed during the course of time. In 1785, James Hutton recognized that the present-day processes have operated throughout the geologic past. In other words, the phenomena of gravity and magnetism operated in the past pretty much as they do so today, i.e., "the present is the key to the past" and that the laws of nature do not change with time. This concept of uniformitarianism is important in geology as it allows us to infer that rock-forming processes have remained more or less unchanged over time. For example, today we can directly observe the formation of current ripples on a sandy river bed. We can measure the size of the sand grains, the velocity of the river currents, and the size and shape of the sand ripples. Given these observations, we can reasonably interpret the conditions and processes that might have operated to produce similar ripples in an ancient river bed.

6.4 Relative Dating Methods

It is the science of determining the relative order of past events, without necessarily determining their absolute ages (Fig. 6.3). It is important to note that relative dating is about deciphering the sequential order in which a series of events occurred, not when they occurred (the exact time; the Numerical age). This latter approach is absolute dating and is discussed later. Relative dating methods use geological principles to place events in a chronological order. These are called Stratigraphic Laws (Fig. 6.4). They enable us to decipher earth's history by carefully looking at how sediment layers are deposited and eroded over time. Hence, they are excellent tools for deciphering the spatial and temporal relationships of rock layers. These laws, largely based upon the works of Nicolaus Steno (Niels Steensen 1638–1686), James Hutton (1726–1797), and William Smith (1769–1839), help in better understanding the relative order in which the layers were formed. However, to understand this order, a big assumption is made, that the geologic processes of the present operated in the same manner as in the present and were constrained by the same laws of physics as they operate today like those of gravity and magnetism. This concept is called the Principle of Uniformitarianism. Interestingly, the other principles, briefly discussed below, have also not changed much in their formulation since their publication in Charles Lyell's book—"Principles of Geology" (1830–1833).

These six fundamental principles are (Fig. 6.4):

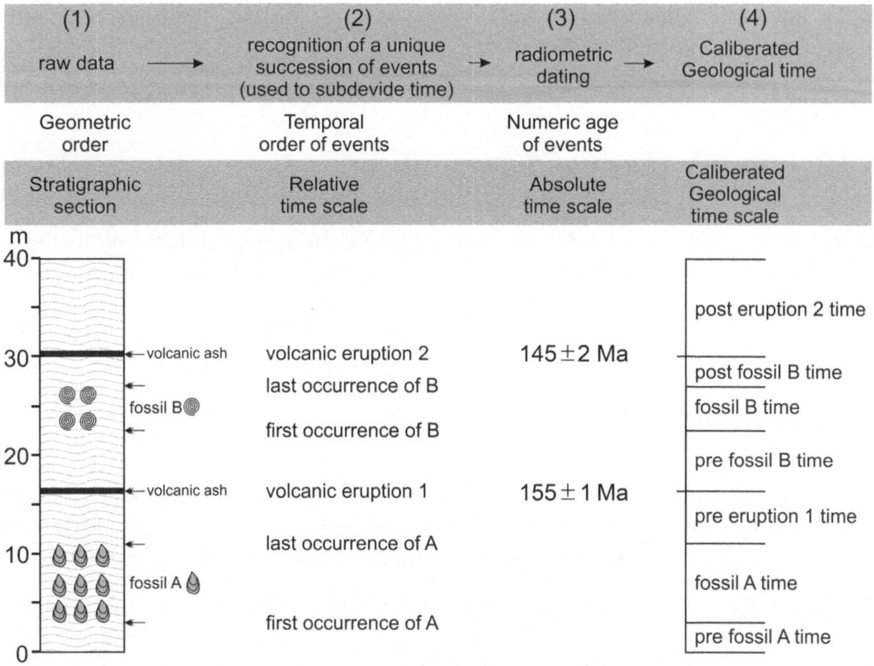

Fig. 6.3 The use of relative and absolute dating methods to construct the geological time scale

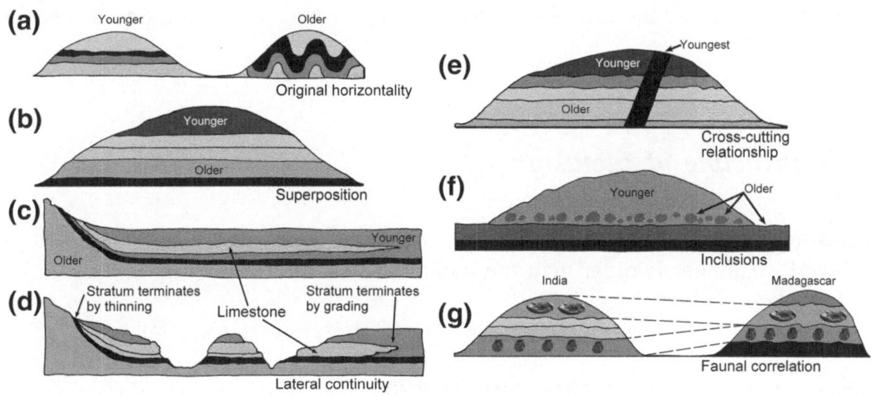

Fig. 6.4 The stratigraphic laws

6.4.1 Principle of Original Horizontality

This is based on the premise that sediment particles deposited from water under the influence of gravity form essentially horizontal sediment layers. Nonhorizontal rocks have been disturbed by an event (such as folding, faulting, etc.) after deposition and lithification (Fig. 6.4a).

6.4.2 Principle (Law) of Superposition

In an undeformed sequence of sedimentary rocks, the youngest beds are always at the top and the oldest at the bottom (this also applies to volcanic rock sequences) (Fig. 6.4b).

6.4.3 Principle of Lateral Continuity

The sediment extends laterally in all directions until it thins, pinches out, or terminates against the edge of the depositional basin. In simple terms, and under normal conditions, deposits originally extended in all directions (Fig. 6.4c, d).

6.4.4 Law of Cross-Cutting Relationships

An intrusion (e.g., igneous) or fault that cuts through another rock is younger than the rock it cuts, i.e., the intruding rock is always younger than the one it invades (Fig. 6.4e).

6.4.5 Principle of Inclusion

Inclusions are older than the rock that contains them, i.e., a structure that is included in another is older than the including structure itself (Fig. 6.4f).

6.4.6 Principle of Faunal Succession

Fossil organisms succeed one another in a definite and determinable order, hence, time period can be recognized by its fossil content (Fig. 6.4g). General evolution pattern is from simple to complex organisms, i.e., specific groups of organisms succeed one another in a definite sequence through earth's history.

However, it is important to note that there are exceptions to these principles. For example: the principle of superposition is primarily based on the concept of gravity where in order for a layer of sediment to be deposited, something has to be beneath it to support it. This implies that the principle of superposition has implications for the relative age of an undisturbed vertical stratal succession; the oldest depositing first. However, this concept fails in cave deposits, where the cave contents are generally younger than both the bedrock below and the suspended roof above. But, a closer examination of the contact between the two (cave infill and the surrounding rock) will reveal the true relative age relationships (by using the principle of cross-cutting relationships). If fragments of the surrounding rock are found within the infill, then the principle of inclusion will be useful for finding the true relative age relationship. Nevertheless, the cave deposits also have distinctive structures of their own (stalactites and stalagmites) that are difficult to mistake from a successional sequence of rock units. The three principles (superposition, cross-cutting relationships, and inclusion), besides being powerful tools providing very high temporal resolution are used commonly to determine the succession of rocks. However, they may not be of much help in finding the amount of time elapsed between events.

6.5 Unconformities

Unconformities are surfaces of erosion or nondeposition of sediments that separate younger rocks from older ones. The time gap in the rock record is known as a Hiatus, which results in an incomplete rock record. There are three types of unconformities (Fig. 6.5):

6.5.1 Disconformity

It is the break in the rock record caused by erosion or nondeposition of sediments. Rocks on either side of the break are essentially parallel and may look like a

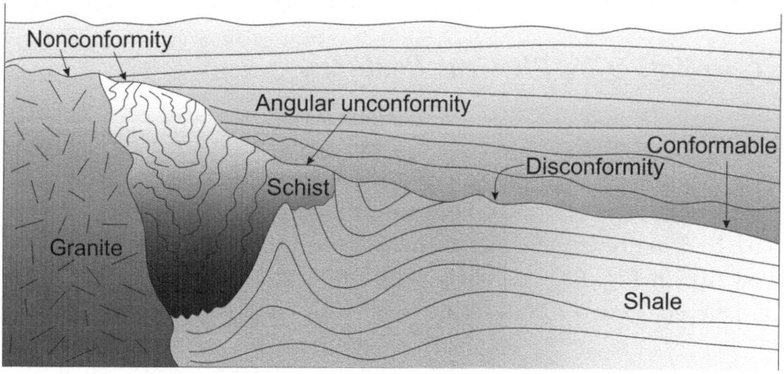

Fig. 6.5 Types of unconformities

bedding plane (Fig. 6.5). Fossils are used to determine the length of break in deposition.

6.5.2 Angular Unconformity

Tilted or folded sedimentary rocks are overlain by more flat-lying strata (Fig. 6.5).

6.5.3 Nonconformity

It is an unconformity between two very different rock types for example the boundary between an eroded igneous intrusion and the overlying sedimentary rock (Fig. 6.5). This may look like an intrusive contact, but there are no heat effects. Inclusions may be useful in distinguishing such nonconformity.

6.6 Correlation of Sedimentary Rocks

Correlation involves matching up rock layers of similar age in different regions. Age correlations using fossils have made it possible to construct a diagram, called the Geologic Column. This was developed based on the principles of superposition and faunal succession in Europe in the 1800s. Later, radiometric dates on igneous rocks provided a numerical scale for absolute ages of the geologic periods. The geological column is made of a hierarchy of relative time subdivisions of the earth's history. The longest units are called Eons, and progressively shorter subdivisions are Eras, Periods, and Epochs. The methods utilized for correlation include:

6.6.1 Correlation by Physical Features

It is the matching of rock units on a small scale by tracing the unit across an outcrop, noting the place of the unit in a sequence of rock strata, or by identifying the same bed in separate areas because of its distinctive lithology/composition (Fig. 6.6). Beds with very distinct characteristics (like concretions in a bed or shell beds; bed five in Fig. 6.6 is a shell bed) are referred to as Marker Beds. These characteristic beds enable both local (basin-wide) and regional level (continental) correlation.

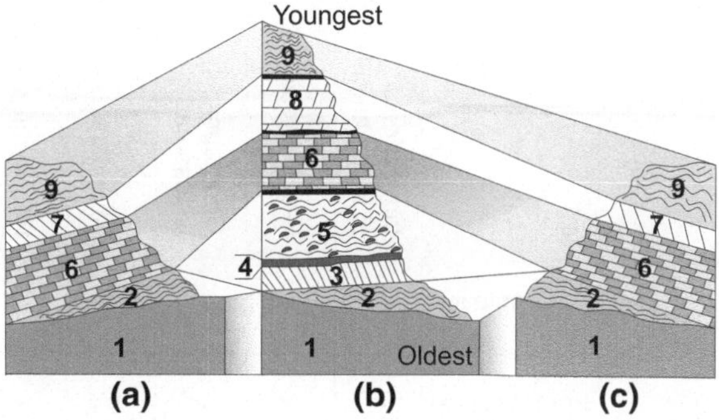

Fig. 6.6 A schematic view of physical correlation that involves the matching of rock units (here beds 1–9) on a small scale by tracing a unit (such as beds 1, 6, and 9) across outcrop or nearby localities (such as a, b, and c)

6.6.2 Correlation by Fossils

Guide fossils are those that are widespread geographically but lived for a very short period of time (i.e., they have a large lateral geographical expansion in a very short-time duration). Hence, this method allows widely separated rocks of different composition to be correlated. Figure 6.7 lists some of the salient requisites for being a good index fossil. Overlapping time ranges of several sets of guide (index) fossils are typically used (Fig. 6.8). A good example is the matching rock units of similar age on a larger scale by using these index fossils such as Ammonites for the Mesozoic Era, and in particular for the Jurassic Period (Fig. 6.9).

6.7 Absolute Dating Methods

Radiometric dating allows dates to be placed on geologic events and ages to be placed on formation of geologic materials.

6.7.1 Radioactivity

This is the process whereby radioactive elements (unstable isotopes) breakdown by nuclear decay into a different element such as from Uranium-238 to Lead-206. This process and its products are illustrated as a flow diagram in Fig. 6.10 (see also Faure 1986).

The products of radioactivity are:

6.7.1.1 Radiation

It is emitted by the nucleus during radioactive decay.

Fossil \ Criteria	Independent of Environment	Fast to evolve	Geographically widespread	Abundant	Readily preserved	Easily recognized	Status as Guide fossil
Graptolite	✓	✓	✓	✓	✓	✓	Good (Ordovician to Silurian)
Foraminifera	✓	✓	✓	✓	✓	✓	Good (Mesozoic to Recent)
Pollen	✓	✓	✓	✓	✓	✓	Good (Cretaceous to Recent)
Coccolith	✓	✓	✓	✓	✓	✓	Good (Mesozoic to Recent)
Ammonite	✓	✓	✓	✓	✓	✓	Good (Devonian to Cretaceous)
Coral	✗	✗	✗	✓	✓	✓	Poor (Carboniferous)
Echinoid	✗	✗	✗	✓	✓	✓	Poor (Cretaceous)
Barnacles	✗	✗	✗	✗	✗	✓	Bad (not used)
Birds	✓	✗	✓	✗	✗	✓	Bad (not used)

Fig. 6.7 The characteristics of a good guide fossil

6.7.1.2 Alpha Particle

This is composed of 2 protons and 2 neutrons. This loss reduces the mass number by 4 and the atomic number by 2 (Fig. 6.11).

6.7.1.3 Beta Particles

An electron produced by the break-up of a neutron (a neutron is the sum of protons and electrons). This loss does not change the mass number, but increases the atomic number by 1. A nucleus can also gain a beta particle (electron-capture), which decreases the atomic number by 1 (Fig. 6.11).

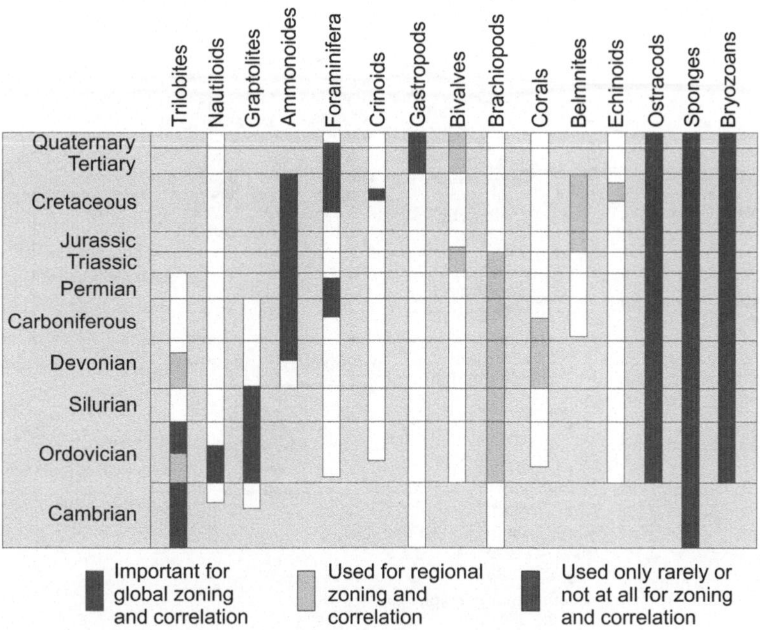

Fig. 6.8 The use of guide fossils through time

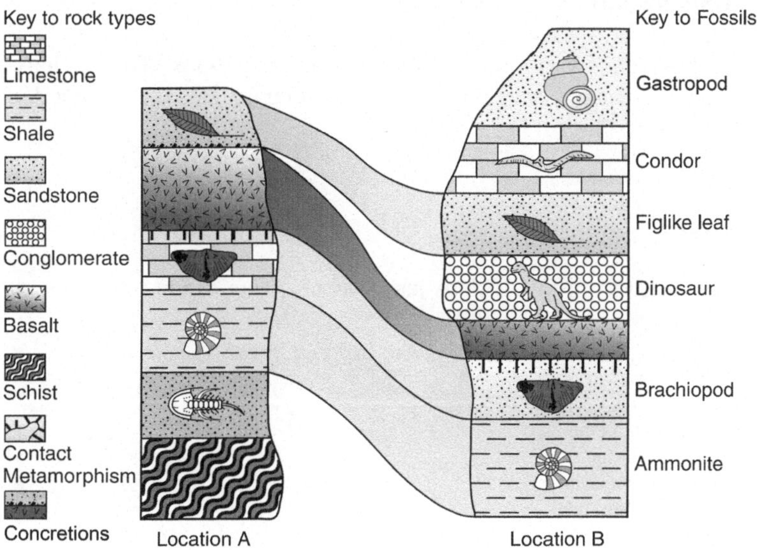

Fig. 6.9 Faunal correlation. A schematic view illustrating the use of fossils and faunal succession to show age equivalency of rocks from widely separated geographic localities (such as Locations A and B). Beds containing the same fossils belong to the same age

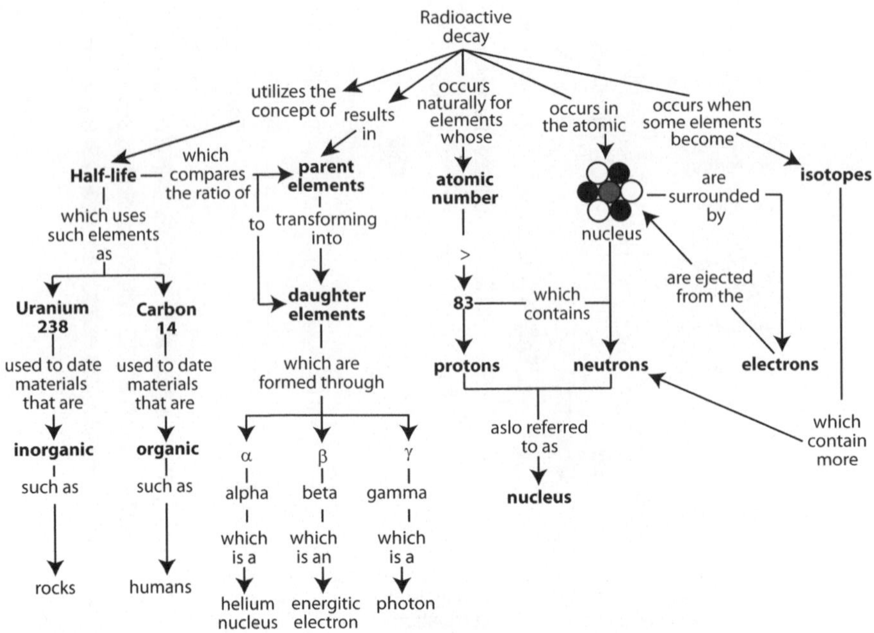

Fig. 6.10 The flow chart of the process of radioactive decay and its products

6.7.1.4 Gamma Ray

These are high energy X-rays, electromagnetic radiations. The nucleus looses energy without a change in its mass or atomic number. This is the most dangerous form of radiation (Fig. 6.11).

6.7.1.5 Heat

Energy is emitted from the nucleus partly as heat (Fig. 6.11).

Fig. 6.11 The atomic number (also known as the proton number), Z, is the number of protons found in the nucleus of an atom and mass number, A, is the total number of protons and neutrons within the nucleus of an atom

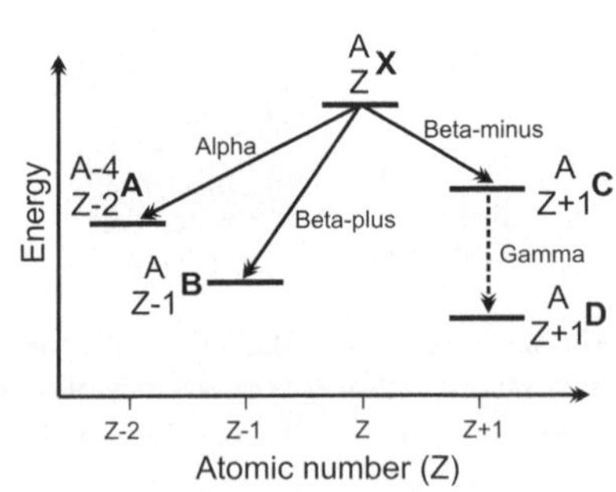

6.7.1.6 Daughter Isotopes

Different elements are produced by changes in the atomic number (Fig. 6.12).

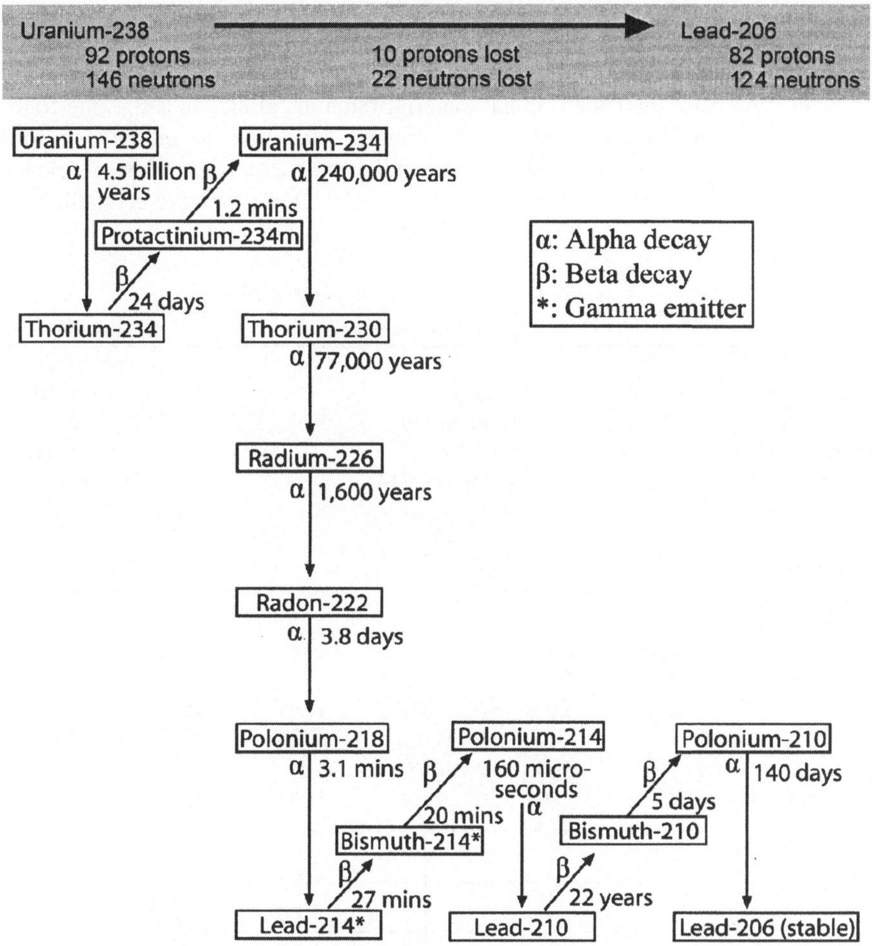

Fig. 6.12 The flow chart of the radioactive decay of Uranium-238 to Lead-206. Radioactive Uranium-238 decays to its stable daughter product, Lead-206, by 8 alpha and 6 beta decay steps. A number of different isotopes are produced as intermediate steps in this decay series. The alpha decay causes the number of protons and neutrons to diminish by 2, whereas beta-negative decay diminishes the number of neutrons by 1 and increases the number of protons by 1. The instability caused by the alpha decay is corrected by the eventual beta decay, leading to the stable nucleus of Lead-206, with its 82 protons and 124 neutrons

6.8 Radiometric Dating

Each unstable parent element decays to a more stable daughter element at a characteristic and fixed geometric rate (Fig. 6.12). The time it takes for one half of the atom to decay is called its half-live. By measuring the amount of parent and daughter isotopes in a sample and knowing the half-life of the parent, an age can be determined for the sample. Half-lives range from fractions of a second to billions of years (Table 6.1).

After 1 half-life, half the original material remains, after 2 half-life one-fourth remains, after 3 half-live, one-eighth remains, etc., and so on and so forth (Fig. 6.13). The rate of decay of a given radioactive isotope is constant. Hence, this provides a clock by which rocks can be dated. Materials tend to crystallize in a

Table 6.1 Radioactive isotopes commonly used in radiometric dating

Parent isotope	Daughter isotope	Half-life (years)
Uranium-238	Lead-206	4.468 billion
Uranium-235	Lead-207	703.8 million
Uranium-234	Thorium-230	245,500
Thorium-232	Lead-208	14.0 billion
Samarium-147	Neodymium-143	106 billion
Rubidium-87	Strontium-87	48.8 billion
Potassium-40	Argon-40	1.25 billion
Carbon-14	Nitrogen-14	5,730

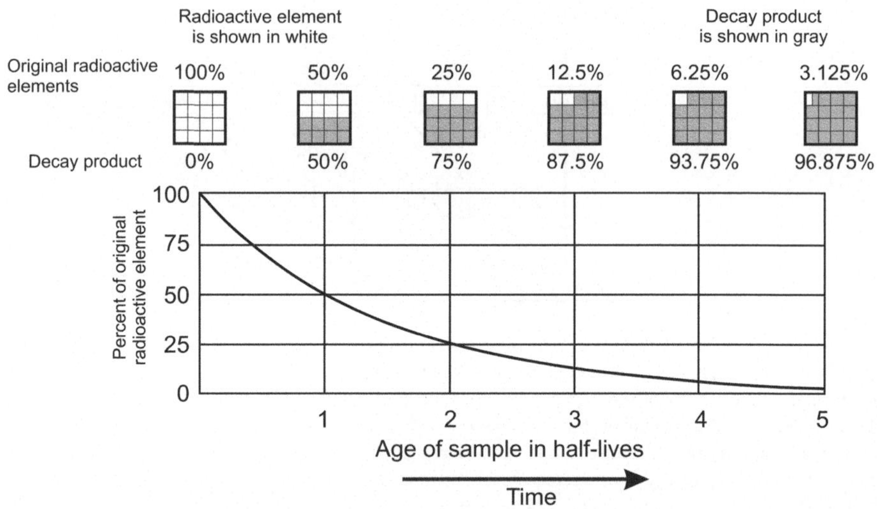

Fig. 6.13 The process of radioactive decay. A geometric radioactive decay curve, in which each time unit represents one half-life, and each half-life is the time it takes for half of the parent element to decay to the daughter element

pure form, so when the crystals of a rock form, they incorporate certain elements and do not incorporate others. For this reason, minerals that incorporate a radioactive isotope tend not to incorporate the stable daughter isotope. Any stable daughter isotope found in a crystal is likely to be the result of decay after the crystal was formed.

The radiometric dates for Sedimentary rocks are largely meaningless as the minerals making them, are parts of other, preexisting rocks, and therefore do not give the age of the formation of the sedimentary rock. The only exceptions to this rule are some sandstones and shales that contain a potassium-bearing mineral are called Glauconite. It only forms at the time of sediment deposition. More often, sedimentary rock ages are bracketed by dating igneous and metamorphic rocks. The most accurate radiometric dates are obtained from igneous rocks as metamorphism affects the parent/daughter ratio. For good dates, there must be no gain or loss of the parent or daughter isotopes (i.e., it should be a closed system). Typically more than one isotope is used for crosschecking. There are six radioactive isotopes that are useful in geologic dating: Carbon-14, Potassium-40, Rubidium-87, Thorium-232, and Uranium-235 and -238. Example of the decay of Uranium-238 to Lead-206 and their respective half-life periods is given in Fig. 6.12.

6.8.1 Long-Lived Radioactive Isotope Pairs

All of the isotopes listed in Table 6.1 (except Carbon-14) have half-lives more than a billion years (Table 6.1). These are used to date very old materials such as meteorites, igneous intrusives, lunar samples, oldest rocks on earth, among others.

6.8.2 Fission-Track Dating

Uranium-238 spontaneously decays by fission. Particles from the nucleus make tracks in rock minerals which can be counted and tied to a number of years. This dating method has the largest useful age range of any radiometric method (Table 6.1).

6.8.3 Radiocarbon Dating

The term "Radiocarbon" is commonly used to denote Carbon-14 (C-14), an isotope of Carbon which is radioactive with a half-life of about 5,730 years. C-14 is produced by cosmic rays in the stratosphere and upper troposphere (see Fig. 4.2). It is then distributed throughout the rest of the troposphere, the oceans, and earth's other exchangeable carbon reservoirs. In the surface atmosphere, about one part

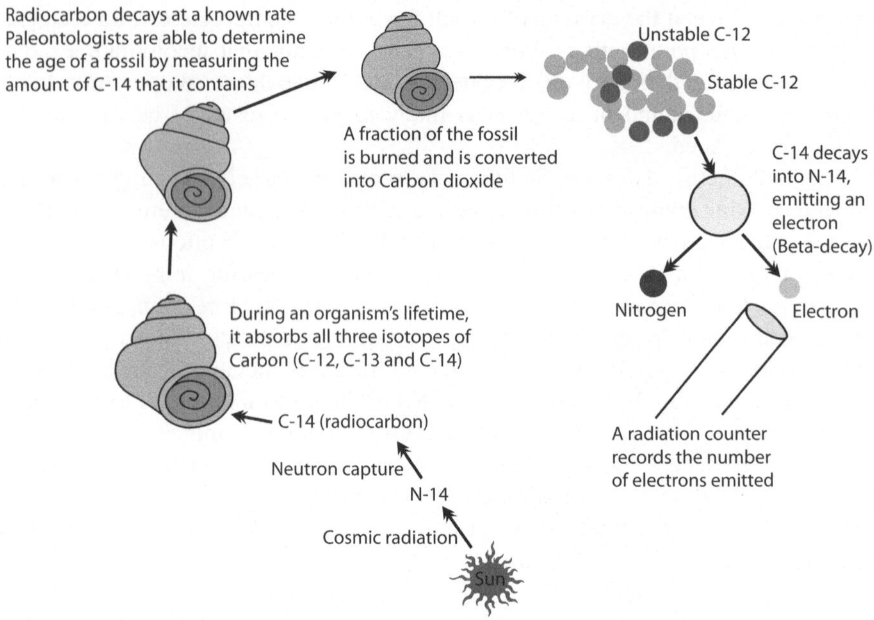

Fig. 6.14 Radiocarbon dating (Carbon-14). All living organisms absorb radiocarbon, an unstable form of carbon that has a half-life of about 5,730 years. During its lifetime, an organism continually replenishes its supply of radiocarbon by breathing and eating. After an organism dies and becomes a fossil, C-14 continues to decay without being replaced. To measure the amount of radiocarbon left in a fossil, scientists burn a small piece of the fossil to convert it into Carbon dioxide (CO_2) gas. Radiation counters are used to detect the electrons given off by decaying C-14 as it turns into Nitrogen (N-14). The amount of C-14 is compared to the amount of C-12, the stable form of Carbon, to determine how much radiocarbon has decayed and to date the fossil

per trillion (ppt) of carbon is C-14. All organisms absorb carbon from their environment (Fig. 6.14). Those that absorb their carbon directly or indirectly from the surface atmosphere have about one ppt of their carbon content as C-14. Such organisms comprise almost all land-dwelling plants and animals. After an organism dies, it stops absorbing C-14. As time passes, C-14 in its tissues is converted into Nitrogen. If we know what the original ratios of C-14 to Carbon-12 were in the organism when it died, and if we know that the sample has not been contaminated by contact with any other carbon since its death, we can calculate when it died by its C-14 to C-12 ratio. Figure 6.15 gives the flow diagram of the Carbon dating process and its products. However, there are limitations to this method. The tiny initial amount of C-14, the relatively rapid rate of decay and the ease with which samples can become contaminated, make radiocarbon dating results for samples "older" than about 5,730 years effectively meaningless. This limit is currently accepted by nearly all radiocarbon dating practitioners. It follows that the older a date is, even within this "limit," the greater are the doubts about the date's accuracy. But with correction factors to account for the variations in the

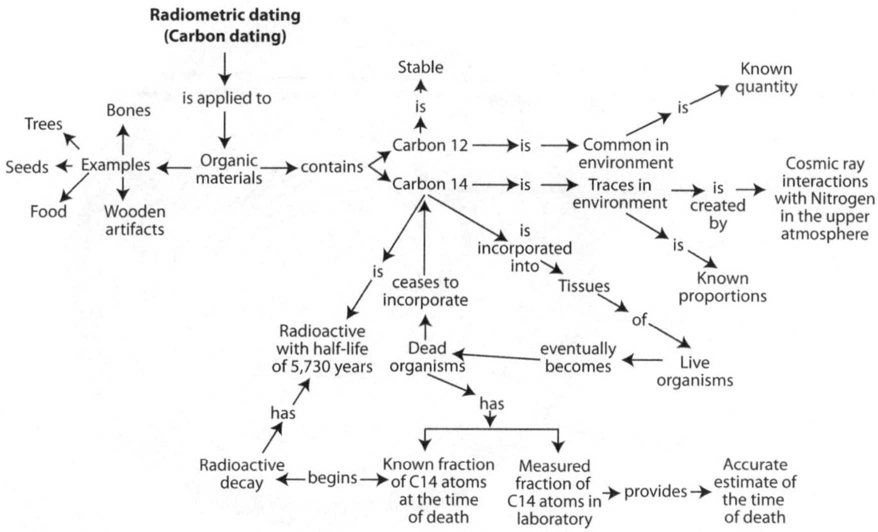

Fig. 6.15 The flow chart for the process and products of radiocarbon dating (see Fig. 6.14 for explanation)

amount of C-12 and C-14 in the atmosphere over time, the dates can be useful. C-14 is also a good match with Tree ring dating methods (up to 11,000 years old).

In general, many rocks cannot be dated by radiometric methods. They are either too young to have incorporated enough daughter products for analysis or are too old and all of the parent isotopes have been decayed. Some even lack sufficient quantities of radioactive elements in them. Quartz is one such example as it does not accommodate radioactive elements in its crystal structure. Hence, it is rarely useful for radiometric dating. Evidences from superposition and fossil content provide an independent check that enable comparing of the relative age with the derived radiometric age of the rocks. Through this system of tests and cross-checks, many reliable radiometric ages have been determined.

6.9 Tree Ring Dating Methods

In temperate climates, the age of trees can be calculated based on their annual growth rings (Fig. 6.16). The thickness and texture of each ring are records of the environment (of temperature, humidity, precipitation, and insect infestations). Even forest fires create distinctive patterns that are recorded in all the trees of an area. Hence, by comparing the pattern of rings from one tree to the other, or from the ones that have died, the chronology of a forested region can be accurately deciphered. By overlapping sections of rings from many different trees, an unbroken tree ring record has been created that extends as far back as 11,000 years (McGovern et al. 1995).

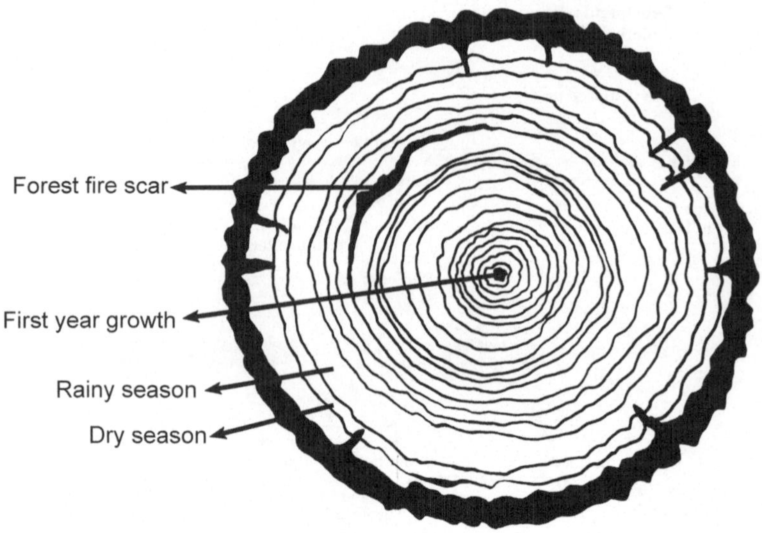

Fig. 6.16 Tree rings. The age of trees can be calculated based on its annual growth rings

6.10 Varves

The rhythmic changes in a sedimentary environment creates a type of clock based on the successions of thin sediment beds. Some of these rhythms correspond to the annual cycle of seasons. Thin layers of clay called Varves (Fig. 6.17), accumulate in still waters of some glacial lakes. Here, each graded layer represents 1 year,

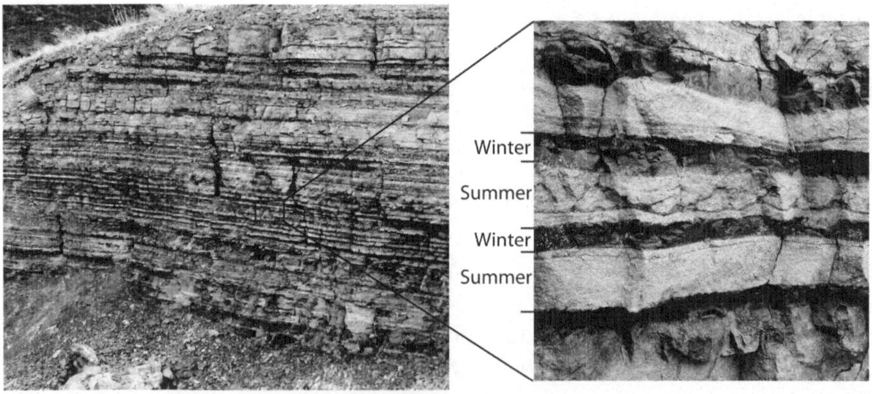

Fig. 6.17 The glacial varves. The dark layers were deposited in winter and the light colored ones in summer when abundant meltwater transported sand and silt to the lake where it settled to the bottom. Typically winter layers are all about the same thickness, whereas summer layers tend to vary more due to differences in weather that influence the amount of meltwater

created by seasonal variations in the amount of sediment-laden meltwater that flowed into a lake. By carefully counting these layers, a record has been created that extends back by ~20,000 years in glaciated regions around the Baltic Sea of Northern Europe (Ojala et al. 2012). Some varves can also be dated by C-14 techniques if they contain sufficient organic material.

6.11 The Geological Column

The currently accepted geologic time scale is based on the standard geologic column, established by faunal succession and superposition, and the numerical radiometric dates of rocks that form precise anchor points (Tables 6.2 and 6.3). The first geologic time scale was proposed by the British geologist Arthur Holmes (1890–1965) in 1913 based on radioactivity and the earth was estimated to be about 4 billion years old (see also York and Farquhar 1972; Hecht 1995; Fortey 1999; Wilde et al. 2001).

6.12 Summary

Archbishop James Ussher (1664) determined the date of creation to be 4004 BC, which required the earth to be no more than about 6,000 years old. This and similar ideas about earth's history dominated the western thinking until the eighteenth century. During the eighteenth and nineteenth centuries, scientific attempts were made to determine the earth's age. Buffon (1707–1788), based on the rate of cooling calculated the age of the earth to be at least 75,000 years from the present. The age of the earth was also calculated from deposition rate of various sediments and the thickness of sediments in the earth's crust. This resulted in ages ranging from 3 million to 1.5 billion years for the earth's origin. John Joly (1857–1933) used ocean's salinity and estimated the age of the earth at about 90 Ma. Problems with these rate-based approaches were that they considered the rates to be uniform. They did not take into account that—the sediment could be removed by erosion, modified by compaction, or sometimes not even deposited at all. Additionally, salt does not come exclusively from continents.

This dating flaw was somewhat removed by Relative dating methods which determines the relative order of past events, without necessarily determining their absolute ages (the latter method is called Absolute dating/Numerical dating). Relative dating methods use geological principles to place events in a chronological order; these are called the Stratigraphic Laws. There are six fundamental principles: Principle of Original Horizontality, Principle (Law) of Superposition, Principle of Lateral Continuity, Law of Cross-Cutting Relationships, Principle of Inclusion, and Principle of Faunal Succession. The three principles (superposition, cross-cutting relationships, and inclusion), provide a very high temporal resolution

Table 6.2 Details about the Phanerozoic Period

Period name	Type locality	Date proposed	Proposed by/Named after	Profession	Author's age	Remarks
Quaternary	France	1829	Jules Pierre François Stanislaus Desnoyers	French geologist and archaeologist	1800–1887	Based on lithology
Tertiary	Italy	1760	Giovanni Arduino	Father of Italian Geology	1714–1795	Originally based on lithology but redefined later on the basis of fossil content
Cretaceous	Paris Basin	1882	Jean Baptiste Julien d'Omalius d'Halloy	Belgian geologist	1783–1875	Based on strata of distinctive chalk beds
Jurassic	Jura Mountains, Northern Switzerland	1795	Friedrich Wilhelm Heinrich Alexander Freiherr von Humboldt	German naturalist, geologist, and explorer	1769–1859	Based on lithology
Triassic	South Germany	1834	Friedrich Von Alberti	German geologist	1795–1878	Based on lithology and fossil content
Permian	Perm province, Russia	1841	Roderick Impey Murchison	English geologist	1792–1871	Based on fossil content
Carboniferous	Central England	1822	William Daniel Conybeare and William Phillips	English geologists and English mineralogist and geologist respectively	1787–1857 and 1775–1828	Based on lithology (coal beds) and fossil content
Devonian	Devonshire, South England	1840	Roderick Impey Murchison	English geologist	1792–1871	Based on fossil content

(continued)

Table 6.2 (continued)

Period name	Type locality	Date proposed	Proposed by/Named after	Profession	Author's age	Remarks
Silurian	Western Wales	1835	Roderick Impey Murchison	English geologist	1792–1871	Based on lithology and fossil content
Ordovician	Western Wales	1879	Charles Lapworth	English geologist	1842–1920	Between Cambrian and Silurian; a unit set up to resolve boundary between the Cambrian and Ordovician
Cambrian	Western Wales	1835	Adam Sedgwick	English geologist and founders of modern geology	1785–1873	Based on lithology

A tip for good remembrance: **Q**uinn **T**old **C**laire **J**ones **T**o **P**ush **P**retty **M**arie **D**odd's **S**cooter **O**ver the **C**liff (**Q**uaternary, **T**ertiary, **C**retaceous, **J**urassic, **T**ertiary, **P**ermian, **P**ennsylvanian, **M**ississippian, **D**evonian, **S**ilurian, **O**rdovician, **C**ambrian)

Table 6.3 The geological time scale

Super-eon	Enothem / Eon	Erathem / Era	System / Period	Series / Epoch	Stage / Age	Age (Ma)	Major events
	Phanerozoic	Cenozoic	Quaternary	Holocene	Atlantic	0.011430 ± 0.00013	**Mount Tambora erupts in 1815, causing the "Year Without a Summer" (1816) in Europe and North America from a volcanic winter.** Last glacial period ends; rise of human civilization. Quaternary Ice Age recedes, and the current interglacial begins. Younger Dryas cold spell occurs, Sahara forms from savannah, and agriculture begins, allowing humans to build cities. Little Ice Age (stadial) causes brief cooling in the Northern Hemisphere from 1400 to 1850. Atmospheric CO_2 levels start creeping from 100 parts per million volume (ppmv) at the end of the last glaciation to the current level of 385 ppmv.
					Boreal		
				Pleistocene	Upper	0.0117	**Last glacial maximum (30,000 years ago)**, last glacial period (18,000-15,000 years ago). Lake Toba supervolcano erupts 75,000 years before present, causing a volcanic winter that pushes humanity to the brink of extinction. Extinction of many large mammals (Pleistocene megafauna). Evolution of anatomically modern humans. Quaternary Ice Age continues with glaciations and interstadials (and the accompanying fluctuations from 100 to 300 ppmv in atmospheric CO_2 levels).
					"Ionian"	0.126 ± 0.005	
					Calabrian	0.781	
					Early	0.500?	
					Gelasian	1.806 ± 0.005	
			Neogene	Pliocene	Piacenzian/ Blancan	2.588 ± 0.005	***Homo habilis* appears.** *Australopithecines*, many of the existing genera of mammals, and recent mollusks appear. Intensification of present Icehouse conditions, Quaternary (present) ice age begins roughly 2.58 Ma ago; cool and dry climate.
					Zanclean	3.600 ± 0.005	
				Miocene	Messinian	5.332 ± 0.005	**First apes appear (*Sahelanthropus tchadensis*).**

(continued)

Table 6.3 (continued)

Super-eon	Enothem / Eon	Erathem / Era	System / Period	Series / Epoch	Stage / Age	Age (Ma)	Major events
					Tortonian	7.246 ± 0.05	Moderate Icehouse climate, punctuated by ice ages; Horse and mastodons diverse. Grasses become ubiquitous. Widespread forests draw in massive amounts of CO_2, gradually lowering the level of atmospheric CO_2 from 650 to ~100 ppmv.
					Serravallian	11.608 ± 0.05	
					Langhian	13.65 ± 0.05	
					Burdigalian	15.97 ± 0.05	
					Aquitanian	20.43 ± 0.05	
			Paleogene	Oligocene	Chattian	23.03	**Major evolution and dispersal of modern flowering plants.** Warm but cooling climate, moving toward Icehouse; Rapid evolution and diversification of mammals.
					Rupelian	28.4 ± 0.1	
				Eocene	Priabonian	33.9 ± 0.1	**First grasses.** Moderate, cooling climate. Archaic mammals (Creodonts, Condylarths, etc.) flourish. Several "modern" mammal families appear. Primitive whales diversify. Decay of seafloor algae drawing in massive amounts of atmospheric CO_2, thus, lowering it from 3,800 to 650 ppmv.
					Bartonian	37.2 ± 0.1	
					Lutetian	40.4 ± 0.2	
					Ypresian	48.6 ± 0.2	
				Paleocene	Thanetian	55.8 ± 0.2	**Modern plants appear**; Mammals diversify following the extinction of dinosaurs. First Climate tropical. Large mammals (Bear or small Hippo size). Indian Subcontinent collides with Asia 55 Ma, Himalayan Orogeny starts between 52 and 48 Ma.
					Selandian	58.7 ± 0.2	
					Danian	~61.1	
		Mesozoic	Cretaceous	Upper	Maastrichtian	65.5 ± 0.3	**Atmospheric CO_2 close to present-day levels.** Flowering plants proliferate, along with new types of insects. Modern teleost fish begin to appear. Ammonites, belemnites, rudist bivalves, echinoids and sponges become common. Many new types of dinosaurs (e.g. *Tyrannosaurs*, *Titanosaurs*, duck bills, and horned dinosaurs) evolve on land, as do *Eusuchia*
					Campanian	70.6 ± 0.6	
					Santonian	83.5 ± 0.7	
					Coniacian	85.8 ± 0.7	

(continued)

Table 6.3 (continued)

Super-eon	Enothem / Eon	Erathem / Era	System / Period	Series / Epoch	Stage / Age	Age (Ma)	Major events
					Turonian	~88.6	(modern crocodilians); and *Mosasaurs* and modern sharks appear in the sea. Primitive birds gradually replace pterosaurs. Monotremes, marsupial and placental mammals appear. Break up of Gondwana.
					Cenomanian	93.6 ± 0.8	
				Lower	Albian	99.6 ± 0.9	
					Aptian	112.0 ± 1.0	
					Barremian	125.0 ± 1.0	
					Hauterivian	130.0 ± 1.5	
					Valanginian	~133.9	
					Berriasian	140.2 ± 3.0	
			Jurassic	Late	Tithonian	145.5 ± 4.0	**Many types of dinosaurs, such as Sauropods, Carnosaurs, and Stegosaurs.** Gymnosperms (especially conifers, Bennettitale and cycads) and ferns common. Mammals' common but small. First birds and lizards. Ichthyosaur and plesiosaurs diverse. Bivalves, ammonites and belemnites abundant. Sea urchins very common, along with crinoids, starfish, sponges, and terebratulid and rhynchonellid brachiopods. Breakup of Pangea into Gondwana and Laurasia. Atmospheric CO_2 levels 4-5 times the present day levels (1200-1500 ppmv, compared to today's 385 ppmv).
					Kimmeridgian	150.8 ± 4.0	
					Oxfordian	~155.6	
				Middle	Callovian	161.2 ± 4.0	
					Bathonian	164.7 ± 4.0	
					Bajocian	167.7 ± 3.5	
					Aalenian	171.6 ± 3.0	
				Early	Toarcian	175.6 ± 2.0	
					Pliensbachian	183.0 ± 1.5	
					Sinemurian	189.6 ± 1.5	
					Hettangian	196.5 ± 1.0	

(continued)

Table 6.3 (continued)

Super-eon	Enothem / Eon	Erathem / Era	System / Period	Series / Epoch	Stage / Age	Age (Ma)	Major events
			Triassic	Upper	Rhaetian	199.6 ± 0.6	**Ceratitic ammonoids abundant.** Archosaurs dominant on land as dinosaurs, in the oceans as Ichthyosaurs and Nothosaurs, and in the air as Pterosaurs. Cynodonts become smaller and more mammal-like, while first mammals and Crocodilia appear. *Dicrodium* flora common on land. Modern corals and teleost fish appear, and many modern insect clades.
					Norian	203.6 ± 1.5	
					Carnian	216.5 ± 2.0	
				Middle	Ladinian	~228.7	
					Anisian	237.0 ± 2.0	
				Lower	Olenekian	~245.9	
					Induan	~249.5	
		Paleozoic	Permian	Lopingian	Changhsingian	251.0 ± 0.4	**Supercontinent Pangea forms.** Permian-Triassic extinction event occurs at 251 Ma: 95 percent of life on Earth becomes extinct, including all Trilobites, Graptolites, and Blastoids. End of Permo-Carboniferous glaciation. Synapsid reptiles (Pelycosaurs and Therapsids) become abundant, while Parareptiles and Temnospondyl amphibians remain common. In the mid-Permian, coal-age flora are replaced by cone-bearing gymnosperms (the first true seed plants) and by the first true mosses. Beetles and flies evolve. Marine life flourishes in warm shallow reefs; Productid and Spiriferid brachiopods, bivalves, foraminifers, and ammonoids become abundant.
					Wuchiapingian	253.8 ± 0.7	
				Guadalupian	Capitanian	260.4 ± 0.7	
					Wordian	265.8 ± 0.7	
					Roadian	268 ± 0.7	
				Cisuralian	Kungurian	270.6 ± 0.7	
					Artinskian	275.6 ± 0.7	
					Sakmarian	284.4 ± 0.7	
					Asselian	294.6 ± 0.8	
			Pennsylvanian (Carboniferous)	Upper	Gzhelian	299.0 ± 0.8	**Highest-ever atmospheric oxygen levels.** Winged insects radiate suddenly. Amphibians become common and diverse. First reptiles and coal forests (scale trees, ferns, club trees, giant horsetails, *Cordaites*, etc.). Goniatites, Brachiopods, Bryozoa,
					Kasimovian	303.4 ± 0.9	
				Middle	Moscovian	307.2 ± 1.0	

(continued)

Table 6.3 (continued)

Super-eon	Enothem / Eon	Erathem / Era	System / Period	Series / Epoch	Stage / Age	Age (Ma)	Major events
				Lower	Bashkirian	311.7 ± 1.1	Bivalves, and Corals abundant.
			Mississippian (Carboniferous)	Upper	Serpukhovian	318.1 ± 1.3	**First land vertebrates**, Large primitive trees, and amphibious sea-scorpions live amid coal-forming coastal swamps. Lobe-finned rhizodonts are dominant big fresh-water predators. Early sharks common and diverse; Echinoderms (especially Crinoids and Blastoids) abundant. Corals, Bryozoa, Goniatite and Brachiopods (Productida, Spiriferida, etc.) very common, but Trilobites and Nautiloids decline.
				Middle	Visean	328.3 ± 1.6	
				Lower	Tournaisian	345.3 ± 2.1	
			Devonian	Upper	Famennian	359.2 ± 2.5	**First Clubmosses, Horsetails and Ferns appear**, as do first seed-bearing plant (Progymnosperms), first trees (the Progymnosperm *Archaeopteris*), and first (wingless) insects. Strophomenid and atrypid brachiopods, rugose and tabulate corals, and crinoids become abundant. Goniatite ammonoids are plentiful, while squid-like coleoids arise. Trilobites and armoured agnaths decline, while jawed fishes (placoderms, lobe-finned and ray-finned fish, and early sharks) rule the seas. First amphibians are still aquatic.
					Frasnian	374.5 ± 2.6	
				Middle	Givetian	385.3 ± 2.6	
					Eifelian	391.8 ± 2.7	
				Lower	Emsian	397.5 ± 2.7	
					Pragian	407.0 ± 2.8	
					Lochkovian	411.2 ± 2.8	
			Silurian	Pridoli	*no faunal stages defined*	416.0 ± 2.8	**First Vascular plants** (the Rhyniophytes and their relatives), first millipede and arthropleurids on land. First jawed fishes, and armoured jawless fish, populate the seas. Sea-scorpions reach large size. Tabulate and rugose corals, brachiopods (*Pentamerida*, Rhynchonellida, etc.), and crinoids all abundant. Trilobites and mollusks diverse.
				Ludlow	Ludfordian	418.7 ± 2.7	
					Gorstian	421.3 ± 2.6	
				Wenlock	Homerian	422.9 ± 2.5	
					Sheinwoodian	426.2 ± 2.4	

(continued)

Table 6.3 (continued)

Super-eon	Enothem / Eon	Erathem / Era	System / Period	Series / Epoch	Stage / Age	Age (Ma)	Major events
				Llandovery	Telychian	428.2 ± 2.3	
					Aeronian	436.0 ± 1.9	
					Rhuddanian	439.0 ± 1.8	
			Ordovician	Upper	Hirnantian	443.7 ± 1.5	**First green plants and fungi on land.** Long straight-shelled cephalopods diversify. Early corals, articulate brachiopods (*Orthida*, *Strophomenida*, etc.), bivalves, nautiloids, trilobites, ostracods, bryozoa, echinoderms (crinoids, cystoids, starfish, etc.), and branched graptolites become common. Conodonts appear. Ice age at end of the period.
					Katian	445.6 ± 1.5	
					Sandbian	455.8 ± 1.6	
				Middle	Darriwilian	460.9 ± 1.6	
					Daapingian	468.1 ± 1.6	
				Lower	Floian	471.8 ± 1.6	
					Tremadocian	478.6 ± 1.7	
			Cambrian	Furongian	Stage 10	488.3 ± 1.7	**Cambrian Explosion.** Many fossils; most modern animal phyla appear. First chordates appear, and a number of extinct, problematic phyla. Reef-building Archaeocyatha abundant; then vanish. Trilobites, priapulid worms, sponges, inarticulate brachiopods (unhinged lampshells), and many other animals become common. Anomalocarids are giant predators, while many Ediacaran fauna die out. Prokaryotes, protists (foraminifer), fungi and algae continue to present day. Gondwana emerges. Atmospheric CO_2 content roughly 20-35 times of present-day (Holocene) levels (6000 as compared to today's 385 ppmv)
					Stage 9	~492	
					Paibian	~496	
				Series 3	Guzhangian	~499	
					Drumian	~503	
					Stage 5	~506.5	
				Series 2	Stage 4	~510	
					Stage 3	~521	
				Terreneuvian	Stage 2	~528	

(continued)

Table 6.3 (continued)

Super-eon	Enothem / Eon	Erathem / Era	System / Period	Series / Epoch	Stage / Age	Age (Ma)	Major events
					Fortunian	542.0 ± 1.0	
Precambrian*	Proterozoic	Neoproterozoic	Ediacaran			630 +5/-30	**Fossils of the first multi-celled animals.** Ediacaran biota flourishes. Presence of simple trace fossils of possible worm-like *Trichophycus*. First sponges and Trilobitomorphs. Enigmatic forms include many soft-jellied creatures shaped like bags, disks, or quilts (like *Dickinsonia*)
			Cryogenian			850	**Possible "Snowball Earth" period.** Fossils still rare. Rodinia landmass begins to break up
			Tonian			1000	**Trace fossils of simple multi-celled eukaryotes.** Rodinia supercontinent persists. First radiation of dinoflagellate-like acritarchs.
		Mesoproterozoic	Stenian			1200	Narrow highly metamorphic belts due to orogeny as Rodinia formed.
			Ectasian			1400	**Green algae colonies in the seas.** Platform covers continue to expand.
			Calymmian			1600	Platform covers expand
		Paleoproterozoic	Statherian			1800	First complex single-celled life: protists with nuclei
			Orosirian			2050	The atmosphere became oxygenic
			Rhyacian			2300	Bushveld Formation formed
			Siderian			2500	Oxygen catastrophe: banded iron formations formed
	Archean	Neoarchean				2800	**Stabilization of most modern Cratons**; possible mantle overturn event
		Mesoarchean				3200	**First stromatolites (probably colonial Cyanobacteria).** Oldest macrofossils
		Paleoarchean				3600	**First known oxygen-producing bacteria.** Oldest definitive microfossils. Oldest cratons on Earth (such as the Canadian Shield and the Pilbara Craton) may have formed during this period.
		Eoarchean				3800	**Simple single-celled life (probably bacteria and perhaps archaea).** Oldest probable microfossils.
	Hadean	Early Imbrian				~3850	This era overlaps the end of the Late Heavy Bombardment of the inner solar system.

(continued)

Table 6.3 (continued)

Super-eon	Enothem / Eon	Erathem / Era	System / Period	Series / Epoch	Stage / Age	Age (Ma)	Major events
		Nectarian				~3920	This unit gets its name from the lunar geologic timescale when the Nectaris Basin and other major lunar basins were formed by large impact events.
		Basin Groups				~4150	**The first Life forms and self-replicating RNA molecules may have evolved on Earth around 4000 Ma during this era.** Oldest known rock (4030 Ma).
		Cryptic				~4570	**Oldest known mineral (Zircon, 4406±8 Ma***).** Formation of Earth (4567.17 to 4570 Ma). Formation of Moon (4533 Ma), probably from giant impact.

Modified after data from Wikipedia.org and International Commission on Stratigraphy (http://www.stratigraphy.org)

[a] Based on isotopic dating, the Precambrian took up 87 % of the geologic time. Hence, the Precambrian is subdivided into three major subdivisions: the Hadean (Hades is, in Greek mythology, the underground place where the dead live; the name alludes to the hell-like nature of earth's early surface), the Archean, and the Proterozoic (Greek for "beginning life"). Each is an Eon, the largest unit of geological time. A fourth, and the youngest, is the Phanerozoic (Greek for "visible life"). The Phanerozoic Eon is the geologic time with an abundant fossil record

[b] After Wilde et al. (2001)

and are commonly used to determine the succession of rocks. However, they may not be of much help in finding the amount of time elapsed between events. Elapsed time is understood by studying Unconformities. These are surfaces of erosion or nondeposition of sediments that separate younger rocks from older ones. The time gap in the rock record is known as a Hiatus, which results in an incomplete rock record. There are three types of unconformities—Disconformity, Angular Unconformity, and Nonconformity.

Correlation involves matching up rock layers of similar age in different regions. Age correlations using fossils have made it possible to construct a diagram, called the Geologic Column. This column is made of a hierarchy of relative time subdivisions of the earth's history; the longest units are called Eons, and progressively shorter subdivisions are Eras, Periods, and Epochs.

The methods utilized for correlation include—correlation by physical features (bed with distinctive lithology/composition) and correlation by fossils (Guide/Index fossils). However, these are Relative methods and do not put a numerical age to the unit. This is done by Absolute dating methods.

Absolute dating methods use radioactive elements. Each unstable parent element decays to a more stable daughter element at a characteristic and fixed geometric rate. The time it takes for one half of the atoms to decay is called its Half-life. By measuring the amount of parent and daughter isotopes in a sample and knowing the half-life of the parent, an age can be determined for the sample. Half-lives range from fractions of a second to billions of years. There are six radioactive isotopes that are useful in geologic dating: Carbon-14, Potassium-40, Rubidium-87, Thorium-232, and Uranium-235 and -238.

In temperate climates, the age of trees can be calculated based on their annual growth rings. The thickness and texture of each ring are records of the environment (of temperature, humidity, precipitation, and insect infestations). Hence, by comparing the pattern of rings from one tree to the other, or from the ones that have died, the chronology of a forested region can be accurately deciphered.

Varves, thin layers of clay, accumulate in still waters of some glacial lakes where each graded layer represents a year. By carefully counting these layers, a record has been created that extends back by $\sim$20,000 years in glaciated regions. Some varves can also be dated by Carbon-14 techniques if they contain sufficient organic material.

The currently accepted geologic time scale is based on the standard geologic column, established by faunal succession and superposition, and the numerical radiometric dates of rocks that form precise anchor points. The first geologic time scale was proposed by the British geologist Arthur Holmes (1890–1965) in 1913 based on radioactivity and the earth was estimated to be about 4 billion years old.

References

Albritton CC (1980) The Abyss of time. Freeman, San Francisco

Berry WBN (1988) Growth of prehistoric time scale. Blackwell, Palo Alto

Brush SG (1982) Finding the age of the Earth by physics or by faith? J Geol Educ 30:34–58

Burchfiel JD (1990) Lord kelvin and the age of the earth. University of Chicago Press, Chicago

Dalrymple GB (1991) The age of the earth. Stanford University Press, Stanford

Faure G (1986) Isotope geology. Wiley, New York

Fortey R (1999) Life: A natural history of the first four billion years of life on earth. Knopf Publishing, New York

Hecht J (1995) The geological timescale. Inside Science 81. New Scientist May 20

McGovern PJ, Sever TH, Myers JW, Myers EE, Bevan B, Miller NF, Bottema S, Hongo H, Meadow RM, Kuniholm PI, Bowman SGE, Leese MN, Hedges REM, Matson FR, Freestone IC, Vaughan SJ, Henderson J, Vandiver PB, Tumosa CS, Beck CW, Smith P, Child AM, Pollard AM, Thuesen I, Sease C (1995) Dendrochronology. Science in Archaeology: A Review. Am J Archaeol 99(1):79–142

Ojala AEK, Francus P, Zolitschka B, Besonen M, Lamoureux SF (2012) Characteristics of sedimentary varve chronologies: A review. Quatern Sci Rev 43:45–60

Wilde SA, Valley JW, Peck WH, Graham CM (2001) Evidence from detrital zircons for the existence of continental crust and oceans on the Earth 4.4 Gyr ago. Nature 409:175–178

York D, Farquhar RM (1972) The earth's age and geochronology. Pergamon Press, Oxford p 178

Part III
The Hydrologic System

Chapter 7
Weathering and Mass Wasting

7.1 Introduction

Weathering is a process of a slow continuous breakdown of rocks into smaller particles that are in equilibrium with the prevailing environment (Bland and Rolls, 1998). This process involves both decomposition (chemical breakdown) and disintegration (physical breakdown) of rocks and minerals. When particles are moved from their place of formation (either by moving water, wind, glaciers, and gravity), the process is called Erosion. Hence, products of weathering are a major source of sediments for both erosion and deposition. Sedimentary rocks are made of sediments that have once been weathered, eroded, transported, and eventually deposited in basins. Additionally, weathering also contributes to the formation of soil by providing mineral particles like sand, silt, and clay. The fact that oceans are saline is also due to the release of ion salts from rocks and minerals caused by weathering.

Weathering and Erosion are often used interchangeably. However, the fundamental processes behind them are very different. Weathering occurs without movement and involves breaking down of the substrate by chemical or physical means whereas in Erosion, movement is the main component. Erosion entails the movement of soil and mineral particles that have been loosened from the substrate, sometimes even by weathering. The other name for Erosion is Mass Wasting. It is simply the down slope movement effected by gravity. Rock falls, slumps, and debris flows are all examples of mass wasting. However, it is erosion, if the rock particle is moved by some flowing agent such as air, water, or ice.

Weathering is largely of two types—physical (mechanical) and chemical. The third, biological, only forms a small fraction. Physical weathering causes the parent rock to break into smaller fragments (reduction in size), but without changing the chemical composition of the parent material. On the other hand, chemical weathering changes the composition of the parent rock through chemical reactions. In it, the mineral constituents react with chemically active reagents such as H_2O, CO_2, O_2, or organic acids to form new minerals and/or to dissolve

S. Jain, *Fundamentals of Physical Geology*, Springer Geology,
DOI: 10.1007/978-81-322-1539-4_7,

elements from minerals (Coleman and Dethier 1986). Both physical and chemical weathering operates together and one often assists the other.

7.2 The Rock Cycle

The Igneous, Sedimentary, and Metamorphic rocks are interrelated to each other by a series of natural processes through the Rock cycle (Fig. 7.1). Igneous rocks form from cooling, crystallization, and solidification of hot molten lava and magma. They further undergo weathering and erosion to form sediments. These are then deposited and lithified by compaction and cementation to form Sedimentary rocks which then get buried deep within the earth, and are thereafter subjected to increased pressure and temperature. This causes them to undergo metamorphism and become Metamorphic rocks. With further burial and heating, Metamorphic rocks begin to melt. As melting progresses, and with increasing temperature and depth of burial (pressure), the rock becomes molten to form Magma. Magma eventually cools and crystallizes to form plutonic Igneous rocks, or is erupted onto the earth's surface as Lava. Lava cools and crystallizes to form volcanic Igneous rocks. These Igneous rocks undergo weathering and erosion to form sediments, and thus this cycle goes on and on (Fig. 7.1).

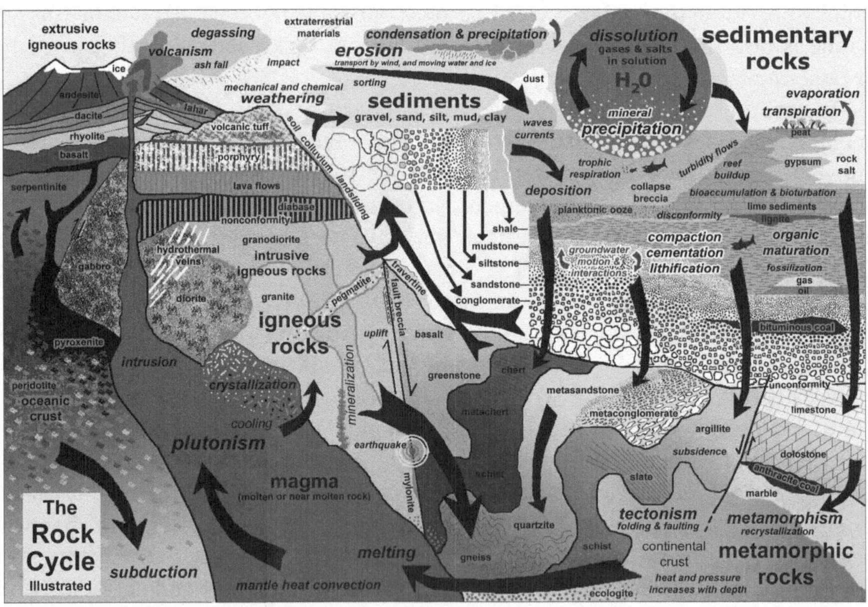

Fig. 7.1 The Rock cycle (reproduced with permission from Dr Phil Stoffer, http://www.geologycafe.com)

7.3 Types of Weathering

The Chemical, Physical, and Biological are the three major categories of weathering that work together to breakdown rocks and minerals into smaller fragments and into minerals that are more stable (and in equilibrium) near the earth's surface or with the prevailing environment.

7.3.1 Physical Weathering

The breakdown of minerals or rock materials by entirely mechanical means characterizes Physical weathering (Thornbury 1969). In this process, the chemical composition of the weathered rock (mineral) remains unchanged. Some of the breaking forces actually originate within the rock or mineral itself, while others are applied externally. The stresses (both internal and external) lead to increased strain that eventually ruptures the rock. This mechanical rupture is largely due to abrasion, crystallization (growth of salt crystals), thermal insolation, unloading (pressure-release), and cycles of wetting and drying. It must also be noted that a rock broken into smaller fragments exposes more surface area of the original rock, thereby also increasing the available channels and chances of weathering. On a smaller scale, mineral grains have boundaries, potential areas of weakness within the rock, and thus potential weathering locations. Sedimentary rocks are often layered and are generally not tightly bound together. Massive rocks have joints, which open as the rocks are exposed to erosion. Physical weathering, with time, also widens these ruptures and fractures. Some of the major factors by which physical weathering occurs, are briefly discussed below.

7.3.1.1 Plants (Roots)

This process is also called Root wedging. It occurs when the root of a plant (especially large trees and bushes) begins to grow into a crack or pore within a rock (Fig. 7.2). As the plant grows larger, so does its root, until the root breaks the rock apart. Plant roots can split even the hardest rocks. This process is commonly noted in city sidewalks and foundations, where tree roots push from underneath, raising the concrete and subsequently cracking it. Additionally, acid-producing microorganisms (Fungi and Lichens) that live on rocks dissolve nutrients (phosphorus, calcium) within rocks. These microorganisms also assist in the breakdown and weathering of rocks.

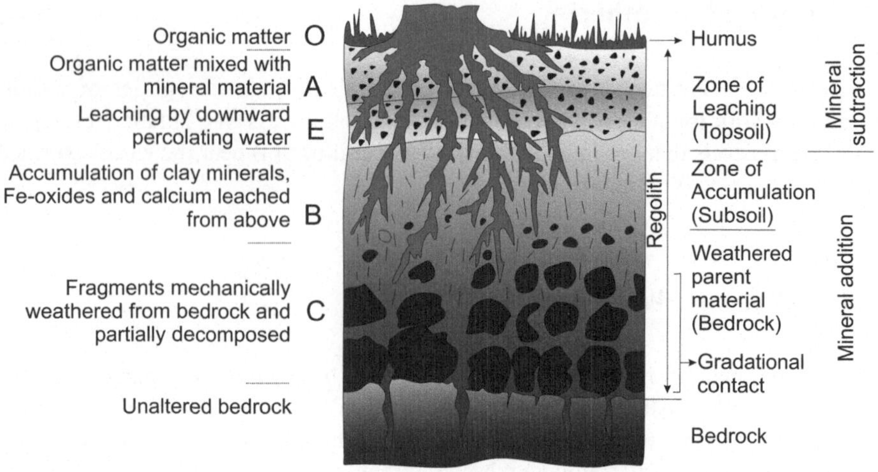

Fig. 7.2 Temperate climate soil profile. It shows the transition from bedrock to regolith through a sequence of layers, or horizons (O–C). The most obvious product of weathering is a blanket of loose, decayed rock debris known as Regolith, which forms a discontinuous cover over the solid, unaltered bedrock. Modified from earth's Dynamic Systems with permission of Eric H. Christiansen

7.3.1.2 Animals

Burrowing by animals (such as worms, termites, reptiles, rodents, etc.) into earth's substrate can move rock fragments and sediments. These movements aid in the disintegration of rocks. Digging by animals or even plowing by humans also results in the slow breaking of rocks into finer particles.

7.3.1.3 Crystallization or Growth of Salt Crystals

Mineral salts develop from mineral crystals as water evaporates moisture from rocks. This is very common in arid climates. The mineral grains are spread apart by these crystal growths. This growth eventually breaks apart the entire rock. Crystallization causes stress that fosters mechanical rupturing of rocks and minerals, alike. Crystal growth causes stress as a result of a compound's or an element's change of physical state with changing temperature. A volumetric change is also caused by the transformation from liquid to the solid crystalline state. This in turn causes the necessary mechanical action for rock rupture. Ice and salt are the two primary types of crystal growth that occurs. The crystallization of salt exhibits volumetric changes from 1 to 5 % depending on the temperature of the rock or mineral surface. Most salt weathering occurs in hot arid regions, but instances are also noted to occur in cold climates.

7.3.1.4 Grinding or Rubbing

The disintegration of rock or soil particles happens by grinding or by rubbing of moving rocks or soil particles against each other.

7.3.1.5 Abrasion

When two rock surfaces come together causing mechanical wearing or grinding of their surfaces, abrasion occurs. This collision normally occurs through the erosional transport of materials by wind, water, and ice. Pure water is not abrasive but the collisions among rock, sand, and silt results in weathering. The wind hurls sand and other small particles against rocks, which results in sandblasting forming unusual and beautiful landforms. Glaciers also cause abrasion as they drag particles ranging in size from clay to boulders across the bedrock. In such a scenario, both the rock fragments (embedded in the ice) and the bedrock beneath are abraded.

7.3.1.6 Unloading or Pressure-Release

The removal of thick layers of sediments overlying deeply buried rocks by erosion or uplift is called Unloading or Pressure-release (Fig. 7.3). Erosion removes the overlying rocks, and thus decreases pressure on the buried ones (Fig. 7.3b). Rocks are slightly elastic, hence, they respond to this pressure reduction by expanding. This results in the formation of fractures (cracks or fissures) parallel to the surface (Fig. 7.3c). With continued erosion, these rocks are exposed on the surface as slabs of rock that break off along the pressure-release fractures (Fig. 7.3c). These results in bare rock surfaces that are more resistant than the surrounding rocks. These are Exfoliation domes (large, rounded masses of rock) and the slabs of rock that break off are called Exfoliation sheets (Fig. 7.3c). Granite commonly fractures by exfoliation, a process in which large plates or shells split away like the layers of an onion (Fig. 7.3d). However, Exfoliation fractures are absent below a depth of 100 m, hence, they seem to be a result of exposure of the granite at the earth's surface.

Unloading plutonic Igneous rocks form depth also creates zones of weakness in them. Hence, when these rocks are exposed, they expand, and the zones of weakness open up as joints. Spalling, the vertical development of fractures, occurs because of the bending stresses of unloaded sheets across a three-dimensional plane.

Although, exfoliation is a form of pressure-release fracturing, however, some scientists have also attributed this to Hydrolysis where feldspars and other silicate minerals react to form clay. This water addition (change from Orthoclase Feldspar to Kaolinite) results in clay having a greater volume than that of the original mineral. Thus, a chemical reaction (hydrolysis) forms clay (Kaolinite), and the mechanical expansion of the clay contributes to exfoliation fractures in onion-shells; Fig. 7.3d.

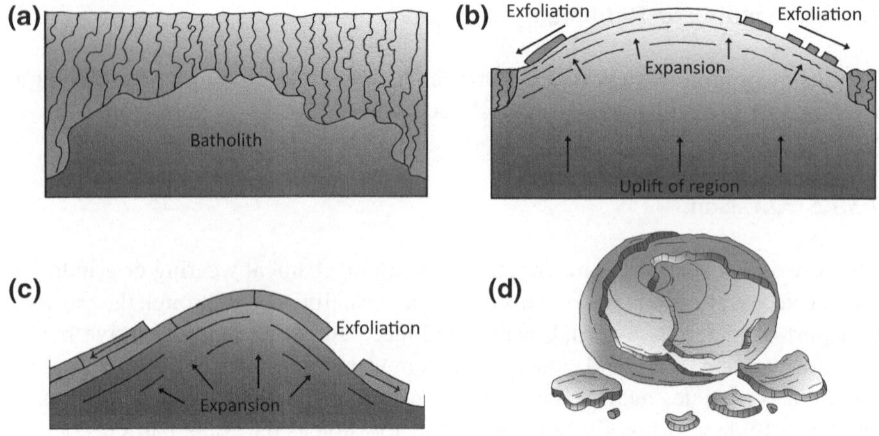

Fig. 7.3 Unloading and Exfoliation. The great white granite domes and cliffs of the High Sierra (like the Half Dome, Yosemite, California, USA), owe their appearance to exfoliation. **a** These rocks were emplaced as molten bodies, or plutons, deep underground, raising the Sierra Nevada range. **b** Erosion unroofed the plutons and took away the pressure of the overlying rock. **c** As a result, the solid rock acquired fine cracks through pressure-release jointing. Mechanical weathering opened up the joints further and loosened these as slabs. **d** Granite commonly fractures by exfoliation, a process in which large plates or shells split away like the layers of an onion

$$2KAlSi_3O_8 + 2H^+ + H_2O \rightarrow Al_2Si_2O_5(OH)_4 + 2K^+ + 4SiO_2$$

Orthoclase Feldspar + Hydrogen ion + Water → Clay (Kaolinite) + Potassium ion + Silica

Excellent examples of exfoliation domes are found in many granitic areas and particularly at the Yosemite National Park in California and at the Stone Mountain in Georgia (USA).

7.3.1.7 Wetting and Drying

Slaking is the alternate wetting and drying of rocks. It is an important contributor to weathering. It occurs by the mechanism of ordered water, which is the accumulation of successive layers of water molecules in between mineral grains of a rock. The rock grains are pulled apart with great tensional stress due to the increasing thickness of water. About 20 cycles of alternating wetting and drying can disintegrate a rock sample. In soils, the disruption by wetting and drying results in its swelling and contracting of soil particles. Abrasion among particles within the soil makes the particles finer. The soil shrinks when dry, and cracks develop, creating an irregular boundary between horizons. As rocks are alternately heated and cooled, they expand and contract; minerals also expand and contract by different amounts, and this differential expansion and contraction stresses the rocks and cracks them open, thereby, facilitating increased weathering.

7.3.1.8 Variation in Temperature

Slowly cooled rocks that have more time to build stronger bonds are more resistant to weathering. Hence, those rocks formed under intense temperature and pressure, but cool slowly, are more stable when exposed to low temperatures and pressures at the earth's surface. Insolation weathering is the physical breakdown of rock by expansion and contraction due to diurnal (daily) temperature changes (which at times on a daily basis can be as large as 30 °C or more). This diurnal temperature change is a notable contributor toward physical weathering. Insolation weathering is an expression of the physical inability of rock to conduct heat well, thus, resulting in differential rates of expansion and contraction. Heat causes expansion, and cooling causes contraction. The surface of the rock expands more than its interior, resulting in stress that eventually causes the rock to rupture. This differential expansion and contraction may also be due to the variance in colors of individual grains of mineral within the rock. Dark colored minerals expand more due to their increased absorptive properties. Hence, in a rock peppered with colored grains, rupturing occurs at differential rates at various mineral boundaries. Different minerals expand and contract at different rates causing stresses along mineral boundaries. Repeated heating and cooling of such rock causes the eventual breakdown of the rock, resulting in enhanced weathering.

7.3.1.9 Freezing and Thawing

Water expands as much as 9 % of its volume as it is converted to ice. Freezing water can exert a pressure of 150 tons per square foot. Freezing and thawing occur on a daily cycle. The expansion force of water as it freezes can split any mineral or rock. Cracks filled with water are forced further apart when it freezes. Frost wedging happens when water filling a crack freezes and expands (Fig. 7.4). The expanding ice presses against the rock and wedges open the crack thereby facilitating weathering (see also Hamblin and Christiansen 2008). Freezing and thawing are most effective in areas where temperatures commonly fluctuates above and below freezing. Good examples are noted in the high mountains of western United States and Canada.

7.3.1.10 Heating and Cooling (Thermal Expansion and Contraction)

Differences in temperature in a rock or soil mass gives rise to differential expansion and contraction. The resulting stresses can fracture minerals. Temperature change can also bring about exfoliation as often noted in granitic terrains. Thermal expansion and contraction are more significant in large outcrops.

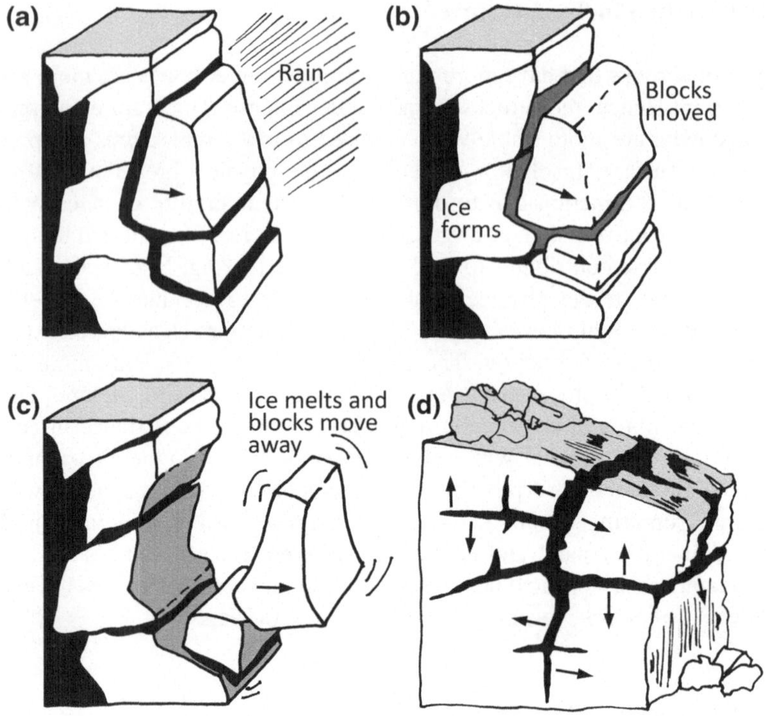

Fig. 7.4 Frost wedging takes place when water seeps into cracks **a**, expands as much as 9 % of its volume as it freezes **b**, and pries loose angular pieces of rock (**c**–**d**). **d** Redrawn from Earth's Dynamic Systems with permission of Eric H. Christiansen

7.3.2 Chemical Weathering

Minerals in rocks are chemically altered by this process which subsequently decompose and decay (Table 7.1). Increasing precipitation (rain) speeds up the chemical weathering process. In fact, water is an essential agent of chemical weathering. Increasing temperature also accelerates chemical reactions causing minerals to degrade (Fig. 7.5). Additionally, climate is another important factor affecting chemical weathering. Climatic conditions control the rate of weathering that takes place by regulating the catalysts of moisture (rain fall/precipitation) and temperature and physical weathering in cooler and drier conditions (Fig. 7.5a). Consequently, chemical weathering is strongest in Tropical wet to Monsoon climates and physical weathering in Subarctic and Tundra (Fig. 7.5b). Weathering, as such is largely a combined function of the effects of precipitation, temperature, and vegetation (Fig. 7.6) with minor inputs based on the nature of topography, rock type, and time that affects the thickness of soil (Fig. 7.7). Hence, tropical weathering rates are three and a half times higher, where temperature, moisture,

Table 7.1 Weathering products of common rock-forming minerals (under the influence of CO_2 and H_2O)

Original material	Main solid products	Additional products (ions in solution)
Amphibole (Ca, Mg, Fe silicate)	Limonite, clay	Ions (K^+, Mg^{2+}, SiO_4^{4-})
Biotite (Fe, Mg, K, Al silicate)	Clay mineral	Ions (K^+, Mg^{2+}, SiO_4^{4-})
Calcite	–	Ions (Ca^{2+}, HCO_3^-)
Ferromagnesian minerals (including biotite)	Clay mineral	Ions (Na^+, Ca^{2+}, K^+, Mg^{2+}), SiO_2, Fe oxides
Gypsum ($CaSO_4$ plus water)	–	Ions (Ca^{2+}, SO_4^{2-})
Halite (NaCl)	–	Ions (Ca^{2+}, Cl^-)
Muscovite (K, Al silicate)	Clay mineral	Ions (K^+, SiO_4^{4-})
Olivine (Mg, Fe silicate)	Limonite, clay	Ions (Mg^{2+}, SiO_4^{4-})
Plagioclase feldspar (Ca, Na, Al silicate)	Clay	Ions (Na^+, Ca^{2+}, SiO_4^{4-})
Potassium feldspar (K, Al silicate)	Clay	Ions (K^+, SiO_4^{4-})
Pyroxene (Ca, Mg, Fe silicate)	Limonite, clay	Ions (Ca^{2+}, Mg^{2+}, SiO_4^{4-})
Quartz (SiO_2)	Quartz grains (sand)	Ion (SiO_4^{4-})

and vegetation are at their maximum, than the rates noted in temperate environments (Fig. 7.7). Contextually, the solubility of minerals depend upon solution (often in water), hydrolysis (the reaction of elements with water) and carbonation (the reaction with HCO_3^{-1}). Chemical weathering is faster where temperatures are higher and water is present (Figs. 7.5–7.7). The most common chemical weathering processes are dissolution and solution, hydration, oxidation, hydrolysis, carbonation and acidification, and reduction. These are briefly discussed below.

7.3.2.1 Dissolution and Solution

Dissolution occurs when rocks and/or minerals are dissolved by water. Carbon dioxide (CO_2) dissolves in water (H_2O) to form Carbonic acid (H_2CO_3) which reacts chemically with minerals. This acid dissolves them or alters them into other minerals. The dissolved material is transported away leaving a space in the rock. This process also forms caves in limestone areas. When the carbonic acid comes in contact with rocks that contain minerals such as calcium, magnesium, and potassium, then these rocks chemically change into carbonates and dissolve in rainwater. Karst topography is a result of this type of chemical weathering and is characterized by numerous sinkholes, caves, and caverns (for details on Karst topography see Chap. 10).

H^+ (Hydron, the positive ion of Hydrogen), a small ion, easily enters crystal structures and releases other ions into water as it is chemically active. Water, and the ions that it carries, moves through and around rocks and minerals, thus,

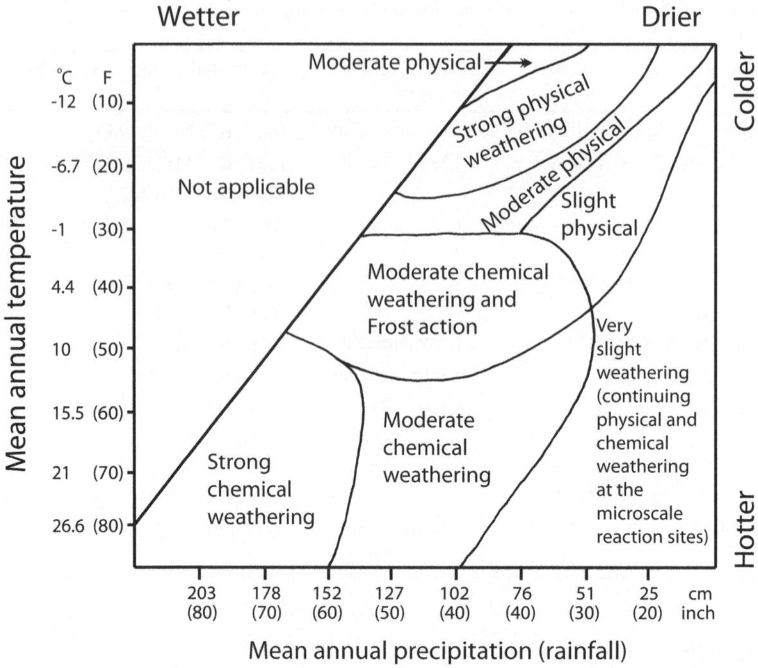
Wetter
Drier
Colder
Hotter
Moderate physical
Strong physical weathering
Moderate physical
Slight physical
Not applicable
Moderate chemical weathering and Frost action
Very slight weathering (continuing physical and chemical weathering at the microscale reaction sites)
Strong chemical weathering
Moderate chemical weathering
Mean annual temperature
°C F
-12 (10)
-6.7 (20)
-1 (30)
4.4 (40)
10 (50)
15.5 (60)
21 (70)
26.6 (80)
203 (80)
178 (70)
152 (60)
127 (50)
102 (40)
76 (40)
51 (30)
25 (20)
cm inch
Mean annual precipitation (rainfall)

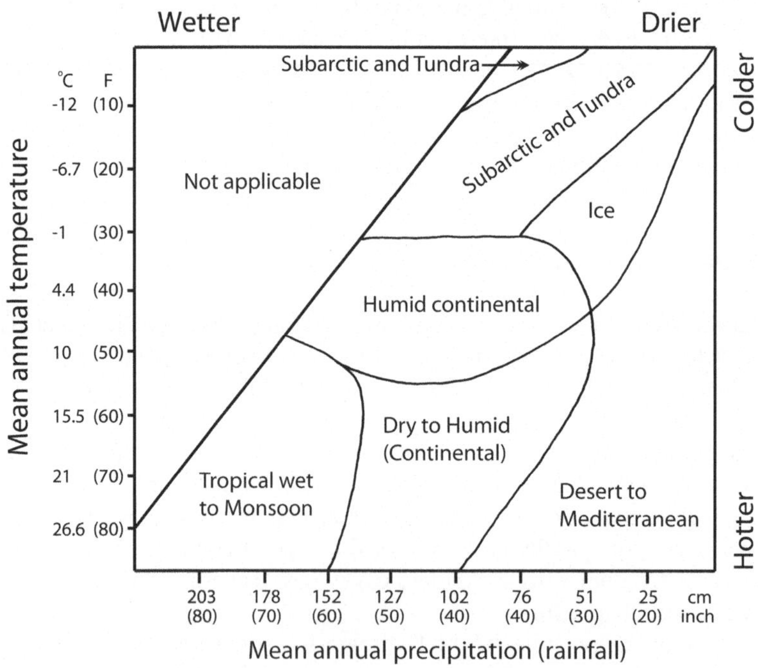
Wetter
Drier
Colder
Hotter
Subarctic and Tundra
Subarctic and Tundra
Not applicable
Ice
Humid continental
Dry to Humid (Continental)
Tropical wet to Monsoon
Desert to Mediterranean
Mean annual temperature
°C F
-12 (10)
-6.7 (20)
-1 (30)
4.4 (40)
10 (50)
15.5 (60)
21 (70)
26.6 (80)
203 (80)
178 (70)
152 (60)
127 (50)
102 (40)
76 (40)
51 (30)
25 (20)
cm inch
Mean annual precipitation (rainfall)

◀ **Fig. 7.5** Climatic conditions control the rate of weathering that takes place by regulating the catalysts of moisture (precipitation) and temperature. **a** At high temperature and precipitation, chemical weathering is strongest, and in cooler and drier conditions, physical weathering. **b** Consequently chemical weathering is strongest in Tropical wet to Monsoon climates and Physical weathering in Subarctic and Tundra. **a** Redrawn from earth's Dynamic Systems with permission of Eric H. Christiansen

furthering the weathering process. This process is called Solution and tends to be most effective in humid and hot climates. Thus, the main agent responsible for chemical weathering reactions is water and weak acids formed in water.

7.3.2.2 Hydration

Some minerals react with water and acid to take up hydrogen and remove other cations. This process is called Hydration. The combination of a mineral or element with water increases the size of the chemical structure, thereby leading to a softer, more stressed, and more easily decomposed mineral Fig. 7.8. Water combines with compounds in rocks, causing a chemical change in the mineral's structure, but more often than not, it physically alters a mineral's grain surface and edges. Good examples of Hydration are the conversions (chemical changes) of Hematite to Limonite (Fig. 7.8) and of Anhydrite ($CaSO_4$) to Gypsum ($CaSO_4 . 2H_2O$), when water is added.

7.3.2.3 Oxidation

Oxidation is a chemical combination of oxygen with a compound. It also brings a change in the oxidation number of the chemical element; electrons are lost in the oxidation process. Oxygen combines with compound elements in rocks to form oxides. Oxidized minerals increase in volume and often become softer. The change of oxidation number of an element also unbalances the mineral's electrical neutrality, making it more susceptible to weathering by water and carbonic acid. Oxidation process is most evident in the weathering of iron-bearing minerals. A good example of this is Rusting. Oxygen reacts with iron in minerals to form iron oxide, like Hematite (Rust). Increased temperatures and the presence of precipitation accelerate the oxidation process. Iron-bearing silicate minerals which also contain Aluminum (such as Pyroxene, Amphibole, and Biotite) undergo both oxidation and hydrolysis, forming both iron oxides and clays.

7.3.2.4 Hydrolysis

Hydrolysis is also one of the most important weathering process that brings changes in the soil profile. The carbonic acid ionizes (breaks down) into two ions,

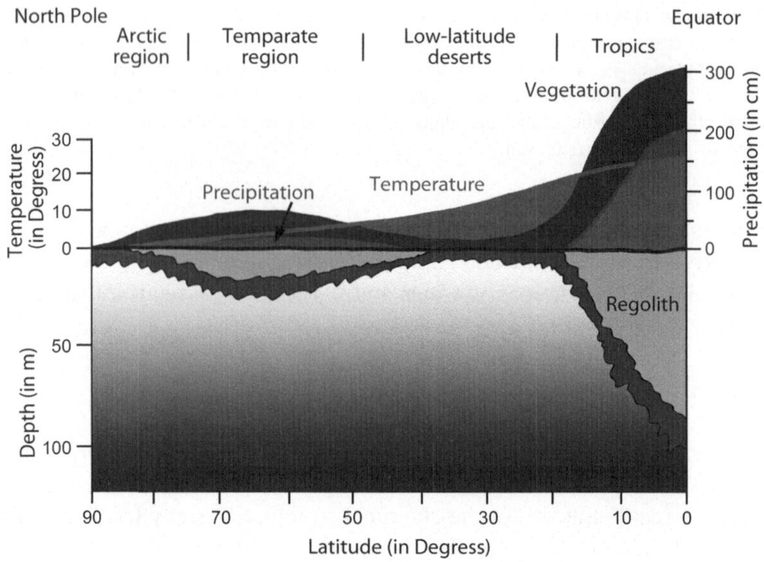

Fig. 7.6 The type and extent of weathering is a function of the prevailing climate. The diagram shows the combined effects of precipitation, temperature, and vegetation. Weathering is most pronounced in the tropics, where precipitation, temperature, and vegetation reach a maximum. Redrawn from earth's Dynamic Systems with permission of Eric H. Christiansen

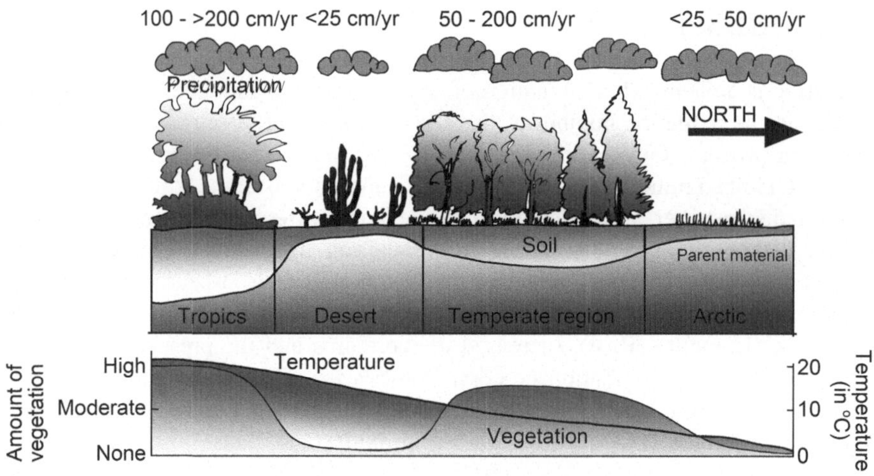

Fig. 7.7 Tropical weathering rates are three and a half times higher where temperature, moisture, and vegetation are at their maximum, than the rates noted in temperate environments (note the higher amounts of eroded soil)

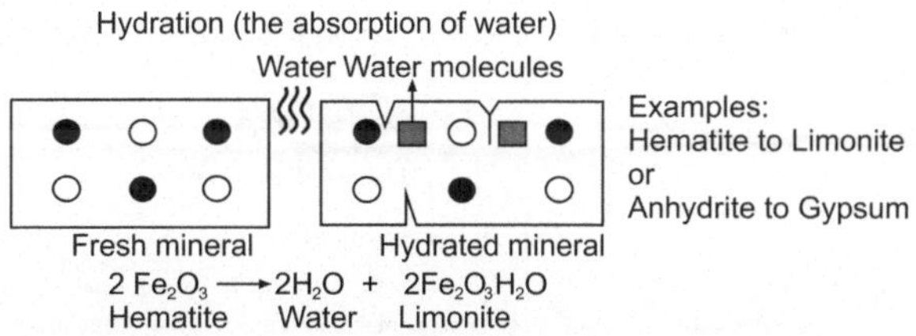

Fig. 7.8 Hydration occurs when some minerals react with water and acids to take up hydrogen and remove other cations. Good examples are the conversions (chemical changes) of Hematite to Limonite. Others example is of Anhydrite ($CaSO_4$) to Gypsum ($CaSO_4$. $2H_2O$), when water is added

hydrogen (H^+) and bicarbonate (HCO_3^{-1}). The free hydrogen ion alters the composition of a mineral by replacing other ions in its atomic structure. It occurs when water (often in the form of precipitation), disrupts the chemical composition and the size of a mineral, thereby creating a less stable mineral. Thus, less stable rock forms will also weather more readily. Hydrolysis also increases the pH of the involved solution through the release of the hydroxide ions. Hydrolysis is especially effective in the weathering of common silicate and alumino silicate minerals because of their electrically charged crystal surfaces. Potassium feldspar weathers to form Kaolinite. In humid climates most of the feldspar in rocks such as Granodiorite (Fig. 7.9) or Granite (Fig. 7.10) will weather to form clay. Nearly all of the minerals in the common rocks of the earth's crust will weather to form clay (with one exception—Quartz). Because of this, clays make up nearly half of the sedimentary rocks on earth.

7.3.2.5 Reduction

Reduction is the addition of one or more electrons. Hence, it is the reverse of Oxidation. Free Oxygen (O_2) reacts with minerals to change the oxidation state of an ion. This is commonly noted in iron (Fe) bearing minerals, as Fe can have several oxidation states, Fe, Fe^{+2}, and Fe^{+3}. Deep in the earth, the most common oxidation state of Fe is Fe^{+2}. In soils, reduction usually takes place when Oxygen is scarce, as in stagnant waters. Reduction in minerals results in electrically unstable compounds, more soluble ones, or more internally stressed ones; these eventually decompose more rapidly. Poorly drained soils will have reduction and oxidation reactions taking place throughout the profile. Reduction is evidenced by the gray color of the soil; red mottles indicate oxidized iron.

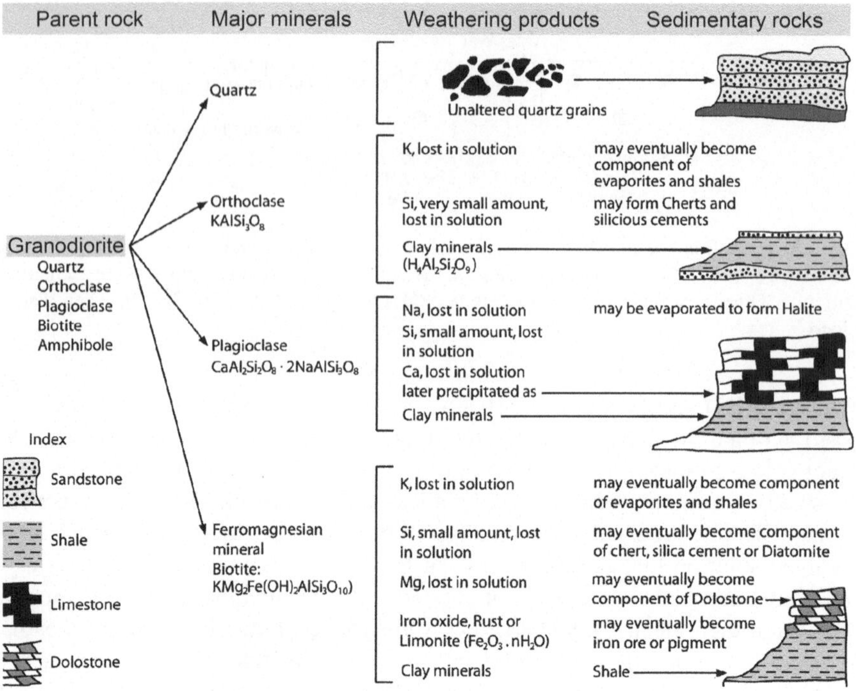

Fig. 7.9 The weathering products of a Granodiorite. The resultant products are used for the formation of various sedimentary rocks

Fig. 7.10 The weathering products of a Granite and the relative rate at which each constituent weathers with progressive chemical weathering. Quartz by and large remains stable

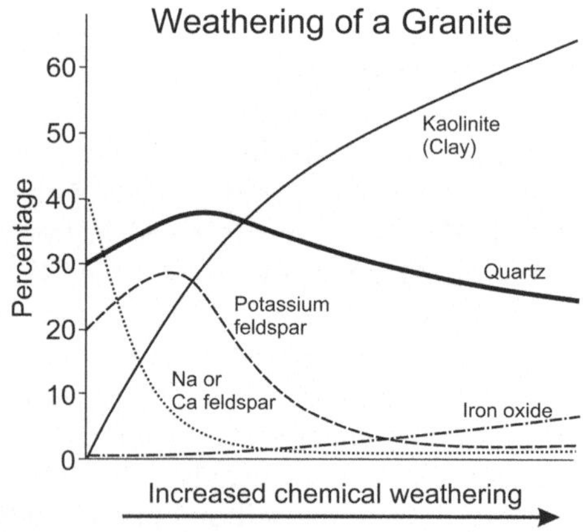

7.3.2.6 Carbonation and Acidification

Carbonation is especially active when the reacting environment is abundant with CO_2. Carbonation is the reaction of carbonate and bicarbonate ions with minerals.

The formation of Carbonic acid (CO_2 and water), is important in the solution of carbonates and the decomposition of mineral surfaces due of its acidic nature. Carbonic acid, a weak acid, is produced when gaseous CO_2 is dissolved in water. The weathering is accelerated by the presence of the hydrogen ion in water, such as that provided by carbonic and organic acids. Hence, acidification is a form of dissolution. Carbonic acid dissolves minerals more readily than water alone and forms the more soluble bicarbonates. In limestone territory, Acidification is responsible for the weathering of limestones and the eventual formation of Sinkholes.

7.3.3 Biological Weathering

Living organisms, ranging from bacteria to plants to animals (including Lichens), breakdown rocks. This process of breakdown and their action is biological weathering (see also Fig. 7.2). Lichen (a crusty, rubbery, light green organic material that grows in patches on rocks as well as on wood) is a combination of fungus and algae that live together in a symbiotic relationship. They can live on bare rocks, and are able to breakdown rocks by secreting acids and other chemicals. The fungal part of the association secretes the acids, which reacts to dissolve minerals that are then used by the algae. Later, water seeps into crevices etched by the acid, and assists in the final breakdown through freezing frost wedging (Fig. 7.4d) and chemical weathering. Biological weathering is also done by Tree roots and Bacteria.

7.3.3.1 Tree Roots

Tree roots grow into cracks and widen them, thus, facilitating physical weathering (Fig. 7.2).

7.3.3.2 Bacteria

Acidic solutions are secreted by some bacteria and other organisms that speed up chemical weathering. Chitons and Limpets also facilitate biological weathering as they bore into beach rocks.

All in all, it must be noted that in a given area, although a single weathering process often dominates, but all three weathering processes (physical, chemical, or biological), often occur simultaneously. Hence, these processes cannot be separated from one another as all three proceed at the same time (though not necessarily at the same rate). Physical weathering helps chemical weathering by breaking rocks into smaller pieces, thus, allowing more surface area to be exposed. With more surface area, chemical reactions occur faster. Chemical weathering

Table 7.2 Summary of the three types of weathering, physical, chemical and biological

Weathering

Physical

Unloading / Pressure-release / Exfoliation
1. Closely associated with the physical disintegration of rocks
2. Rocks experience great confining pressure; any reduction in it results in decreased rock-strength and elastic expansion forming fractures, cracks and fissures
3. Sheets joints, pseudo-bedding planes and Exfoliation fractures are widely noted in granitic rocks
4. Expansion is at 90 degrees to the ground surface forming a series of cracks parallel to the surface. This vertical development of fractures is called Spalling
5. Formation of pressure-release joints (line of weakness) causes eventual rock disintegration [both chemical (Hydrolysis) and physical weathering act upon it]

Freezing and Thawing (Frost weathering)
1. Most common type
2. Occurs in high-latitudes, high-altitudes, mid-latitude and in desert regions
3. Results from Frost-wedging / Splitting
4. Water penetrates joints and bedding planes, and then undergoes change from liquid to ice; this change expands the volume by 9% thereby creating fissures
5. The growing ice crystals exert pressure, detaching joint bounded blocks or breaking into small clasts.
6. Fissures are wide-spread resulting in large blocks to be detached
7. Average frost shattering force can be as high as 14kg/cm^3

Variation in Temperature, (Insolation weathering)
1. Occurs when there is a large diurnal range in temperature (>30 degrees)
2. Rock types most likely to be fractured are dark colored, fine grained and heterogenous rocks. Dark colored rocks have low albedo and thus absorb maximum solar energy.
3. Heterogenous rocks (containing different minerals) possess different coefficients of expansion and contraction, hence are most susceptible to breakdown
4. Changes in surface temperature, aided by pressure-release results in tiny exfoliation cracks, causing fracturing and eventual rock disintegration

Crystallization or growth of salt crystals (Salt weathering)
1. Essentially a physical process but also involves chemical weathering
2. Crystallization of super-saturated solutions of salt occupy fissures and pore spaces within the rock
3. As the crystal grows, expansive stresses are applied to joints resulting in rock disintegration
4. Acts uniformly over rock surfaces or is concentrated at particular sites
5. It is mainly a hot-desert process (low-rainfall and high temperatures)
6. Other forms of salt-weathering include expansion of salts within confining spaces by heating and by the stresses imposed by the hydration of salts
7. Salt-weathring has an important effect on building materials in desert regions

Heating and Cooling (Thermal expansion and contraction)
1. Temperature differences in a rock give rise to differential expansion and contraction thus causing exfoliation.
2. Common in large outcrops

Wetting and Drying
1. Slaking is the alternate wetting and drying of rocks. An important contributor to weathering.
2. The rock grains are pulled apart with great tensional stress due to the increasing thickness of water. About twenty cycles of alternating wetting and drying can disintegrate a rock sample.

Chemical

Dissolution and Solution
1. Dissolution occurs when rocks and/or minerals are dissolved by water
2. Effectiveness governed by the acidity and alkalinity of the groundwater
3. Good example of this is Karst topography

Hydration
1. Results due to the capacity of certain minerals to absorb water
2. Volumetric changes take place, setting up physical stresses and causing physical disintegration

Oxidation
1. A process of combining with Oxygen
2. Oxidation usually involves water in which Oxygen is dissolved
3. Iron minerals are very prone to Oxidation [Rusting; Iron minerals are oxidized]
4. Commonly noted in free-draining, well-aerated tropical soils and regoliths
5. Oxidation is responsible for its red and reddish brown color
6. Iron-bearing silicate minerals undergo both oxidation and hydrolysis, forming iron oxides and clays.

Carbonation and Acidification
1. Calcium carbonate is changed to Calcium bi-carbonate by rainwater containing dissolved Carbon-di-oxide
2. Bicarbonates are readily removed from solution
3. Most effective in weathering common silicate and alumino-silicate minerals. Potassium feldspar weathers to form Kaolinite

Hydrolysis
1. Chemical reaction between H+ and OH- ions of water and the ions of the mineral
2. An important process that initiates the decomposition of Feldspars
3. Feldspars weathers to Clay (Kaolinite)

Reduction
1. A process of disassociating from Oxygen i.e. an addition of one or more electrons; the reverse of Oxidation
2. It results in gray color soil; red mottles indicate oxidized iron
3. Reduction results in electrically unstable compounds, more soluble ones, or more internally stressed ones; that eventually decompose more rapidly.

Biological

Living organisms, ranging from plants to animals (including Lichens) to bacteria, break down rocks.

Tree roots: Tree roots grow into cracks and widen them, facilitating physical weathering

Bacteria: Acidic solutions are secreted by some bacteria and other organisms that speed up chemical weathering

helps physical weathering by weakening the mineral grains that binds the rock together. Biological weathering helps both physical and chemical weathering. Trees fracture rocks with their roots, which make the rocks easier to break up physically, thereby, exposing more surface area for chemical weathering. Bacteria secretes acid solutions which speed up the chemical weathering process. Thus, working in concert, these three weathering processes can reduce a once resistant rock into nothing (as in case of limestones) or to easily erode into weaker materials (such as clays).

Table 7.2 gives a summary of the characteristics of the three types of weathering discussed above.

7.4 Weathering Products

Weathering byproducts include (a) the complete loss of particular atoms or compounds from the weathered surface, (b) the addition of specific atoms or compounds to the weathered surface, and (c) a breakdown of one mass into two or more masses, with no chemical change in the mineral or rock. Thus, the residue of weathering consists of chemically altered and unaltered materials. The most common unaltered residue is Quartz which is resistant to chemical decay, and therefore, only undergoes size reduction during transportation (Figs. 7.9–7.10). Other minerals such as feldspar, olivine, augite, and hornblende react with chemical reagents to produce various products. Generally the Si, Al, and Fe ions from these minerals are used up in generating secondary solid products like quartz, muscovite, kaolinite, hematite, and goethite. Many of the chemically altered products of weathering become very simple small compounds or nutrient ions. These residues can then be dissolved or transported by water, released to the atmosphere as a gas, or taken up by plants for nutrition. Some of the products of weathering, like the less resistant alumino silicate minerals, become clay particles.

7.5 Weathering and Associated Rocks

When rock weathers, they usually do so by working inward from a surface that is exposed to the weathering process. This results in Spheroidal weathering, Exfoliation, and Weathering rinds.

7.5.1 *Exfoliation*

These are shells of weathering that form on the outside of a rock and become separated from the rock with the release of stress that results from changes in the volume of minerals that occur as a result of the formation of new minerals. This process is also called Onion-skin weathering (Fig. 7.3d).

7.5.2 *Spheroidal Weathering*

Fig. 7.11 Weathering rind. Note the presence of an outer weathered zone (weathering rind) and an inner unweathered zone. As weathering progresses, the thickness of the weathering rind increases. It is, thus, sometimes also used as an indicator of the amount of time the rock has been exposed to the weathering process

If joints and fractures in the rock beneath the surface form a three-dimensional network, the rock will be broken into cube-like pieces separated by fractures. Water penetrates through cracks in rocks and dissolves the cement that binds particles together and also erodes sharp edges and corners of the rock, making it appear spheroidal in shape. The unaltered rock is in the center and the weathered one toward the outside. Such progression of weathering is referred to as Spheroidal weathering.

7.5.3 Weathering Rinds

Often, a Weathering rind is an indicator of the amount of time a particular rock was exposed to the weathering process. In the initial stages of weathering, a rock shows an outer weathered zone and an inner unweathered zone (Fig. 7.11). This outer zone is called a Weathering rind and its thickness increases as weathering progresses. Ferromagnesian silicate minerals like olivine, pyroxene, hornblende, and biotite rust to produce hematite and other oxide minerals.

7.6 Rates of Weathering

Weathering rates are a function of the rock type, slope (topography), structure, and the prevailing climate (Figs. 7.5–7.7). Rocks that are most resistant are composed of minerals that are relatively unaffected by chemical weathering. Quartz is a classic example which is unaffected by dissolution, hydrolysis and oxidation. Therefore, rocks composed almost exclusively of Quartz are more resistant than any other rock types. Chemical weathering occurs fastest at the sharp edges of rocks as they have a large surface area and less volume. This results in faster chemical reactions, thereby, converting the sharp edges into rounded ones. On the other hand, physical weathering breaks rocks into smaller pieces. Hence, more surfaces are exposed to chemical weathering. Often, both physical and chemical weathering occurs together. Weathering rates are differential. Some general statements can be made about the factors that influence the rate of weathering. These are enumerated below.

7.6.1 Time

Time is a crucial factor in weathering as it determines the rate of weathering. Rate is how fast something occurs in a given amount of time. Rates of weathering vary from rapid to extremely slow and are largely a function of slope, climate, and the type of rock. If all things are equal, the longer a rock is exposed at the surface, the more weathered it will be.

7.6.2 Slope (Topography)

On steep slopes, rain may quickly wash the weathering products away. However, on gentle slopes the weathering products accumulate. For example, on gentle slopes the water may stay in contact with rock for longer periods of time, and thus, result in higher weathering rates.

7.6.3 Climate

Chemical reactions progress faster in the presence of high amounts of water and higher temperatures. Hence, in warm humid climates, there are more weathered rocks, and subsequently, the rates of weathering are much higher in comparison to cold dry climates (Figs. 7.5–7.7). Thus, a limestone in a dry desert climate is very resistant to weathering, but in a tropical setting, it weathers very rapidly. Heavy rainfall and high temperatures promote chemical weathering. Thus, a rock in a desert will show less weathering than the one in a tropical rain forest (Fig. 7.5). Hence, the type of weathering depends upon climatic conditions that it is being worked upon (Figs. 7.5–7.7). A temperature range that varies between freezing and thawing conditions and with plentiful supply of water is essential for physical weathering. Warm temperatures and a plentiful supply of water are necessary for chemical weathering. Areas with little water or low temperatures have relatively slow rates of chemical weathering (Fig. 7.5). Physical weathering is most common in cold climates where frost action occurs. Frost action is most prevalent when temperatures are low and moisture is abundant (Figs. 7.5–7.7).

7.6.4 Type of Rock (Mineral Composition)

The susceptibility of a rock type to weathering is largely a function of its constituent minerals (Table 7.1). Sandstone consists exclusively of Quartz, a mineral that is very stable on the earth's surface. It will not weather at all in comparison to a Limestone that is composed entirely of Calcite, a mineral that easily dissolves in a wet climate (Table 7.1). The more soluble a mineral is, the more easily weathered it will be. Calcite, Halite and, Olivines are highly soluble, while Quartz dissolves very slowly, as it is a tough framework structure (Table 7.1). Additionally, the earlier a mineral crystallizes from the melt, the less stable it is at surface conditions (Fig. 7.12).

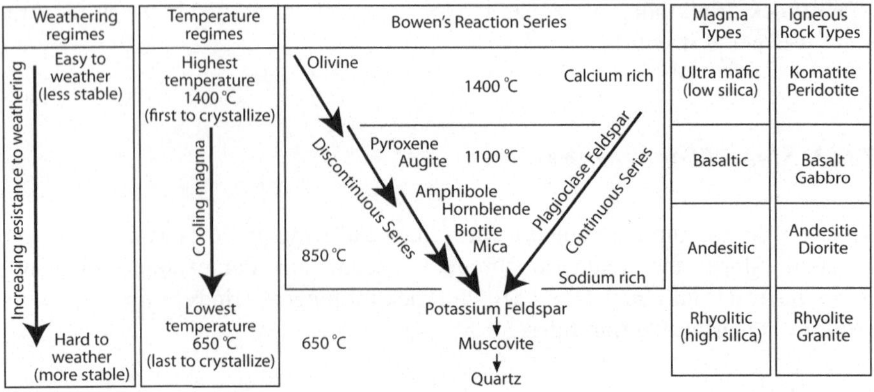

Fig. 7.12 The inter-relationship between weathering, temperature regimes, magma and igneous rock types and the Bowen's Reaction Series. The latter series consists of a discontinuous branch along which a succession of ferromagnesian silicates crystallize as the magma's temperature decreases, and a continuous branch along which plagioclase feldspars with increasing amounts of sodium crystallize. Notice also that the composition of the initial mafic magma changes as crystallization takes place along the two branches. The earlier a mineral crystallizes from the melt, the less stable it is at surface conditions and the more easily it is weathered

7.6.5 Exposure

Materials that have a high surface area exposed to air, water, or organisms have higher rates of weathering.

7.6.6 Particle Size

As particle size decreases, more surface area is exposed, thereby, increasing the chances of weathering.

7.6.7 Rock Structure

Joints, Fractures, Fissures, and Bedding planes provide potential surfaces for both physical and chemical weathering processes to act. Such features also provide easy pathways for the entry of water. Thus, the more massive the rock is, the slower it will weather physically. Hence, rocks that contain them are weathered more rapidly than equivalent unfractured ones. Additionally, as soils hold moisture and organisms; rocks weather more quickly when they are in contact with a soil.

7.6.8 *Gravity*

It is the primary agent for the movement of weathered material from the site of weathering to the site of deposition. But, gravity always acts in tandem with the running water.

7.6.9 *Animals*

Burrowing organisms like rodents, earthworms, and ants, bring material to the surface were it can be exposed to the various agents of weathering.

7.7 The Stability of Minerals

Rocks, as a result of uplift and/or erosion, when they arrive near the surface, encounter conditions very different from those under which they were originally formed, i.e., deep within the earth. Due to these contrary conditions, minerals in rocks react with their new environment to produce new minerals that are stable under the prevailing conditions (near the surface). Table 7.3 lists minerals in order of most stable to least stable.

Those minerals that crystallize at a high temperature from the magma are most unstable, i.e., higher the temperature of crystallization, the less stable they will be at low temperatures (Fig. 7.12). Low temperature conditions prevail near the earth's surface. Rocks that are formed under intense temperature and pressure cool rapidly to form crystalline structures in minerals and are less stable when exposed to low temperatures and pressures at the earth's surface. Hence, these weather more rapidly. However, rocks that are formed under intense temperature and pressure, but cool more slowly and occur later in the volcanic magma cooling process, are more stable even when exposed to the low temperatures and pressures at the earth's surface. Bonds holding atoms together determine mineral hardness. Rocks that have cooled more slowly have time to build stronger bonds, hence, are more resistant to weathering.

7.8 Weathering of Common Rocks

In the weathering processes most of the rocks form clay minerals (Figs. 7.9–7.10). Quartz and muscovite remain as residual minerals because they are resistant to weathering. The ferromagnesian minerals present in Igneous rocks are largely weathered to clay, highly soluble ions, and suspended matter (Figs. 7.9–7.10).

Table 7.3 Distribution of minerals in order of their stability with respect to weathering

Slow weathering → Rapid weathering

Least stable	Stable in atmosphere	Unstable in atmosphere	Dissolve and re-precipitate
			Halite
			Gypsum
			Calcite
		Pyrite	Dolomite
		Olivine	
		Ca-plagioclase	
		Pyroxene	
		Amphibole	
		Biotite	
		Na-plagioclase	
		K-feldspar	
		Muscovite	
	Quartz		
	Clay		
	Aluminum oxide		
Most stable Iron oxide			

Mineral calcite and dolomite of sedimentary rocks (limestone), weather to highly soluble ions. The iron oxides and sulfides, salts and gypsum, volcanic debris, and organic material are weathered to clay, highly soluble ions, and suspended matter. The common rocks and their weathering products are given in Table 7.4 and weathering reactions for common minerals in Table 7.5.

An example of Granodiorite, an Igneous rock, is given to illustrate the weathering products of major minerals and their subsequent uptake for the eventual making of sedimentary rocks. The unweathered granodiorite contains minerals such as plagioclase feldspar, potassium feldspar, quartz, and small

Table 7.4 Weathering of common rocks

Rock	Primary minerals	Residual minerals	Leached ions
Granite	Feldspars	Clay minerals	Na^+, K^+
	Micas	Clay minerals	K^+
	Quartz	Quartz	–
	Fe–Mg minerals	Clay minerals + hematite + goethite	Mg^{+2}
Basalt	Feldspars	Clay minerals	Na^+, Ca^{+2}
	Fe–Mg minerals	Clay minerals	Mg^{+2}
	Magnetite	Hematite, goethite	–
Limestone	Calcite	None	Ca^{+2}, CO_3^{-2}

Table 7.5 Weathering reactions for common minerals

Original mineral	General formula	Weathering reactions	dissolved ions	Residual minerals
Gypsum	$CaSO_4\ 2H_2O$	Dissolution by water	Ca, SO_4	
Halite	NaCl	Dissolution by water	Na, Cl	
Olivine	(Mg, Fe)$_2$SiO$_4$	Oxidation		Fe oxides
		Dissolution by acid	Mg, Fe	
Pyroxene	Ca (Mg, Fe)Si_2O_6	Oxidation	–	Fe oxides
		Dissolution in acid	Mg, Fe, Ca	
Amphiboles	NaCa (Mg, Fe)$_5$AlSi$_7$O$_2$2(OH)$_2$	Oxidation		Fe oxides
		Partial solution by acid	Na, Ca, Mg	Clay
Plagioclase	$NaAlSi_3O_8$ to $CaAl_2Si_2O_8$	Partial solution by acid	Na, Ca	Clay
K-feldspar	$KAlSi_3O_8$	Partial solution by acid	K	Clay
Muscovite	$KAl_3Si_3O_{10}(OH)_2$	Partial solution by acid	K	Clay
Biotite	K(Mg, Fe)$_3$AlSi$_3$O$_{10}$(OH)$_2$	Oxidation		Fe oxides
		Partial solution by acid	K, Mg	Clay
Quartz	SiO_2	Resists dissolution		
Calcite	$CaCO_3$	Dissolution by acid	Ca	
Dolomite	$CaMg(CO_3)_2$	Dissolution by acid	Mg, Ca	
Pyrite	FeS_2	Oxidation	SO_4	Fe oxides

amounts of biotite, amphibole, orthoclase and sometimes even muscovite (Fig. 7.9). Resultant weathering under warm, humid conditions forms a rock called Saprolite (rotten rock). The weathered rock fragments are one of the constituents of soil. The feldspars undergo hydrolysis to form clay (Kaolinite) and sodium and potassium ions. The Na and K ions are removed through leaching and carried in solution along with the running water. The biotite and/or amphibole undergo hydrolysis to form clay, and oxidation to form iron oxides. Quartz (and if present muscovite) remain as residual minerals as they are resistant to weathering. The weathered products are subjected to various physical processes such as transportation and deposition. Quartz grains are eroded, and incorporated into sedimentary rocks or just become Quartz sand. This is ultimately transported to the sea (as bed load), where it accumulates to form beaches. Clays ultimately are eroded and washed out to the Sea. The fine-grained clay remains suspended in the water column (as suspended load) and is subsequently deposited in quiet waters. Dissolved ions are transported by rivers to the Sea (as dissolved load), and eventually become part of the salts in the Sea. A schematic diagram of minerals and their weathering products from a Granodiorite is given in Fig. 7.9 and for fresh Granite that primarily contains quartz and feldspar and they change to clay, quartz, and iron oxides as given in Fig. 7.10.

7.9 Mass Wasting

Gravity-initiated down slope movement of Regolith (loose particles of soil and rock) without the aid of a transporting medium (such as water, ice, or wind) is called Mass Wasting. The term is often used interchangeably with Mass Movement and comes within the gamut of erosional processes, between weathering and stream transport. The loose particles of soil and rock (Regolith) are picked up by the transporting agent and then moved to a site of deposition (such as an ocean basin or a stream bed). For the regolith to move on its own (the basic criterion for a mass wasting process), it must be on a slope allowing gravity to affect and cause motion. The slope at which loose and unconsolidated material sits at rest is called the Angle of Repose and it is typically between 25 and 40°. Besides slope, another factor that plays a key role in mass wasting events is water. It lubricates the material and adds weight, thereby resulting in increased instability and greater motion.

7.10 The Mass Wasting Process

The classification of mass wasting process is somewhat difficult as most types (of mass wasting) grade into one another. Many classifications of types have been proposed. However, it is now universally agreed that for such classification to be successful, it should be based on the following criteria: (a) the type of material in motion (particle size, degree of coherence); (b) the nature of motion (falling, toppling, sliding, flowing, etc.), and (c) the speed of motion. But, the mode, speed, and volume of down slope mass movements vary enormously. Some types of commonly recognized landforms based on the nature of motion include creep debris, landslide, avalanche debris, mudflow, flow debris, rock fall, slide, rock

Table 7.6 Classification of mass wasting (see Fig. 7.14 for illustration)

	Velocity →					
Materials	Nature of motion	Slow 1 cm/year Low water content		Moderate 1 km/year High water content	Fast 5 km/year High water content	
Rocks	Flow					Rock avalanche
	Slide or fall			Rockslide		Rockfall
Unconsolidated materials	Flow	Earth creep	Earthflow	Debris flow		
				Mudflow		
	Slide or fall		Slump		Debris slide	Debris avalanche

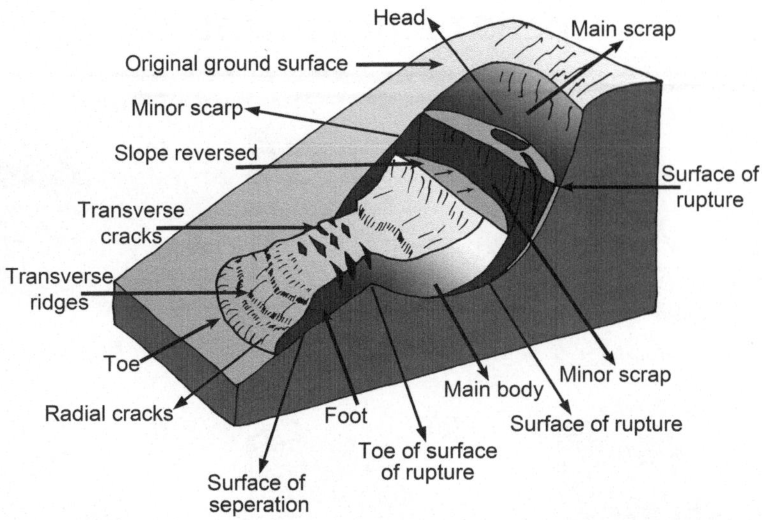

Fig. 7.13 Slump terminology used in the chapter. Modified from (USGS, 2004)

slide, earthflow, slump, gelifluction, solifluction, and lahar. It is to be noted that most of them grade into one another, without any clear boundaries between them, complicating the very basis of classifying them into types. Landslide is, thus, a better all-encompassing term that is best suited to describe all the above landform types. All down slope movement of materials (bedrock, regolith, or a mixture of these) are therefore, referred to as Landslide. Broadly, the mass wasting processes can be divided into two categories—Flow and Slide (Table 7.6).

The terminology used in the text ahead is diagrammatically illustrated in Fig. 7.13.

7.10.1 Rockfall

This might involve either one or few rocks (Fig. 7.14a). Generally, the rocks detach from part of a steep slope, drop, and bounce as they move very rapidly down slope. They are quite dangerous as they can occur without warning, and more so as the rocks are traveling at high velocity. Talus, an accumulation of fallen material, is quite a characteristic feature of Rockfalls. The Yosemite Valley, California (USA), is a good example of rockfalls. Here, the rockfalls range in size from individual boulders to moderate-sized block with volumes close to 100,000 m^3 (Matthes, 1930; Wieczorek et al., 1992; Wieczorek and Jäger, 1996).

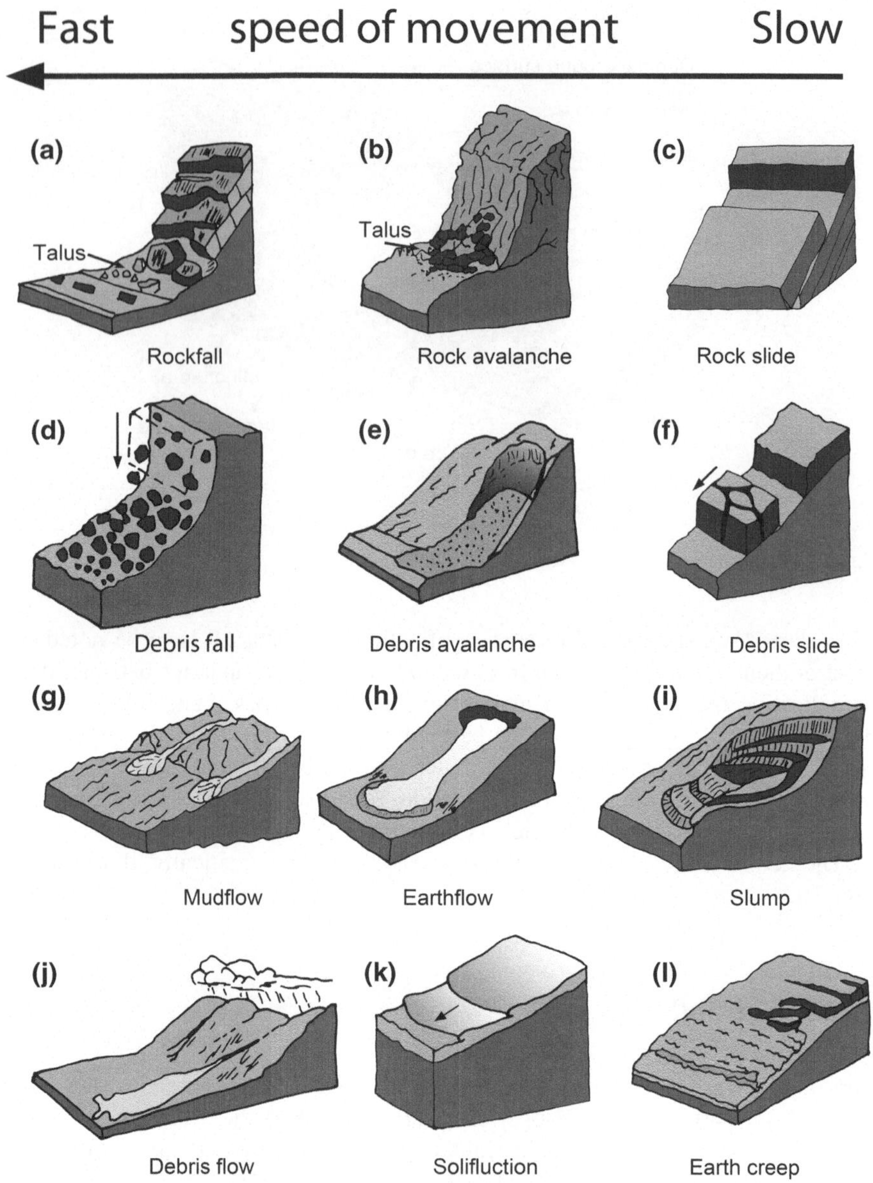

Fig. 7.14 Diagrammatic representation of Mass wasting classification discussed in the chapter. Modified from (USGS, 2004)

7.10.2 Rock Avalanche

Rock avalanches are often caused by heavy rain, melting snow, or quite often by earthquakes (Fig. 7.14b). Here steep slopes are involved. As a fast-moving fall touches the base of the mountain, it breaks into thousands of fragments that continue tumbling down slope at high velocity, this is Rock avalanche. Among all mass movements, it is one of the most destructive. The Denali earthquake (M = 7.9) in November 2002 triggered three large rock avalanches onto Black Rapids Glacier in the central Alaska Range of interior Alaska (Harp et al., 2003; Jibson et al., 2006). Estimates of the total volume of landslide debris emplaced ranges from 25×10^6 to $37 \times 10^6\ m^3$ (Jibson et al., 2004, 2006) and about 11 km^2 of the glacier was covered by debris. Another good example is from the Sherman Glacier in the Chugach Mountains of southern Alaska (USA) where in March 1964, the great Alaska ('Good Friday') earthquake (M = 9.2) triggered thousands of landslides across southern Alaska (Keefer, 2002).

7.10.3 Rockslide

This occurs when a tabular mass of rock glides down a slope, which is usually underlain by more of the same rock, with planes of weakness parallel to the slope (Fig. 7.14c). These planes of weakness include either a bedding plane or a joint surface. The Sierra Nevada of California and Nevada (USA) is a good example of rockslides. This region has undergone large prehistoric and historical rockslides triggered either by strong seismic shaking or by long periods of unusually wet weather. At lower elevations on the non-glaciated slopes of the Sierra Nevada, rockslides commonly occur within more weathered granitic rocks, where the strength of the rock mass is typically affected by joint weathering and alteration of the intact rock to Saprolite (Wieczorek, 2002).

7.10.4 Debris Fall

It is the collapse of weathered rock material and/or soil from a steep slope or cliff (Fig. 7.14d).

7.10.5 Debris Avalanches

These are very high velocity flows of large volume mixtures of rock and regolith that results from the complete collapse of a mountainous slope (Fig. 7.14e). They move down slope and travel for considerable distances along relatively gentle slopes. They are often triggered by earthquakes and volcanic eruptions. A good example of debris avalanche is the 300,000 and 360,000 year ago deposit that extended 43 km NW from the base of Mount Shasta across the floor of Shasta Valley, California (USA), where it covered an area of at least 450 km^2. The estimated volume was of about 26 km^3, the largest known Quaternary landslide on earth (Crandell et al., 1984).

7.10.6 Debris Slide

When blocks of rock, or masses of unconsolidated material slide down a slope it is called Debris slide (Fig. 7.14f). These are very destructive mass movement types and are generally triggered by heavy rain, melting snow, or earthquakes. The hilly areas of Hawaii (USA) have experienced immense destruction, as do areas of extreme northern California, Idaho, Oregon, and Washington (USA).

7.10.7 Mudflows

It is high velocity mixture of sediment and water with a consistency of a wet concrete (Fig. 7.14g). It differs from a Debris flow in that fine-grained (sand, silt, and clay) material is predominant. Mudflows usually result from heavy rains in areas where there is abundance of unconsolidated sediment that can be picked up by streams. Mudflows can travel for long distances over gently sloping stream beds. Because of their high velocity and long distance of travel they are potentially very dangerous. The mudflows triggered by the November 1985 eruption of Nevado del Ruiz (Colombia, Central America) killed more than 25,000 people—resulting in the worst volcanic disaster in the 20th century since the catastrophe at Mont Pelee in 1902. The eruption of Mount St. Helens in Washington State (northwest America) on June 12, 1980 produced a 25 feet thick bed that was deposited in less than a day.

7.10.8 Earthflow

These form in humid areas on hillsides following heavy rain or melting snow, in fine-grained materials (clay and silt). They also occur at the toe of slumps when associated with heavy rains (Fig. 7.14h) and remain active for long periods of time. They generally tend to be narrow tongue-like features that begin at a scarp or on a small cliff. Good examples of both large and small earthflows are found in the Palos Verdes Hills of Los Angeles County, California (USA).

7.10.9 Slumps

Slumps are sliding mass of soil or other loose material along a curved, rotational surface (Fig. 7.14i). The upper surface of each slump block remains relatively undisturbed, as do the individual blocks. Slumps leave arcuate scars or depressions on the hill slope. Heavy rains or earthquakes usually trigger slumps. At the bottom (or toe) of the slump, Earth flow, or flow of soil, occurs. The Trollwood Park in north Fargo, Dakota (USA) provides good examples of Slump and earth creep.

7.10.10 Debris Flow

Debris flow results from heavy rains. This causes saturation of the soil and regolith with water, followed by its flow (Fig. 7.14j). Debris flow sometimes starts with slumps and then flows downhill forming lobes with an irregular surface consisting of ridges and furrows. Those occurring in volcanic areas are called Lahars. The Sierra Nevada of California and Nevada (USA) are good examples of debris flow (Wieczorek, 2002). The mountains of Colorado and the Sierra Nevada of California have experienced debris flows in areas receiving high rates of rainfall, rapid snowmelt, or a combination of both (Highland et al., 1984; see also USGS Fact Sheet 176–97).

7.10.11 Solifluction

It occurs in areas underlain by permafrost within the periglacial environment. A periglacial environment is defined as any place where the geomorphic processes related to the freezing of water occurs. Solifluction produces distinctive lobes on hill slopes (Fig. 7.14k). These occur in areas where the soil remains saturated with water for long periods of time and the sediment flows over an impermeable material (permafrost—ground that remains frozen for many years). The flow rates are in the order of a cm per year. Good examples are noted in Serbia and Canada (especially in Newfoundland). Deposits are also noted in Germany that date back to the Younger Dryas.

7.10.12 Earth Creep

It is an unusually slow continuous movement of regolith and soil downhill. Creep occurs on almost all slopes, but the rates vary. Evidence for creep is often seen in bent trees, offsets in roads and fences, and inclined utility poles (Fig. 7.14l).

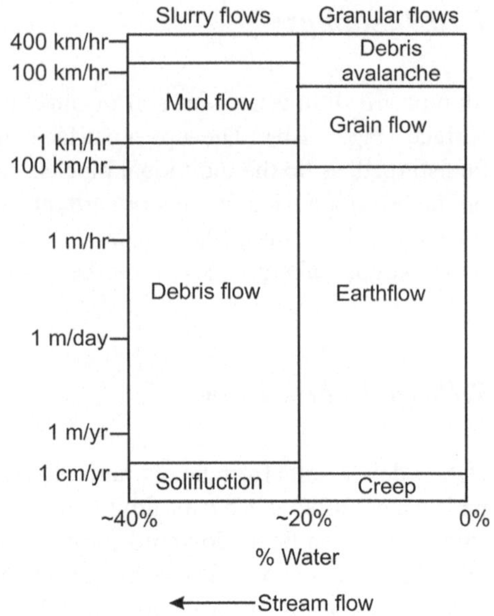

Fig. 7.15 A sediment flow is a mixture of rock, regolith and water. These are grouped into two types—Slurry and Granular Flows, depending on the amount of water present

7.10.13 Sediment Flows

Sediment flow occurs when sufficient force is applied to rocks and regolith that they begin to flow down slope. It is a mixture of rock, regolith, and water. Sediment flows can be grouped into two types depending on the amount of water present—Slurry and Granular Flows (Fig. 7.15).

7.10.13.1 Slurry Flows

These contain between 20 and 40 % water (Fig. 7.15). As the water content increases beyond 40 %, slurry flow grades into a stream (Fig. 7.15).

7.10.13.2 Granular Flows

They occur with little or no water (Fig. 7.15). Their fluid-like behavior is due to their mixing with air.

7.10.14 Grain Flows

Grain flow usually forms in relatively dry material, such as a sand dune, on a steep slope. A small disturbance sends the dry unconsolidated grains moving rapidly down slope.

7.11 Factors Responsible for Mass Wasting

Many factors (such as slope, water, geology, and relief, precipitation, slope modification, etc.) affect Mass Wasting. However, among them, slope and water are prime.

7.11.1 Slope

Regolith is always stable on a flat surface. If the slope is steep, weathering will take place at a higher rate in comparison to in-flat areas. On a steep slope, the tangential component of gravity increases and produces a shear stress parallel to the slope. This helps to move the loose particles of soil and rock in the down slope direction.

7.11.2 Water

Water, inspite of the fact that it is not a direct medium of transportation, plays an important role in the mass wasting process. Slightly wet unconsolidated materials exhibit a very high angle of repose as the surface tension between the water and the grains tends to hold the grains in place. But, as loose particles of soil and rock become saturated, the angle of repose is reduced and the material tends to flow like a fluid. A mass wasting event occurs as soon as the slope becomes unstable. Sometimes, as in earth Creep or Solifluction (Fig. 7.12k-l), the slope is unstable all the time and hence, the process is always continuous.

7.11.3 Geology and Relief

Both geology and relief of an area affects the mass wasting process indirectly. Rocks such as limestone and granite are particularly susceptible to weathering primarily due to their mineral composition and hence, have relatively high rates of weathering. And if these rocks, that generally dominate in regions of high relief, witness high precipitation, mass wasting in such areas is most likely to occur.

7.11.4 Precipitation

Heavy rains saturate regolith thereby reducing the grain-to-grain contact. This changes the angle of repose, thus, triggering a mass wasting event.

7.11.5 Slope Modification

Any change in slope either by humans or by natural causes will change the slope angle and this will eventually alter the angle of repose. A mass wasting event will occur until the slope is not restored to its equilibrium angle.

7.11.6 Undercutting

Streams eroding their banks or strong surf action along a coast can also undercut a slope making it unstable.

7.11.7 Shocks

An earthquake triggers slope instability. Additionally, aftershocks can also further damage the landscape thereby creating new areas for potential mass wasting events. Shocks, such as those produced by heavy trucks going down the road, or by manmade explosions can also trigger mass wasting events.

7.11.8 Volcanic Eruptions

These produce shocks similar to those generated by explosions and earthquakes. They cause snow to melt or even empty crater lakes, thereby triggering mass wasting events.

7.11.9 Submarine Slope Failures

Rapid deposition of sediment that does not allow trapped water between grains escape leads to slope failure and a mass wasting event. Similarly, the generation of methane gas from the decay of organic materials, increases pressure between unconsolidated grains and reduces grain-to-grain contact, leading to slope failure.

7.11.10 Frost Heaving

In cold climates, when water-saturated soils freeze, they expand, pushing rocks and boulders on the surface upward perpendicular to the slope. When the soil thaws, the boulders move down vertically resulting in a net down slope movement, thus, causing a mass wasting event.

7.11.11 Gelifluction

This process occurs in cold climates when the upper layer of the soil thaws during warmer months. This thaw produces a water-saturated soil that slowly moves down slope. It is similar to Solifluction, though restricted to cold regions. Gelifluction is the flowage of wet, unfrozen soil down slope over a frozen substrate (Baulig 1956). The latter may be either seasonal or perennial frost (permafrost).

7.11.12 Rock Glaciers

It is a lobe of ice-cemented rock debris that moves slowly downhill. Good examples of rock glaciers are noted in the high Sierra Nevada south of the Lake Tahoe region (USA).

7.12 Summary

Weathering is a process of the slow continuous breakdown of rocks and minerals into smaller particles and involves both decomposition (chemical breakdown) and disintegration (physical breakdown). Weathering occurs without movement and involves breaking down of the substrate by chemical of physical means. On the other hand, when particles are moved from their place of formation, it is called Erosion. The other name for Erosion is Mass Wasting.

Weathering is largely of two types—physical (mechanical) and chemical. The third, biological, only forms a small fraction. Physical weathering causes the parent rock to break into smaller fragments (reduction in size), but chemical weathering changes the composition of the parent rock through chemical reactions.

Both physical and chemical weathering operate together and one often assists the other.

Some of the major factors by which physical weathering occurs include—Plants (Root wedging), Animals, Crystallization or growth of salt crystals, Grinding or Rubbing, Abrasion, Unloading or Pressure-release, Wetting, and Drying, Variation in Temperature, Freezing and Thawing, and Heating and Cooling (Thermal expansion and contraction).

Some of the major factors by which chemical weathering occurs include—Dissolution and Solution, Hydration, Oxidation, Hydrolysis, Reduction and Carbonation, and Acidification.

Biological weathering is the breakdown of rocks by living organisms, ranging from bacteria to plants to animals (including Lichens).

Weathering byproducts include (a) the complete loss of particular atoms or compounds from the weathered surface, (b) the addition of specific atoms or compounds to the weathered surface, and (c) a breakdown of one mass into two or more masses, with no chemical change in the mineral or rock.

When rock weathers, they usually do so by working inward from a surface that is exposed to the weathering process. This results in Spheroidal weathering, Exfoliation, and Weathering rinds.

Weathering rates are a function of the rock type, structure, and the prevailing climate. Weathering rates are differential. Rocks that are most resistant are composed of minerals that are relatively unaffected by chemical weathering. Some factors that influence the rate of weathering are—time, slope, climate, type of rock (mineral composition), exposure, particle size, rock structure, gravity, animals, and the stability of minerals within the rock.

Mass Wasting is the gravity-initiated down slope movement of Regolith (loose particles of soil and rock) without the aid of a transporting medium (such as water, ice, or wind). The term is often used interchangeably with Mass Movement and comes within the gamut of erosional processes, between weathering and stream transport.

The classification of mass wasting process is somewhat difficult as most types (of mass wasting) grade into one another. Mass wasting processes can be divided into two categories—Flow and Slide. These include—rock avalanche, rockfall, debris avalanches, debris slides, rockslide, debris flow, mudflows, earthflow, slumps, earth creep, sediment flows (including slurry and granular flows), solifluction, and grain flows.

Many factors affect Mass Wasting. These include slope, water, geology and relief, precipitation, slope modification, undercutting, shocks, volcanic eruptions, submarine slope failures, frost heaving, gelifluction, and rock glaciers. However, among them, slope and water are the most important ones.

References

Baulig H (1956) Vocabulaire de géomorphologie. Les Belles Lettres, Paris

Bland W, Rolls D (1998) Weathering: an introduction to scientific principles. Hodder Arnold, New York

Hamblin WK, Christiansen CH (2008) earth's dynamic systems. Printice Hall, USA 767p

Colman SM, Dethier DP (eds) (1986) Rates of chemical weathering of rocks and minerals. Academic Press, Orlando 603p

Crandell DR, Miller CD, Glicken HX, Christiansen RL, Newhall CG (1984) Catastrophic debris avalanche from ancestral mount Shasta volcano, California. Geology 12:143–146

Harp EL, Jibson RW, Kayen RE, Keefer DK, Sherrod BL, Carver GA, Collins BD, Moss RES, Sitar N (2003) Landslides and liquefaction triggered by the M 7.9 Denali fault earthquake of 3rd November 2002. GSA Today, August 4–10

Highland LM, Ellen SD, Christian SB, Brown III WM (1984) Where do debris flows occur? debris flows in the Western United States. US Geological Survey Fact Sheet 176–97 (http://pubs.usgs.gov/fs/fs-176-97/fs-176-97.pdf)

Jibson R, Harp EL, Schulz W, Keefer DK (2004) Landslides triggered by the 2002 Denali fault, Alaska, earthquake and the inferred nature of the strong shaking. Earthq Spectra 20:669–691

Jibson RW, Harp EL, Schulz W, Keefer DK (2006) Large rock avalanches triggered by the M 7–9 Denali fault, Alaska, earthquake of 3rd November 2002. Eng Geol 83:144–160

Keefer DK (2002) Investigating landslides caused by earthquakes—a historical review. Surv Geophys 23:473–510

Matthes FE (1930) Geologic history of the Yosemite Valley: U.S. Geological Survey Professional Paper 160, p 137

USGS (2004) Landslides Types and Processes. US Geological Survey Fact-Sheet 2004-3072. (http://pubs.usgs.gov/fs/2004/3072). Open source, retrieved on 3/3/2012.

Thornbury WD (1969) Principles of geomorphology, 2nd edn. Wiley, New York, p 594

Wieczorek GF (2002) Catastrophic rockfalls and rockslides in the Sierra Nevada USA. In: Evans SG, DeGraff JV (eds.) Catastrophic landslides effects, occurrence, and mechanisms: Geol Soc Am Rev Eng Geol. Boulder, Colorado 25:126

Wieczorek GF, Jäger S (1996) Triggering mechanisms and depositional rates of postglacial slope-movement processes in the Yosemite Valley, California. Geomorphology 15:17–31

Wieczorek GF, Snyder JB, Alger CS, Isaacson KA (1992) Rock falls in Yosemite Valley, California: US Geol Surv Open-File Rep. 38:92–387

Chapter 8
Streams

8.1 Introduction

A large stream is a river and earth's lifeblood. While flowing to the oceans, they drain about 68 % of the earth's land surface; the remainder is either covered by ice or drains into closed basins. Areas draining to the sea are common in humid regions, whereas those draining to the interior closed basins occur in arid regions or in areas of active tectonic subsidence. Streams are one of the main transporting mediums in the formation of sedimentary rocks, carrying millions of tons of sediment, dissolved ions and products of chemical weathering, into oceans, thus, making the sea, salty (Fig. 8.1) (see also Morisawa 1985; Leopold et al. 1995; Robert 2003; Hamblin and Christiansen 2008). They are also a major part of the erosional process, working in conjunction with weathering and mass wasting (for details see Chap. 7).

8.2 Development Stages

The pattern of river discharge and the shape of the channel changes with time along with the landscape that it flows through. A stream that flows along a clearly defined path is called a channel. Stream gradient also changes over time from very uneven (as in lakes and waterfalls) to gentle where it ends (Fig. 8.2). Thus, the stream gradient decreases progressively from its upper to lower reaches and with time, attains a smooth graded profile. Once this profile has been attained, it is maintained while getting progressively lowered, closer to the base level. A stream always operates in or toward a state of total equilibrium. Those streams close to a graded condition have a tendency to develop meanders and vertical erosion surfaces (characteristic of ungraded streams), being replaced with lateral valley development and widening of the floodplain (Fig. 8.3). Thus, the evolution of a stream and its valley goes through various stages of development from initial

S. Jain, *Fundamentals of Physical Geology*, Springer Geology,
DOI: 10.1007/978-81-322-1539-4_8,

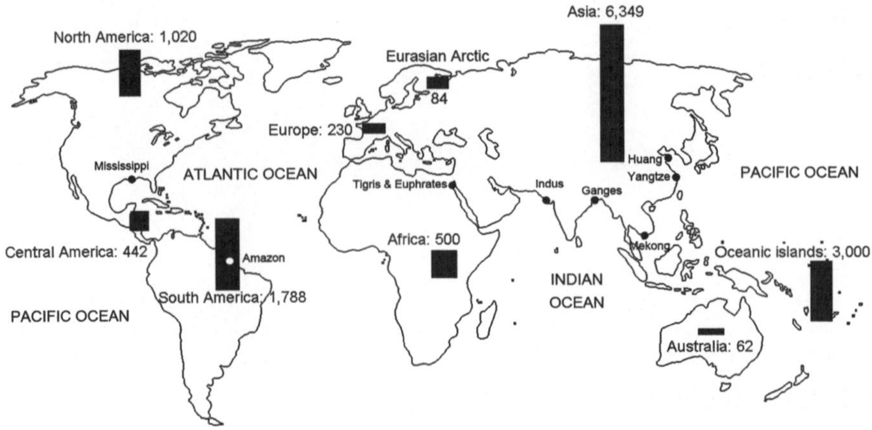

Fig. 8.1 Figure shows the suspended sediment discharge (in million tons per year) into the oceans by rivers

(young rivers), intermediate (mature), and terminal (old rivers) (Fig. 8.3). These three stages are discussed briefly.

8.2.1 Initial (Youthful Stage; Upper Course)

The initial stage is typically found in rivers that occur in the highlands or mountainous regions (Fig. 8.4). Erosion is their primary work. At this stage, a stream usually erodes its bed more rapidly than it erodes its banks, thus, producing a V-shaped valley with steep sides (Figs. 8.3, 8.4). Waterfalls and rapids are quite common. Youthful rivers usually have relatively few tributaries, and hence carry a small volume of water and much of the precipitation falling on the watershed that does not reach the main stream; instead, forms lakes at high altitudes. The initial stage of a river is thus, characterized by:

(a) steep gradient
(b) channel is deeper than it is wider and down cutting is greater than lateral erosion (hence, V-shaped valley)
(c) water velocity is strong and is capable of moving all sediment sizes (from ions to boulders)
(d) cliffs/flanks are steep sided
(e) drainage network is poorly developed
(f) waterfalls and rapids are common
(g) erosion dominates over deposition.

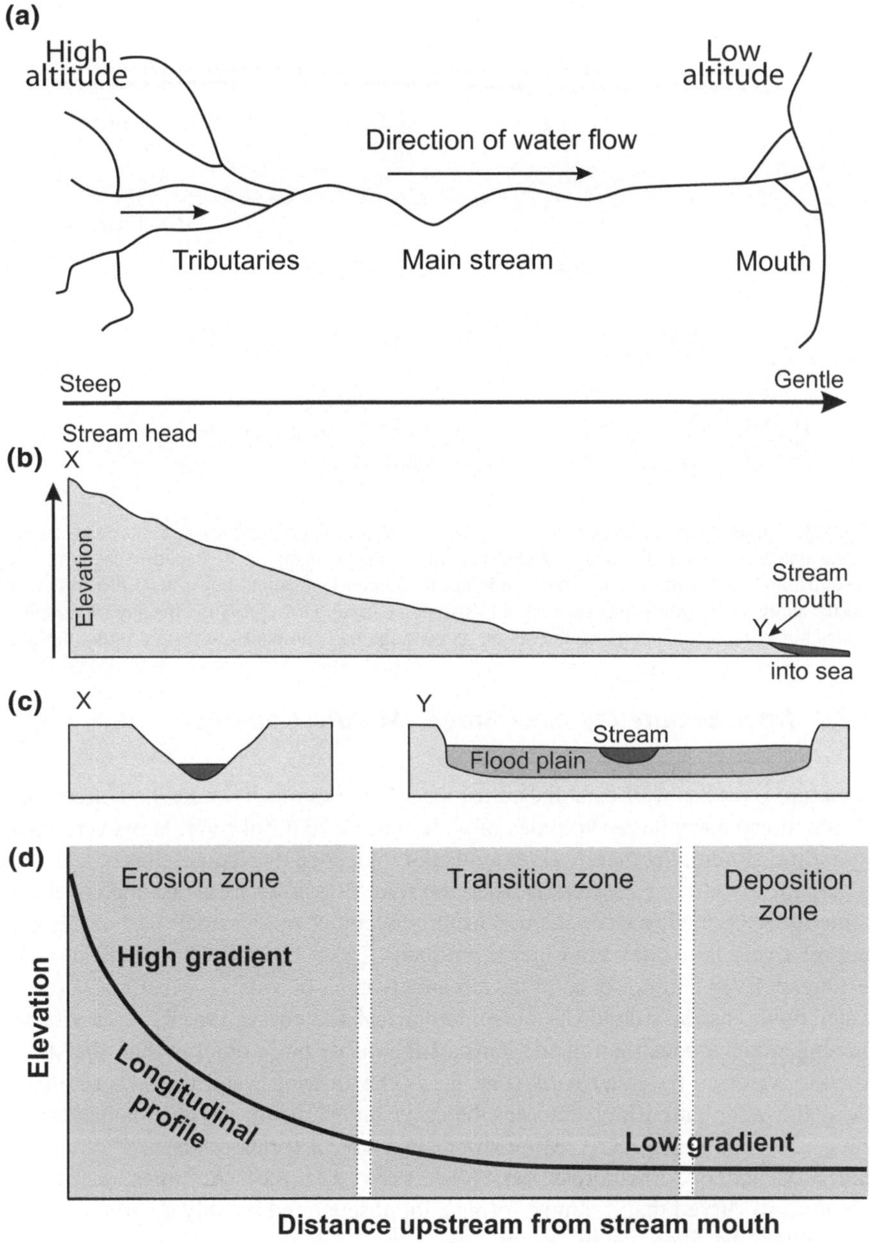

Fig. 8.2 Stream profile and stream gradient. **a** Longitudinal stream profile, **b** vertical section of a stream from head to mouth, **c** cross-sections of a stream from head to mouth, and **d** stream gradient and characteristics from stream's head to mouth

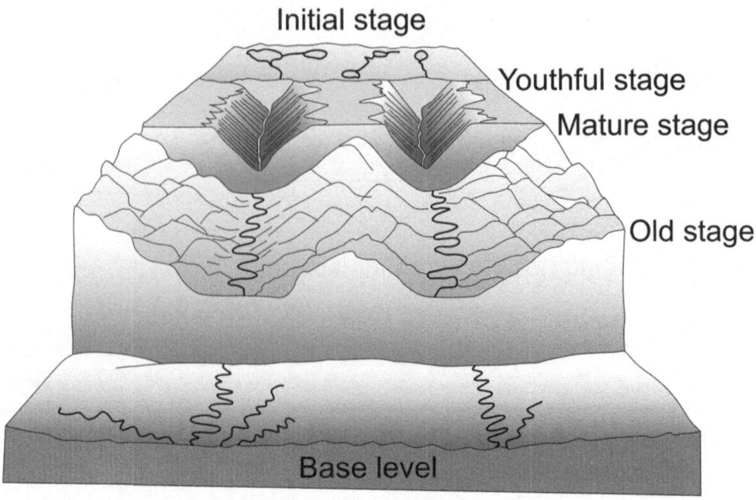

Fig. 8.3 Three stages of stream development. *Youthful stage*: At this stage, a stream usually erodes its bed more rapidly than it erodes its banks, thus, producing a V-shaped valley with steep sides. *Mature stage*: the mature river tends not to deepen its channel but instead erosion occurs mostly along valley walls thus forming a U-shaped channel. *Old age*: Now, the stream no longer erodes but in fact deposits its sediments on its own channel and banks to form a broad flat plain

8.2.2 Intermediate (Mature Stage; Middle Course)

A mature river has well-established tributaries and drains its watershed effectively. Hence, it can carry larger volumes of water than a youthful river. However, unlike a youthful stream, the mature river tends not to deepen its channel. Instead, erosion occurs mostly along valley walls when the river overflows its banks and covers the valley floor. A mature river channel usually occupies only a small part of the wide and relatively flat valley floor that it produces (Figs. 8.3, 8.4). The gradient is also less steep. Even a minor bend in the stream channel, becomes a wider curve (as the water flows fastest around the outside edges of the curve; Fig. 8.5). This faster-flowing water erodes the outside bank of the curve more quickly than the slower moving water on the inner bank (Fig. 8.5). The slowing water deposits sediments along the inner bank which enlarges the curve and shifts the stream channel toward the outside bank (Fig. 8.5). More often than not, a series of these wide curves, called Meanders form across the valley floor (Fig. 8.6). At times, a meander becomes so curved that it almost forms a loop, separated by only a narrow neck of land. When the river eventually cuts across this neck, it deposits sediments at both ends of the meander and eventually abandons it. The meander is, thus, isolated from the river. If the water remains in the abandoned meander, an Oxbow lake is formed (Figs. 8.6, 8.7).

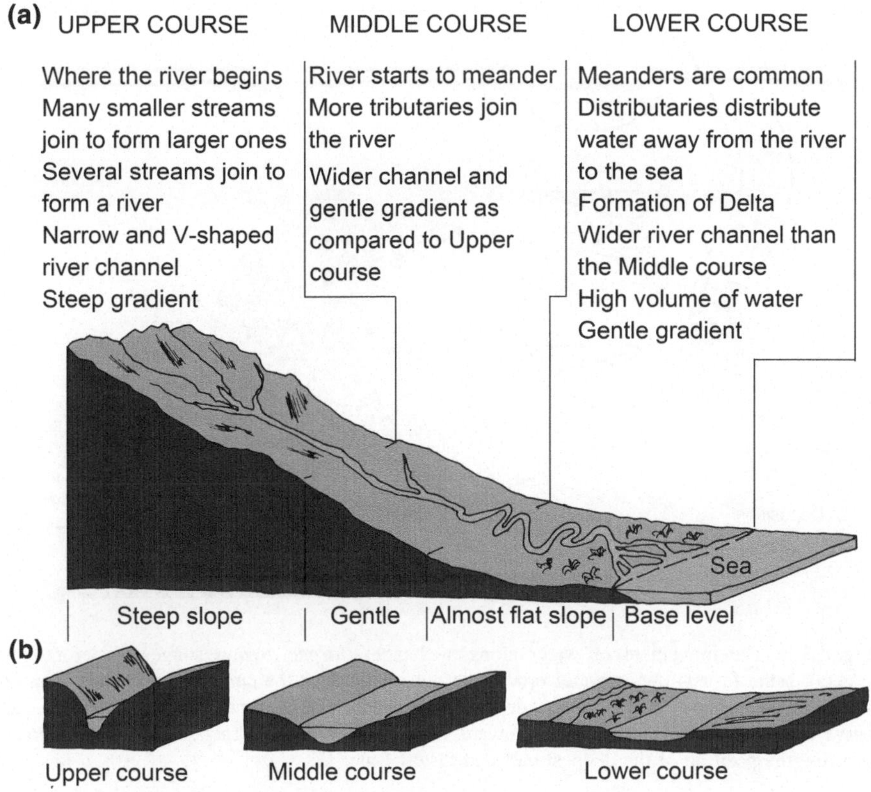

Fig. 8.4 **a** Characteristics of stream stages (course) and **b** their profile sections

For now, the mature river has reached a stage of equilibrium (i.e., its erosion and deposition are nearly in balance), thereby allowing maximum amount of water to flow in a fairly efficient manner. Now the mature river is characterized by:

(a) a moderate gradient
(b) U-shaped channel; wider than the youthful stage
(c) moderate down cutting along with lateral erosion
(d) slower velocity than the youthful stage
(e) moderately strong water velocity
(f) stream is capable of moving all sediment sizes (except larger ones like boulders)
(g) beginnings of the Flood Plain formation
(h) erosion present but deposition also occurs (of sand and gravel bars)
(i) stream channel has more water, i.e., greater discharge than the youthful stage (it carries more sediments)
(j) has well-established tributaries.

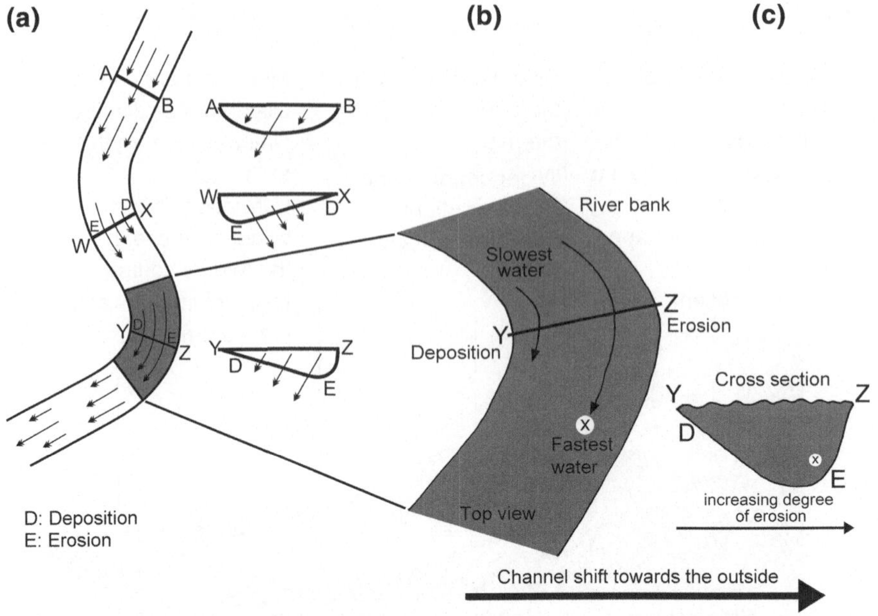

Fig. 8.5 **a** The movement of water along a channel (*larger arrows* suggest faster moving waters), **b** the faster-flowing water erodes the outside bank of the curve more quickly than the slower moving water. The latter, thus, deposits sediments along the inner bank which enlarges the curve and shifts the stream channel toward the outside bank. **c** Cross-section of the stream showing the position of the deepest part and fastest water (*x*)

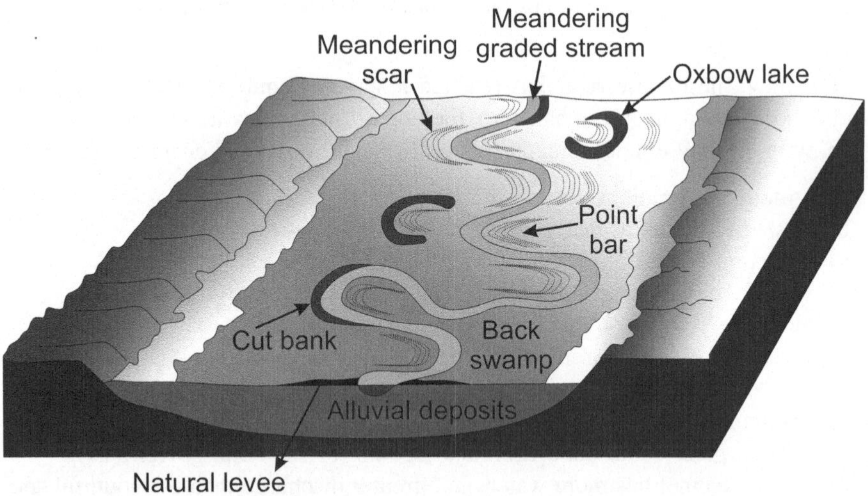

Fig. 8.6 Mature stage stream landforms. Redrawn from Earth's Dynamic Systems with permission of Eric H. Christiansen

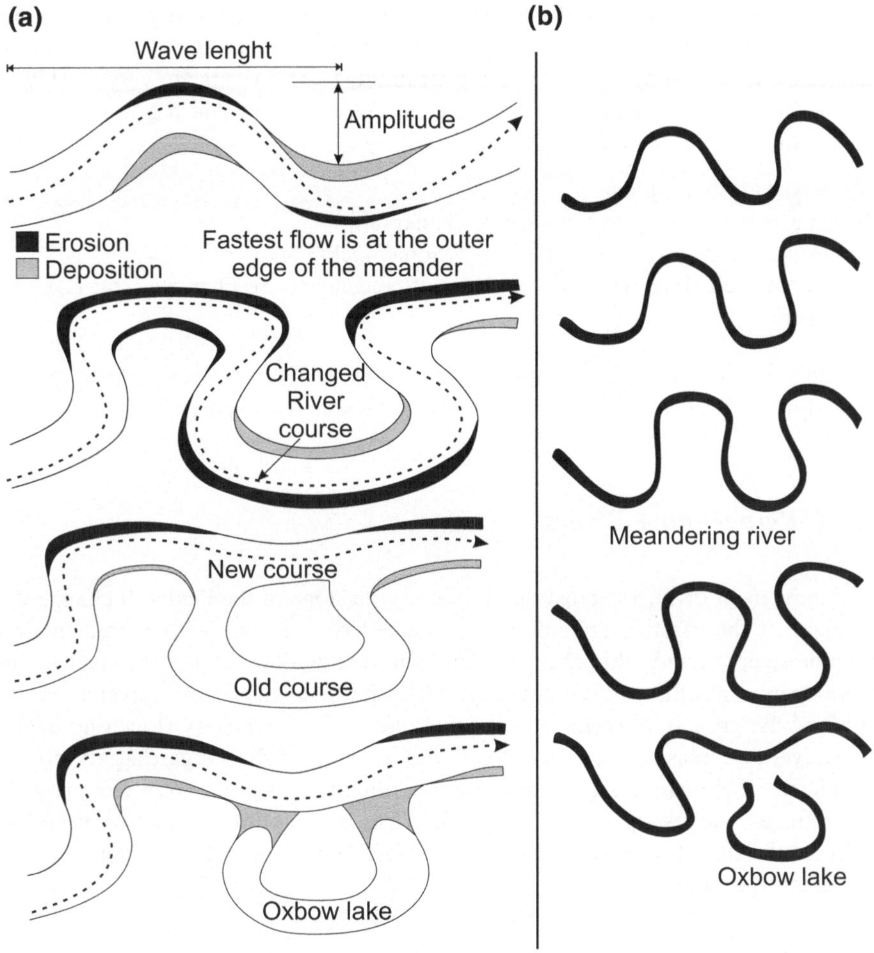

Fig. 8.7 Meandering river and the eventual formation of an Oxbow lake. **a** A series of wide curves are called Meanders which from across the valley floor. At times, these becomes so curved that they form a loop, separated by only a narrow neck of land. When the river eventually cuts across this neck, it deposits sediments at both ends of the meander and eventually abandons it. The meander is, thus, isolated from the river. If the water remains in the abandoned meander, an Oxbow lake is formed. **b** Simplified version of an Oxbow lake formation

8.2.3 Terminal (Old Stage; Lower Course)

As the river continues to age, its gradient and velocity decreases correspondingly. Now, the stream no longer erodes but in fact deposits its sediments on its own channel and banks to form a broad flat plain. More meanders develop, and there are fewer tributaries. Smaller tributaries merge and become larger ones. At this stage, the river becomes flat, with almost no slope (Figs. 8.3, 8.4) and consequently with very little momentum. Such old stage rivers are characterized by very elaborate and intricately meandering courses. Often swampy areas appear around old rivers due to lack of slope, and hence leads to very poor drainage. Old rivers are very muddy as large amounts of suspended material are carried by a river that has low velocity. Thus, at this stage, the river is characterized by:

(a) very wide flood plains
(b) land is worn down to flat surface (Peneplain)
(c) resistant rocks form residual hills (Monadnocks) and
(d) pronounced river meanders and cut-off meanders are characteristic (Oxbow lakes).

Table 8.1 and Fig. 8.8 summarize the three stages of stream development indicating their characteristic features and processes involved.

8.2.4 Rejuvenated Rivers

Any movement of earth's crust that increases the slope of the land will change the gradient of the existing stream. A rejuvenated river is one whose gradient has become steeper due to this change. This increased gradient allows the river to cut more deeply into the valley floor (Fig. 8.9). The main causes of rejuvenation are uplift of the ground, lowering of sea level due to glacial eustasy (lowering of the base level), decrease of sediment load, and increase of discharge (largely due to the effects of climate change). Rejuvenation often results in the formation of step-like terraces, suggesting that the valley floor has been uplifted and a new floor has been cut through (Fig. 8.9). Incised river, antecedent river, and river capture occur due to rejuvenation.

8.3 Classification of Streams

Streams are classified based on their constancy of flow, condition of the groundwater table and the stream equilibrium.

8.3.1 Constancy of Flow

On this basis, the rivers are classified into three types:

Table 8.1 Summary of stream development along with their characteristic features

Characteristics	Youth	Maturity	Old Age
Gradient	Steep, irregular	Moderate, smooth	Low, smooth
Valley profile	Narrow (V-shaped)	Broad, moderately U-shaped	Very broad
Valley depth	Deep	Deepest to moderate	Shallow
Meanders	Absent	Common	Extremely common
Floodplain	Absent or small	Equal width of meander belts	Wider than width of the meander belt
Natural levees	Absent	May be present	Abundant
Tributaries	Few but small	Many	Few, large
Velocity	High	Moderate	Sluggish
Waterfalls	Many	Few	None
Erosion	Downward cutting	Downward and lateral cutting is equal	Lateral cutting
Deposition	Absent or transitory	Present, but partly transitory (narrow flood plains)	Much and fairly permanent (broad flood plains)
Culture	Steep-walled valleys and barriers to roads and railroads	Flat valley floors are good transportation routes	Large rivers and nearby swamps are barriers
Regional			
Dissection	Partial	Complete	None
Divides	High, flat, and broad	High, rounded, and narrow	Few, low, and broad
Valley development	Youthful to mature	Mostly mature	Old age
Number of streams	Few	Maximum	Few
Relief	Great	Maximum	Minimum
Tributaries	Small	Many	Few
Drainage	Poorly developed	Excellent	Poor and swampy
Topography	Rugged canyons, flat plateaus	Hilly, land mostly in slopes	Plane surface (penelplane)

8.3.1.1 Perennial

The water flows throughout the year.

8.3.1.2 Intermittent

These flow seasonally.

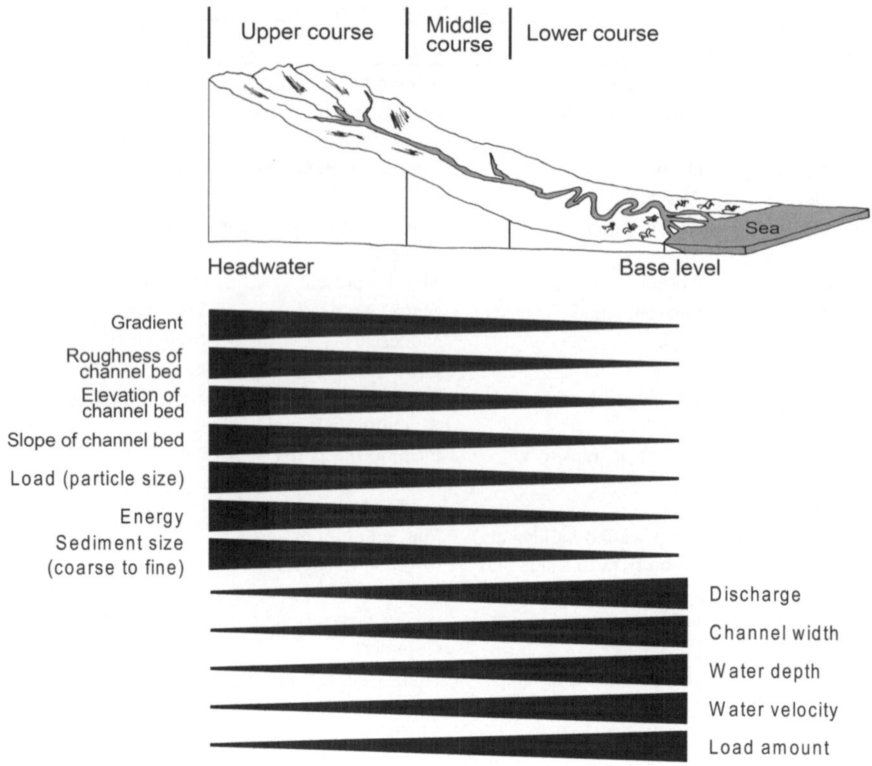

Fig. 8.8 Summary of processes involved in the stream development

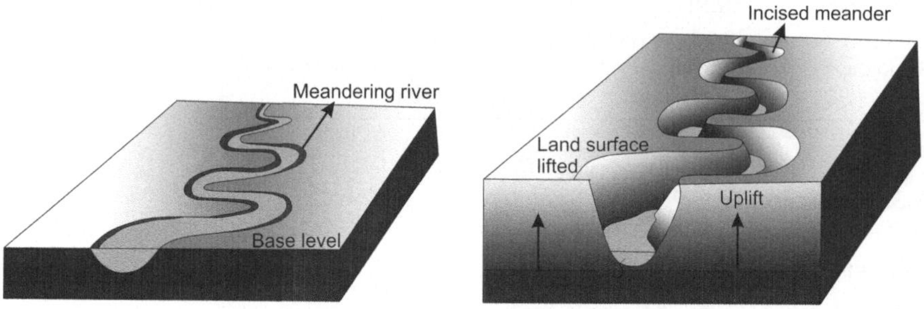

Fig. 8.9 Formation of an incised meander. If the land rises (uplift), the meandering river will be lifted further above base level and will renew its erosion into the land surface, forming incised meanders. This is a case of rejuvenation. Classic examples of incised meander are preserved throughout the Colorado Plateau (Western USA), documenting the widespread uplift of the Plateau

8.3.1.3 Ephemeral

The water flows only in direct response to precipitation (rain).

8.3.2 Groundwater Table Conditions

On this basis, the rivers are classified into two types:

8.3.2.1 Influent

The river contributes water to the groundwater table and the flow of the water decreases downstream. It is, thus, also known as a loosing stream.

8.3.2.2 Effluent

The river receives water from the groundwater table and the water flow increases downstream. It is, thus, called a gaining stream.

8.3.3 Stream Equilibrium Conditions

On the basis of downstream changes in flow, rivers are classified into two types:

8.3.3.1 Graded Stream (Steady State)

A stream that has regulated its various parameters (depth, width, slope, velocity, base level, discharge, channel shape or size, and sediment load) for optimum flow conditions and sediment transport is called a graded stream. Such streams maintain a steady-state condition where neither erosion nor sedimentation occurs and its gradient also does not vary. No falls or basins exist within the channel profile and no net erosion or deposition occurs along its channel. Graded stream is capable of handling all sediment introduced to it from its tributaries. On an average, a graded stream is neither eroding nor depositing sediment but simply transporting it. Hence, it is in complete equilibrium or balance.

8.3.3.2 Nongraded Stream

It contains Falls and Basins. Falls result in the concentration of energy, thereby promoting erosion whereas the formation of Basins results in the decrease of energy, thus, promoting deposition. Hence, the energy within the system is not evenly distributed along the profile.

8.4 Genetic Classification of Streams

Based on this, there are five groups:

8.4.1 Consequent Streams

This stream is formed as a direct consequence of the original slope (regional or local) of the surface, i.e., such streams follow the original slope of the land.

8.4.2 Subsequent Streams

They are secondary and possess a structurally controlled drainage pattern. They develop independently of the primary consequent drainage and generally develop after the original stream. Their course is determined by selective headward erosion along the weak strata (i.e., their flow is determined by the weak rock belts).

8.4.3 Obsequent Streams

They flow locally in opposite direction to the regional slope, and thus in opposite direction of the consequent drainage.

8.4.4 Resequent Streams

They follow along the original relief, but at a lower level than the original slope (i.e., they flow down a course determined by the underlying strata in the same direction). They develop later and are generally a tributary to a subsequent stream.

8.4.5 *Insequent Streams*

They are a product of headward erosion and have an almost random drainage; often forming dendritic patterns. Typically, these are tributaries that have been developed by headward erosion on a horizontally stratified belt or on homogeneous rocks.

8.4.6 *Superimposed Stream*

It is a stream that keeps its course through different preexisting rock sequences and structures or a folded stratum. They erode downward into the underlying rock.

8.4.7 *Antecedent Stream*

Due to crustal movement, when a part of the drainage basin is uplifted, rejuvenation of a channel occurs. If this uplift occurs at a slower pace than undercutting, then the channel keeps its course and a gorge is formed. A river that forms a gorge along a meandering course is called an antecedent stream.

8.4.7.1 Nickpoint

If the uplift movement is faster than the undercutting, the riverbed becomes discontinuous on either side of the uplifted area. The upper reaches of the river are dammed up, and a Nickpoint appears on the downstream side. The Nickpoint (the source of the truncated river), moves back toward the upper course, often creating a waterfall. A Nickpoint is also formed when the rate of erosion of rocks is larger in the lower reaches than it is in the upper (Fig. 8.10). A typical example of a Nickpoint is the Niagara Falls that straddles the international border between Canada (Ontario State) and the USA (New York State).

8.4.8 *Incised River*

If a meandering channel has been rejuvenated, a channel begins to undercut a riverbed while keeping the meandering course. As a result, a meandering valley is formed, and the river running in the valley is an incised river (Fig. 8.9).

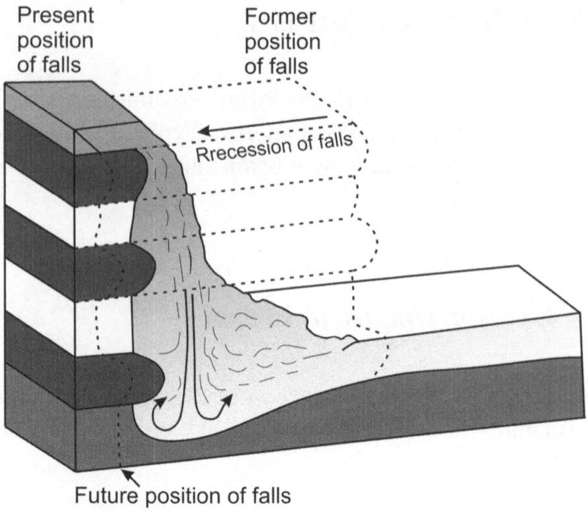

Fig. 8.10 Nickpoint. It is formed when the rate of erosion of rocks is larger in the lower reaches than it is in the upper

8.5 Stream Channels

8.5.1 Geometry

The volume of water passing at any point on a stream is called discharge. It is measured in units of volume per time (m^3/second).

8.5.1.1 Cross-Sectional Shape

The cross-sectional shape of the channel is a function of the position it is in the stream and its discharge. The deepest parts of the channel occur where the stream velocity is the highest. As discharge increases the cross-sectional shape changes and the stream becomes deeper and wider.

8.5.1.2 Long Profile

It is a plot of elevation versus distance and usually shows a steep gradient near the source of the stream and a gentle one as the stream approaches its mouth (Fig. 8.2).

8.5.1.3 Longitudinal Profile

Channel slope decreases with increasing distance downstream (Fig. 8.2).

8.5.1.4 Base Level

Base level is defined as the limiting level below which a stream cannot erode its channel (Fig. 8.3). Local base levels occur where the stream meets a resistant body of rock, or where a natural or artificial dam impedes further channel erosion, or where the stream empties into a lake. When a natural or artificial dam impedes stream flow, the stream adjusts to the new base level by adjusting its long profile.

8.5.2 Dynamics of Discharge

8.5.2.1 Velocity

A stream's velocity is a function of the irregularities (caused by resistant rocks) in the stream channel and the stream gradient. Stream flow can be either laminar, in which all water molecules travel along similar parallel paths, or turbulent, in which individual particles take irregular paths. Turbulent flow can keep the sediment in suspension longer than the laminar flow and thus aid in greater erosion of the stream bottom. The average linear velocity is generally greater in laminar flow than in the turbulent flow. Stream velocity is greatest in the upper central part of the channel flow, due to the effect of friction against channel sides and bottom.

8.5.2.2 Discharge

It is the amount of water passing at any point in a given time within a stream.

Discharge (m^3/sec) = cross-sectional Area (width × average depth) (m^2) × average Velocity (m/sec) or $Q = A \times V$.

As the amount of water in a stream increases, the stream adjusts its velocity and the cross-sectional area toward equilibrium. Discharge increases as more water is added by rainfall, tributary streams, or from groundwater seeping into the stream. As discharge increases, stream width, depth, and velocity also increase. Stream discharge typically increases downstream, but can decrease as well (if a river enters an arid area and loses its waters through evaporation and infiltration). However, over short stretches, the channel discharge remains relatively constant (assuming that the channel receives no tributaries over this stretch), and therefore both cross-sectional area (*A*) and average velocity (*V*) are inversely related. However, this is valid only for ideal conditions. Average velocity is defined as the time it takes for a given particle of water to traverse a given distance. In reality, stream flow is very turbulent. Even in straight channels, the actual flow of water is not straight. There are multiple short-lived currents moving in all possible directions within the overall flow, thus, giving the stream flow considerable complexity. Differences in flow velocity are especially pronounced for very sinuous and meandering rivers.

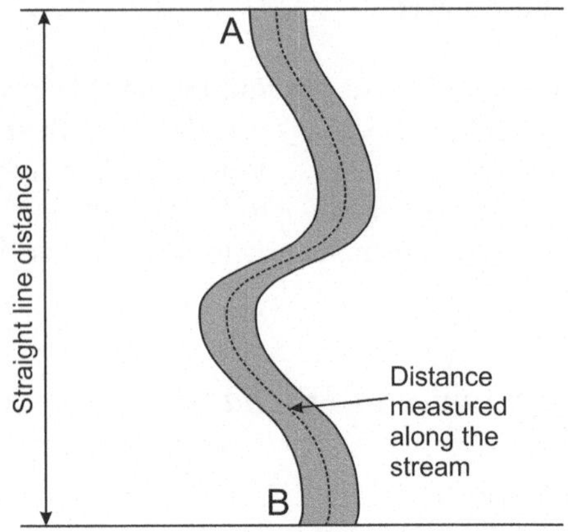

$$\text{Sinuosity ratio} = \frac{\text{Distance measured between two points (A and B) along the stream}}{\text{Straight line distance between two points}}$$

Fig. 8.11 The sinuosity ratio. A sinuosity ratio is used to determine whether a channel is straight or curved (meandering). This ratio is the distance between two points on the stream measured along with the channel divided by the *straight-line* distance between the two points. If the sinuosity ratio is 1.5 or greater, the channel is considered to be a meandering one

8.6 Channel Morphology

Streams are partially confined bodies of water flowing downhill, carrying rock fragments, as bed load, suspended load and in solution, within a well-defined path called a Channel. There are four major types of stream channels: straight, sinuous, meandering, and braided.

Straight channels are relatively rare and short. Their presence indicates strong control of the underlying geologic structure. Straight channels are generally noted when a channel develops along a master joint in a hard rock terrain. A sinuosity ratio is used to determine whether a channel is straight or curved (meandering) (Fig. 8.11). This ratio is the distance between two points on the stream measured along the channel divided by the straight-line distance between the two points (Fig. 8.11). If the sinuosity ratio is 1.5 or greater, the channel is considered to be a meandering one. Sinuous channels are the most common, i.e., between straight and meandering ones. Meandering channels develop on mature floodplains, indicating that the river is close to equilibrium. Braided channels are a series of interwoven and interconnected shallow channels, flowing on roughly parallel courses. Stream channel has a slope called gradient, which is the angle between the channel surface (or a channel bed) and a horizontal plane. Each channel types are briefly discussed below.

8.6.1 Straight Channels

As mentioned, straight channels are generally controlled by a linear zone of weakness in the underlying rock. This could be a fault or a joint system. In these rare channels, water flows in a sinuous manner. The deepest part of the channel changes from one bank to the other and the velocity is highest in the zone overlying the deepest one. In these areas, sediment is transported readily resulting in pools. Where the velocity of the stream is low, sediment is deposited to form Bars (Fig. 8.6). The bank closest to the zone of highest velocity is usually eroded and this results in a cutbank (Fig. 8.6).

8.6.2 Meandering Channels

It is the most common channel variety. Rivers develop alternating bends with an irregular spacing along the trend of the valley (Fig. 8.6). Meandering channels (with sinuosity ratio $\geq$1.5) form where streams are flowing over a relatively flat landscape with a broad floodplain (Fig. 8.5). They are characteristic of plains and lowlands. The stream is close to its base level, so there is little or no down cutting and therefore lateral erosion dominates. Aggradation is common, resulting in broad, flat floodplains over which water flows during floods. Channels in these streams are characteristically U-shaped and actively migrate over the extensive floodplain (Fig. 8.12). If lateral erosion dominates over vertical erosion, meanders will develop even in an initially straight stream.

In streams flowing over low gradients with easily eroded banks, straight channels will eventually erode into meandering channels. Erosion will take place on the outer parts of the meander bends where the velocity of the stream is highest. Sediment deposition will occur along the inner meander bends where the velocity is low (Fig. 8.5). Such deposition of sediment results in exposed bars, called Point bars (Fig. 8.6). Because meandering streams are continually eroding on the outer meander bends and depositing sediment along the inner meander bends, meandering stream channels tend to migrate back and forth across their flood plain.

If erosion on the outside meander bends continues to take place, eventually a meander bend can become cut off from the rest of the stream. When this occurs, the cutoff meander bends will collect water and form an Oxbow lake (Fig. 8.7).

8.6.3 Braided Channels

These have interwoven network of channels, separated by the deposition of coarse alluvium (Fig. 8.13). They occur where sediment load is high and both discharge and velocity fluctuates rapidly. In streams that have a highly variable discharge

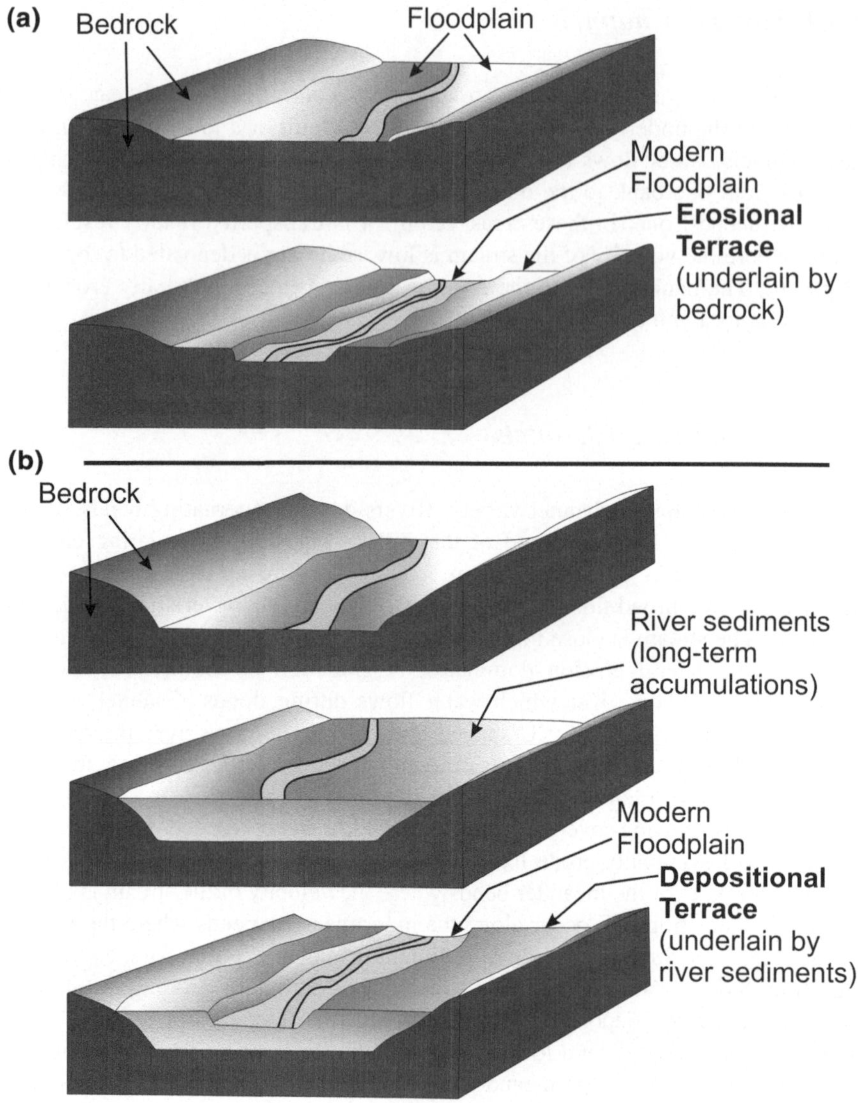

Fig. 8.12 Formation of erosional and depositional terraces. **a** A stream has formed a broad floodplain. **b** Local tectonic uplift or climatic change causes the stream to downcut into its bed. As the stream cuts downward, the old flood plain becomes a terrace above the new stream level. **c** A new floodplain forms at a lower level

and easily erodable banks, the deposited sediment forms bars and islands that are exposed during periods of low discharge. In such a stream, the water flows in a braided pattern around islands and bars, dividing and reuniting as it flows downstream (Fig. 8.13). This is a Braided channel. During times of increased

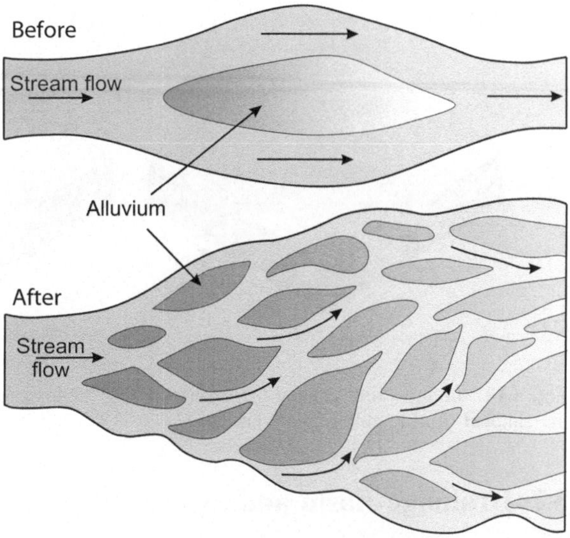

Fig. 8.13 Braided channel. These have interwoven network of channels, separated by the deposition of coarse alluvium. They occur where sediment load is high and both discharge and velocity fluctuates rapidly

discharge, the stream channel contains water and the islands are then covered to become submerged bars. During such high discharge, some of the islands may erode, but the sediment would be re-deposited as the discharge decreases, forming new islands or submerged bars again. Braided rivers tend to have steeper gradients, more variable discharge, coarser sediment loads, and lower sinuosity than meandering streams. Their channels also tend to be relatively wide and shallow. Best examples are noted in semi-arid regions (or those that are subject to irregular flooding), or in regions of glacial outwash, with seasonal melt water.

8.6.4 Stream Capture

Stream capture or stream piracy is common in the youthful stage of a stream and occurs when rapid head erosion proceeds into an adjacent drainage basin; the valley head eventually works its way toward another channel, and it becomes connected with the upper reaches of the formerly separate basin (Fig. 8.14). The place where river capture occurs is called Elbow of capture. The lower reaches of the captured river are deprived of the headwaters and a dry valley, a Beheaded river or a Wind gap, remains. On the other hand, if there is an increase in the discharge of the river, the drainage basin is enlarged and under-cutting is accelerated to form a Gorge. Stream capture can happen several times in the evolution of a drainage basin.

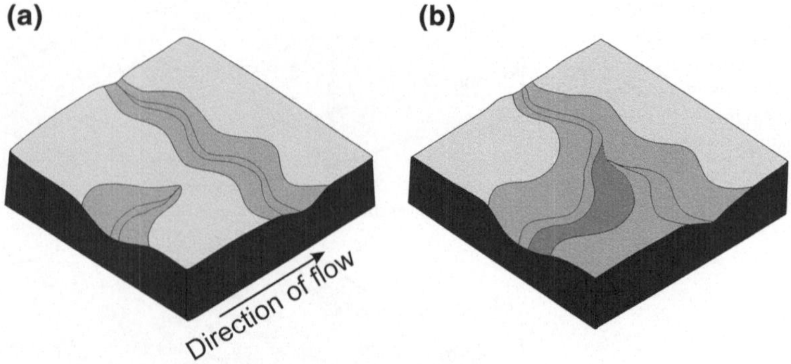

Fig. 8.14 Stream capture. Note the diversion of a stream's flow from its original channel (**a**) to the channel of a neighboring stream (**b**)

8.7 Drainage Basin and System

The area drained by a stream and its tributaries is called a Drainage basin. It is grouped by its size, shape, and arrangement of tributaries, drainage density, and morphometry. It is a system where sediment is produced, transported, and deposited.

8.7.1 Drainage Basins

Each stream in a drainage system drains a certain area, called a Drainage basin (Fig. 8.15). In a single drainage basin, all water falling in the basin drains into the same stream. Drainage basins can range in size from a few sq. km (for small streams), to extremely large areas (for large streams such as the Ganga and Bhamputra Rivers basins in India and the North American Mississippi-Missouri Rivers in the USA; see Fig. 8.15).

8.7.2 Drainage Divide

A divide separates each drainage basin (Fig. 8.16). Continents can be divided into large drainage basins that empty into different ocean basins such as for the African Nile and Congo rivers. These are called Continental divides or the Great Basin Divide, USA (Fig. 8.15). The Continental divides usually run along high mountain crests that were either formed recently or have not been eroded. Thus, major continental divides and the drainage pattern in major basins reflect the recent geologic history of the continents. However, sometimes, waters on each side of the divide never meet, but do flow into the same ocean (for example, the divide

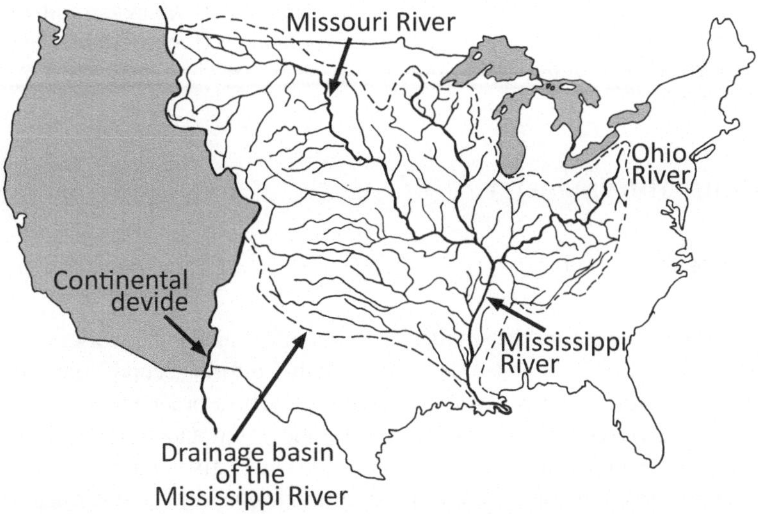

Fig. 8.15 A drainage divide is a divide that separates each drainage basins. Drainage basins can range in size from a few sq. km (for small streams), to extremely large areas such as the one shown here, the Great Basin Divide of the North American Mississippi-Missouri Rivers

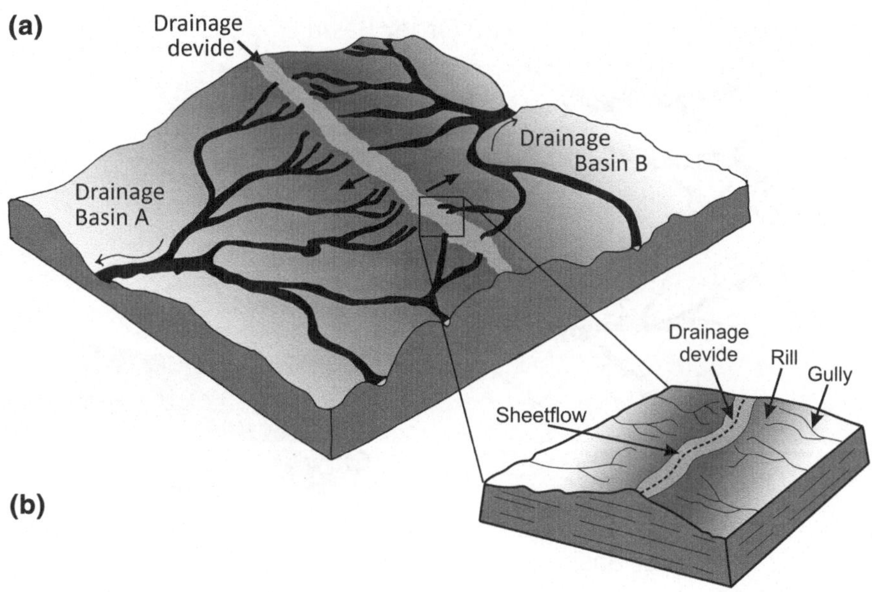

Fig. 8.16 **a** An ideal drainage divide marked by a high ridge and the subsequent formation of Sheetflow, Rills, and Gullies (**b**)

between the Yellow River basin and the Yangtze, China) whereas in other cases, waters part, but eventually rejoin at a river confluence (for example, the Mississippi and Missouri divides, USA; Fig. 8.15).

8.8 Basin Morphometry

8.8.1 Stream Order

The first-order streams are the smallest streams in a drainage network and have no tributary (Fig. 8.17). Two first-order streams unite to form second-order ones. The latter only has first-order streams as tributaries. Third-order streams have only second- and first-order streams as tributaries, and so on and so forth. The stream originating in the uppermost part of a drainage basin is given the first order. If two first-order streams join, the stream order becomes second in the lower reaches, i.e., if two *n*th order streams join, a $(n + 1)$th stream is born. Where a stream joins with another stream having a lower order, the stream order does not change.

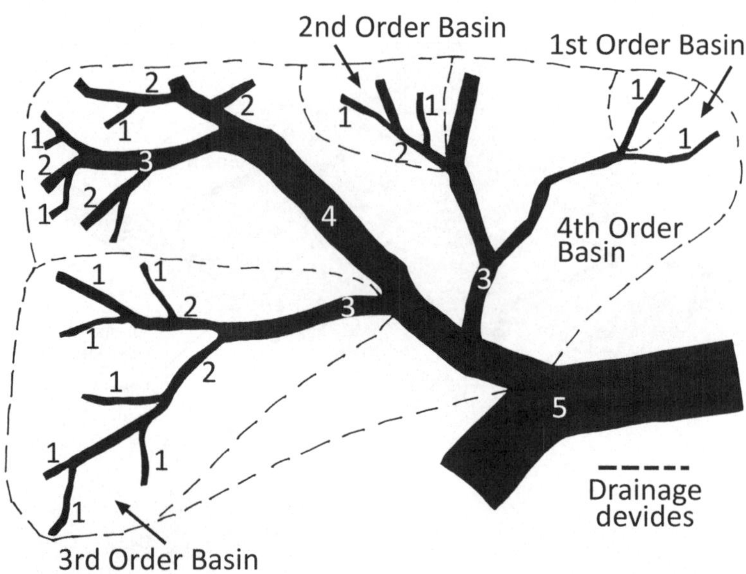

Fig. 8.17 Stream order. The first-order streams are the smallest in a drainage network and have no tributary. Two first-order streams unite to form second-order one. The third-order streams only have second- and first-order streams as tributaries, and so on and so forth

8.8.2 Drainage Patterns

Characteristics of a drainage basin are best exemplified by its drainage pattern (Fig. 8.18) and its drainage texture (Fig. 8.19). It is possible to deduce the geology of the basin, the strike and dip of the deposited rocks, existence of faults, and other information about geological structures by studying the drainage pattern of an area. The drainage texture reflects climate, permeability of rocks, vegetation, and the relief ratio. Various types of drainage patterns develop within a region, reflecting the structure of the rock. The patterns also relate to large-scale conditions of climate and tectonics that can only be appreciated on a global perspective. In general, a drainage pattern is largely a function of basin geology (structure, lithology, and overburden), climate, developmental history, and slope. Eight major types of drainage pattern are recognized (Fig. 8.18) and these are:

8.8.2.1 Dendritic

It has a tree-like drainage pattern possessing a main stream. It is the most common drainage pattern. It develops in areas where there are uniform type of rocks, slope, and geologic features with an impervious soil underneath (Fig. 8.18a). The longer the time of formation of a drainage basin, the more easily this pattern is formed.

8.8.2.2 Trellis

This is the modified version of the Dendritic pattern (Fig. 8.18b). This pattern develops where sedimentary rocks of varying resistance have been tilted (or have tilted and alternating hard and soft strata), folded (parallel), or faulted strata. Hence, for this pattern to develop, local geology plays a strong structural control upon its formation. The tributaries generally have an orientation at right angles to the main stream which run parallel following the lowlands.

8.8.2.3 Subdendritic

This pattern is intermediate between dendritic and trellis drainage pattern (Fig. 8.18c).

8.8.2.4 Pinnate

A dendritic drainage pattern in which the main stream receives many closely spaced, subparallel tributaries that joins it at acute angles (Fig. 8.18d). Thus, the pattern resembles a feather in plan view.

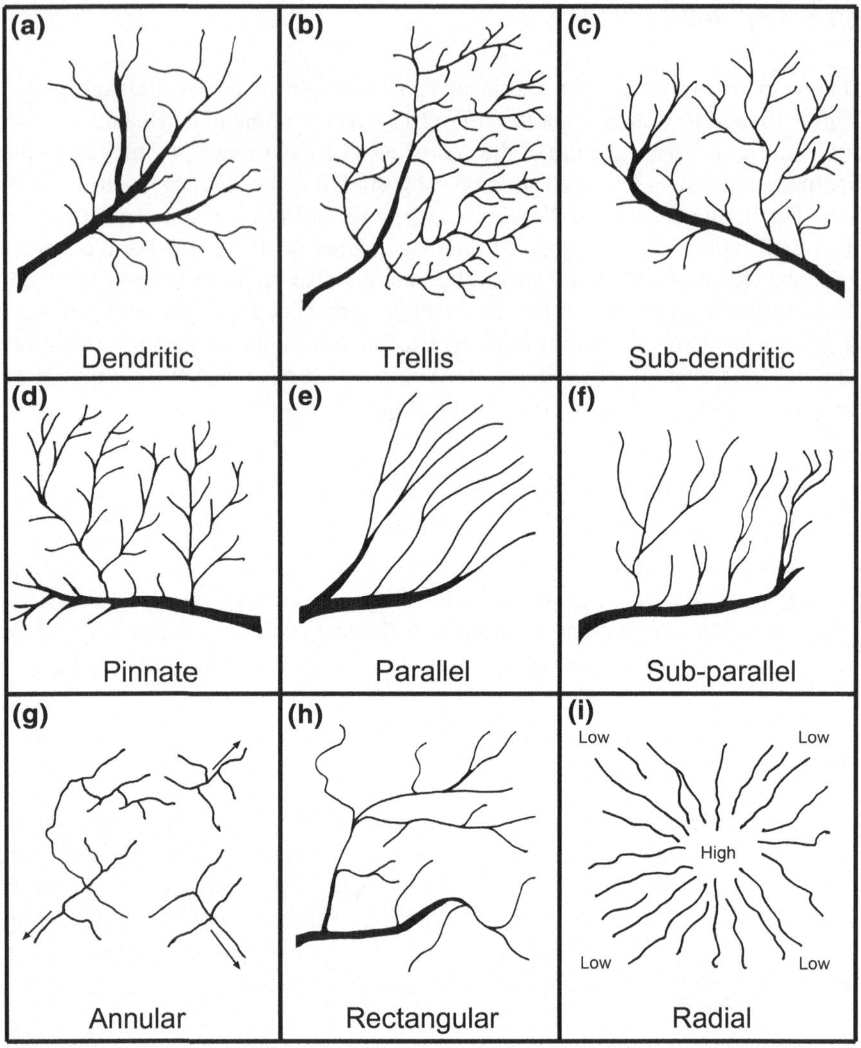

Fig. 8.18 Stream pattern

8.8.2.5 Parallel

This pattern develops in regions with moderate to steep slopes and in areas where there are parallel (and elongate) landforms (Fig. 8.18e). Hence, most of the streams run in the same direction. The tributaries often join the main stream at approximately the same angles. Parallel stream pattern is suggestive of a gently dipping bed or of a uniformly sloping topography such as a sloping basal flow or a young coastal region.

8.8.2.6 Subparallel or Linear

It forms typically on a newly uplifted surface that has not yet developed a dendritic pattern (Fig. 8.18f). Most streams run in the same direction.

8.8.2.7 Annular

It is similar to Trellis, but is curved around domal structures; concentric circular joints surrounding an uplifted dome of sedimentary rocks (Fig. 8.18g). The streams form in the weaker strata of the dome with an approximately circular or annular pattern. Often fractures control the flow of tributaries that are generally at right angles to the main stream.

8.8.2.8 Rectangular (Angular)

It develops in the region with geologic faults, joints, and fractures within the bedrock and so the drainage takes on a grid-like or a rectangular pattern (Fig. 8.18h). Hence, it is often also called rectangular dendritic or angular dendritic and is but a modified version of the dendritic pattern discussed above. Similar to the latter pattern, it also shows abrupt (close to 90°) changes in stream direction and very pronounced acute or obtuse angles of stream juncture. The rectangular drainage pattern is often associated with massive, intrusive igneous and metamorphic rocks.

8.8.2.9 Radial

This drainage pattern develops around a newly formed volcanic mountain, peak, or a cone or a dome (Fig. 8.18i). Channels form along maximum gradient. Thus, here streams emanate from a central focal point and flow outward radially.

8.8.2.10 Deranged

This pattern represents uncoordinated drainage and is mostly found in immaturely eroded areas such as those recently disturbed by events like glacial activity or volcanic deposition. The prerequisite for this drainage pattern to develop is a high water table and a flat or gently undulating topographic surface (such as in a glacial till plain). Few tributaries develop here.

Table 8.2 summarizes some of the salient points of major drainage patterns discussed above.

Table 8.2 Some of the major drainage patterns and their characteristics

Pattern	Origin	Characteristics	Geology
Dendritic and subdendritic	Insequent	Random, acute-angle junctions. Horizontal sediments or uniformly resistant crystalline rocks; gentle regional slope at present or at time of drainage inception	Homogenous, horizontal beds
Trellis	Subsequent	Parallel streams, high-angle junctions. Dipping or folded sedimentary, volcanic, or low-grade metasedimentary rocks; areas of parallel fractures	Heterogenous, tilted beds
Annular	Subsequent	Circular patterns. Structural domes and basins, diatremes, and possibly stocks	Heterogenous, breached domes
Rectangular/angular	Subsequent	High-angle junctions, high-angle bends in tributaries. Joints and/or faults at right angles; streams and divides lack regional continuity	Jointed rocks
Radial	Consequent	Streams flowing in all directions from central high area. Volcanoes, domes, and residual erosion features	Volcanic or intrusive domes

8.8.3 Drainage Texture and Drainage Density

Drainage texture is qualitative expressed as coarse or fine (Fig. 8.19). Texture is a function of drainage density and frequency. Drainage density is defined as the summation of channel lengths per unit area. It is an indication of the runoff potential and the degree of landscape dissection. The drainage density is controlled

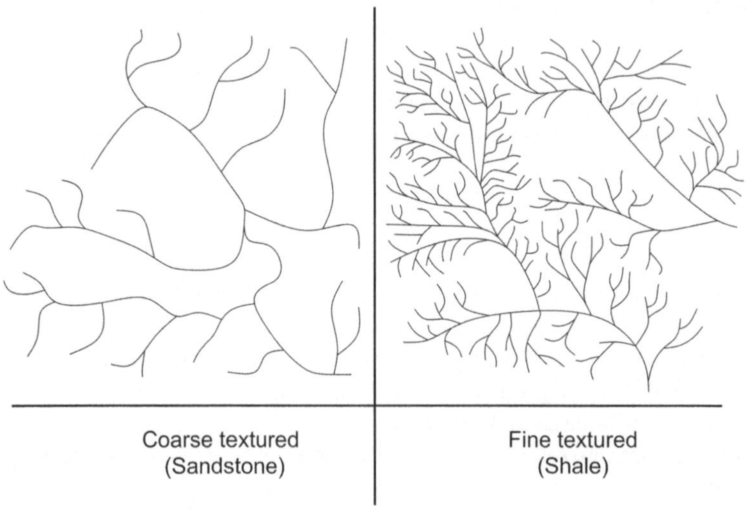

Fig. 8.19 Drainage texture

by permeability, erodibility of surface materials, vegetation, slope, and time. Drainage frequency is the total number of streams per area of the drainage basin and is inverse of drainage density. Although two drainage basins may show similar drainage density, their drainage frequency may not be the same. Conversely, drainage frequency does not mean similar drainage density. Coarse drainage density may appear in areas of permeable rocks and low rainfall intensity, whereas fine drainage density develops in badlands, due to impermeable rocks, thin vegetation cover, and heavy rainfall. A relationship between drainage density, texture, and their characteristics is given in Table 8.3.

8.9 Geological Work of Streams

The work of streams can be divided into three activities: Erosion, which creates particles, Transportation, which moves particles and Deposition, which deposits particles. The crushing and grinding of particles, rocks, and boulders when carried by a stream is Abrasion, and is the principle means for eroding the bedrock. Table 8.4 gives an overview of river processes, work of streams and fluvial landforms discussed in this chapter.

8.9.1 Stream Erosion

Running water carries out four major processes—Hydraulic action (Erosion), Abrasion/Corrosion, Solution/Corrosion, and Attrition.

Erosion is a hydraulic action, derived from the energy of running water. It kicks off a frictional drag action on particles of stream bed. But this action is only capable of removing small particles such as sand and fine gravel. However, as velocity increases, there is more turbulence and energy, thereby, lifting more sand grains from the stream bed. The river banks are often undermined by hydraulic action, and they eventually collapse into the river. Erosion makes a channel broader (lateral erosion) and deeper (deepening erosion). Lateral erosion forms a broader riverbed. But if deepening erosion dominates, a Canyon is formed.

Abrasion/Corrosion results in the removal of materials (the load of a river) by grinding the particles against the sides and the bed of the river channel. This results in the widening and deepening of the channel. Most down cutting of a river channel is done by this process. Abrasion is effective when the river has a large load. Potholes are a common feature in regions of rapid abrasion. They are cylindrical holes drilled into the rock by turbulent high velocity flows. Vertical eddies rotate a pebble, thereby grinding a depression in the rock. Very large potholes are found underneath glaciers where water velocities are quite high. Thus, the speed of abrasion is a function of streams load; the larger the particles in the load the quicker the erosion.

Table 8.3 Drainage density, texture and characteristics

Texture	Drainage density (km/ km sq.)	Characteristics
Coarse (low)	<8	Permeable or resistant rock, humid, and well vegetated
Medium	8–20	Permeable rocks, heavy rainfall, and well vegetated
Fine (high)	20–200	Impermeable surface, Low rainfall, and Little vegetation
Ultra-fine	>200	Impermeable surface, Low rainfall, easily erodible rocks, and little vegetation

Table 8.4 Drainage density, texture and characteristics

River processes	**Erosion**	**Transportation**
	Corassion	Saltation
	Attrition	Traction
	Solution	Solution
	Hydraulic action	Suspension
Work of Streams	Erosion	
	Transportation	
	Deposition	
Fluvial Landforms	**Erosional**	**Depositional**
	Rill	Colluvial Fan
	Gully	Alluvium fans and cones
	V-shaped valleys	Bajada
	Gorges	Bars
	Waterfalls	Point bars
	Potholes	Channel bars
	Cut-bank	Dunes and Ripples
	Pediments	Crevasse splays
	Escarpment	Floodplain
	Hogbacks and Cuestas	Crevasse
	Mesa and Butte	Levees
	Peneplain	Aggradation
	River terraces	Floodplain deposits
	Badlands topography	Abandoned channels
		River Terraces
		Deltas
		Alluvial Fans

In Solution/Corrosion, stream water acts as a solvent and dissolves soluble minerals such as a limestone. Water reacts chemically with rocks and dissolves them. In limestone regions, solution results in a characteristic Karst topography (for details see Chap. 10). Here, the CO_2 accelerates the solution of the limestone ($CaCO_3$) due to the formation of Carbonic acid that dissolves the rock and carries it away in solution. Almost all rocks are soluble to some extent.

Attrition makes the carried materials smaller, rounder, and smoother resulting from colliding into each other during transportation. Hence, all particles moving downstream undergo a progressive reduction in their size. The sharp edges and

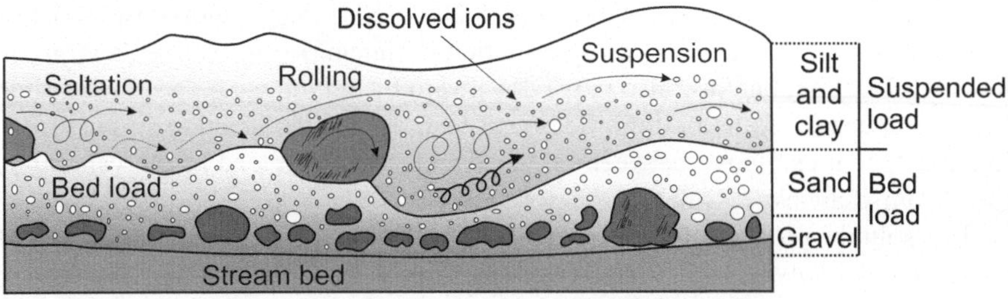

Fig. 8.20 Stream transportation. Fine materials such as clay, silt, and fine sand are transported in the water or on the water surface without contact with the riverbed through a process called Suspension. Those materials that are carried in suspension are called Suspended load. The larger rock particles (sand, gravel, and boulders) are too heavy to be suspended in the water flow, and thus remain on the bed of the stream and move by either Traction (rolling, sliding, or dragging) or by Saltation (bouncing). Sediment load carried by Traction is called Bed load. Sediment can move between bed load and suspended load as the velocity of the stream changes. Redrawn from Earth's Dynamic Systems with permission of Eric H. Christiansen

angles of the particle become more continuously rounded. Thus (in general), the upstream parts of a river have large angular sediments, whereas the downstream has fine rounded ones. However, this is not the norm as the angularity and size of particles are a function of prevailing lithology, length of time they travel in the river and distance covered by them in the stream.

8.9.2 Stream Transport

Stream transport is the movement of eroded particles downstream. The sediment load is transported by three different processes: Solution/Corrosion, Suspension, and Traction/Saltation (Fig. 8.20).

Rock particles and dissolved ions carried by the stream are called Stream load (Fig. 8.20). Material transported by streams can be in the form of solution, suspension, and bed load (top to bottom) (Fig. 8.20). The size of the fragments that is transported is a function of stream velocity and the nature of flow, laminar, or turbulent. Turbulent flow can keep fragments in suspension longer than in the laminar flow.

Corrosion corrodes rocks and brings them into solution. Fine materials such as clay, silt, and fine sand are transported in the water or on the water surface without contact with the riverbed through a process called Suspension (Fig. 8.20). Those materials that are carried in suspension are called Suspended load. Suspended load creates turbidity (the muddiness of a river) and it is also the largest fraction of the load (Fig. 8.20). The muddy appearance of a stream during a flood or after a heavy rain is largely due to a large suspended load. The larger rock particles (sand, gravel, and boulders) are too heavy to be suspended in the water flow, and thus remain on the bed of the stream (Fig. 8.20). These move either by Traction (rolling, sliding, or dragging) or by Saltation (bouncing) (Fig. 8.20). Sand grains move by Traction, but they also move downward by Saltation, i.e., a series of short leaps or bounces off the bottom. Sediment load carried by Traction is called Bed load. Sediment can move between bed load and suspended load as the velocity of the stream changes (Fig. 8.20).

The soluble products of the chemical weathering processes can make up a substantial Dissolved load (dissolved ions; Fig. 8.20) in a stream. This load is invisible as the ions are dissolved in the water. The load consists of ions like sodium (Na^+), potassium (K^+), calcium (Ca^{+2}), magnesium (Mg^{+2}), carbonate (CO_3^{-2}), bicarbonate (HCO_3^-), chloride (Cl^-), and sulfate (SO_4^{-2}). These ions are eventually carried to the oceans and give the oceans their saltiness. Streams that have a deep underground source generally have higher dissolved load than those whose source is at or near the surface.

8.9.2.1 Stream Capacity

Stream capacity is the maximum load of solid material (both suspended and bed load; Fig. 8.20) that can be carried by a stream at a given discharge. Stream capacity is function of the velocity of flow; higher stream velocity increases the capacity of stream to hold the material in suspension and carry the bed load. Generally, bed load capacity increases 3 to 6 power of the prevailing velocity. If velocity increases two folds, then the amounts of bed load that the stream can transport increases 8–64 times. For small and high velocity streams, most of the load is the bed load with little suspended load. However, for large and slow streams, much of the load is carried in suspension. Stream capacity is of great importance as it determines whether a stream will cause erosion or deposition. If the stream is transporting much less sediment than its capacity, it has free energy to erode its bed and the sides of a channel. Conversely, when a stream is carrying a full load, any reduction in its flow will cause deposition. As stream capacity is related to the discharge of water, hence, if more water is flowing in the stream, it will carry more sediment. When flow increases, stream load capacity increases and stream starts widening and deepening the channel. When the water flows with decreased velocity, stream load capacity decreases and deposition occurs. Hence, fluctuations in the rates of discharge will create a cycle of erosion and deposition.

8.9.2.2 Stream Flow and Work

The relationship between particle size and flow velocity for the three types of geologic work performed by water (Erosion, Transportation, and Deposition) is illustrated in Fig. 8.21.

For Erosion, it is quite difficult to erode materials from the channel than it is to transport or deposit them. For finest particles like clay, a very high velocity of flow

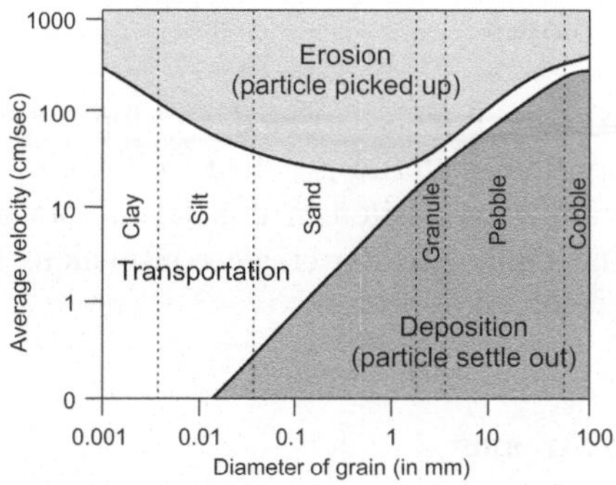

Fig. 8.21 The Hjulstrøm curve. This diagram illustrates the relationship between particle size and flow velocity for the three types of geologic work performed by water (erosion, transportation and deposition). For finest particles like clay, a very high velocity of flow is required to dislodge them from the bed of the channel due to their strong molecular bonding. On the other end of the spectrum, the largest particles (cobble) also require equally high flow velocities to erode them. For transportation, note that once clays have been eroded they can be transported over a wide range of velocities, even very slow ones, before being deposited

is required to dislodge them from the bed of the channel due to their strong molecular bonding. On the other end of the spectrum, the largest particles (cobble) also require equally high flow velocities to erode them.

For Transportation, note that once clays have been eroded they can be transported over a wide range of velocities, even very slow ones, before being deposited. The area of the graph for transportation narrows as you move to larger particle sizes. Larger sized materials are harder to transport due to their weight (Fig. 8.21).

For Deposition note that it does not extend over to the very smallest size on the bottom left of the diagram (Fig. 8.21). The very smallest particles are easily transported even under low flow velocities and will not settle out. The minimum velocity for deposition (the line that separates transportation and deposition) climbs steadily up to the right. This indicates that as particle size and weight increases, it is more difficult to transport material and deposition will occur even with a slight drop in stream velocity (Fig. 8.21).

8.9.3 Deposition

The deposition processes is divided into two broad categories: deposition on slopes (dominantly in overland flow) and deposition in streams (channel flow).

8.9.3.1 Deposition in Slopes

Materials eroded by overland flow is carried down slope and deposited. It is a very slow process. This deposited material is called Colluvium. As this is laid by the overland flow, Colluvium is deposited in thin layers, not always obvious to an observer. Both Rill and Gully erosion create conspicuous accumulations of the Colluvium.

8.9.3.2 Deposition in Streams

It is the accumulation of transported particles due to decreased or sudden changes in current velocity. Large particles are deposited first, and the smaller ones are carried further and deposited later. Within a stream profile, velocity varies with position. If the sediment gets moved to the part of the stream where velocity is lower, then the sediments comes out of suspension and are deposited. Stream dynamics is also affected by sudden changes in its velocity. Hence, during flood discharge, stream velocity suddenly increases and it overtops its banks and flows onto the flanks forming Levees and Floodplains (Fig. 8.22). If the gradient of the stream suddenly changes by emptying into a flat-floored basin (a lake or an ocean basin), the velocity of the stream will suddenly decrease, resulting in the deposition of sediments that can no longer be transported. This results in the deposition of Alluvial fans (Fig. 8.23) and Deltas (Fig. 8.24).

8.10 Fluvial Landforms

The landforms created by running water, i.e., by fluvial processes are fluvial landforms. These landforms are divided into two broad groups: Erosional and Depositional (Table 8.4). The former are sequential landforms shaped by the progressive removal of particles and the latter are created by the accumulation of transported sediments that have been removed from sequential landforms.

8.10.1 Erosional Landform

8.10.1.1 Rill

These are small closely spaced microchannels best seen along roadsides (road cuttings). They range in size in width between 50 and 300 mm and can be as deep as 30 mm. They are common in sparsely vegetated dry lands where intense rainfall occurs on a regular basis (Fig. 8.16).

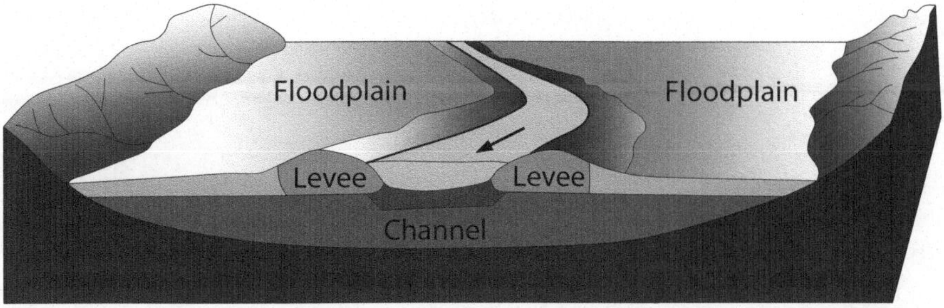

Fig. 8.22 Levees, channels, and floodplains. If the sediment gets moved to the part of the channel where velocity is lower, then the sediments come out of suspension and are deposited. Hence, during flood discharge, channel velocity suddenly increases and it overtops its banks and flows onto the flanks forming levees and floodplains. Redrawn from Earth's Dynamic Systems with permission of Eric H. Christiansen

Fig. 8.23 Alluvial fans. As gradient changes, the velocity of the stream also decreases causing the sediments to deposit which can no longer be transported. This results in the formation of an alluvial fan

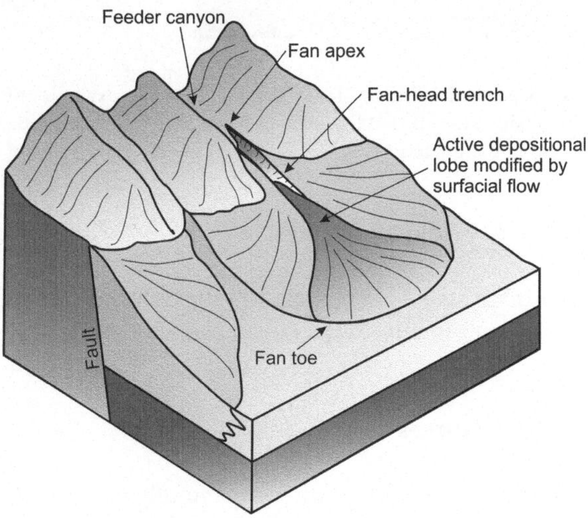

8.10.1.2 Gully

These look like steep-walled trenches with actively growing upper ends (Fig. 8.16). Their larger version is a stream channel. They commonly have a depth range between 0.5 and 25 m. However, distinction between them and stream channels (which closely share the same dimensions) is arbitrary. These are the larger version of Rills.

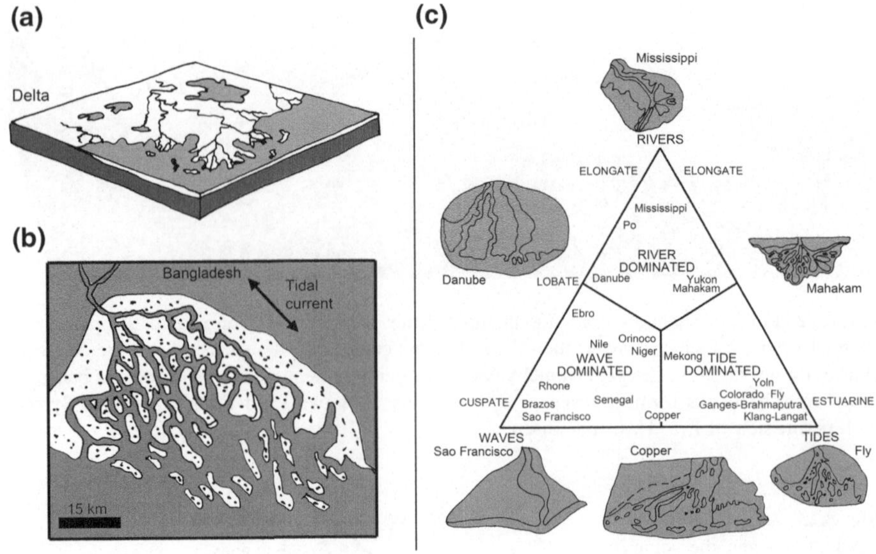

Fig. 8.24 Deltas. **a** A Tidal delta and **b** an example of a Tide-dominated Ganges delta. **c** Classification of delta types based on their formation (i.e., river-dominated, wave-dominated, and tide-dominated)

8.10.1.3 V-shaped Valleys

These are steep sided, narrow valleys formed due to vertical erosion. The valley walls join directly to the streambed (Fig. 8.4).

8.10.1.4 Gorges

These are deep valleys often structurally controlled by the bedrock. There is some confusion in the use of "Gorge" and "Canyon". They both are the same, except that the word Canyon comes from the French, whereas Gorge comes from Spanish. A gorge is a steep-sided valley where the sides have often cliff-like appearance. A gorge commonly has a river flowing in it and is generally formed by a fast-flowing river that, with time, cuts through the surrounding landscape. Waterfalls can also form gorges as they cut backwards over time. Water erosion is the principal cause of gorge formation, although weathering can also help the process. The Grand Canyon in Arizona is arguably the most famous gorge of the world.

8.10.1.5 Waterfalls

These are generally formed when a more resistant unit of bedrock juts, thereby creating a nickpoint in the stream's longitudinal profile. The weaker rocks below the resistant unit are preferentially eroded, undercutting the resistant rock, and maintaining the vertical face of the waterfall as it migrates upstream (Fig. 8.10).

8.10.1.6 Potholes

These are caused by the abrasion of the bedrock by pebbles, sand, etc., caught in small depressions within the bedrock. If a rock carried by water is trapped in a small depression within the stream bed, it is turned by the flowing water and slowly grinds a pothole.

8.10.1.7 Cut-Bank

They form on the outside curve of a meandering stream. Higher water velocity causes erosion on the outside of the curve (Fig. 8.6).

8.10.1.8 Pediments

They are eroded and gently inclined (usually ~2–3°, but <10–15°) bedrock surfaces near the mountain base. They are usually formed by lateral planation by ephemeral streams or by Sheetwash flooding.

8.10.1.9 Escarpment

An escarpment is an edge of a plateau usually with a steep and jagged cliff (Fig. 8.25). The cliff is generally long that results either from erosion of the softer layer beneath or by faulting. The less resistant rocks erode faster, retreating until the point they are overlain by more resistant ones.

When the dip of the bedding is gentle, a Cuesta is formed (Fig. 8.26), however, with steeper dips (>30–40°) Hogbacks are formed (Fig. 8.26).

8.10.1.10 Hogbacks and Cuestas

Hogbacks get their name because they look like the topside of a razorback hog (Fig. 8.26). When the rocks are uplifted by the mountain building process and are tilted in one direction, erosion of the softer, less resistant rock occurs leaving a

Fig. 8.25 Escarpment. It is an edge of a plateau usually with a steep and jagged cliff

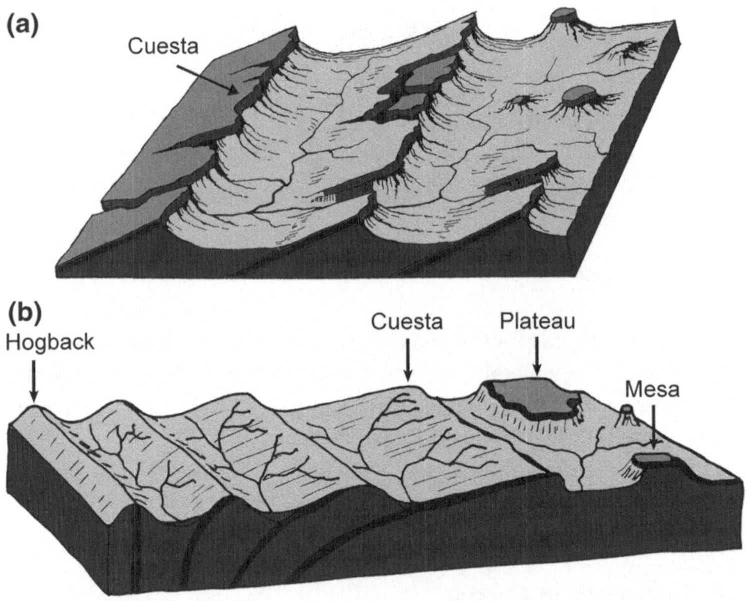

Fig. 8.26 Cuestas and Hogbacks. When the rocks are uplifted by the mountain building process and are tilted in one direction, erosion of the softer, less resistant rock occurs leaving a ridge of the more resistant rock. If the angle of dip is low, the ridges are called Cuestas (a), but if the rocks dip at a high angle, the ridges are Hogback (b)

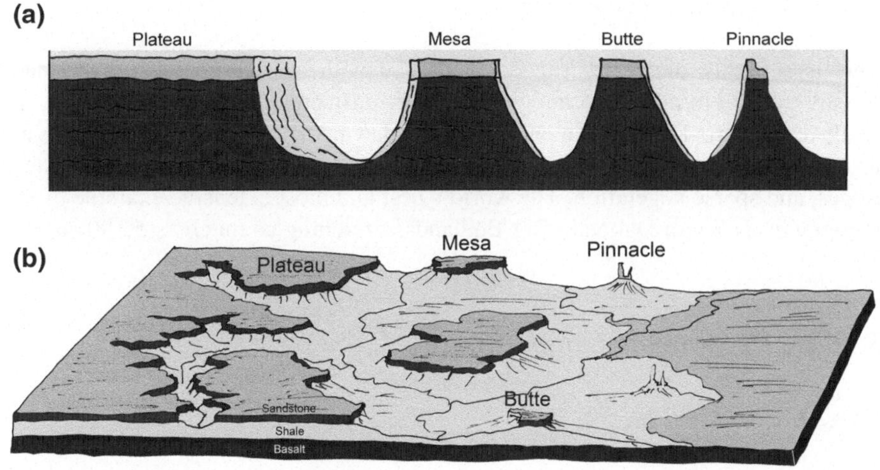

Fig. 8.27 Butte, Mesa, Pinnacle, and Plateau. **a** *Cross-sectional view* of a terrain. **b** *Aerial view* Butte is an isolated hill with steep, often vertical sides and a small, relatively flat top. Mesa is an elevated land with a flat top and steep cliffs sides. **b** Redrawn from Earth's Dynamic Systems with permission of Eric H. Christiansen

ridge of the more resistant rock. If the angle of dip is low, the ridges are called Cuestas, but if the rocks dip at a high angle, the ridges are Hogbacks (Fig. 8.26).

8.10.1.11 Mesa and Butte

Mesa is an elevated land with a flat top and steep cliffs sides, thus displaying a characteristic table-top shape (Fig. 8.27). Butte, on the other hand, is an isolated hill with steep, often vertical sides and a small, relatively flat top; it is smaller than a Mesa (Fig. 8.27).

8.10.1.12 Peneplain

It is a more or less level land surface representing an advanced stage of erosion undisturbed by crustal movements.

8.10.1.13 River Terraces

Part of an old flood plain that has been left perched on the side of a river valley. It results from rejuvenation, a renewal in the erosive powers of a river (Fig. 8.12).

8.10.1.14 Badlands Topography

This is a highly dissected, rugged landscape with an extremely high drainage density (>100 km/mile). Such regions are generally underlain by impermeable and easily eroded rocks or sediments that are also poorly vegetated. Badlands are usually found in semiarid climates and are characterized by countless gullies, steep ridges, and sparse vegetation. The world's best and most extensive example of this topography is South Dakota's Big Badlands, stretching as much as 5,180 sq. km.

8.10.2 Depositional Landform

8.10.2.1 Colluvial Fan

It is a loose deposit of rock debris accumulated through the action of gravity at the base of a cliff or slope.

8.10.2.2 Alluvium Fans and Cones

The sediments are deposited by a stream (Fig. 8.23).

8.10.2.3 Bajada

They are formed by the lateral merging of a series of alluvial fans (Fig. 8.28).

8.10.2.4 Bars

They develop due to the reduction in stream discharge and are commonly formed in elevated areas. Streams carrying coarse sediments develop sand and gravel bars as in braided streams. There are two types of Bars: Point and Channel bars.

Point Bars

These form on the inside curves of a meandering stream (Fig. 8.6). They develop where stream flow is locally produced either due to decrease in stream velocity and/or reduced water depth.

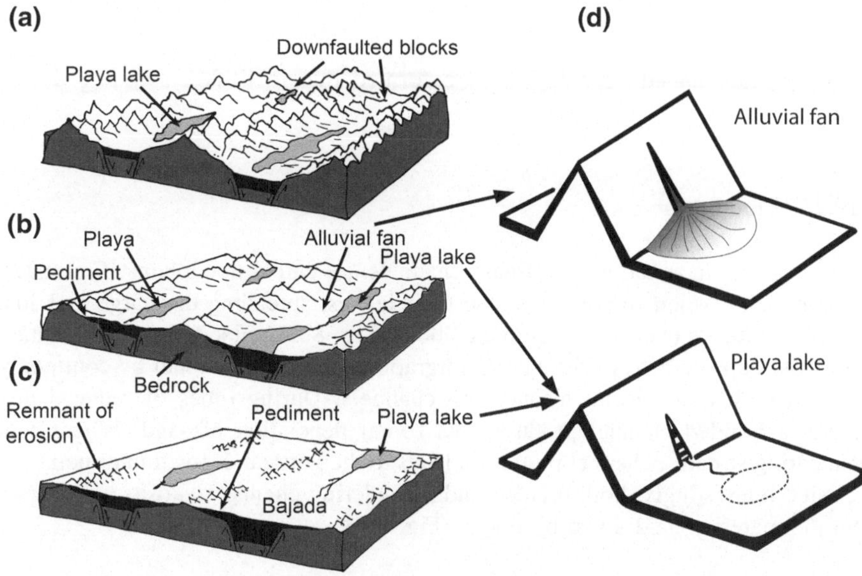

Fig. 8.28 Formation of Bajada, Pediment, Playa Lake, and other landforms from the initial stage (**a**) to the final stage of erosion (**c**). **d** Plan view of an Alluvial fan and the Playa Lake. **a-c** Redrawn from Earth's Dynamic Systems with permission of Eric H. Christiansen

Channel bars

These form in braided streams (Fig. 8.13) and generally contain relatively coarse-grained sediments. They are similar to Point bars in that they are also formed as a result of decrease in stream velocity. However, channel bars occur in the interior of the channel rather than along the edges.

8.10.2.5 Dunes and Ripples

Stream channels that are mainly composed of sand and silt, Dunes and Ripples form as primary sedimentary structures. Streams with higher velocities form dunes. Dunes are about 10 or more cm in height and are spaced a meter or more, apart. Slow moving streams with fine textured beds form ripples. They are only a few cm in height and spacing. Both dunes and ripples move over time, migrating downstream.

8.10.2.6 Crevasse Splays

These are fan-shaped deposits that are formed due to bank crevassing during floods.

8.10.2.7 Floodplain

Relatively flat areas alongside stream channels are called floodplains (Fig. 8.22). These develop when streams over-top their levees; spreading discharge and suspended sediments over the land during floods (Fig. 8.22). Floodplains also contain sediments deposited from the lateral migration of a river channel as commonly noted in both braided and meandering channels. During times of reduced discharge, a braided channel produces horizontal deposition of sand. Whereas in meandering streams, channel migration leads to the vertical deposition of point bar deposits. Interestingly, both braided and meandering channel deposits are coarser than the materials laid down by floods (Fig. 8.13).

8.10.2.8 Crevasse

These are narrow gaps intersecting the levees. These features allow for the movement of water to the floodplain and back during floods. Topographical depressions are found scattered around the floodplain. Depressions contain some of the finest deposits on the floodplain because of their elevation.

8.10.2.9 Levees

During a flood, as a stream overtops its banks, the velocity of the flood is first very high, but it decreases as water flows out over the gentle gradient of the floodplain. This decrease in velocity causes the coarser grained suspended sediments to be deposited along the riverbank, forming a Levee. Generally, in diameter, a Levee is approximately one-half to four times the width of a channel (Fig. 8.22).

8.10.2.10 Aggradation

Upon retreat of the floodwaters, stream velocities are reduced causing deposition of the alluvium. Repeated flood cycles over time result in the deposition of many successive layers of such alluvial material thereby raising the elevation of the stream bed. This process is called aggradation.

8.10.2.11 Floodplain Deposits

Floodplains have a great variety of depositional materials such as colluvium (debris from valley sides), channel deposits (sand and gravel), and vertical accretion deposits (clay and silt deposited by overbank flows).

8.10.2.12 Abandoned Channels

These are depressions on the floodplain that were former meander bends.

8.10.2.13 River Terraces

When a channel rejuvenates, it deepens to form a new gorge. Thus, the surface of the former alluvial plain becomes higher than the river bed, and the river water ceases to overflow onto it. The former alluvial plain then becomes a river terrace (Fig. 8.12). Erosional terrace are formed when either a stream or river downcuts through the bedrock. As it is continues to downcut, the flattened valley bottom composed of bedrock (overlain with a possible thin layer of alluvium) is left above either a stream or river channel. These bedrock terraces are erosional in nature (Fig. 8.12a). Depositional terrace are formed as the valley is filled with alluvium, the alluvium is incised, and the valley fills again with material but now to a lower level than before. The terrace that results for the second filling is depositional in nature (Fig. 8.12b).

8.10.2.14 Deltas

When a stream enters a standing body of water such as a lake or ocean, there is a sudden decrease in velocity and the stream deposits its sediment in a deposit called a Delta (Fig. 8.24).

8.10.2.15 Alluvial Fans

When a steep mountain stream enters a flat valley, there is a sudden decrease in both its gradient and velocity. Hence, alluvial fans develop when streams carrying a large load reduce their velocity as they emerge from mountainous terrain to a nearly horizontal plain (Fig. 8.23). An alluvial fan is characterized by coarse sediments; the fines are deposited more toward the downstream and the very coarse sediments are deposited closer to the mountains.

8.11 Factors Affecting Fluvial Systems

Lithology and local to regional structure and tectonics affect fluvial systems. Strata that are horizontal, tilted, and folded also have a strong bearing on fluvial systems.

8.11.1 Structure and Tectonics

Drainage, either adjusts passively to the varying resistance of geologic structures, or is actively induced to follow a particular course by the prevailing tectonics (Fig. 8.29). The Ganges–Brahmaputra delta region is a good example that shows influence of faulting. Structural and tectonic controls are exhibited well for streams that emerge from mountain fronts onto surrounding plains. Where mountain fronts are erosional because of a complex interplay of geomorphic variables, they may develop flanking surfaces of planation called Pediments (Fig. 8.28). Deposition at

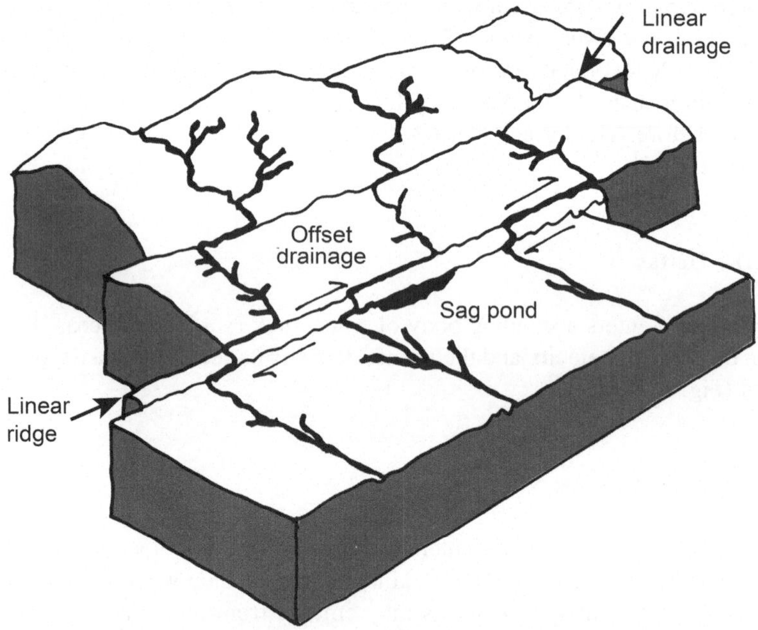

Fig. 8.29 Prevailing tectonics and the varying resistance of geologic structures to water erosion and flow defines the drainage pattern of an area. The area adjusts passively or is actively induced to follow a particular course defined by the prevailing tectonics and geologic structures. Notice that the movement along the fault has offset the stream drainage pattern on either side and resulted in the formation of sag ponds along the fault. Redrawn from Earth's Dynamic Systems with permission of Eric H. Christiansen

the mountain front produces alluvial fans (Fig. 8.23) because of the tremendous increase in width as a stream emerges from a mountain canyon.

8.11.2 Lithology

Varied rates of erosion result in the formation of small hills or mountains of resistant rocks; the less resistant ones are eroded to a lesser extent.

8.11.2.1 Horizontal Strata

Differential weathering and erosion of horizontal strata results in terraced slopes with steep, cliff-like edges in hard strata and gentle slopes covered by weathered material in the soft strata. The resistant rocks tend to have near vertical faces and are maintained this way by undermining of the weaker strata beneath. As erosion progresses, the plateau of resistant cap rock become smaller and eventually is reduced to isolated remnants called Mesas (Figs. 8.26, 8.27). It is called a Butte (Figs. 8.26, 8.27), if it is reduced until its diameter is less than its height.

8.11.2.2 Tilted Strata

Streams cut down into tilted softer strata, leaving the more resistant strata as ridges. In areas where the strata are gently tilted, the ridges are distinctly asymmetrical, consisting of a steep, erosional escarpment, or scarp slope, with a gentle dip slope. These are called Cuestas (Fig. 8.26). However, when the strata are steeply inclined, the ridges are more symmetrical; these are then called Hogbacks or Razorbacks (Fig. 8.26).

8.11.2.3 Folded Strata

Initially, streams will flow along the synclines, with the anticlines forming parallel ridges. Due to fractures and joints developed during folding, the arched rock strata are structurally weaker, while the synclines are hardened by compression. Given time, the synclines tend to form ridges, with the streams occupying anticlinal valleys. Thus, an inversion of relief is formed.

8.12 Summary

A large stream is a river and earth's lifeblood. While flowing to the oceans, they drain about 68 % of earth's land surface; the remainder is either covered by ice or drains into closed basins. The pattern of river discharge and the shape of the channel changes with time along with the landscape that it flows through. Stream gradient also changes over time from very uneven (as in lakes and waterfalls) to gentle where it ends. Thus, the evolution of a stream and its valley goes through various stages of development from initial (young rivers), intermediate (mature), and terminal (old rivers). The initial stage is typically found in rivers that occur in the highlands or mountainous regions. Erosion is their primary work. A mature river tends not to deepen its channel. Instead, erosion occurs mostly along valley walls when the river overflows its banks and covers the valley floor. For the mature river, a stage of equilibrium is reached where its erosion and deposition are nearly in balance. In the terminal stage, as the river continues to age; its gradient and velocity decreases correspondingly. Now, the stream no longer erodes but in fact deposits its sediments on its own channel and banks to form a broad flat plain.

A rejuvenated river is one whose gradient has become steeper due to this change. This increased gradient allows the river to cut more deeply into the valley floor. incised river, antecedent river, and river capture occur due to rejuvenation.

Streams are classified based on their constancy of flow (Perennial, Intermittent, and Ephemeral), condition of the groundwater table (Influent and Effluent), and the stream equilibrium (graded and non-graded stream). The genetic classification of streams includes: consequent streams, subsequent streams, obsequent streams, resequent streams, insequent streams, superimposed stream, sntecedent stream (Nickpoint), and incised River.

Streams are partially confined bodies of water flowing downhill, carrying rock fragments, as bed load, suspended load and in solution, within a well-defined path called a channel. There are four major types of stream channels: straight, sinuous, meandering and braided. Straight channels are controlled by a linear zone of weakness in the underlying rock; this could be a fault or a joint system. The meandering channels (with sinuosity ratio $\geq$1.5) are the most common channel variety. In streams flowing over low gradients with easily eroded banks, straight channels will eventually erode into meandering channels. Braided channels have interwoven network of channels, separated by the deposition of coarse alluvium. They occur where sediment load is high and both discharge and velocity fluctuates rapidly. In such a stream, the water flows in a braided pattern around islands and bars, dividing and reuniting as it flows downstream.

Stream capture or stream piracy is common in the youthful stage of a stream and occurs when rapid head erosion proceeds into an adjacent drainage basin; the valley head eventually works its way toward another channel, and it becomes connected with the upper reaches of the formerly separate basin.

The area drained by a stream and its tributaries is called a basin. A drainage basin is grouped by its size, shape, and arrangement of tributaries, drainage density and morphometry. It is a system where sediment is produced, transported, and deposited. A drainage divide separates each drainage basin.

Basin morphometry is a function of stream order and drainage patterns. The first-order streams are the smallest streams in a drainage network and have no tributary. Two first-order streams unite to form second-order ones. The latter only have first-order streams as tributaries. Third-order streams have only second- and first-order streams as tributaries, and so on and so forth.

Characteristics of a drainage basin are best exemplified by its drainage pattern and its drainage texture. Eight major types of drainage pattern are recognized—Dendritic, Trellis, Subdendritic, Pinnate, Parallel, Subparallel or Linear, Annular, Rectangular (Angular), Radial, and Deranged.

Drainage texture is qualitative expressed as coarse or fine. Texture is a function of drainage density and frequency. Drainage density is defined as the summation of channel lengths per unit area. It is an indication of the runoff potential and the degree of landscape dissection. The drainage density is controlled by permeability, erodibility of surface materials, vegetation, slope, and time. Drainage frequency is the total number of streams per area of the drainage basin and is inverse of drainage density.

The work of streams can be divided into three activities: Erosion, which creates particles, transportation, which moves particles and deposition, which deposits particles. The crushing and grinding of particles, rocks, and boulders when carried by a stream is Abrasion, and is the principle means for eroding the bedrock.

Stream capacity is the maximum load of solid material (both suspended and bed load) that can be carried by a stream at a given discharge. Stream capacity is function of the velocity of flow; higher stream velocity increases the capacity of stream to hold the material in suspension and carry the bed load. Stream capacity is of great importance as it determines whether a stream will cause erosion or deposition.

The landforms created by running water, i.e., by fluvial processes are fluvial landforms. These landforms are divided into two broad groups: erosional and depositional landforms. The former are sequential landforms shaped by the progressive removal of particles and the latter are created by the accumulation of transported sediments that have been removed from sequential landforms.

The erosional landform include—Rill, Gully, V-shaped valleys, Gorges, Waterfalls, Potholes, Cut-bank, Pediments, Escarpment, Hogbacks and Cuestas, Mesa and Butte, Peneplain, River terraces, and Badlands topography.

The depositional landform include: Colluvial Fan, Alluvium fans and cones, Bajada, Bars (Point bars and Channel bars), Dunes and Ripples, Crevasse splays, Floodplain, Crevasse, Levees, Aggradation, Floodplain deposits, Abandoned channels, River terraces, Deltas, and Alluvial fans.

Factors that affect fluvial systems include the prevailing structure and tectonics and lithology (Horizontal strata, Tilted strata, and Folded strata).

References

Hamblin WK, Christiansen CH (2008) Earth's Dynamic Systems. Printice Hall, Upper Saddle River 767 p

Leopold LB, Wolman MG, Miller JP (1995) Fluvial Processes in Geomorphology. Dover Publications Inc., New York, USA (Reprinted)

Morisawa M (1985) Rivers. Longman Inc., New York 222 p

Robert A (2003) River processes: an introduction to fluvial dynamics. Hodder Arnold, London

Chapter 9
Groundwater

9.1 Introduction

Groundwater originates as rainfall or snow (precipitation), and then moves through the soil into the groundwater system, where it eventually makes its way back to surface as streams, lakes, or oceans (Fig. 9.1) (see also Price 1996). Groundwater makes up about 0.64 % of the total water on earth (Table 9.1); its total volume is equivalent to a 55 m thick layer spread out over the entire surface of the earth. Both surface and subsurface waters are related through the Hydrological cycle (Fig. 9.1), where a staggering 100 million billion gallons of water is moved per year by different paths such as evaporation, transpiration, precipitation, volcanism, surface runoff, infiltration, percolation, etc. However, inspite of this, the total amount of water in the earth system remains relatively constant.

The openings between grains of soil and sediments and within joints and fractures contain a large volume of water which moves and collects as groundwater or underground water. Much of the precipitation (rainfall and snowfall) that falls does not flow directly into rivers but seeps into the ground and moves, underground, slowly to stream beds. Hence, groundwater is a reservoir that can sustain streams even during periods of extreme droughts. The freshwater reservoir is about 28 million km^3 and a major part ($\sim$85 %) of this is in the form of polar ice and glaciers. Groundwater, as such, represents the largest reservoir of freshwater (14 %) that is readily available to us (Table 9.2; see also Fig. 9.2).

9.2 Vertical Distribution

Transpiration allows the recycling of small amount of water that infiltrates into the ground, back into the air almost immediately. The rest of the water either stays, or slowly works its way down to the Vadose zone (the Zone of Aeration; Fig. 9.3). Here, the pores between rocks and minerals are not filled with water. But this is so, in the zone below, the Zone of Saturation (Fig. 9.3). This subsurface occurrence of

S. Jain, *Fundamentals of Physical Geology*, Springer Geology,
DOI: 10.1007/978-81-322-1539-4_9, © Springer India 2014

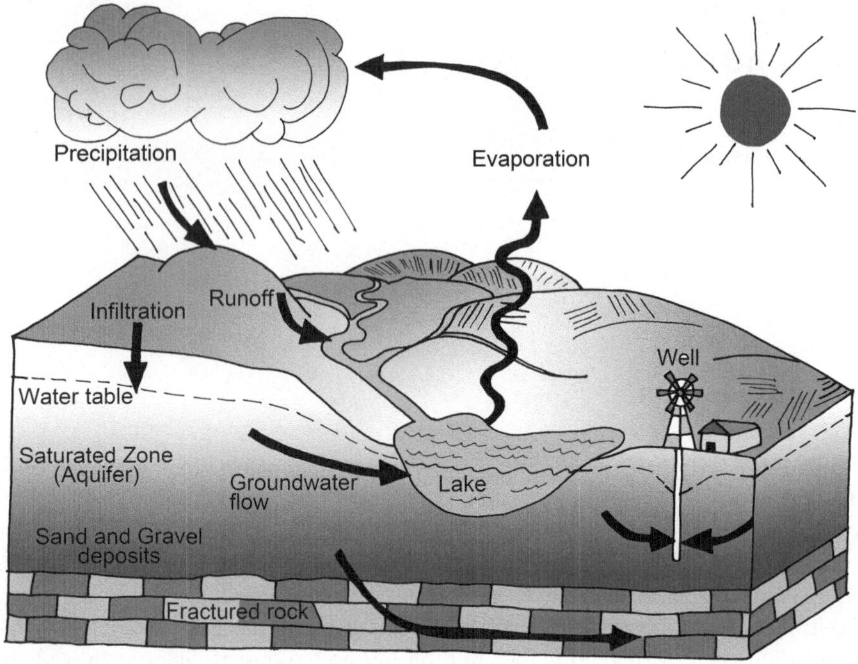

Fig. 9.1 The Hydrological cycle. The groundwater originates as rainfall or snow (precipitation), and then moves through the soil into the groundwater system, where it eventually makes its way back to surface as streams, lakes, or oceans

Table 9.1 The distribution of water

Reservoir	Percentage of Water
Oceans	97.3
Glaciers and polar ice	2.05
Soil moisture and underground aquifers	0.64
Lakes and rivers	0.016
Atmosphere	0.001
Biosphere	0.00004

Table 9.2 Distribution of freshwater reservoirs

Freshwater reservoir	Volume (%)
Ice	85
Groundwater	14
Lakes	0.5
Soil moisture	0.3
Atmosphere	0.05
Rivers	0.004

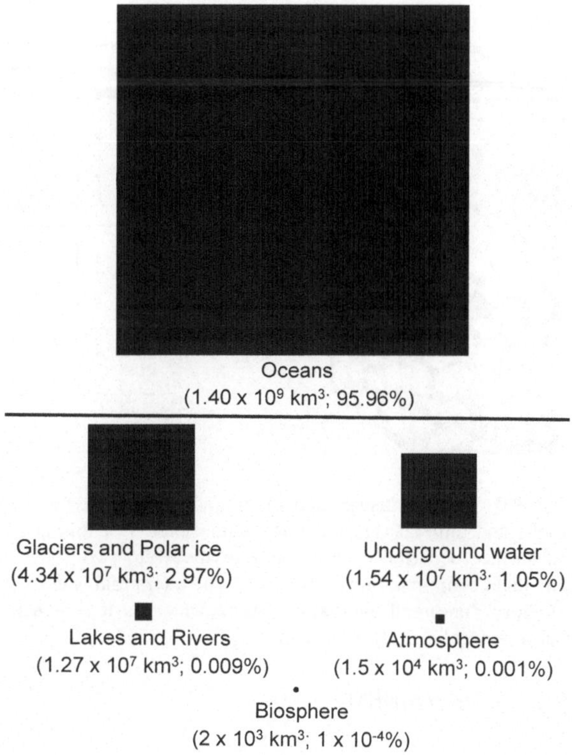

Fig. 9.2 Earth's inventory of water

water can be divided into two zones: Interstitial water and the water in chemical combination with the rock (Internal water). Often, groundwater is used interchangeably for interstitial water.

9.2.1 Zone of Aeration

The interstices or pores (open spaces) are partially occupied by water and partially by air (Fig. 9.3). The water in this zone is called the Suspended or Vadose water. In arid regions, this zone may be more than 300 m thick, but in moist areas, it may be entirely absent. The Zone of Aeration is divided into three zones namely Soil water (moisture) zone, Intermediate zone, and the Capillary zone (Fig. 9.3).

9.2.1.1 Soil Water (Moisture) Zone

It is the topmost part where water is present in the form of soil moisture. Plants mostly utilize this water.

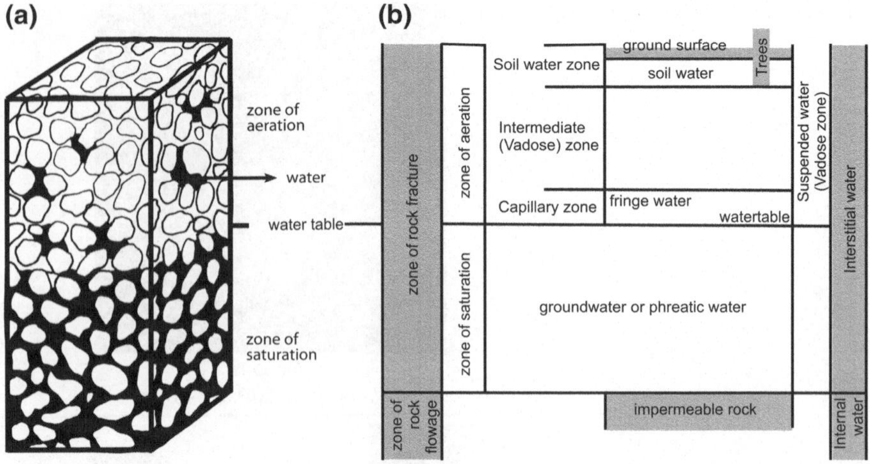

Fig. 9.3 Vertical distribution of groundwater. **a** At the Zone of Aeration, the pores between rocks and minerals are not filled with water. But this is so in the Zone of Saturation. **b** The subsurface occurrence of water is divided into two zones: Interstitial water and the water in chemical combination with the rock. The interstitial water is further divided two zones: Aeration (Vadose Zone) and Saturation (Groundwater). **a** Redrawn from Earth's Dynamic Systems with permission of Eric H. Christiansen

9.2.1.2 Intermediate Zone

This zone is commonly called as the Vadose zone.

9.2.1.3 Capillary Zone

It is the zone between the Vadose and upper part of the groundwater (the Zone of Saturation). Here, ground-water rises due to capillary action.

9.2.2 Zone of Saturation

All interstices (open spaces) within the rock are filled with water (Fig. 9.3). Water in this zone is known as Groundwater or Phreatic water). The upper part of the zone is called the Water table (Fig. 9.3). The groundwater zone merges at some depth into a zone of dense rock. Here, water does not migrate as the pores are not connected. The depth at which these zones will merge varies with the prevailing geology of the terrain. In crystalline intrusive and metamorphic rocks, the depth of this zone may start at 3,000 m, whereas in sedimentary basins, it may be nearly 15,000 m. At great depths, as much as 30,000 m, the pores of the rock are closed due to high pressure and temperature. Here, water is present only in a chemical combination with other material.

9.3 Hydrological Property of Rocks

9.3.1 Porosity

It is the percentage of the volume of open spaces (pore spaces) within the rock (Fig. 9.4). Open spaces determine the amount of water that a rock can hold. In sedimentary rocks, porosity is a function of grain size, shape of the grain, degree of sorting, and the degree of cementation. Well-rounded coarse-grained sediments have higher porosity (as the grains do not fit well together) in comparison with fine-grained sediments (Fig. 9.4). Poorly sorted sediments usually have lower porosity as the fine-grained sediments tend to fill in open spaces. Since cement tends to fill in pore spaces, hence, highly cemented sedimentary rocks have lower porosity. In Igneous and Metamorphic rocks, porosity is usually low as most minerals tend to intergrow, leaving little free space. However, highly fractured igneous and metamorphic rocks possess high porosity (Fig. 9.4).

There are four main types of pore spaces or voids in rocks. These are spaces between mineral grains, between fractures, between solution cavities, and between vesicles. In sand and gravel deposits, pore space can make up as much as 20–30 % of the total volume (Fig. 9.4). However, porosity is greatly reduced, if grain sizes vary, as the smaller grains fill spaces in-between the larger ones. Sometimes significant amounts of cementing materials fill the spaces between grains, thereby

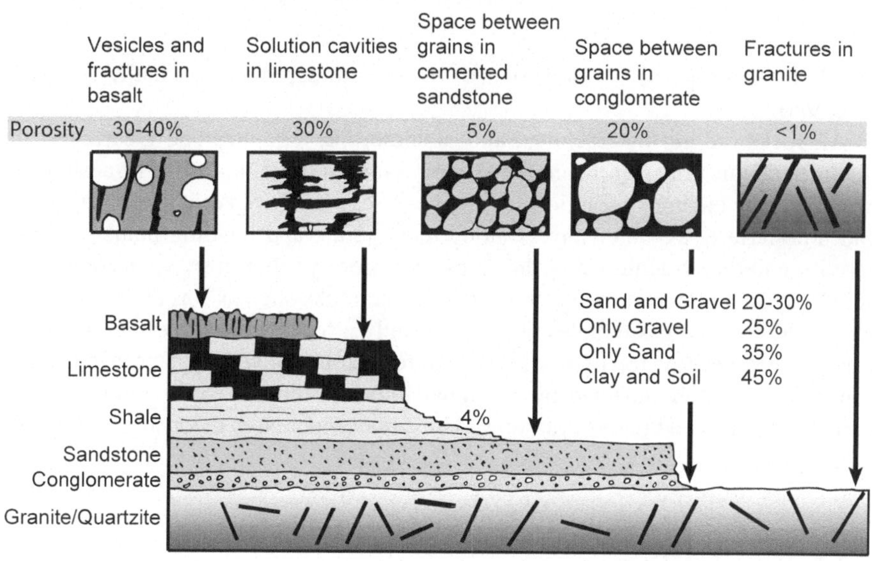

Fig. 9.4 Pore spaces and the flow of groundwater. The open pore spaces determine the amount of water that a rock can hold. Well-rounded coarse-grained sediments have higher porosity in comparison with fine-grained sediments. Redrawn from Earth's Dynamic Systems with permission of Eric H. Christiansen

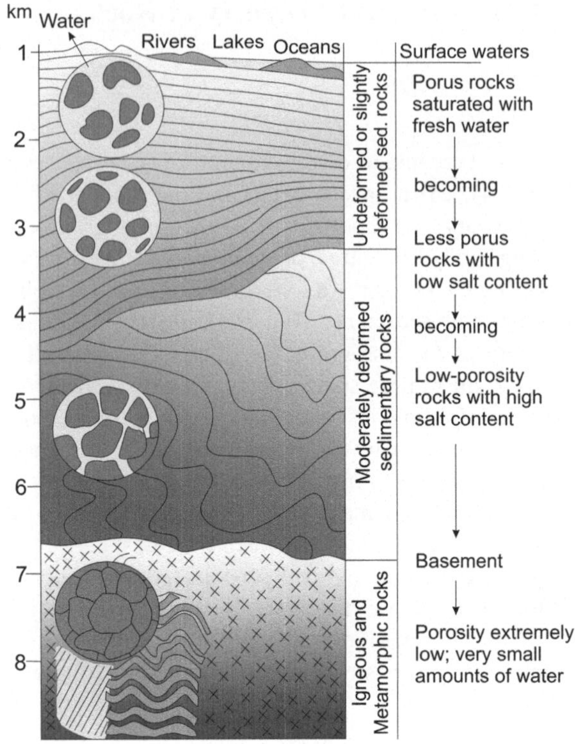

Fig. 9.5 Relationship of porosity with depth

greatly reducing porosity. Porosity also varies with depth and gradually decreases (Fig. 9.5).

All rocks are cut by fractures, and in some, like Granites and Quartzites, the fractures are the only significant pore spaces available (Fig. 9.4). In a Limestone (or in a carbonate territory), solution activity removes soluble materials resulting in pits and holes. Here, as water moves along these joints and bedding planes solution activity enlarges fractures and develops passageways that grow to become sinkholes, caves, and caverns (Fig. 9.6). In volcanic rocks such as Basalts, vesicles are formed by trapped gas bubbles that significantly affect porosity. Vesicles are often concentrated near the top of a lava flow and form zones of very high porosity. These zones can be interconnected by columnar joints or through voids in cinders and rubble at the top and base of the flow. This greatly increases the available porosity of the rock.

9.3.2 Permeability

The degree with which pore spaces are interconnected is Permeability (Fig. 9.7). Hence, permeability measures the rate of water passage through rocks, while

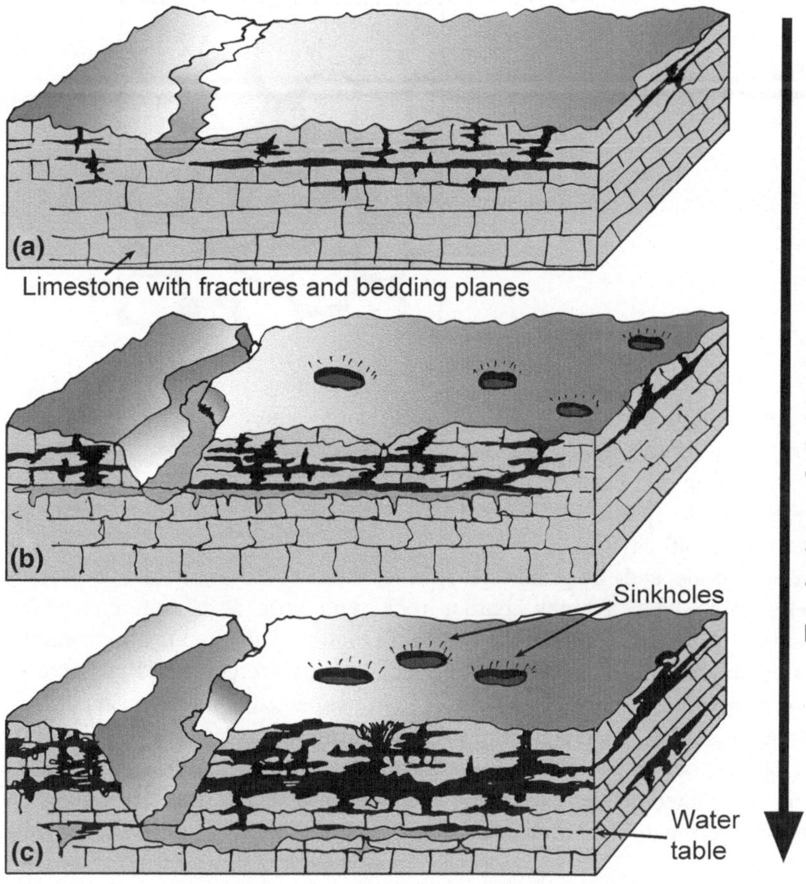

Fig. 9.6 In a Limestone, the solution activity removes soluble materials. As water moves along these joints and bedding planes solution activity enlarges fractures and develops passageways that grow to become sinkholes, caves and caverns. **a** Early stage: The water seeps through fractures and bedding planes. Groundwater seeps downward to the water table and then moves toward the surface streams. Soluble minerals are dissolved and the flow paths of the groundwater become enlarged. **b** Intermediate stage: Surface streams erode the valley floor. Water table drops. Surface water seeps through the zone of Aeration and enlarges the existing joints and caves. A system of horizontal caverns is developed. Sinkholes start to develop. **c** Mature stage: River erodes a deeper valley. Water in the underground channels seeks a new path to this lowered river level. Early stage starts all over again leading to a new, lower system of horizontal caverns. The older, higher caverns continue to enlarge and finally collapse to form sinkholes, or cave deposits. Redrawn from Earth's Dynamic Systems with permission of Eric H. Christiansen

porosity is a measure of open spaces in rock volume. The size of the interconnections further facilitates increased permeability. Low porosity usually results in low permeability, but high porosity does not necessarily imply high permeability. A rock is permeable if the fluids pass through it (Fig. 9.7), and it is impermeable if the flow is negligible. Rocks that have high permeability are conglomerates,

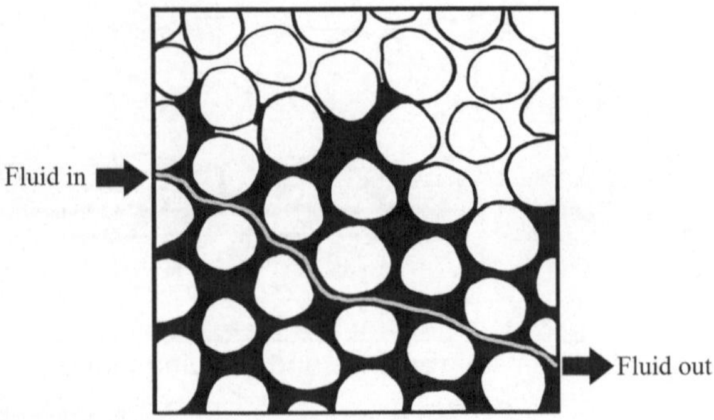

Fig. 9.7 Permeability. A rock is permeable if the fluids pass through it (i.e., fluid in and fluid out), and it is impermeable if the flow is negligible

sandstones, basalt, and certain limestones. Permeability in sandstones and conglomerates is high because of the relatively large, interconnected pore spaces between grains. Basalt is permeable because it is often extensively fractured by columnar jointing and because the top part of most flows is vesicular. Fractured limestones are also permeable, as are limestones in which solution activity has created many small cavities.

Permeability depends on grain size, sorting, shape of the grain, and its packing. It is possible to have a highly porous rock with little or no interconnections between pores. A good example of a rock with high porosity and low permeability is a vesicular volcanic rock, where the bubbles that once contained gas give the rock high porosity, but since these holes are not connected to one another, the rock has low permeability. Coarse-grained rocks are usually more permeable than fine-grained ones. Rocks that have low permeability are shale, unfractured granite, quartzite, and other dense, crystalline metamorphic rocks. Table 9.3 gives the porosity and permeability of major sediments and rocks.

Table 9.3 Porosity and permeability of sediments and rocks

Sediment		Porosity (%)	Permeability
Gravel		25–40	Excellent
Sand		30–50	Good to excellent
Silt		35–50	Moderate
Clay		35–80	Poor
Glacial till		10–20	Poor to moderate
Rock			
Conglomerate		10–30	Moderate to excellent
Sandstone	Well-sorted with little cement	20–30	Good to very good
	Average	10–20	Moderate to good
	Poorly sorted and well cemented	0–10	Poor to moderate
Shale		0–30	Very poor to poor
Limestone, Dolomite		0–20	Poor to good
Cavernous limestone		up to 50	Excellent
Crystalline rock	Unfractured	0–5	Very poor
	Fractured	5–10	Poor
Volcanic rock		0–50	Poor to excellent

9.4 Aquifers and Aquicludes

The ability to hold and supply groundwater enables their categorization into Aquifers and Aquicludes.

9.4.1 Aquifers

An Aquifer is a large body of permeable material where groundwater is present in the saturated zone. Hence, it is a water-bearing rock. Good aquifers have high permeability. Examples include poorly cemented sands, gravels, and sandstones or highly fractured rocks. Aquifers are of two types—Unconfined and Confined (Fig. 9.8a).

9.4.1.1 Confined Aquifers

They occur when an aquifer is confined between two layers of impermeable strata (Fig. 9.8). A special kind of confined aquifer is an Artesian system. These are desirable as they result in free flowing artesian springs and artesian wells.

An Artesian well (Fig. 9.8a) forms when an aquifer is slanted between two aquitards (beds of low permeability along an aquifer) such that groundwater rises above the level of the aquifer and if the opening is below the pressure surface (the level of the water table in the recharge region), the well will flow freely. When the water table intersects the earth's surface and flows out of the ground, it forms a Spring. Spring forms because the aquitard blocks the downward movement of groundwater and forces the water to move laterally (Fig. 9.8b), an area where the water table intersects the surface and the water flows out of the ground. Alternatively, springs occur when an impermeable rock (called an aquiclude) intersects a permeable rock that contains groundwater (an aquifer). Such position between permeable and impermeable rock can occur along geological contacts and fault zones. When the water table intersects at the earth's surface for a large region, it forms a Swamp (Fig. 9.8b).

9.4.1.2 Unconfined Aquifers

The water table is exposed to the earth's atmosphere through the Zone of Aeration (Fig. 9.8b). It is the most common type of aquifer.

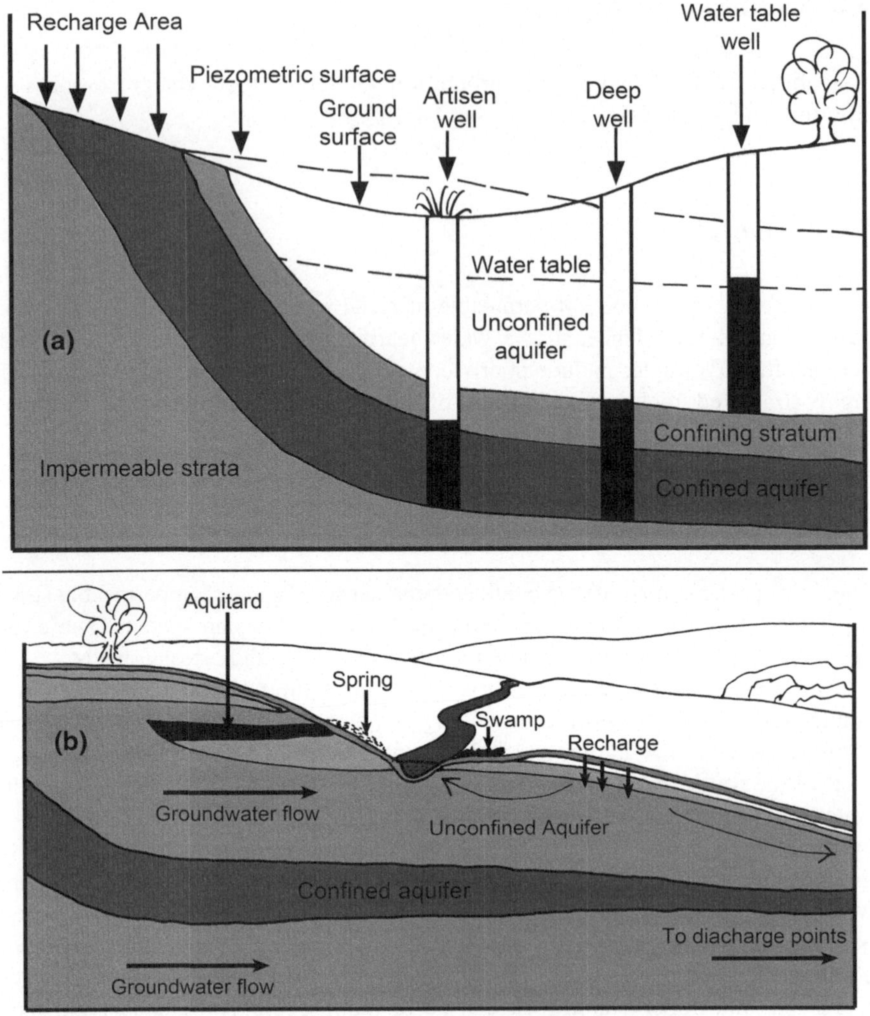

Fig. 9.8 Unconfined and Confined aquifers. **a** A schematic cross-section of an unconfined and confined aquifers, **b** Aquitard, Spring, Unconfined and Confined aquifers

9.4.2 Aquicludes

They have little or no porosity or permeability and hence, are water-excluding. Clay is a good example of this type.

Fig. 9.9 shows various conditions for an ideal Aquifer, Aquitard, and an Aquiclude.

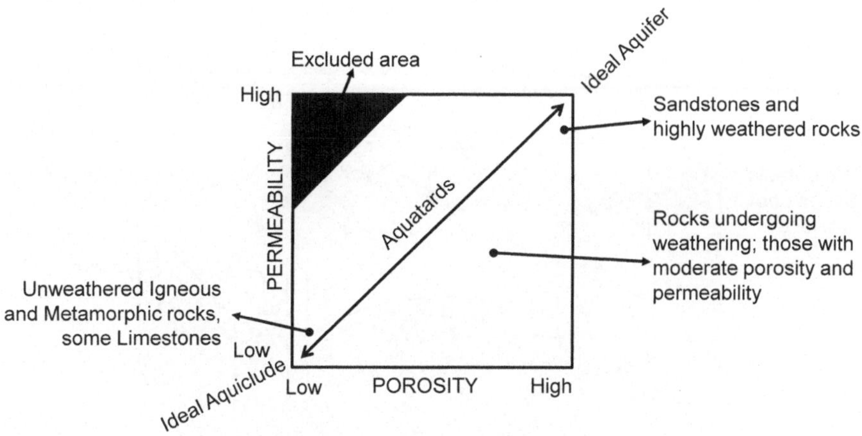

Fig. 9.9 Figure shows the optimum operating conditions for an Aquifer, Aquitard, and an Aquiclude

9.5 The Water Table

It is the top of the Zone of Saturation (Fig. 9.3). Below this point, all of the pore spaces are filled with water. Water table is not flat; it is higher under hills and lower under valleys. Where the water table intersects (or lies above) the ground surface, springs, lakes, swamps, or rivers are formed (Fig. 9.8). The position of the water table may fluctuate with droughts. During dry seasons, the depth to the water table increases (goes down from the surface), but, during wet seasons, the depth decreases (i.e., it lies closer to the surface).

9.6 Groundwater Movement

A small fraction of rainwater adheres to grains in the soil by molecular attraction, and some of it is absorbed by plant roots, while some seeps down into the saturated zone. Here, in the saturated zone, groundwater is in constant motion. It migrates through pores within sediments and moves downward under the force of gravity and upward toward the zones of lower pressure. Porosity and Permeability of the rock determines the rate of groundwater flow. It is very slow in rocks that possess very low porosity and permeability (such as Granites) and it is high in high porosity and permeability sediments such as Sand (Fig. 9.4).

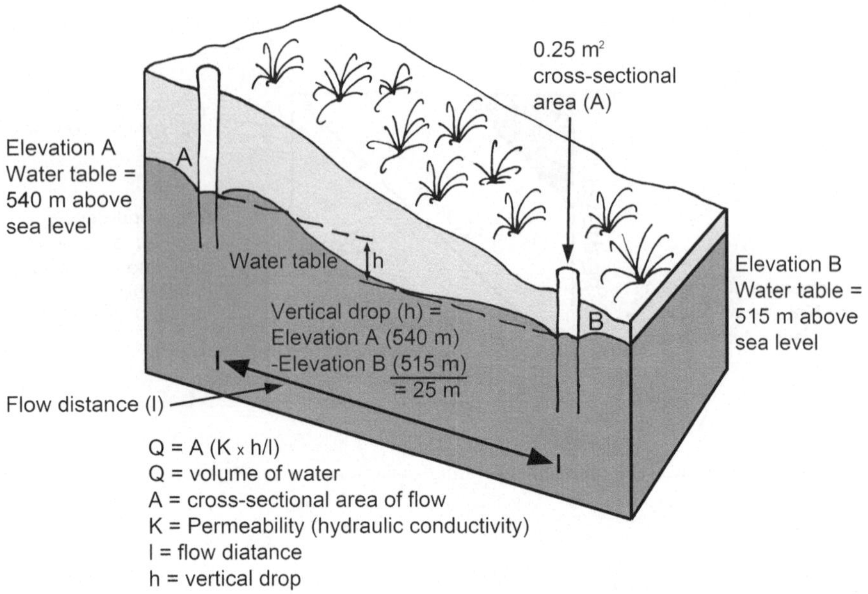

Fig. 9.10 The Darcy's law provides an accurate description of the flow of ground water in almost all hydrogeological environments

9.6.1 Darcy's Law

The Permeability of the rock and the Hydraulic gradient determine the rate at which groundwater moves through the saturated zone. The latter is defined as the difference in elevation divided by the distance between the two points on the water table (Fig. 9.10). For almost all hydrogeological environments, the flow of ground water is best explained by Darcy's law.

9.7 Geologic Activity of Groundwater

9.7.1 Caves, and Caverns

When, at the subsurface level, large areas of limestone are dissolved by the action of groundwater, they form Cavities. These eventually become larger and develop into Caves and Caverns; both may possess many interconnected chambers (Fig. 9.6). Once a cave is formed, it is open to the atmosphere and the water percolating through it can precipitate new materials like Stalactites (those hanging from the ceiling) and Stalagmites (those growing from the floor upwards) (Fig. 9.11). Both of these eventually unite to form Columns (Fig. 9.11). Water percolating from a

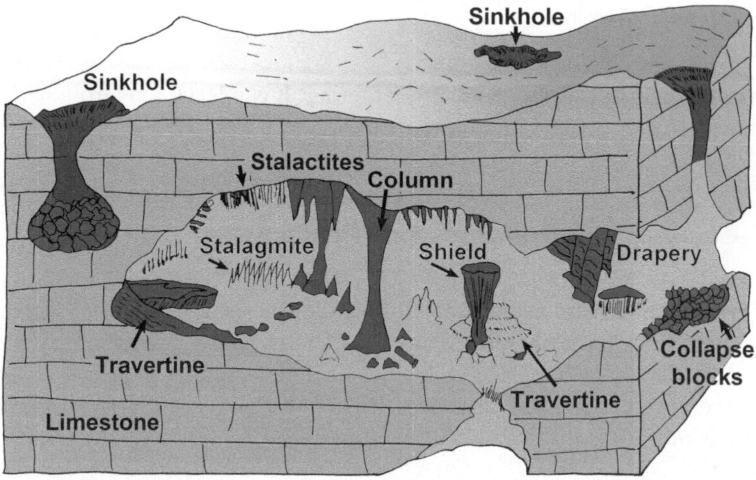

Fig. 9.11 Cave deposits. Many varieties of cave deposits are shown in this idealized diagram. Most are composed of calcite deposited by water that seeps into the open cave and then loses carbon dioxide as the water evaporates. Stalactites (icicle-shaped pendants hanging from the ceiling) and Stalagmites (cone-shaped deposits growing up from the floor) are impressive deposits formed by the dissolution of carbonate rocks by weakly acidic groundwater. As a stalactite grows downward and a stalagmite grows upward, they eventually join to form a Column. Water percolating from a fracture along a slanting ceiling may form a thin, vertical sheet of rock known as Drapery (named so because of its shape). Pools of water on the cave floor flow from one place to another, and as they evaporate, calcium carbonate is deposited on the floor, forming terraces made of travertine—a layered cave or hotspring rock composed of calcium carbonate. Redrawn from Earth's Dynamic Systems with permission of Eric H. Christiansen

fracture along a slanting ceiling may form a thin, vertical sheet of rock called Drapery (Fig. 9.11). Pools of water on the cave floor flow from one place to another, and as they evaporate, leading to the deposition of calcium carbonate ($CaCO_3$) on the floor. These deposits form terraces made of Travertine—a layered cave or hot spring rock composed of $CaCO_3$ (Fig. 9.11). Groundwater dripping from a cave's roof contains more CO_2 than the surrounding air. In an attempt to reach equilibrium with the air, CO_2 diffuses out of the water droplet. This diffusion reduces the amount of carbonic acid as well as the amount of calcite that can be dissolved in the water. As a result, the water becomes saturated with $CaCO_3$ and precipitates out. These deposits are collectively called Speleothems (= Cave deposits).

9.7.2 Sinkholes

If the roof of a cave or cavern collapses, it forms a Sinkhole (Fig. 9.11). Sinkholes are common in regions underlain by Limestones. For more details on Sinkholes, see Chap. 10.

9.7.3 Dissolution

Water is the main agent of chemical weathering. It can leach ions from rocks, and, in carbonate rocks (as in Limestone), it dissolves it completely.

9.7.4 Chemical Cementation and Replacement

Water is also the primary agent acting during diagenesis. It carries in dissolved ions that precipitate to form chemical cements thereby holding sedimentary particles together. Water can also replace molecules on a molecule-by-molecule basis, thus, often preserving the original structure. This is best noted in the fossilization of a petrified wood.

9.8 Karst Topography

In regions where the primary type of weathering is dissolution (as in Limestone terrains), the formation of caves, caverns, and sinkholes, their collapse and coalescence, and the presence of cave deposits and disappearing streams together

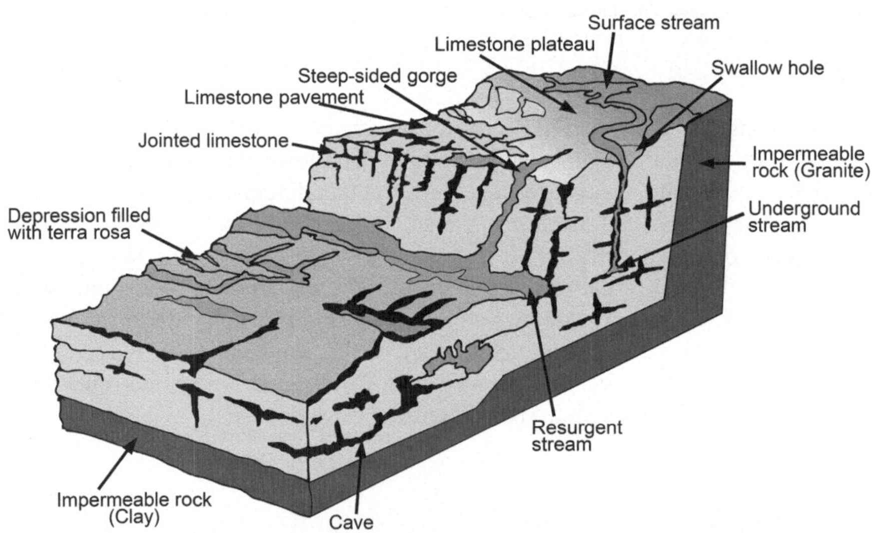

Fig. 9.12 Karst topography develops largely by groundwater erosion in areas underlain dominantly by limestone and other readily soluble rocks. The name Karst is derived from the plateau region of the border area of Slovenia, Croatia, and northeastern Italy where this type of topography is well developed. In the United States, regions of karst topography include large areas of Illinois, Indiana, Kentucky, Tennessee, Missouri, Alabama, and Florida. The Carlsbad Caverns in New Mexico (USA) it has the largest cave which is taller than the U.S. Capitol building and can accommodate 14 football fields. For details on Karst topography, see Chap. 10

form a highly irregular topography called Karst topography (Fig. 9.12). Details of this topography are given Chap. 10.

9.9 Summary

Groundwater originates as rainfall or snow, and then moves through the soil into the groundwater system, where it eventually makes its way back to surface as streams, lakes, or oceans. Groundwater makes up about 0.64 % of the total water on earth. Transpiration allows the recycling of small amount of water that infiltrates into the ground, back into the air almost immediately. The rest of the water either stays, or slowly works its way down to the Vadose zone (the Zone of Aeration). Here, the pores between rocks and minerals are not filled with water. But this is so, in the zone below, the Zone of Saturation. The water in the Zone of Aeration is called the Suspended or Vadose water and is divided into three subzones namely Soil water (moisture) zone, Intermediate zone, and the Capillary zone. The water in the Zone of Saturation is known as Groundwater. The upper part of the zone is called the Water table.

The hydrological property of rocks is a function of its Porosity and Permeability. Porosity is the percentage of the volume of open spaces (pore spaces) within the rock. The open spaces determine the amount of water that a rock can hold. In sedimentary rocks, porosity is a function of grain size, shape of the grain, degree of sorting, and the degree of cementation. In Igneous and Metamorphic rocks, porosity is usually low as most minerals tend to intergrow, leaving little free spaces. Highly fractured Igneous and Metamorphic rocks, however, have high porosity. There are four main types of pore spaces or voids in rocks—spaces between mineral grains, spaces between fractures, spaces between solution cavities, and spaces between vesicles. Permeability is the degree with which pore spaces are interconnected. Hence, permeability measures the rate of water passage through rocks, while porosity is a measure of open space in rock volume. Permeability depends on grain size, sorting, shape of the grain, and its packing.

The ability to hold and supply groundwater enables categorization into Aquifers and Aquicludes. An Aquifer is a large body of permeable material where groundwater is present in the saturated zone. Hence, it is a water-bearing rock. There are two types of Aquifers: Unconfined and Confined. Aquicludes have little or no porosity or permeability and hence, are water-excluding. An Artesian well forms when an aquifer is slanted between two aquitards (beds of low permeability along an aquifer) such that groundwater rises above the level of the aquifer and if the opening is below the pressure surface (the level of the water table in the recharge region), the well will flow freely and vice versa if the opening is above the pressure surface. When the water table intersects the earth's surface and flows out of the ground, it is called a Spring. When the water table intersects at the earth's surface for a large region, it forms a Swamp.

The rate at which groundwater moves through the saturated zone is determined both by the permeability and the Hydraulic gradient of the rock. The latter is defined as the difference in elevation divided by the distance between the two points on the water table. For almost all hydrogeological environments, the flow of ground-water is best explained by Darcy's law.

The geologic activity of groundwater creates unique landforms. These are caves and caverns, sinkholes, dissolution structures, chemical cementation and replacement, and the Karst Topography.

References

Hamblin WK, Christiansen CH (2008) Earth's Dynamic Systems. Printice Hall, New Jersey, Web Edition 1.0, 767p

Price M (1996) Introducing groundwater. Chapman and Hall, London, 278p

Chapter 10
Karst

10.1 Introduction

The Karst region is characterized by a landscape that is dotted with sinkholes, underlain by caves and with many large springs that discharge into stream valleys. It is a terrain with distinctive characteristics of relief and drainage arising primarily from the solution of soluble bedrock by natural waters (Fig. 10.1) (see also Sweeting 1973, 1981; Jennings 1985; Ford 1988; Ford and Williams 2007; Palmer 2009). Rainwater (and snowmelt) seeps through a relatively thin soil cover, into fractured and soluble bedrock such as a limestone or a dolostone and the water moves through these fractured rocks, slowly dissolving and enlarging pathways along fractures and bedding planes (Fig. 10.2). Once these underground drainage pathways have been established in the bedrock, surface-water drainage is diverted underground (Fig. 10.2). Hence, Karst areas often lack the network of surface streams as noted in most other terrains. The surface runoff drains into sinkholes and flows through solution-enlarged conduits (caves) in the underlying rock. But, eventually the waters do return to the land surface, as springs (Fig. 10.3).

Karstification, is thus, defined as the differential chemical and mechanical erosion by water on soluble bodies of rock, such as limestone, dolomite, gypsum, or salt (see also Waltham 1971; Sweeting 1973, 1981; Hamblin and Christiansen 2008). Regions with thick and fractured rocks and those that possess pure limestones in a humid environment excellently exhibit the process of Karstification. Here, the resulting Karst morphology is characterized by dolines (sinkholes), hums (towers), caves, and a subsurface drainage system that is quite complex (Figs. 10.3 and 10.4). The Karstified limestones cover approximately one-tenth of the land surface of the earth, and around 25 % of the world's population lives in these regions (which are mostly in southern China, large areas of Central and Southern Europe and much of Central America). It must be noted that only about 20 % of earth's surface has major limestone sequences exposed at the surface and so the development of Karst topography is limited to these regions.

S. Jain, *Fundamentals of Physical Geology*, Springer Geology,
DOI: 10.1007/978-81-322-1539-4_10, © Springer India 2014

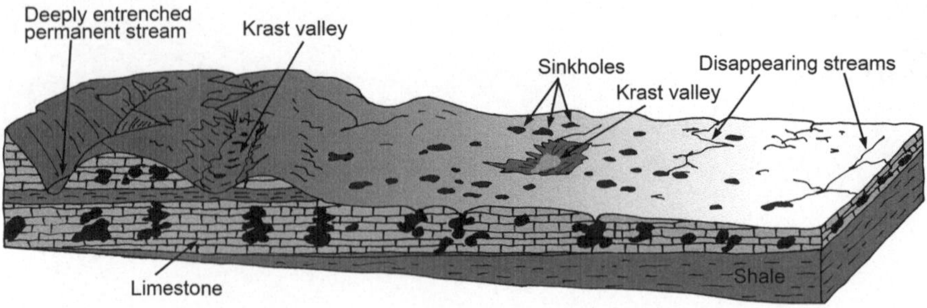

Fig. 10.1 Various features of Karst topography

10.2 Factors Affecting Karst Development

The development of a Karst terrain is a function of several factors in varying degrees. These include lithology, structure, topographic relief, hydrology, climate, and vegetation. Each are discussed briefly (see also Trudgill 1985; White 1988).

10.2.1 Lithology

Lithology largely determines where Karstification occurs. Karst development is primarily affected by the carbonate content or other soluble mineral content of the rock, the mechanical strength of the rock, and the absence of primary porosity within a rock unit. Besides these, the composition and thickness of individual beds, the nature of interbeds (especially the ones with shale), and lateral facies variations affect both the style and degree of Karstification. Thus, Karst morphology is a function of the solubility, composition, porosity, and thickness of the rock. Karst morphology development is somewhat reduced in carbonate rocks that have high primary porosity or possess high proportions of insoluble minerals. However, such rocks are likely to have some degree of dissolution enlargement of joints, bedding planes, and other voids and may include areas with considerable Karst development. Although, Karstification dominates in limestones and dolomites but several other soluble rocks such as rock salt and gypsum are also susceptible for Karst development.

10.2.2 Structure

Structure and stratigraphy determine the spatial arrangement of Karst and non-Karst rocks.

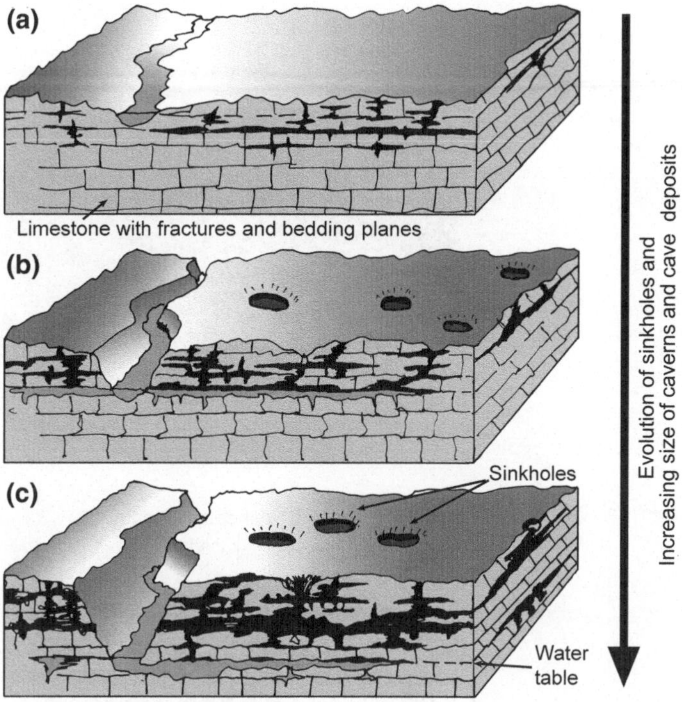

Fig. 10.2 Cave development. In a limestone, the solution activity removes soluble materials. As water moves along these joints and bedding planes solution activity enlarges fractures and develops passageways that grow to become sinkholes, caves, and caverns. **a** Early stage: The water seeps through fractures and bedding planes. Groundwater seeps downward to the water table and then moves toward the surface streams. Soluble minerals are dissolved and the flow paths of the groundwater become enlarged. **b** Intermediate stage: Surface streams erode the valley floor. Water table drops. Surface water seeps through the zone of aeration and enlarges the existing joints and caves. A system of horizontal caverns is developed. Sinkholes start to develop. **c** Mature stage: River erodes a deeper valley. Water in the underground channels seeks a new path to this lowered river level. Early stage starts all over again leading to a new, lower system of horizontal caverns. The older, higher caverns continue to enlarge and finally collapse to form sinkholes, or cave deposits. Redrawn from Earth's Dynamic Systems with permission of Eric H. Christiansen

10.2.2.1 Fracture

Limestones and dolomites at or near the surface tend to deform by brittle fracture. This tendency to form complex joint sets is directly responsible for the secondary permeability required for the development of subsurface solution drainage and subsequent cave developments. Vertical fractures usually manifest themselves at the surface and focus the solution processes along them.

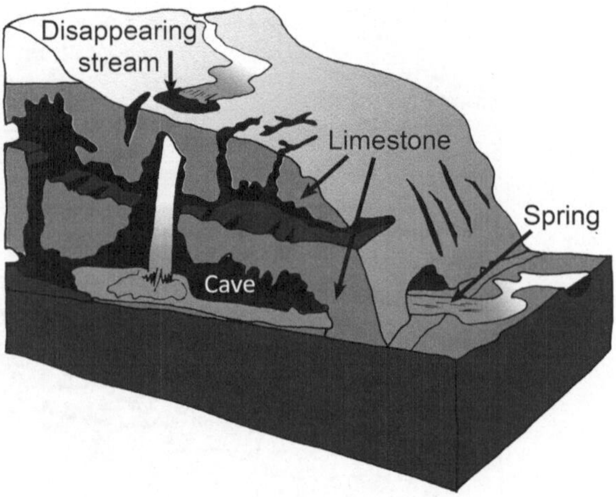

Fig. 10.3 Karst topography. Sinkholes and caverns form in limestone. Streams commonly disappear into sinkholes and flows through caverns to emerge elsewhere

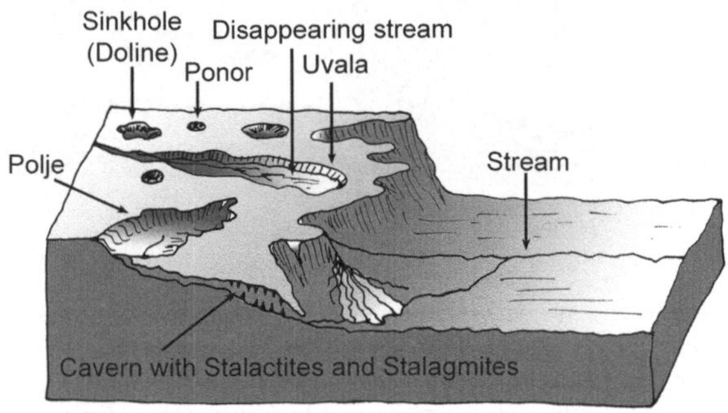

Fig. 10.4 Karst landforms

10.2.2.2 Folds

The direction of the ground-water flow is determined by regional fold systems. Water flow is more constrained in down-dip or along the strike of the fold system.

10.2.2.3 Faults

Regional folding generally accompanies numerous smaller faults. These offset confining units thereby allowing hydraulic connection between stratigraphically separate Karst aquifers.

10.2.2.4 Large-Scale Structures or Tectonics

These not only form Poljes (large, flat-floored depressions within the Karst limestone; Fig. 10.4) but also influence the rate and degree of Karstification. Most Poljes are associated with boundary faults.

10.2.3 Topographic Relief

Lithology and geologic structures outwardly manifest as topographic relief and is the elevation difference between the highest and lowest points on the surface for a given area. The hydraulic gradient drives the water through the aquifer. Steep topography or entrenched rivers produce high hydraulic gradient. Hence, greater the driving force, greater the potential for Karst development. This is assuming that both geologic and climatic factors remain constant.

10.2.4 Hydrology

Karstification not only involves the solution of carbonate rocks such as limestones but also involves mechanical erosion. The dissolution of carbonate rocks involves three components—CO_2, water, and $CaCO_3$. Atmospheric CO_2 diffuses into the moisture within the air or soil and simultaneously becomes hydrated to form carbonic acid: $CO_2 + H_2O <-> H_2CO_3$ (carbonic acid). For limestones, carbonic acid dissolves Calcite ($CaCO_3$).

$$\underset{\text{Limestone}}{\mathrm{CaCO3 + CO2 + H2O}} \quad < - > \quad \underset{\text{dissolved Limestone}}{\mathrm{Ca^{++} + 2HCO_3-}}$$

Thus, water undersaturated with respect to dissolved carbonate is capable of dissolving limestones. Most limestone dissolution results from an influx of fresh aggressive water. Dissolution often occurs within minutes to a few hours. Additional acids, such as organic acids from the soil and sulfuric and nitric acids from the acid rain also contribute to the dissolution process of carbonate rocks. This dissolution process is episodic in nature with maximum peaks corresponding to surges in the Karst water flow. These flood events can be from a single-storm or seasonal. Macro-porosity within the soil such as cracks, root channels, animal burrows, and other visually discernible openings with dimensions on the order of 0.1 cm to several centimeters also affect the dissolution process. Such openings are present in all soils and in most unconsolidated materials. Plant growth, animal activity, shrinking, and swelling in response to changes in temperature or moisture content also leads to the formation of macropores, hence accelerating Karst development.

10.2.5 Climate

Both Tropical and Temperate environments are ideal grounds for Karstification. The rates are intense in Tropics as higher temperatures lead to both greater rainfall and CO_2 production in soils. Cooler waters absorb larger quantities of CO_2 and hence, they have a higher potential to become more acidic. The 10 °C water dissolves twice as more CO_2 than the water at 30° C. If the water temperature rises from 0° to 35 °C, the CO_2 saturation level decreases from one-third to two-thirds. Therefore, waters in cold climates hold greater quantities of CO_2. However, in cold regions there is much less CO_2 available to dissolve in the water due to low biochemical activity. High temperatures increase biochemical activity, so more CO_2 and organic acids are formed in tropical or temperate environments.

However, more important than temperature is the partial pressure of CO_2. Water under pressure can dissolve more CO_2 and therefore can hold more calcium carbonate ($CaCO_3$) in solution. The release of pressure will result in the deposition of $CaCO_3$ from water previously under hydraulic pressure. The pressure of CO_2 at sea level is relatively constant at 0.035 %. The partial pressure of CO_2 in soil air can exceed that of the atmosphere from ten to several hundred times. Therefore, the presence and nature of soil cover is a far more potent factor in contributing to the intensity of Karstification, rather than atmospheric air.

10.2.6 Vegetation

Often, where vegetation is dense, highest concentration of moisture, biogenic activity, and production rates of CO_2 occur. Under these conditions and with free water circulation, the solution erosion rates are highest. The organic wastes produced by birds and bats, strongly corrode limestones. The humus generated by vegetation, increases the acidity of the water and the partial pressure of CO_2, thus, releasing organic acids. The algae generate acid solvents and thus, produce an intricately pitted, sharp-edged topography in limestone terrains called Phytokarst.

10.3 Karst Landforms

10.3.1 Sinkholes (Dolines)

It is a naturally occurring, cone or bowl-shaped depression formed because of the collapse of soil cover into a crevice in the underlying bedrock, or the collapse of a cave roof and its overlying rock and soil cover into the cavity below (Fig. 10.5). The sinkholes are divided into various types (Fig. 10.5).

10.3.1.1 Solution Sinkholes

These are funnel shaped sinkholes formed by solution along a joint or along the intersection of several joints. Regolith drapes the floor of these sinkholes (Fig. 10.5a).

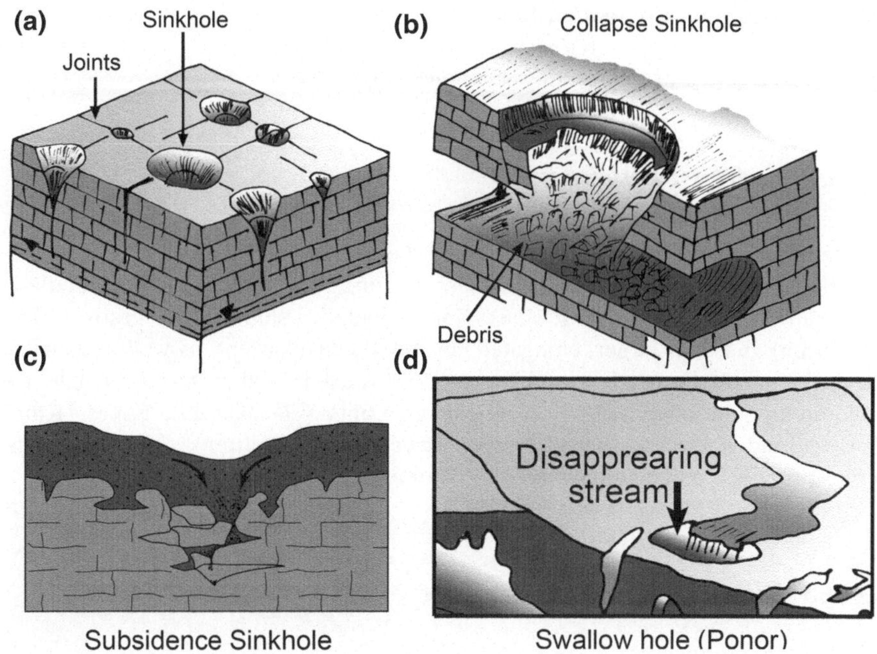

Fig. 10.5 Types of sinkholes

10.3.1.2 Collapse Sinkholes

These are steep sided sinkholes formed by subterranean corrosion. Large cave forms, overlying rock no longer is supported and eventually collapses (Fig. 10.5b).

10.3.1.3 Subsidence Sinkholes

These are similar to solution sinkholes (dolines) but the overlying soil has washed into the cave system (Fig. 10.5c).

10.3.1.4 Swallow Hole (Ponors)

These are limestone sinks into which a stream disappears (Fig. 10.5d).

10.3.1.5 Subjacent Karst Collapse Sinkholes

These sinkholes are formed by the collapse of the overlying noncarbonate rocks into a limestone cavern.

10.3.1.6 Uvaluas (Compound Sinkholes)

These are a series of intersecting sinkholes (Fig. 10.4).

10.3.2 Karren Features

Karren is a general term used to describe the total complex of superficial micro-solutional features of soluble rocks (such as limestone and gypsum), particularly on limestone pavements (Fig. 10.6). Karren includes sharp-ridged grooves (Rillenkarren) and their larger, elongated cousins (Rinnenkarren), as well as rounded Runnels formed beneath a soil cover (Rundkarren). Other forms include the ubiquitous solutional hollows (Kamenitsa), clints (Flachkarren), grikes (Kluftkarren), and horseshoe-shaped stepped structures (Trittkarren). Of these, clints, grikes, runnels, pits, and pans are quite common.

10.3.3 Clints and Grikes

Limestone is characterized by divisions into blocks called Clints (Fig. 10.7) and these are bounded by deep vertical fissures called Grikes (Fig. 10.7). Clints and grikes form under relatively deep soil cover where water, carrying carbonic acid (formed from the dissolved CO_2 as well as organic acids from decaying vegetation), picks out vertical lines of weakness (joints) in the rock. Over time, these fissures widen as the acidic water preferentially attacks along the lines of maximum weakness. Grikes take many thousands of years to form as the rate of solution is slow. Some of the material lost into the Grikes is washed deep into the drainage systems of the limestone pavements through connecting fissures, often leaving open Grikes that are as big as a meter or more in depth.

10.3.4 Runnels, Pits, and Pans

These take on varied forms depending on the structure of the limestone pavement on which they form. Runnels are gutter like channels eroded out of the surface of

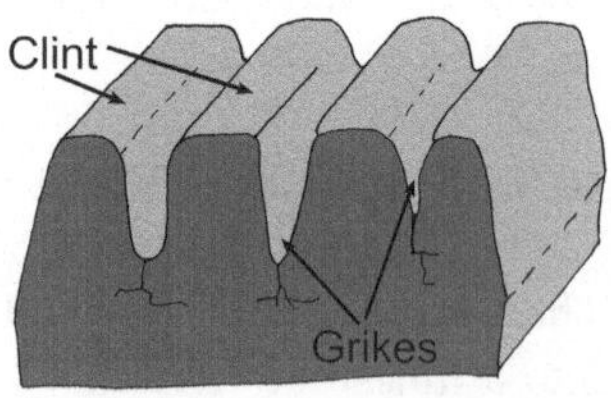

Fig. 10.6 Clint and Grikes

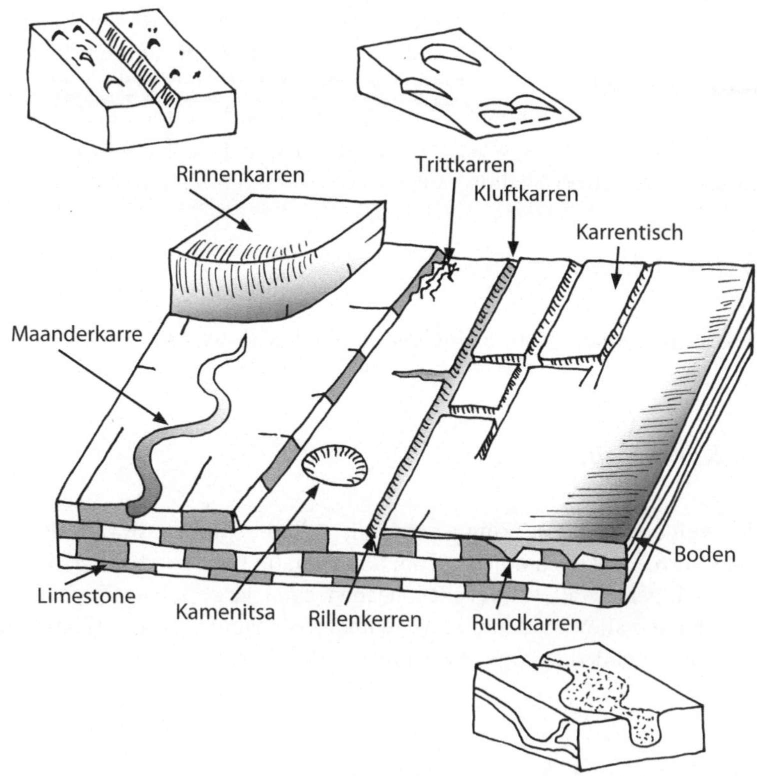

Fig. 10.7 Various Karren features

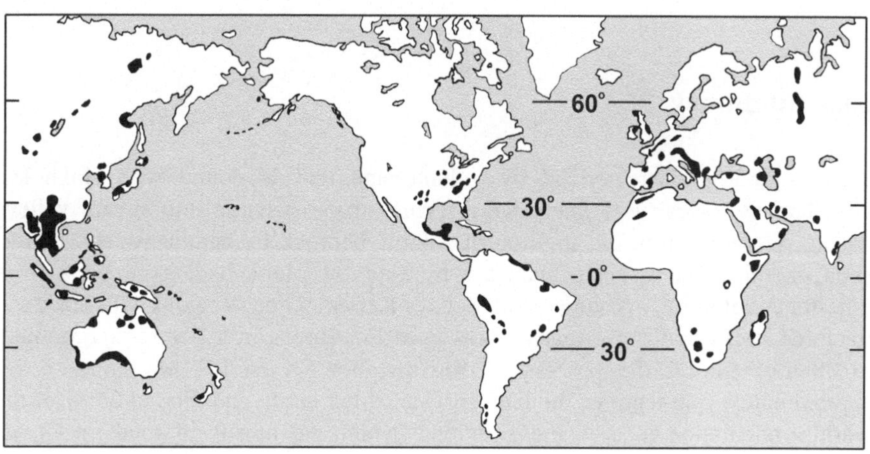

Fig. 10.8 Karst regions of the world

the limestone, which drain into Grikes. The formation of Runnels takes place under a shallow layer of soil. Pits and Pans are small-scale solution features (i.e. formed by water and acids dissolving the limestone) found on the tops of Clints. Pits are deep and free draining into the subterranean limestone drainage system. Pans are shallow, and possess an impervious base and hold water. Both of these features also form under shallow soil cover.

10.3.5 Karst Lakes

Karst lakes are formed when a sinkhole is filled with water.

10.3.6 Karst Valleys

A well-developed valley lacking a stream is called a Blind valley. A deep gorge-like valley formed as stream flows from a nonkarstic region into a karstic region is called as Allogenic valley. A valley headed by a large spring is called Pocket valley. A large valley or depressions with broad valley floors oriented along a tectonic trend is called a Poljes (see Fig. 10.4).

10.4 Major Karst Areas of the World

Major Karst areas found on all continents (Fig. 10.8) (except Antarctica; see also Herak and Stringfield 1972; Jennings 1985; White 1988) are listed in Table 10.1:

10.5 Summary

Karst region is characterized by a landscape that is dotted with sinkholes, underlain by caves, with many large springs that discharge into stream valleys arising primarily from the solution of soluble bedrock by natural waters. Differential chemical and mechanical erosion by water on soluble bodies of rock, such as limestone, dolomite, gypsum, or salt defines Karstification. Regions with thick and fractured rocks and those that possess pure limestones in a humid environment excellently exhibit the process of Karstification. Karstified limestones cover approximately one-tenth of the land surface of the earth, and around 25 % of the world's population lives in these regions (which are mostly in southern China, large areas of Central and Southern Europe and much of Central America). Only about 20 % of earth's surface has major limestone sequences exposed at the surface and so the development of Karst topography is limited to these areas.

Table 10.1 Major Karst areas of the world (except Antarctica)

Madagascar	Anjajavy Forest, western Madagascar
	Ankarana Reserve
	Madagascar dry deciduous forests, western Madagascar
	Tsingy de Bemaraha Strict Nature Reserve
Asia	Phong Nha Cave in Phong Nha-Ke Bang, Vietnam
China	Area around Guilin and Yangshuo
	Jiuzhaigou and Huanglong National Park
	South China Karst, World Heritage Site
	Stone Forest
	Zhangjiajie National Forest park
Georgia	Arabika Massif, Abkhazia, Georgia
Indonesia	Bantimurung
	Gunung Sewu
Israel	Ofra region, Israel
Japan	Akiyoshi Plateau
	Atetsudai and Taishakudai Plateaus
	Shikoku Karst
	Hiraodai Plateau
	Okinoerabujima Island and other islets of Nansei Islands
Laos	Vang Vieng
Lebanon	Dunnieh Mountains, North Lebanon
	Jeita Grotto
	Parts of Mount Lebanon
Malaysia	Gunung Mulu National Park
	Kilim Karst Geoforest Park, Langkawi
	Kinta Valley, Perak
	Perlis State Park, Perlis
	Batu Caves, Selangor
Philippines	El Nido, Palawan
	Coron, Palawan
	Sagada, Mountain Province
	Chocolate Hills, Bohol
	Negros and Gigante Islands, Negros Oriental
	Virac, Catanduanes
Poland	Polish Jura Chain (Jura Krakowsko-Częstochowska)
	Holy Cross Mountains (Góry Świętokrzyskie) with the Jaskinia Raj
	Tatra Mountains including the Jaskinia Wielka Śnieżna (the Great Snowy Cave)
Romania	Apuseni Mountains, Romania
Serbia	Dinaric Alps region
	MeroKarst of eastern Serbia
Scotland	Assynt, southeast Skye and near Kentallen in Scotland, United Kingdom
Slovakia	Slovak Paradise, Slovak Karst, and Muránska planina, Slovakia

(continued)

Table 10.1 (continued)

Slovenia	Region of Inner Carniola, Goriška, Upper Carniola, and Lower Carniola
	Kras, a plateau in southwestern Slovenia and northeastern Italy
Spain	El Torcal (Antequera—Spain)
	Picos de Europa and Basque mountains, northern Spain
	Larra-Belagua, Navarre, northern Spain
	Cadí mountain range, Spain
	Garraf Natural Park area, Spain
	Ciudad Encantada in the Cuenca province, Castilla-La Mancha
	El Torcal de Antequera nature preserve, southern Spain
Switzerland	19 % of the surface (~ 7,900 sq km) of Switzerland is Karst
Ukraine	Podolia and Bukovina regions in the northeastern edge of the Carpathian Mountains.
Wales	Southern region of the Brecon Beacons National Park, Wales, United Kingdom
North America: Canada	Marble Canyon, British Columbia
	Monkman Provincial Park, British Columbia
	Northern Vancouver Island, British Columbia
	Niagara Escarpment, Ontario
	Port au Port Peninsula, Newfoundland
	Nahanni region in the Northwest Territories
	Wood Buffalo National Park in Alberta and the Northwest Territories
	Avon Peninsula, Nova Scotia

The development of a Karst terrain is a function of several factors in varying degrees. These include lithology, structure (fracture, fold, faults, large-scale structures, or tectonics), topographic relief, hydrology, climate, and vegetation.

Landforms in a Karst terrain include: sinkholes (Dolines) [Solution sinkholes; Collapsed sinkholes; Subsidence sinkholes; Swallow hole (Ponors); Subjacent Karst collapse sinkholes and Uvaluas (Compound sinkholes)], karren features [Karren: Rillenkarren, Rinnenkarren, Rundkarren, Kamenitzas, Clints (Flachkarren), Grikes (Kluftkarren) and Trittkarren], Runnels, Pits and Pans, and Karst Lakes and Karst valleys.

References

Ford DC (1988) Characteristics of dissolutional cave systems in carbonate rocks. In: James NP, Choquette PW (eds) Paleokarst. Springer, New York, pp 25–57

Ford DC, Williams P (2007) Karst hydrogeology and geomorphology. Wiley, New York, ISBN 978-0-470-84996-5

Hamblin WK, Christiansen CH (2008) Earth's dynamic systems. Printice Hall, New Jersey, Web Edition 1.0, 767p

Herak M, Stringfield VT (1972) Karst, important karst regions of the northern hemisphere. Elsevier Publishing Company, New York, p 551

Jennings JN (1985) Karst geomorphology, 2nd edn. Blackwell, New York, ISBN 0-631-14032-8

Palmer AN (2009) Cave geology, 2nd edn. Cave Books, Dayton, ISBN 978-0-939748-66-2

Sweeting MM (ed) (1981) Karst geomorphology, benchmark papers in geology, vol. 59. Hutchinson Ross Publishing Company, Stroudsburg, pp. 1–448

Sweeting MM (1973) Karst landforms. Columbia University Press, New York. ISBN 0-231-03623-X, 362 p

Trudgill S (1985) Limestone geomorphology. White Plains, Longman, New York

Waltham AC (1971) Controlling factors in the development of caves. Trans Cave Res Group of Great Br 13:73–80

White WB (1988) geomorphology and hydrology of Karst terrains. Oxford University Press, New York, p 464 p

Chapter 11
Glaciers

11.1 Introduction

A glacier is a thick mass of ice moving downhill under the pull of gravity. It is an accumulation of large quantities of ice, air, water, and sediments (rock debris) that flows with gravity due to its own massive mass. Glaciers make up the Cryosphere, the part of the earth that remains below the freezing point of water and covers about 10 % of the earth's land surface (about 15 million km^2; Hambrey 1994; French 1996; Ben and Evans 1998; Anderson 2004). On a global basis, glaciers cover one-tenth of the earth's surface, and store 68 % of the world's freshwater supply (Fig. 11.1). If all of the global supply of land ice locked up in glaciers were to melt, then sea level would rise $\sim$70 m.

The glacial ice can be as large as a continent, such as the ice sheet covering Antarctica or it can be a valley glacier, just filling a small valley between two mountains. Most glacial ice today is found in the Polar Regions, above the Arctic and Antarctic circles (in Iceland, the Arctic, Antarctica, Alaska, and Canada) (Flint 1971; Ives and Barry 1974; Hughes 1998; Martini et al. 2001). Glaciers flow very slowly, from tens of meters to thousands of meters per year. The fastest moving glacier is the Ilulissat (Jakobshavn), located in Greenland (Fig. 11.2) that moves $\sim$20 m in 1 day. The longest glacier in the world is the Lambert glacier and is located in Antarctica; it is about 65 km wide and 710 km long. Figure 11.3 illustrates the global distribution of glaciers.

11.2 Formation of Glaciers

Glaciers form where snowfall in the winter exceeds the amount of snow and ice that melts in the following summer as in areas of high elevation and or high latitudes (Fig. 11.4). The size and extent of glaciers are determined by the climate of a region. The balance between ice accumulation and melting is called the glacier mass balance. Accumulation that occurs high up on the glacier forms the

S. Jain, *Fundamentals of Physical Geology*, Springer Geology,
DOI: 10.1007/978-81-322-1539-4_11,

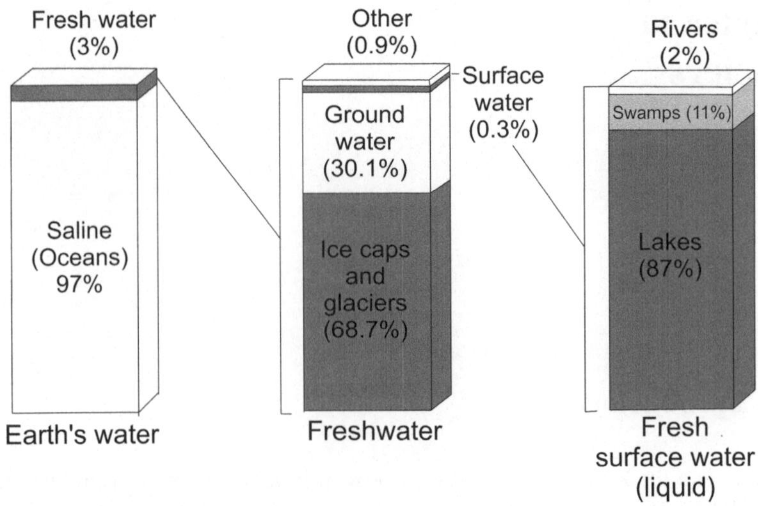

Fig. 11.1 Distribution of earth's water

Zone of Accumulation and where glacier melts, the Zone of Ablation (Fig. 11.5). The line that separates these two zones is called the Equilibrium line (Fig. 11.5). It is also called the Firn line. The Equilibrium line divides the glacier into distinct accumulation and ablation zones whose areas lie in a definite ratio to each other. Various ratios (Accumulation: Ablation) have been suggested for Alpine glaciers ranging from 8:9 to 3:4 to 2:3 (see also Meier 1962; Deynoux et al. 2004). In the accumulation zone, snow doesn't melt even during summers but as glacier move down, they eventually melt in the ablation zone. The elevation of the equilibrium line varies each year depending on the temperatures of that year and the amount of snowfall received. If a glacier has more accumulation than ablation for several years, then, the glacier advances. If more ablation occurs than accumulation, then, it retreats. At all times, ice is continually moving down the glacier, even when the terminus is stable for several years. The line above which, snow forms and remains round the year is called the Snowline and glaciers can form only at latitudes or altitudes above the snowline (Fig. 11.6). At present, the snowline lies at the sea level in Polar latitudes but rises up to 6,000 m in tropical regions (23.5° North to 23.5° South; see Fig. 11.4).

11.3 Glacier Movement

Glaciers move slowly. The rate of flow varies from 0.01 to 0.1 m per day for large Continental glaciers to 0.1–2 m per day for Alpine glaciers. During a period of rapid glacier movement, an alpine glacier can even flow at the rate of 50–100 m per day. Flow rates are greatest in the center of the ice mass, and minimum at rock

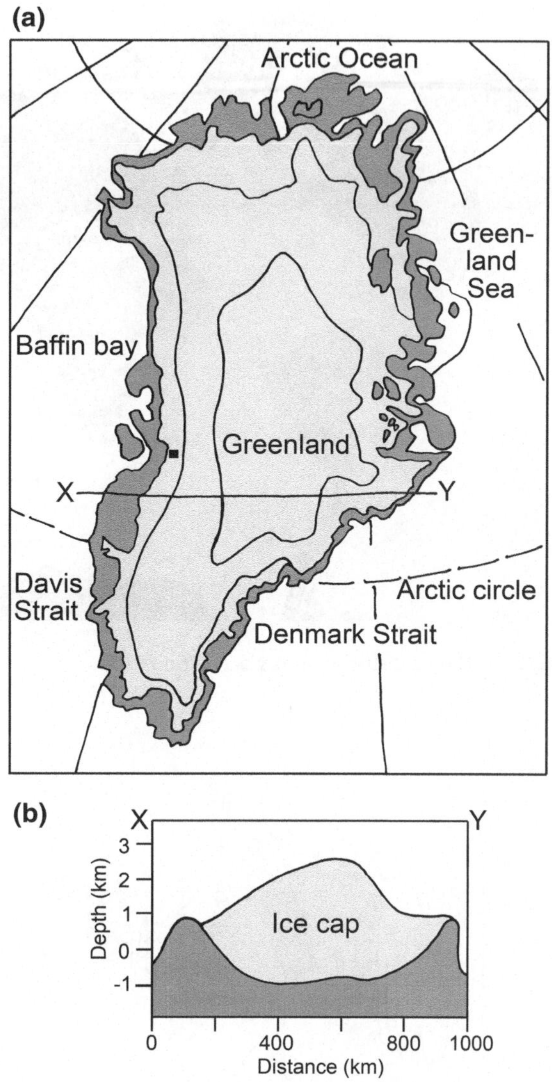

Fig. 11.2 The Greenland ice cap. The position of the fastest moving glacier in Greenland, the Ilulissat (Jakobshavn) is shown as a *solid black square* (center left). It moves ~20 m in a day

contacts. Glacier movement is by basal slip and plastic flow in the lower part (Fig. 11.7a), with upper brittle part riding on the lower. Striations and chatter marks (horseshoe-like indentations in the bedrock) are evidence of basal slip. As long as ablation (loss) of ice is less than accumulation, glaciers advance. The flow of glacial ice produces vertical to nearly vertical, wedge-shaped cracks called Crevasses (Fig. 11.7b). These may range in size from a few cm to over 10 m in width and up to about 40 m in depth. Glaciers move to lower elevations under the force of gravity by two different processes: Internal Flow (Plastic flow) and Basal slip (Sliding).

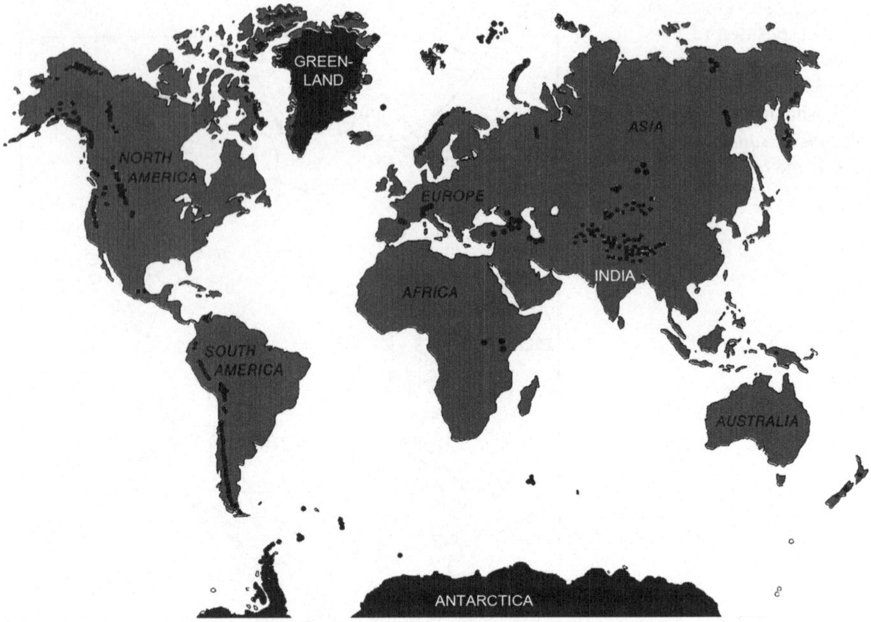

Fig. 11.3 Global distribution of glaciers (in *black*)

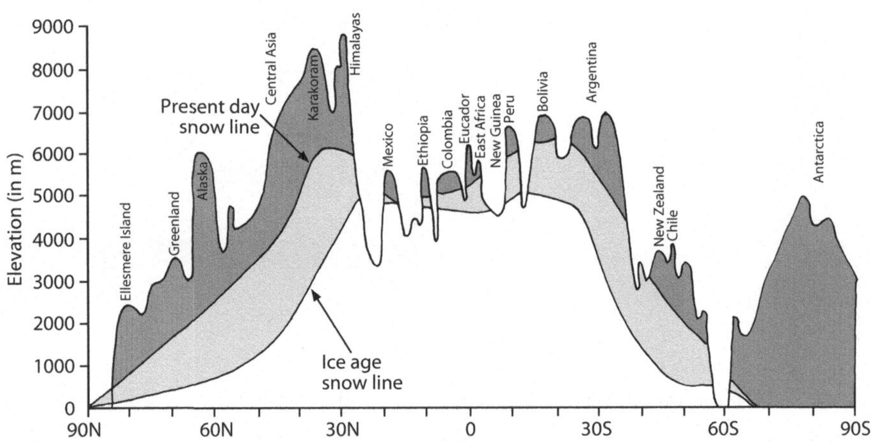

Fig. 11.4 Altitude of present and ice age snowlines

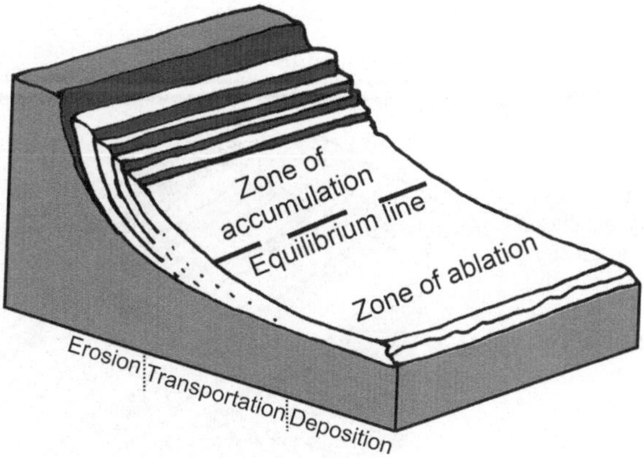

Fig. 11.5 Accumulation and ablation zones. Where accumulation of snow occurs it is called the accumulation zone and where glaciers melt it is called the ablation zone

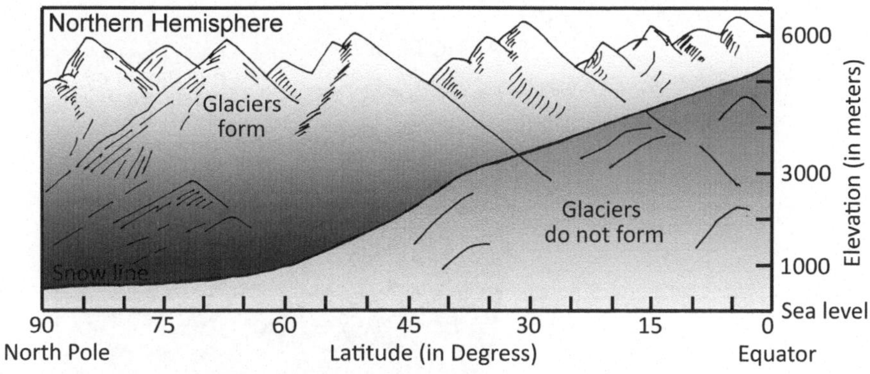

Fig. 11.6 The snowline. The line above which, snow forms and remains round the year is called the snowline and glaciers can form only at latitudes or altitudes above the snowline

11.3.1 Basal Slip (Sliding)

The melt water at the base of the glacier reduces friction by lubricating the surface and allows the glacier to slide across its bed (Fig. 11.7a). Polar glaciers are usually frozen to their bed and are thus, too cold for this mechanism to occur.

11.3.2 Internal Flow (Plastic Flow)

The ice crystals slide over each other like deck of cards, i.e., there is displacement between them. This type of movement generally occurs in polar glaciers, but

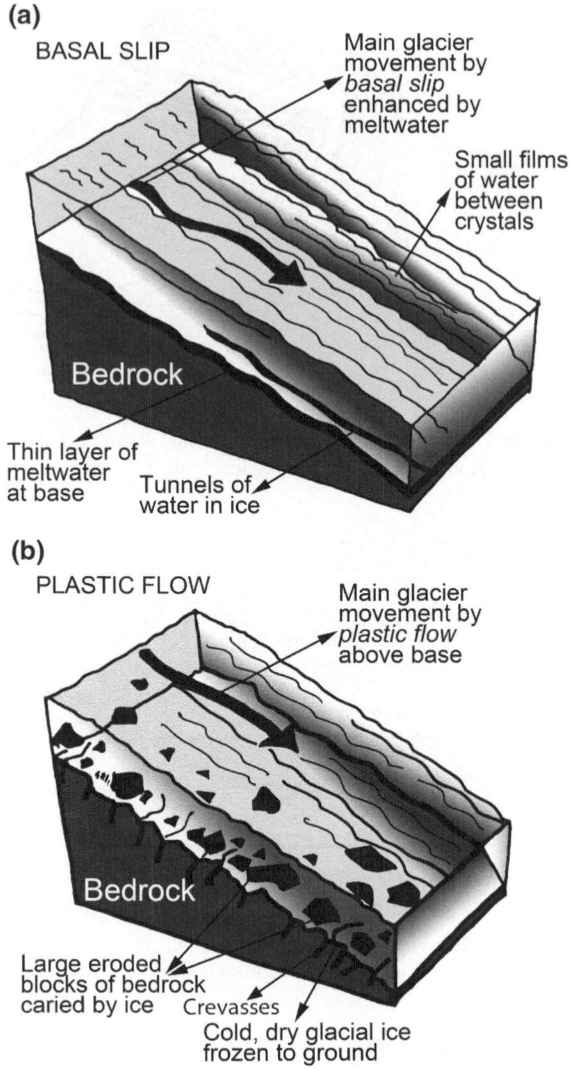

Fig. 11.7 The two mechanisms of glacier movement: Basal slip (sliding) and Plastic flow. **a** Basal slip is sliding over the underlying surface. If a glacier is solidly frozen to its bed, it moves only by Plastic flow. The *top* of the glacier moves farther in a given time than the *bottom* does. **b** Plastic flow involves internal deformation within the ice

sometimes also in temperate glaciers. The upper portions of glaciers are brittle, when the lower portion deforms by internal flow (plastic flow), the upper portion fractures form cracks called Crevasses (Fig. 11.7b). Often, these occur when the lower portion of a glacier flows over a sudden change in topography or when accumulating snow and ice reach a critical thickness of about 40 m; the stress on the ice at depth is great enough to induce plastic flow. Plastic flow is the dominant form of movement where all parts of the glacier are below freezing, including the base.

11.4 Glacier Types

The classification of glaciers is based on their size and their relationship to topography. The smallest glaciers are confined to mountain valleys and are called Valley glaciers or Alpine glaciers. Larger masses of ice may cover an entire mountain range or a volcano. These are called Ice caps and cover several thousand square kilometers.

11.4.1 Glacier Classification

Glacier classification is based on two types—internal temperature and locale and size. Glaciers based on their internal temperature include Temperate and Polar glaciers (see also Sharp 1992).

11.4.1.1 Temperate Glaciers

Ice in a temperate glacier is at a temperature near its melting point.

11.4.1.2 Polar Glaciers

Ice in a polar glacier always maintains a temperature well below its melting point. Glaciers, based on their locale and size, can be divided into two types—Alpine (Mountain/Valley) and Continental glaciers. There are several sub categories of each types.

Alpine (Mountain/Valley) Glaciers

These originate on a mountain or in a mountain range (Fig. 11.8a) and are much smaller than continental glaciers.

Cirque Glacier

These glaciers are formed in a depression (Fig. 11.8) and usually occur at the head of valleys. They occupy hollows or bowl-shaped depressions on the sides of mountains.

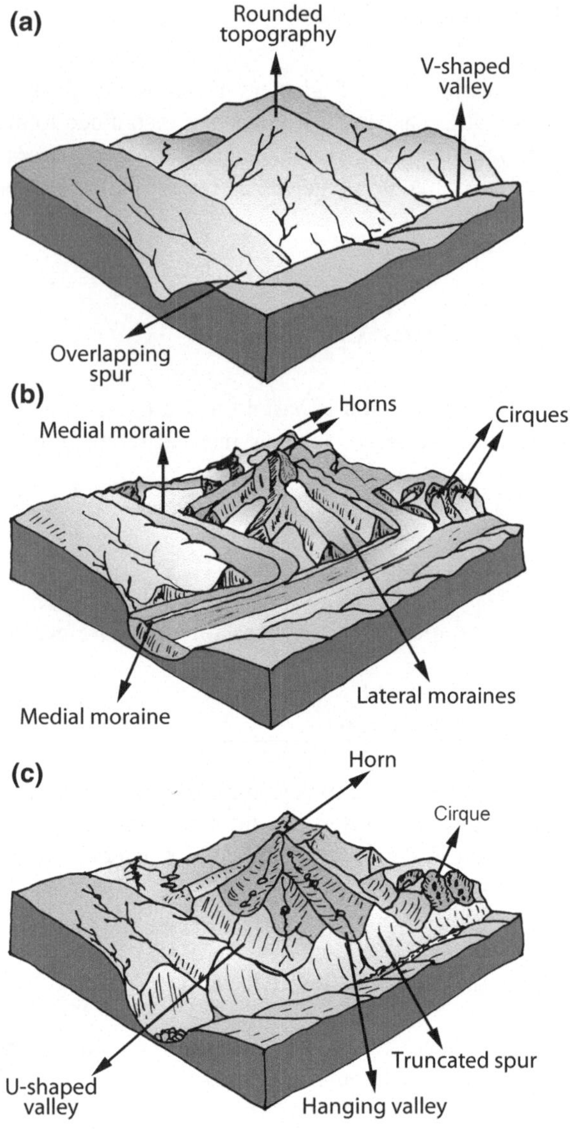

Fig. 11.8 Landforms of Valley glaciers. **a** Topography before glaciation, **b** Formation of Moraines and topography during glaciation, **c** Topography after glaciation. Redrawn from Earth's Dynamic Systems with permission of Eric H. Christiansen

Valley Glacier

It is the prototypical river of ice. As the cirque glacier grows larger, they spread out into the valley and flow down it, as a Valley glacier (Fig. 11.8). The prevailing topography defines the path of these valley glaciers. Valley glaciers are common in the mountains of western North America, especially Alaska and Canada, as well as the Andes in South America, the Alps in Europe, the Alps of New Zealand, and the Himalayas in Asia.

Piedmont Glaciers

These extend from their mountain origin all the way on to the plain. Hence, if a valley glacier extends down a valley and covers a gentle slope beyond the mountain range, it becomes a Piedmont glacier. These are formed by two or more coalescing Alpine glaciers when entering a flatter area.

Fjord Glaciers

If a valley glacier extends down to the sea level, it carves a narrow valley into the coastline. These are then called Fjord glaciers (Fig. 11.9).

Continental Glaciers

Most of the glacial ice in the world lies in Continental glaciers (Fig. 11.10). Here, the ice can be hundreds to thousands of meters thick. Greenland and Antarctica are good examples of Continental glaciers. Both Greenland and Antarctica ice sheets contain 99 % of the world's ice and about three-fourths of earth's fresh water. The Greenland sheet, at places, is more than 2.7 km thick and covers around 1.8 million sq. km, whereas, the Antarctic ice sheet blankets about 13 million sq. km, almost 1.5 times the size of the United States. During the last Ice age, Continental glaciers covered nearly 32 % of the total land area of the planet. Continental glaciers, based on their size, are subdivided into two main types—the smaller ones are Ice caps and the larger ones are Ice sheets.

Ice Cap

An Ice cap covers less than 50,000 km^2 of land area. These are not constrained by topographic features. Hence, they lie over the top of a mountain or a mountain range. Ice caps can be divided into groups: (i) Confluent glacier: these merge into two or more glaciers and (ii) Outlet glaciers: these valley glaciers drains an Ice cap or an Ice sheet.

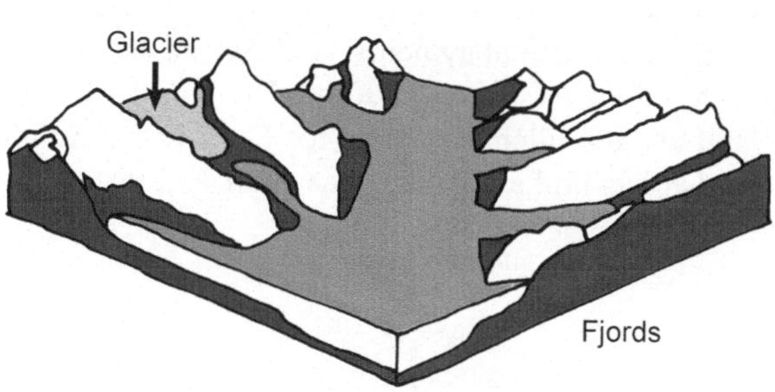

Fig. 11.9 Fjord glacier. Redrawn from Earth's Dynamic Systems with permission of Eric H. Christiansen

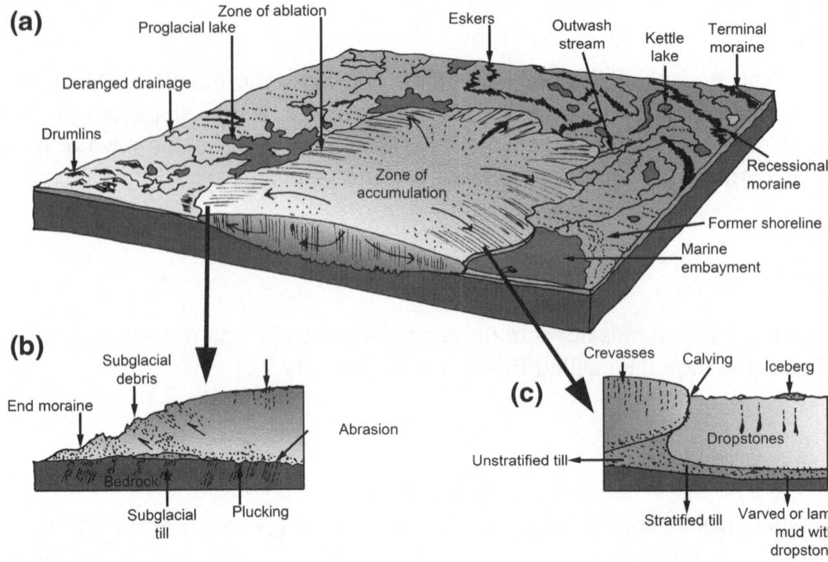

Fig. 11.10 Glacial process and landforms of continental glaciers. **a** Overview. (**b** and **c**) Details of glacial landforms. Redrawn from Earth's Dynamic Systems with permission of Eric H. Christiansen

Ice Sheets

An Ice sheet that covers more than 50,000 km^2 of land area possesses the largest glacier on earth. They cover an entire land mass or a continent. These are not controlled by topographic features such as valleys and hills. They cover everything in the landscape, except at the margins, where they are thin. The Antarctic and Greenland ice sheets are the only existing ice sheets today and together they make up $\sim$99 % of all the glacial ice currently on earth with an estimated volume of $\sim$30 million km^3.

Ice Shelves

Ice shelves are sheets of ice floating on water and attached to land. They usually occupy coastal embayments, and extend hundreds of kilometers from land and reach a thickness of $\sim$1,000 m. The world's largest ice shelves are the Ross Ice Shelf and the Filchner-Ronne Ice Shelf in the Antarctica. An iceberg is a piece of ice that has broken from an ice shelf and that floats in the Ocean.

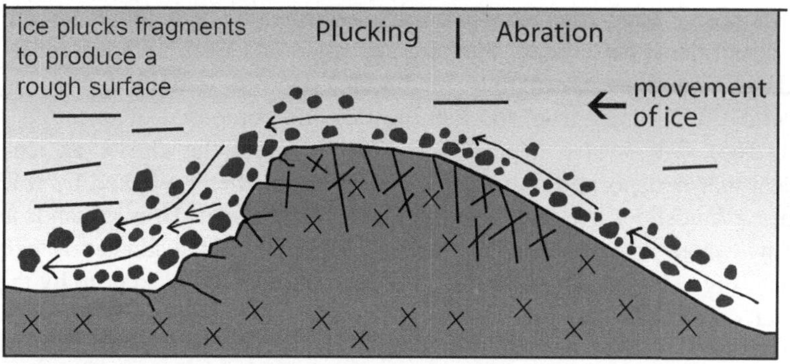

Fig. 11.11 Abrasion, plucking and the movement of ice

11.5 Glacial Processes

Rocks and sediments are eroded and incorporated into a glacier by Abrasion, Plucking, and Freezing-on (Fig. 11.11).

11.5.1 Abrasion

Abrasion occurs when debris-rich ice slides over the bedrock and abrades it (Fig. 11.11). This process is similar to using a sand paper on a block of wood. Abrasion is a particle-by-particle erosion process and produces large amounts of silt-sized (0.002–0.0625 mm) sediments.

11.5.2 Plucking

Plucking occurs when ice flows into or refreezes in fractures in the bedrock (Fig. 11.11). The ice wedges and removes blocks of bedrock and incorporates it into the glacier. This process is most rapid in areas where water is refreezing near the bed.

11.5.3 Freezing-on

It usually occurs in the leeward (down side) of the bedrock and/or in the presence of other obstacles to the ice flow. Glacier erosion is most rapid when (i) ice flows at high velocity, (ii) the material is freezing to the bed, and (iii) there is plenty of water. In most glaciers, this usually occurs downstream and below the equilibrium line. As sediments get eroded, they are deposited and form a variety of features. The glacial sediments are mainly melt water deposits (stratified drift) and Till (unstratified drift-direct from the melting ice, and composed of unsorted debris) (Fig. 11.10c). Till is also deposited at the bottom of the glacier as ice melts. Sediment that is deposited on top of a glacier and is then reworked by water and becomes a landslide, it is called a Supraglacial sediment. Gravel, sand, and silt sorted by water and deposited by streams in front of the glacier is called an Outwash (Fig. 11.10a). Some of this finer material may be picked up by the wind and later deposited as Loess. Commonly small lakes and ponds form near the margin of glaciers and the material is deposited as lake sediment in these areas. Glaciers usually deposit their load in the Ablation zone where ice is melting (Fig. 11.10a). All sediment deposited as a result of glacial erosion are called Glacial Drift. There are two types of glacial drift—Glacial marine drift and Stratified drift (Fig. 11.10c).

11.5.4 Glacial Marine Drift

Glaciers that reach the ocean or a lake, calve off into large icebergs which then floats on the water surface until they melt. Upon melting, the rock debris they contain becomes deposited on the seafloor or lakebed as an unsorted chaotic deposit. Sometimes single large rock fragments fall out on the floor of the water body, and are then called as Dropstones (Fig. 11.10c).

11.5.5 Stratified Drift

Glacial drift can be picked up and moved by meltwater streams. This is later deposited. Such materials are called Stratified Drift (Fig. 11.10c).

11.6 Erosional Landforms

Erosional landforms are formed by two types of glaciers—one by Alpine or Valley glaciers and the other by Continental Ice sheets and Ice caps.

11.6.1 Alpine Glaciers

Glacial erosion on mountains produces some of the most remarkable natural features. The most characteristic is the formation of U-shaped valleys (Figs. 11.8 and 11.12). Other erosional features include Hanging valleys, Arêtes, Cirques, Cols, Horns, Tarns, and Roches moutonnée (Fig. 11.13). All these are briefly described below.

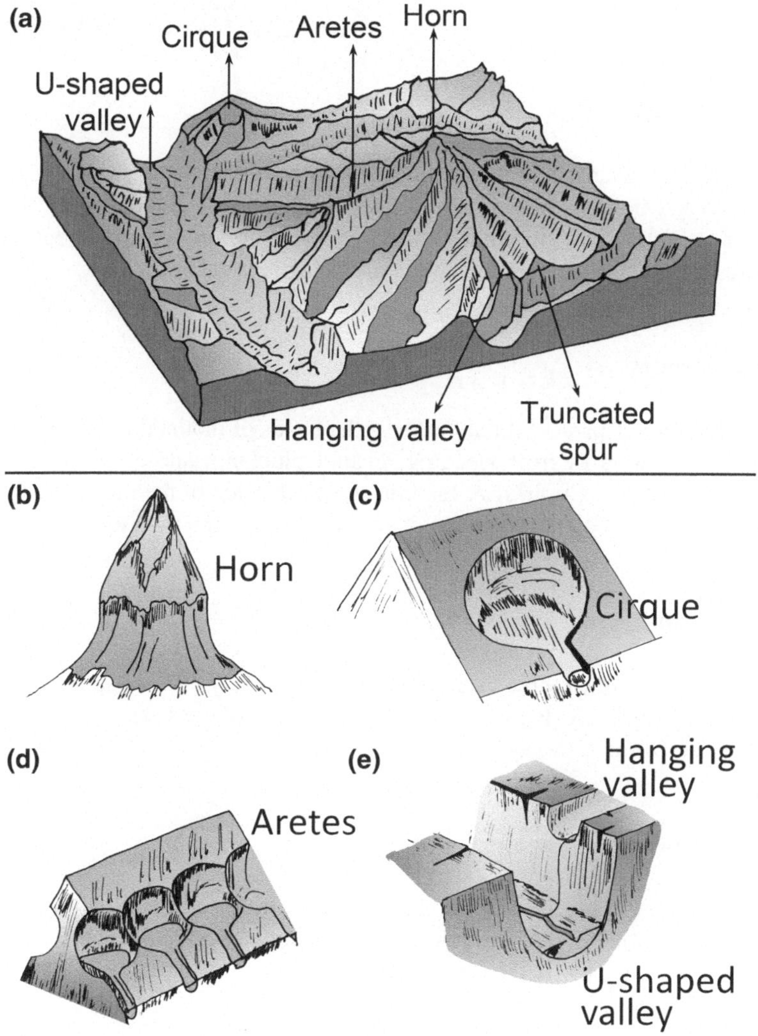

Fig. 11.12 Alpine Glaciers. **a** Characteristic landforms, **b–e** Details of Alpine glacier landforms. **a** Redrawn from Earth's Dynamic Systems with permission of Eric H. Christiansen

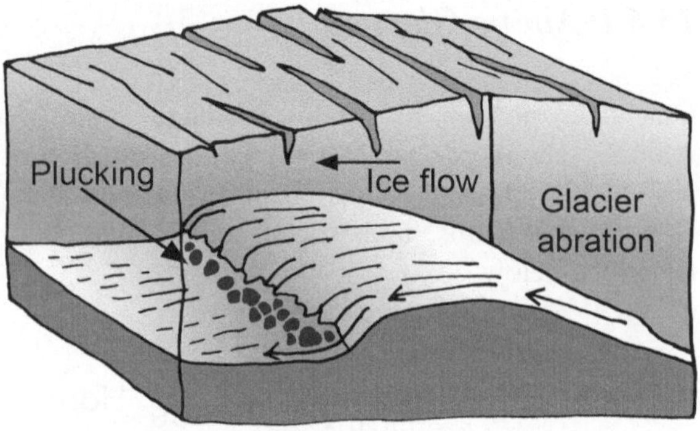

Fig. 11.13 Roche moutonnée

11.6.1.1 Glacial Valleys

The weight of a glacier and its scouring action results in steep sided valley walls and almost flat bottoms producing U-shaped valleys (Fig. 11.12). This U-shaped valley is a distinctive morphology, and if it occurs in areas without glaciers anymore, it is indicative that there was a glacier present in the past.

11.6.1.2 Cirques

The bowl shaped depressions that occur at the heads of mountain glaciers resulting from a combination of frost wedging, glacial plucking, and abrasion are called Cirques (Fig. 11.12). Cirques are features where the side of the mountain has been scooped out, as if a giant ice cream scoop has been taken out (Fig. 11.12).

11.6.1.3 Horns

Horns are formed when three or more cirques are steeply carved out of a mountain producing a sharp peak (Figs. 11.8 and 11.12b). The Matterhorn in the Alps is the most famous Horn.

11.6.1.4 Arêtes

These are groups of horns in a row (Fig. 11.12d). If two adjacent valleys are filled with glacial ice, the ridges between the valleys are carved into a sharp knife-edge ridge (Fig. 11.12). This is then called an Arête.

11.6.1.5 Tarn

These are small lakes that occur at the bottom of a Cirque (Fig. 11.12).

11.6.1.6 Hanging Valleys

Hanging valleys are formed where two or more glaciers or former valleys intersect at different elevations (Fig. 11.12a and e). When a glacier occupying a smaller tributary valley meets the larger one, the tributary glacier usually does not have the ability to erode its base to the floor of the main valley. Thus, when the glacial ice melts the floor, the tributary valley hangs above the floor of the main valley and is then called a Hanging valley (Fig. 11.12). Waterfalls occur where the hanging valley meets the main valley (Fig. 11.12e).

11.6.1.7 Fjords

These are narrow inlets along the seacoast that were once occupied by a valley glacier (Fig. 11.9).

11.6.1.8 Roche Moutonnée

These are eroded and smoothed bedrock hills (Fig. 11.13). The gentle side faces up ice and has been smoothed by glacial abrasion. The down ice side has a steep plucked face, where pieces of bedrock have been plucked out by the glacier ice. Hence, Roche moutonnée are smooth upstream, with rough and steep downstream resistant rocks (Fig. 11.13). The term Roches Moutonnée is derived from the French words *roche* for "rock" and *mouton*, for "sheep". Clusters of Roches Moutonnée resemble herds of grazing sheep. Both alpine and continental glaciers form Roches moutonnée.

11.6.1.9 Glacial Striations

They are long deep parallel scratches and grooves produced at the bottom of glaciers by rocks embedded in the ice that scrap against the rock. These are the small scale erosional features. These markings show the direction of ice movement. Hence, glacial striations are used to map the flow directions of glaciers.

11.6.1.10 Glacial Polish

These are small-scale features where the glacier acts like a sandpaper on the underlying surface, and as a result, produces a smooth surface. The same small-scale abrasional features such as striations and glacial polish can occur beneath ice caps and ice sheets, particularly in temperate environments. The land surface beneath a moving continental ice sheet can be molded into smooth streamlined elongate hills called Drumlins (Fig. 11.14a). These elongated forms form when a glacier flows over a mound of sediment and the flow of the ice creates the streamlined shape, elongated in the same direction as the glacial flow. Drumlins are usually about 1–2 km long and about 15–50 m high. Most are made of till and some partly of bedrock. Good examples of Drumlins are seen in the northern United States, more so in the rolling farmland of Wisconsin (USA).

Continental ice sheets (Antarctica and Greenland), have over a km thick ice slowly moving seaward over a landmass. They override low hills, producing a cleaned off, polished, and striated bed rock region with U-shaped valleys and truncated hanging tributaries. Occasional peaks sticking through the ice (Nunataks) show alpine features.

11.7 Depositional Landforms

There are two types of depositional landforms—Till and Stratified drift deposits (Fig. 11.10).

11.7.1 Till Deposits

These include Till, Erratics, and Moraines.

11.7.1.1 Till

It is a nonsorted glacial drift ("drift" is a generic term for all glacially deposited sediments) deposited directly from ice (Fig. 11.10). It is made up of a random mixture of different sized fragments of angular rocks in a matrix of fine grained, sand-to clay-sized fragments that were produced by abrasion within the glacier. This fine-grained material is often called Rock flour. A Till that has undergone diagenesis and is now rock hard is called a Tillite.

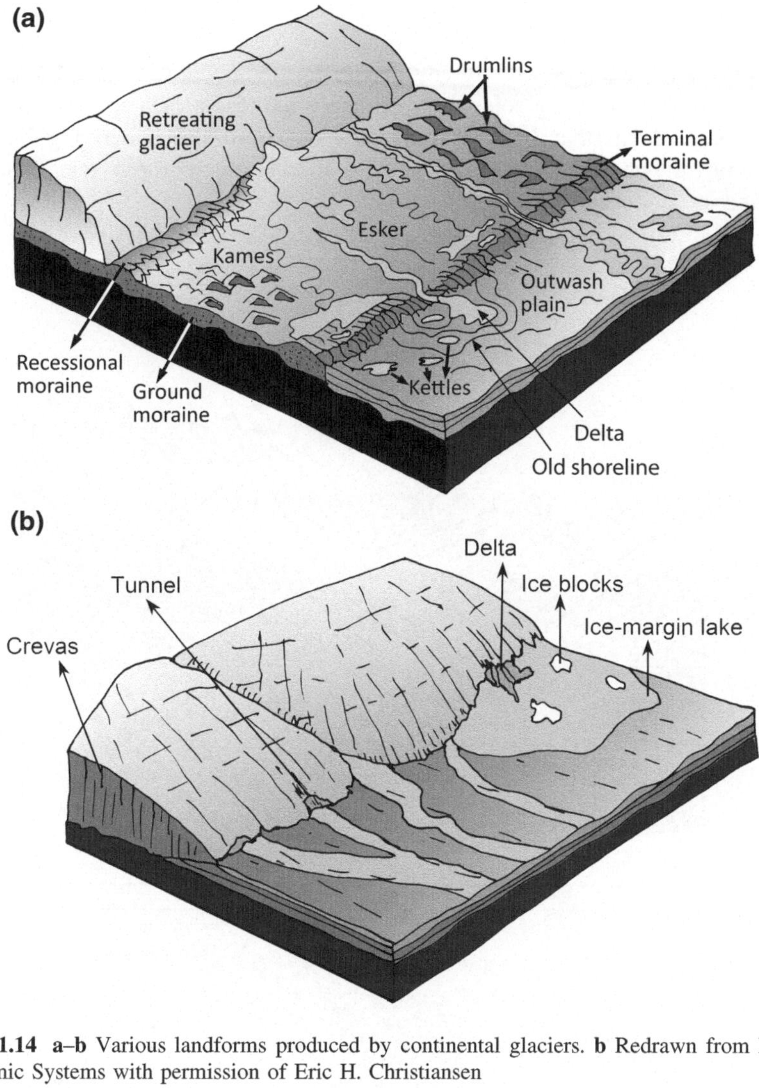

Fig. 11.14 **a–b** Various landforms produced by continental glaciers. **b** Redrawn from Earth's Dynamic Systems with permission of Eric H. Christiansen

11.7.1.2 Erratics

These are large boulders that have been transported by glaciers, and often deposited at a considerable distance from their origin. These are left behind when the ice melts. By mapping the distribution pattern of Erratics, one can determine the flow direction of ice that carried them to their present location. This also enables to plot past ice movements across large areas.

11.7.1.3 Moraines

This is a French word describing any glacier-formed accumulation. Moraine is a material transported by a glacier and then deposited (Figs. 11.8 and 11.15). These are deposits of till that have a form different from the underlying bedrock. Depending on where moraines are formed in relation to the glacier, they are classified as:

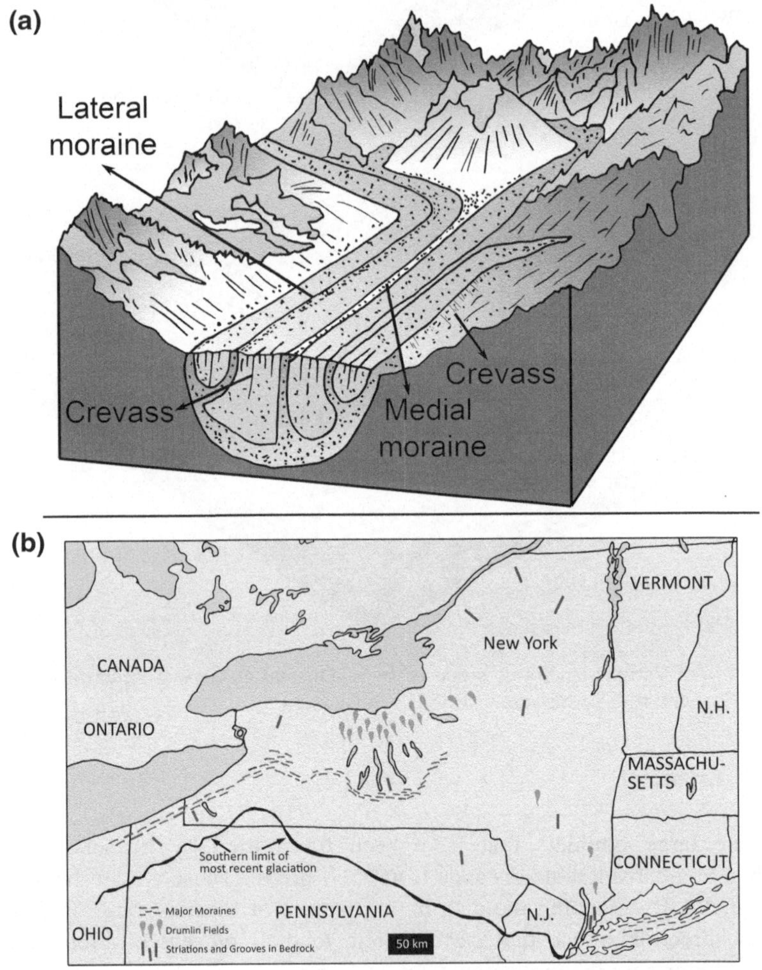

Fig. 11.15 Moraine types **a** and their presence on continental United States **b**. **a** Redrawn from Earth's Dynamic Systems with permission of Eric H. Christiansen

Ground Moraine

These are deposited beneath the glacier and result in a hummocky topography with lots of enclosed small basins (Fig. 11.15a). Till deposited as a relatively thin layer over a broad area forms a Ground Moraine. These fill old stream channels and low lying areas. Thus, this leveling process disturbs drainage patterns. Good examples are the swamps in the northern Great Lakes region (USA) that were formed when the most recent continental glaciers receded.

End, Terminal and Recessional Moraines

These are piles of material dropped by the glacier at its terminus (Fig. 11.15a). Hence, their presence marks the furthest advance of a glacier. If there are warmer conditions, then a glacier recedes. If the glacier stabilizes again during its retreat and the terminus remains in the same place for a year or more, a new End moraine, called a Recessional moraine is formed. Both End and Ground moraines are characteristic of alpine and continental glaciers.

Lateral Moraines

These are deposited along the sides of mountain glaciers (Fig. 11.15a).

Medial Moraines

When two valley glaciers meet to form a larger glacier, the rock debris along the sides of both glaciers merges to form a medial moraine (Fig. 11.15a). These occur as black streaks in an active glacier.

11.7.2 Stratified Drift Deposits

Geologists define drift as all rock or sediment transported and deposited by a glacier, whereas stratified drift deposits are sediments deposited by glacial melt-water that are sorted and layered. These include the following:

11.7.2.1 Kettle and Kettle Lakes

If depressions form underneath a glacier and remain after the glacier is melted, then water filling these depressions become small lakes (called Kettles) where fine-grained sediment are deposited (Figs. 11.10 and 11.14a). Melted chunks of ice left by a retreating glacier forms Kettle lakes.

11.7.2.2 Kames and Kame Terraces

Streams and lakes formed on top of a stagnant ice may deposit stratified sediments on top of a glacier. When the glacier melts these deposits are set down on the ground surface. These former lake deposits become Kames, and former stream deposits become Kame terraces. Kames are hills of sediment that are commonly found near Kettles (Figs. 11.10 and 11.14a).

11.7.2.3 Eskers

Eskers (Fig. 11.14a) are long sinuous ridges of sediment (sand and gravel) deposited by streams that ran under or within a glacier. The sediment deposited by these streams becomes an Esker after the ice has melted. Most Eskers are sinuous, with a height of a few to several tens of meters. The longest ones continue for several kms. However, most are shorter or discontinuous. Eskers are often broad and flat-topped, or may have a single crest or they split into parallel ridges. They are often formed at the margin of warm-based glaciers or ice sheets during ice retreat.

11.7.2.4 Outwash Plains

These form in front of a glacier as the meltwater carries silt, sand, and gravel away from it (Fig. 11.14a). Streams running off the end of a melting glacier are usually choked with sediment and form braided streams (flowing in multiple channels), which deposit poorly sorted stratified sediment in an outwash plain. These deposits are often referred to as outwash. These are often silt-rich, and are reworked by wind to form Loess. Outwash plains are characteristics of continental glaciers.

11.7.2.5 Outwash Terraces

If the outwash streams cut down into their outwash deposits, the banks form river terraces called Outwash terraces.

11.8 Summary

A glacier is a thick mass of ice moving downhill under the pull of gravity. Glaciers cover one-tenth of the earth's surface, and store over 75 % of the world's fresh-water supply. Most glacial ice today is found in the Polar Regions, above the Arctic and Antarctic circles. Mostly glaciers are in Iceland, the Arctic, Antarctica, Alaska, and Canada.

Glaciers form where snowfall in the winter exceeds the amount of snow and ice that melts in the following summer as in areas of high elevation and or high latitudes. The balance between ice accumulation and melting is called the glacier Mass balance. Accumulation that occurs high up on the glacier forms the Zone of Accumulation and where glacier melts, the Zone of Ablation. The line that separates these two zones is called the Equilibrium line. It is also called the Firn line. The line above which, snow forms and remains round the year is called the Snowline and glaciers can form only at latitudes or altitudes above the snowline. At present, the snowline lies at the sea level in Polar latitudes but rises up to 6,000 m in tropical regions.

Glaciers move slowly. The rate of flow varies from 0.01 to 0.1 m per day for large Continental glaciers to 0.1–2 m per day for Alpine glaciers. Glacier movement is by basal slip and plastic flow. The upper portions of glaciers are brittle, when the lower portion deforms by internal flow (plastic flow), the upper portion fractures to form cracks called Crevasses. Plastic flow is the dominant form of movement where all parts of the glacier are below freezing, including the base. The melt water at the base of the glacier reduces friction by lubricating the surface and allows the glacier to slide across its bed by Basal Slip.

Glacier classification is based on two types—internal temperature and locale and size. Glaciers based on their internal temperature include Temperate and Polar glaciers.

Glaciers, based on their locale and size, can be divided into two types—Alpine (Mountain/Valley) [Cirque, Valley, Piedmont, and Fjord glaciers] and Continental glaciers (Ice cap, Ice Sheets, and Ice Shelves). There are several subcategories of each types.

Rocks and sediments are eroded and incorporated into a glacier by Abrasion, Plucking, and Freezing-on.

Glaciers that reach the ocean or lake, calve off into large icebergs which then floats on the water surface until they melt. Upon melting, the rock debris they contain becomes deposited on the seafloor or lakebed as an unsorted chaotic deposit called Glacial marine drift. Dropstones are single large rock fragments that fall out on the floor of the water body. Glacial drift that are moved by meltwater streams and deposited called Stratified Drift.

The erosional landforms formed by glaciers include Glacial valleys, Cirques, Horns, Arêtes, Tarn, Hanging Valleys, Fjords, Roche moutonnée, Glacial striations, and Glacial polish.

There are two types of depositional landforms—Till deposits (Till, Erratics, and Moraines) and Stratified drift deposits (Kettle and Kettle Lakes, Kames and Kame Terraces, Eskers, Outwash Plains, and Outwash Terraces). Moraines include Ground, End, Terminal, and Recessional Moraines (Lateral and Medial Moraines).

References

Anderson D (2004) Glacial and periglacial environments. Hodder Arnold, London

Ben DI, Evans DJA (1998) Glaciers and glaciation. Wiley, New York

Deynoux MJ, Miller MG, Domack EW, Eyles N, Fairchild I, Young GM (eds) (2004) Earth's glacial record. Cambridge University Press, Cambridge

Flint RF (1971) Glacial and quaternary geology. Wiley, New York

French HM (1996) The Periglacial environment. 2nd edn. Longman, London

Hambrey M (1994) Glacial environments. B.C., UBC Press, Vancouver

Hughes TJ (1998) Ice sheets. Oxford University Press, New York

Ives JD, Barry RG (1974) Arctic and alpine environments. Menthuen, London

Martini IP, Brookfield ME, Sadura S (2001) Principles of glacial geomorphology and geology. Prentice Hall, New Jersey

Meier MF (1962) Proposed definitions for glacier mass budget terms. J Glaciol 4(33):252–263

Sharp RP (1992) Living ice. Cambridge University Press, New York

Chapter 12
Ocean

12.1 Introduction

Oceans determine earth's weather and climatic patterns and hence, play a crucial role in earth's habitability. Most of the solar energy that reaches earth is stored in the ocean, which helps power oceanic and atmospheric circulation. Additionally, about 200 Ma of earth's recorded geologic and biologic history is found in the ocean's floor. By studying this ocean data (through oceanic sediments), we learn about our ancient climate, and how it changed, enabling us to better predict our own, in future. The average ocean depth is about 3,798 m, and the deepest point is 11,033 m located in the Mariana Trench in the Pacific Ocean. Contrast this depth with that of the tallest mountain, the Mount Everest, that only measures 8,846 m! Hence, if Mount Everest were to be placed into the Mariana Trench it would be covered with sea water more than 1.5 km deep. On earth, there is 527,864,832 km^3 of seawater, covering approximately 71 % of its surface. At 3.89 °C, the temperature of the deep ocean is only a few degrees above freezing. Interestingly, the present sea levels have largely remained constant for the past 5,000 years and more than 66 % of the world's population lives within 100 km of the coastline.

12.2 Features of an Ocean

The Ocean basins are divided into Continental shelf, Continental slope, Continental rise, Abyssal plains, Oceanic ridges, Trenches, Seamounts, Guyots, Transform faults, and Rift zones (Fig. 12.1) (see also Defant 1961; Dietrich 1963; Fairbridge 1966; Trenhaile 1997; Bird 2000; Masselink and Huges 2003). These are briefly mentioned below.

S. Jain, *Fundamentals of Physical Geology*, Springer Geology,
DOI: 10.1007/978-81-322-1539-4_12, © Springer India 2014

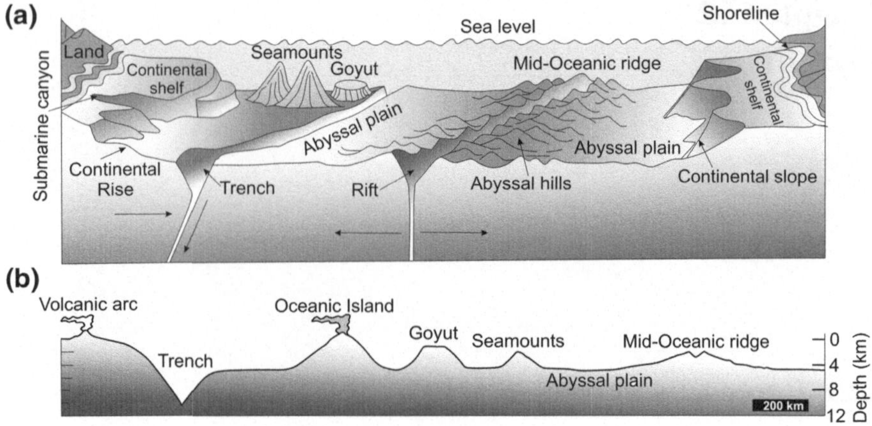

Fig. 12.1 Oceanic features. **a** Idealized diagram of an ocean basin and various submarine features. **b** Actual depths at the Pacific basin (vertical exaggeration is more than 20x)

12.2.1 Continental Shelf

The Continental shelf (Fig. 12.1) is a gently sloping (0.1–0.5°) and submerged part of the continent. It can be as wide as 1,500 km (like the Siberian Shelf in the Arctic Ocean) or just a few kilometers wide, as is near the Pacific coast of North America; the average width is only ~200 km. The water depth at the seaward edge of the Continental shelf is between 100 and 200 m deep; the average depth is ~150 m.

12.2.2 Continental Slope

Continental slope is the boundary between the continental and oceanic crust (Fig. 12.1). As compared to the shelf, it slopes more steeply. The average slope is about 3–5°, but this can be as low as 1° or as high as 10–25°. Continental slope is usually ~20 km wide and about 100–200 m deep. The Continental shelf and slope are together called the Continental margin.

12.2.3 Continental Rise

It is at the base of the continental slope (Fig. 12.1) and constitutes a wedge of sediments that extends from the lower part of the continental slope to the deep-sea floor. The continental rise, typically slopes between 0.5 and 1° (average is at about

0.5°) and ends in a flat abyssal plain at depths of about 5 km. The slope angle decreases gradually and the slope is hundreds of kilometer wide. The continental rise rests upon the oceanic crust.

12.2.4 Abyssal Plains

They occur at the base of the Continental rise and are almost flat (<1 m of vertical drop for every 1,000 m of horizontal distance) ocean floor formed by horizontally deposited sediments (Fig. 12.1) with depths ranging from 3 to 6 kms (average depth is 5 km) (see also Bruun 1957). The Abyssal plains cover more than 50 % of the earth's surface. The Canary Abyssal plain (Canary Islands, North Atlantic) are the largest, covering 900,000 km^2.

12.2.5 Oceanic Trenches

Oceanic trenches are the deepest part of the ocean and often exceed 10,000 m in depth (Fig. 12.1). The Mariana trench in the Pacific Ocean is the deepest (11,033 m). The continental slope forms the landward wall of the trench, which steepens with increasing depth. This slope is typically 4–5° on the upper side, increasing to about 10–15° near the lower side of the trench. The oceanic trenches are elongate, often 8–10 km deep and occur at Subduction zones where the oceanic crust subducts downward into the mantle, associated with earthquakes and volcanoes.

12.2.6 Seamounts

These are volcanic peaks (conical undersea mountains) under the sea that rise 1,000 m or more above the sea floor (Fig. 12.1). They form along the Mid-Ocean Ridges (MOR) or over Hot spots. The western Pacific seafloor has the highest concentration of seamounts; one estimate puts the number at 10,000. Globally, a compilation of bathymetric and altimetric data suggested ~125,000 seamounts (height > 1000 m) but could be between 45,000 and 350,000; while smaller seamounts (height < 100 m) could be 25 million (8–80 million) (Wessel et al. 2010). Large seamounts may hinder plate subduction, trigger slides, and earthquakes and also generate tsunamis (Gonzalez 1999).

12.2.7 Guyots

When the seamounts are eroded to a flat top, they are called Guyots (Fig. 12.1). These are mostly found in the western Pacific Ocean where they stand to minimum height of 900 m. They also commonly occur in the islands of Hawaii. Seamounts are often ringed by coral reefs, these are then called Atolls (circular reef surrounding a lagoon over a now-submerged volcanic peak) (discussed later in the chapter).

Many Seamounts and Guyots are aligned in chains. Such volcanic chains, along with ridges on the seafloor, are called Aseismic ridges, i.e., submarine ridges not associated with earthquakes.

12.3 The Ocean Sediments

Rivers deposits large volumes of sediments in the ocean (Fig. 12.2). These are of four types.

12.3.1 Terrigenous Sediments

Terrigenous sediments (often interchanged with Lithogenous sediments) are non-organic sediments from land, blown out into the sea (Fig. 12.2a). These include mineral grains from weathered continental rocks, fine-grained sediment like clay and mud and volcanic ash. These sediments accumulate slowly and may take anything between 5,000 and 50,000 years to deposit a single cm of layer within the ocean. The Terrigenous sediments largely make up the continental rise and the abyssal plains.

12.3.2 Biogenous Sediment

These are of biological origin (i.e., produced directly by an organism; Fig. 12.2b). The biogenous sediments are primarily made of shells and skeletons of microscopic planktons. They also include bones, teeth, clams, and corals (as commonly noted in warm and shallow seas). They are quite common deposits in open oceans and deep seas. Biogenous sediments in the open-ocean are called "Oozes." Ooze is made up of microscopic shells of calcareous and siliceous algae and/or protozoans. By definition, the biogenous oozes contains greater that 30 % of biogenic component and the remainder is made up of nonbiogenic sediments, such as lithogenous mud. The biogenous sediments are divided into three types:

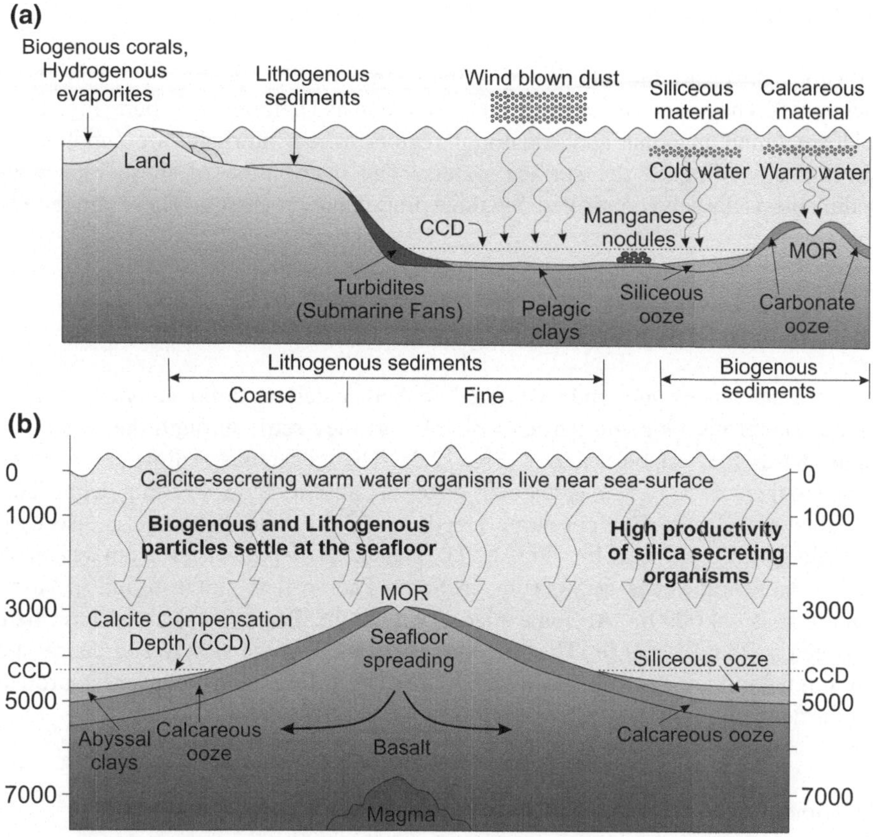

Fig. 12.2 Distribution of sediments on the Continental margin. **a** A generalized view of sediment accumulation and various types of oozes. **b** A more detailed view of sediment accumulation and various types of oozes. The particles making up the Calcareous ooze are skeletons of Foraminifera (floating single-celled organisms) and Coccolithophores (floating single-celled plants), whereas Siliceous ooze is composed of skeletons of Radiolarians (single-celled floating organisms) and Diatoms (single-celled floating plants)

12.3.2.1 Calcareous Oozes

Calcareous Oozes are remains of molluscs, corals, snails (pteropods), foraminifera, plants, algae (seaweed), and coccolithophores (those covered with plates are called coccoliths) (Fig. 12.2). These form Calcium Carbonate ($CaCO_3$) deposit such as Chalk and are widespread in occurrence in relatively shallow areas of the deep sea (Fig. 12.2b).

12.3.2.2 Siliceous Oozes

Siliceous Oozes are the remains of animals (amoeba like Radiolarians) and plants (Diatoms). They form Silica (SiO_2) deposits such as Diatomite or Chert (Fig. 12.2) and are found in polar and equatorial regions where nutrients are supplied by vertical upwelling to the surface waters. The distribution of these biogenous sediments is largely controlled by three important processes. These are briefly mentioned.

Dissolution in Deep Waters

Deep-ocean waters are undersaturated in both calcium carbonate and opalline silica. Therefore, biogenic particles dissolve as they settle through the water column and as they sit on seafloor (Fig. 12.2). The calcareous sediments are more prone to this effect. Hence, calcareous oozes are absent below a certain depth. This is called the Calcite Compensation Depth or CCD (often Carbonate is also used interchangeably for Calcite; Fig. 12.2). The CCD depth varies from ocean to ocean; in Atlantic it is at 5,000 m, in South Pacific it is 4,500 m and in North Pacific it is <4,000 m. At some places within the Pacific, it can be anywhere between 500 and 1,500 m. The siliceous particles dissolve slowly and hence, are not limited in their distribution by depth. Their distribution is, however, controlled by nutrient supply.

Dilution

In regions where the input of terrigenous sediment is very high, such as along continental margins, the calcareous and siliceous components are diluted to less than 30 % of the total sediment volume (and hence, do not make the "biogenous ooze"; Fig. 12.2). Here, surface productivity is also high and dissolution is maximum. The higher influx of terrigenous sediments does not allow the biogenous oozes to form.

Production in Surface Waters

The growth of marine algae (algae forms the base of the oceanic food chain) is controlled by the availability of nutrient elements, namely Nitrogen and Phosphorus. These are supplied to surface waters by deep waters that "upwell" to the surface. Hence, biological productivity is high in areas of strong upwelling, i.e., along the equator, in certain coastal regions, and in the Southern Ocean around Antarctica.

12.3.3 Hydrogenous Sediment

These are authigenic or diagenetic minerals that precipitate from ions in seawater by chemical reactions. Near hydrothermal vents, metal ions are released into the water, which oxidize or combine with silica and precipitate out as dark, metal-rich sediments. Manganese nodules (Fig. 12.2) are good examples of hydrogenous sediments. Other examples include inorganic limestones and phosphorites. The hydrogenous sediments are less common than the terrigenous or biogenous ones and are generally never the dominant sediment type also.

12.3.4 Cosmogenous Sediments

These are extraterrestrial in nature and are generally like miniature meteorites. These sediments are the remains of impacts of large bodies of space material (such as comets and asteroids). They are comprised of silicates and mixtures of different metals and, as one might imagine, they are not incredibly common to find. This is rather surprising because there is a constant "rain of these materials that falls to earth daily." The amounts of such sediments also leads researchers to wonder if these space-driven events might have been responsible for mass extinction and thus these sediments hold several possible keys to future understanding of ancient life on earth.

Cosmogenous sediment consists of two main types:

(a) Microscopic spherules that are microscopic globular masses composed of silicate rock material (called Tektites) or composed mostly of iron and nickel. Although they were once thought to be the product of meteors, it is now believed they are produced by collisions between asteroids, which produce microscopic space dust particles that drift harmlessly through earth's atmosphere.
(b) Macroscopic meteor debris that forms near an impact site when meteors collide with earth at great speeds.

12.4 Shore Processes

12.4.1 Ocean Waves

Waves move through water in an oscillatory manner; floating objects move up and down rather than forward. Wave height in the open ocean can be over 10 m, and the largest observed is about 34 m in the North Pacific in 1933. Waves are generated by winds that blow over the surface of Oceans. In a wave, water travels in

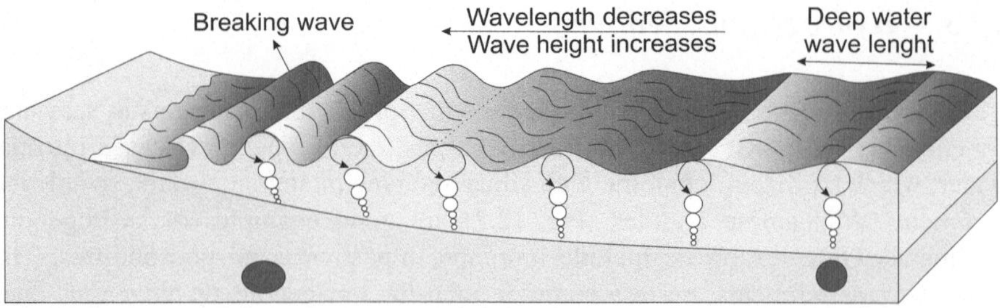

Fig. 12.3 Ocean wave motion. When a wave reaches the shore, the circular motion flattens out and becomes elliptical (*in red*), the wavelength shortens, and the wave steepens until they finally break, creating surf. Redrawn from Earth's Dynamic Systems with permission of Eric H. Christiansen

loops (Fig. 12.3). But since the area affected is the top (surface), the diameter of the loops decreases with depth (and changes in shape from a circular to an ellipse; Fig. 12.3).

Waves are measured using Frequency (f), the number of crests passing a point in given length of time; Height (h), the distance from crest to trough; Period (P), the time for two successive crests to pass a given point; Velocity (V), the distance traveled by a wave in a unit of time and Length (L), the distance from crest to crest. The speed of a moving wave is determined as follows: Speed of wave (C) = Wavelength (L)/Period (P).

The effectiveness of wind in generating a wave depends on three factors: (a) its average speed, which determines its force, (b) its duration, and (c) the extent of open water across which it blows (this is also called the Fetch). When gusty winds blow for a long time and cover large distances of the open water, waves of great height are formed (as much as 20 m as in the Pacific). However, the motion of waves is only effective in moving objects where the water depth is equal to one half of its wavelength (L/2) (Fig. 12.3). Waters deeper than that do not move objects. Thus, waves cannot erode the bottom or move sediments where the water is deeper than L/2. This depth is known as the Wave base. In the Pacific Ocean, wavelengths up to 600 m are noted; hence, waters deeper than 300 m will not feel the passage of this wave. However, the outer parts of the continental shelves are not so deep (average depth is ~200 m), hence, considerable erosion takes place at the edge of the continental shelf from long wavelength waves. When waves approach the shore, water depth decreases and the wave starts feeling the bottom. Because of friction, wave velocity (= L/P) decreases, but its period (P) remains the same. Hence, wavelength (L), decreases. Furthermore, as the wave "feels the bottom," the circular loops of water motion changes to elliptical, as loops get deformed by the bottom (Fig. 12.3). As wavelength (L) shortens, wave height (h) increases (Fig. 12.3). Eventually, the steep front portion of the wave cannot support the water as the rear part moves over, and the wave breaks. This results in the turbulent water of the surf, where the incoming waves meet back the flowing water.

12.4.1.1 Tsunami

Tsunamis are super destructive seismic sea waves triggered by massive changes to the sea floor (Fig. 12.4) (see also Fred 1990; Gonzalez 1999). They are caused by earthquakes, volcanic eruptions, landslides, explosions, underwater nuclear tests and, more rarely, by the impact of cosmic bodies, such as meteorites or asteroids. Hence, any disturbance of appreciable magnitude, above or below the sea floor can trigger a Tsunami. A Tsunami usually forms when the sea floor abruptly deforms; Tectonic earthquakes are a characteristic underwater disturbance that displaces the overlying water mass. Subduction earthquakes are vertical movements of earth's crust at plate boundaries. The slipping of oceanic plates under continental plates also triggers a Tsunami (Fig. 12.4). Large earthquakes are often followed by submarine landslides that disturb the overlaying water. Tsunamis may also be generated by submarine volcanic eruptions as well as by the collapse of volcanic edifices. Underwater nuclear testing can also trigger a Tsunami. The water above the deformed area is displaced and causes the sea to rise vertically as high as 30 m. However, the first wave is the most dangerous. The waves that reach the shore can be several minutes to an hour apart and the danger from a Tsunami can last for several hours after the arrival of the first wave. For a Tsunami, in the open ocean, the wave heights are low (<1 m), periods and wavelengths are long (10–20 min; ~200 km), but velocities are extremely high (~1,000 km/h). A Tsunami is not just a single wave but "a wave train"—a series of waves that can be as long as 100 km. Such long waves gather early and produce very high breakers (<30–60 m) that are very destructive to coastal areas.

12.4.1.2 Tides

Tides cause the rise and fall of sea level to a meter or so, once or twice per day. They are caused by the attraction of both Moon and Sun on the Earth (Fig. 12.5) (see also Cartwright 1999). As Moon is closer to Earth than Sun, it has a larger effect and causes the Earth to bulge toward itself. At the same time, a bulge occurs on the opposite side of the Earth due to inertial forces. Tides at certain times of the month are unusually lower or higher than at others. The reason for this has to do with the position of both the Sun and the Moon relative to the Earth. Tidal effects are greatest when the configuration of the coastline includes narrow openings, such as estuaries, bays, straits, etc.

The highest high tide is called the Spring tide and occurs when the Sun, Moon and the Earth are in line on New and Full moon days (Fig. 12.5). During the moon's quarter phases, the Sun and Moon work at right angles, causing the bulges to cancel each other. The result is a smaller difference between high and low tides. Such a tide is called a Neap tide. These are weak tides and occur when the gravitational forces of Moon and Sun are perpendicular to one another (with respect to the Earth). The highest tides occur in the Bay of Fundy, Nova Scotia (Canada) where the tidal range is 15–16 m. Usually there are two high tides and

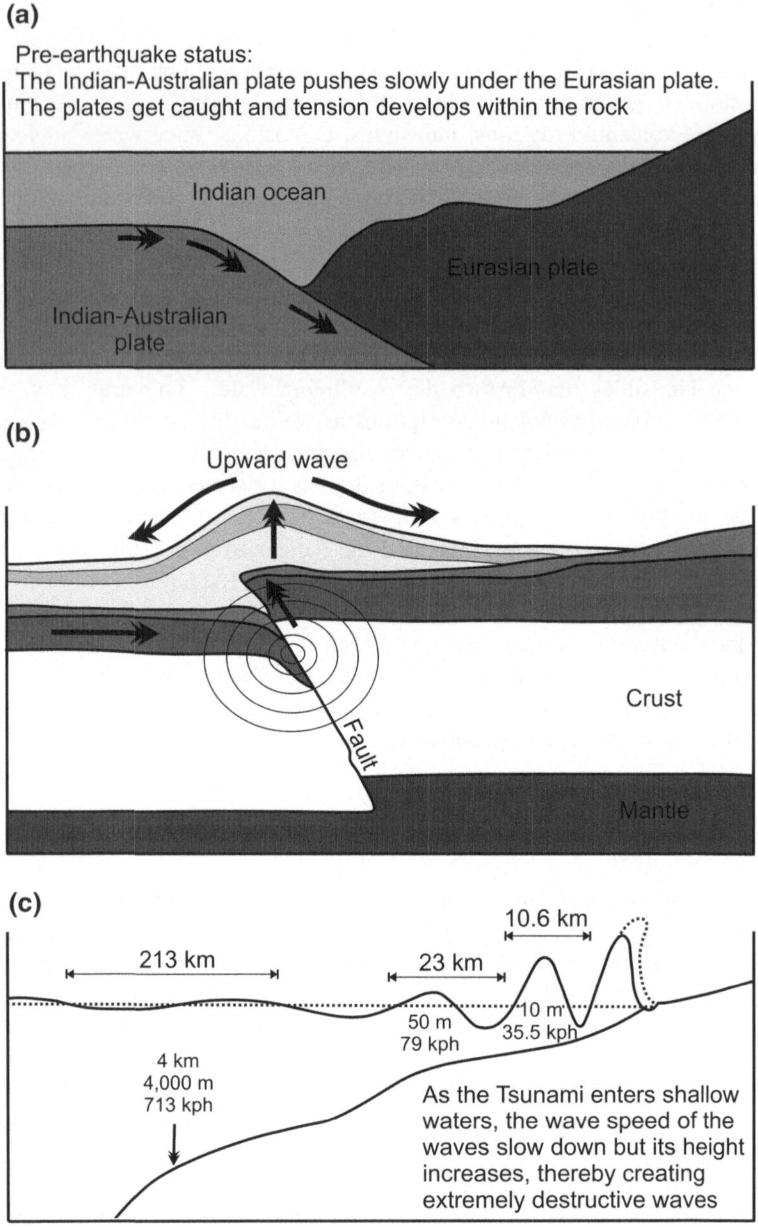
(a)
Pre-earthquake status:
The Indian-Australian plate pushes slowly under the Eurasian plate.
The plates get caught and tension develops within the rock
Indian ocean
Eurasian plate
Indian-Australian plate
(b)
Upward wave
Crust
Fault
Mantle
(c)
213 km
23 km
10.6 km
50 m
79 kph
10 m
35.5 kph
4 km
4,000 m
713 kph
As the Tsunami enters shallow waters, the wave speed of the waves slow down but its height increases, thereby creating extremely destructive waves

◀**Fig. 12.4** Formation of Tsunami. Most Tsunamis are caused by strong submarine earthquakes (Subduction earthquakes), like the one that struck the Indian Ocean on 26th December 2004 (**a**) or by earthquakes close to the coasts (submarine landslides, often activated by earthquakes cause only local destruction). Tsunamis with high magnitude and shallow focus are capable of producing vertical displacement of the sea bottom. So when a strong submarine earthquake occurs, the sea bottom undergoes a vertical uplift displacing the water column and producing waves (**b**) with a small amplitude (tens of cm) but with high wavelength (**c**). In the open sea Tsunami waves are usually unnoticed as their height generally does not exceed more than a meter, though; their wavelengths do reach hundreds of km (**c**). **c** the relationship of depth, wave velocity and wave length. Note that with decreasing depth (as the Tsunami nears the shore), wave velocity decreases but wave length increases dramatically. This increase height causes massive damage. Note that in deep waters, the tsunamis move very fast (between >700 km/h at depths between 4,000 and 7,000 m), with very long wave lengths ($\sim$200 km)

two low tides each day, so the tides must come in within about a 6 hour period. Along most coasts, the tide range is about 2 m, but in narrow inlets, the tidal currents can be strong and fast, causing variations in sea level as much as up to 16 m. The rising tides produce Flood currents, while falling tides produce Ebb currents. Areas that are alternately submerged and exposed by the rising and falling of tides are called Tidal Flats.

12.4.2 Refraction

Waves generally do not approach the shoreline parallel to the shore but bend as they enter shallow waters (primarily due to changes in their velocity). As the wave base touches the bottom, waves slow down. The part of a wave in shallow water moves slower than the part in deeper waters. Hence, when the depth under a wave crest varies with respect to the crest, waves bend. This is Refraction. In simple terms, the refraction of the ocean wave occurs as the trough of the wave scrapes the bottom of the shallow coastline, thereby, causing the waves to slow down (due

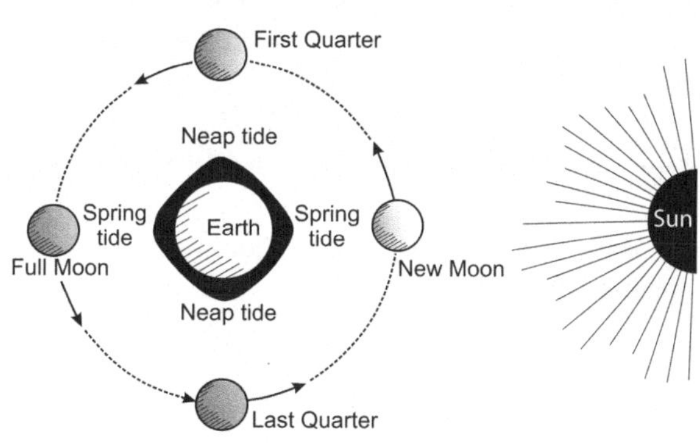

Fig. 12.5 Diagram shows how tides are cause by the gravitational attractions of the Moon and the Sun

to friction). Hence, the oncoming (faster) waves impact the back of the waves and push them causing Refraction.

12.4.3 Ocean currents

The unidirectional flow of water is called a Current. Ocean currents are controlled by winds (like trade winds) or by solar heating of the earth. The currents are acted upon by the Coriolis Effect which, due to earth's rotation, has the tendency for currents in the Northern hemisphere to turn right, and left in the Southern hemisphere.

There are several types of currents. Surface currents result from the drift of the upper 50–100 m of the Ocean, caused by the drag of the wind. Thus, these currents follow the same pattern as the atmospheric circulation and tend to form large circular gyres. In the Northern hemisphere, these gyres rotate clockwise, and in the Southern hemisphere, counterclockwise. Surface currents are found in the upper 400 m of the Ocean and make up about 10 % of all the water in the Ocean. The speed of surface currents is greatest when closer to the Ocean's surface and decreases at about 100 m below the surface.

Deep water currents are also called Thermohaline circulation and are found below 400 m. These make up about 90 % of the Ocean. Like surface currents, gravity plays a role in the creation of deep water currents. However, besides gravity, density differences in water, is the principal cause for the formation of currents. Density difference is a function of changes in both water salinity and temperature. Warm waters hold less salt than cold ones; hence, warm water is less dense and rises toward the surface, whereas, the cold, salt laden water sinks.

Contrary to the upper few hundred meters of wind-driven circulation, the density-driven circulation that dominates below is called the Thermohaline circulation. This circulation is influenced by heating, cooling, freshening, and salinification of the water that produces regional density differences within the ocean. This circulation, for the most part, is an 'overturning' circulation in which warm waters flow pole ward near the surface and is subsequently converted into cold water that sinks and flows equator ward in the interior. Radiocarbon measurements show that the Thermohaline circulation turns over all the deep water in the Ocean every 600 years or so. Hence, this circulation is also known as the Global Conveyor Belt.

12.5 Coastal Erosion and Sediment Transport

Rigorous erosion of sea floor occurs in the surf zone, i.e., between shoreline and breakers (Fig. 12.6).

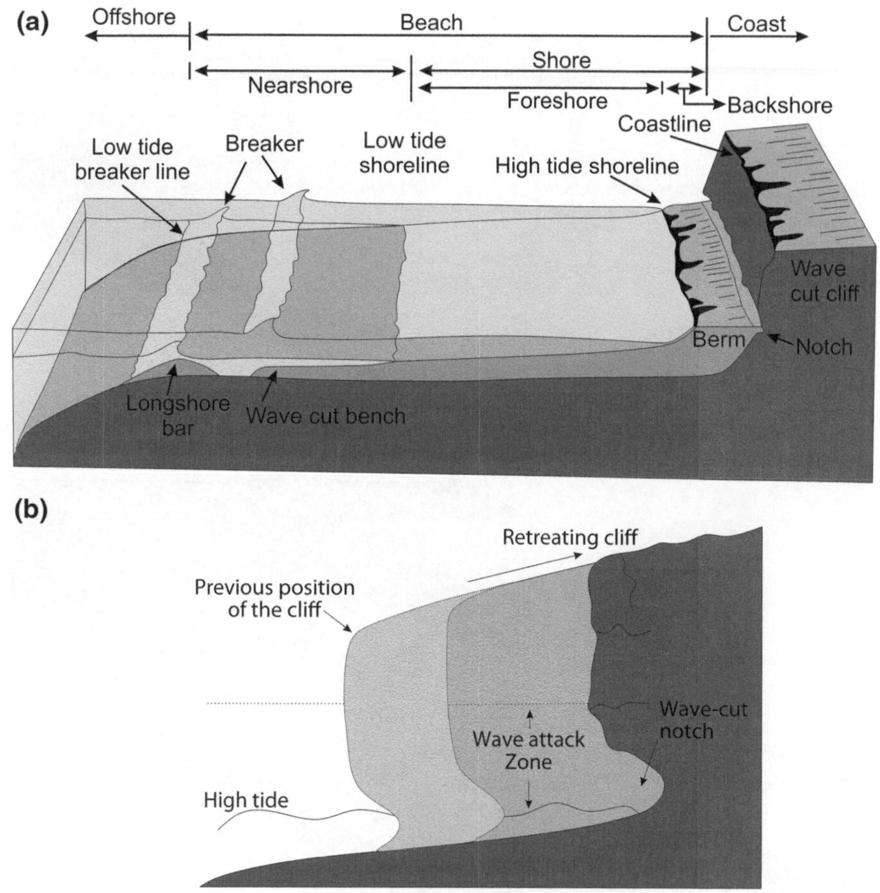

Fig. 12.6 Landforms of the coastal region (a) and retreating cliff (b)

12.5.1 Erosional Features

A Coast is the boundary between the sea and land (Fig. 12.6a). When waves hit the coast, they erode by breaking rocks into finer particles and abrading other rocks by flinging rocks, sand and water against them. A rocky coast is formed when a wave has had enough time to lower the coastline to the sea level. Due to resistance to erosion, a Wave cut bench and Wave cut cliff develops (Fig. 12.6b). If subsequent uplift of the Wave-cut bench occurs, a Marine terrace is formed (Fig. 12.7a). In areas of differential erosion, undercutting initially produces Sea caves. If Sea caves, from opposite sides of a rocky headland meet, then a Sea arch is formed (Fig. 12.7b). Gradual weakening of the Sea arch can lead to its collapse, thus, forming a Sea stack (Fig. 12.7b).

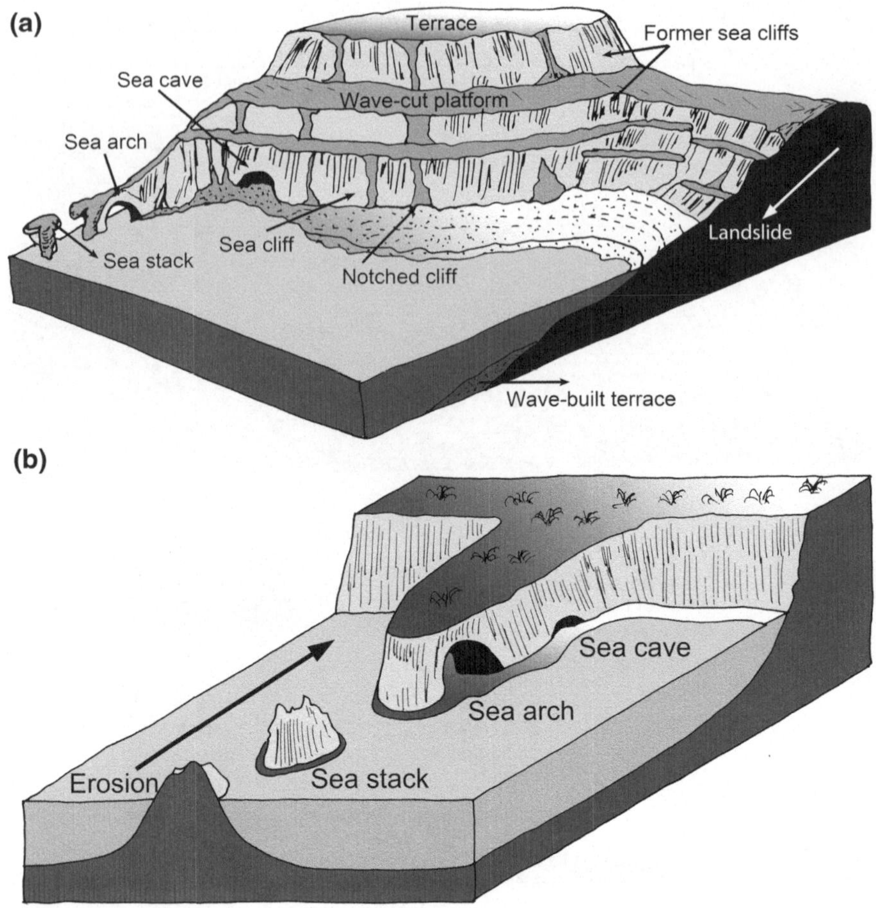

Fig. 12.7 Erosional features of the coast. **a** Broad view of the coast. **b** A more detailed view showing Sea stack, Sea arch and a Sea cave

Sediments are transported along the beach by waves. Waves rush onto the beach at an angle, but retreat perpendicular to the shore line due to gravity resulting in the swash of the incoming wave moving the sand up the beach in a direction perpendicular to the incoming wave crests and the backwash moving the sand down the beach perpendicular to the shoreline (Fig. 12.8). Thus, with successive waves, the sand will move along a zigzag path along the beach (Fig. 12.8). This is the Longshore current (Fig. 12.8).

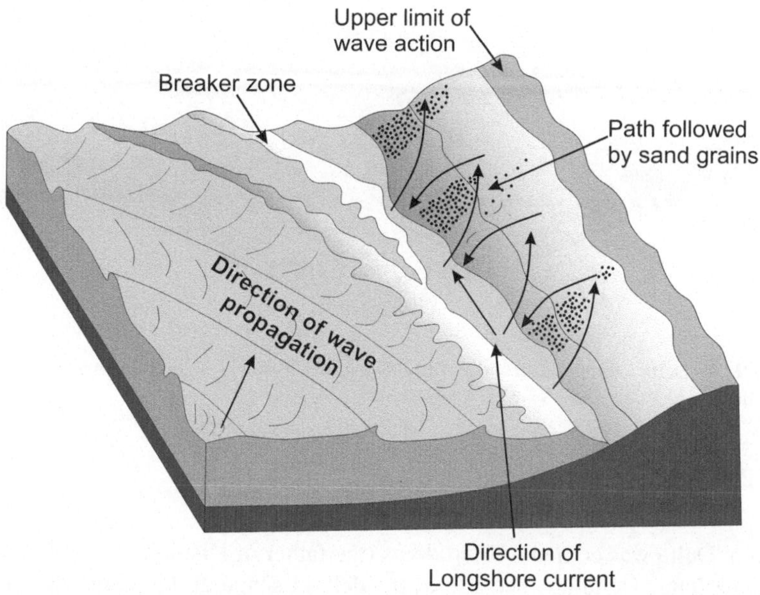

Fig. 12.8 The Longshore current. Redrawn from Earth's Dynamic Systems with permission of Eric H. Christiansen

12.6 Depositional Features Formed by Waves

The fine balance between wave energy and sediment supply defines the coastlines. Equilibrium is maintained if these two remain constant. However, if any one of these changes, shoreline also adjusts. For example, the winter storms increase wave energy. However, if the sediment supply remains constant, then the fine grained beach sand is carried offshore, resulting in the formation of a pebble or a cobble beach. Other depositional landforms formed by waves include marine Deltas that form due to the input of sediments from a river. Beach and Longshore drift currents also form Spits, Bay barriers, and Tombolos (see also Snead 1982). All these are briefly discussed below.

12.6.1 Beach

Beaches occur when sand is deposited along the shoreline (Fig. 12.6) (see also Hardisty 1990). A beach is divided into a foreshore zone (equivalent to the swash zone), and a backshore zone that is commonly separated from the foreshore by a distinct ridge, called a Berm (Fig. 12.6). Behind the backshore is a zone of cliffs, marshes, or sand dunes.

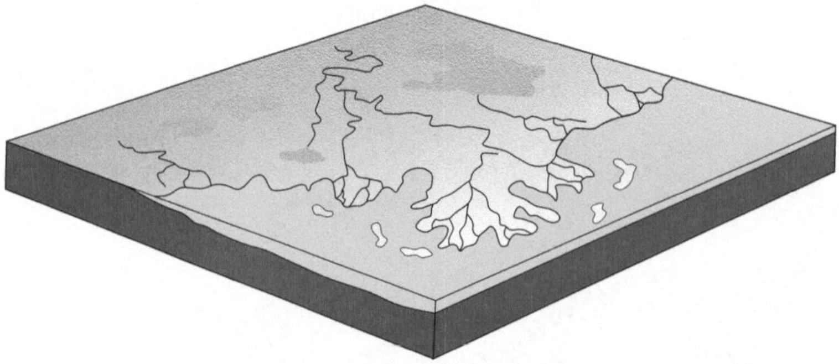

Fig. 12.9 A Delta. Redrawn from Earth's Dynamic Systems with permission of Eric H. Christiansen

12.6.2 Deltas

The term Delta was coined by Herodotus (the father of History, 484–425 BC) after the Greek letter Delta (Δ) because of the deltoid-shape at the mouth of the River Nile. L. D. Wright (1978) defined delta as coastal accumulations, both subaqueous and subaerial, of river-derived sediments adjacent to, or in close proximity to, the source stream, including the deposits that have been secondarily molded by various marine agents, such as waves, currents, or tides. Delta forms when sediment supply is greater than the ability of waves to remove (Fig. 12.9). They form where rivers meet the sea. The rivers carry a lot of sediment and these are deposited on the sea to form deltas. The deltas have three common characteristics: (a) presence of a large catchment, or drainage, basin (the area where all run-off water drains to the river), (b) they are all formed at the mouth of large river systems that carry

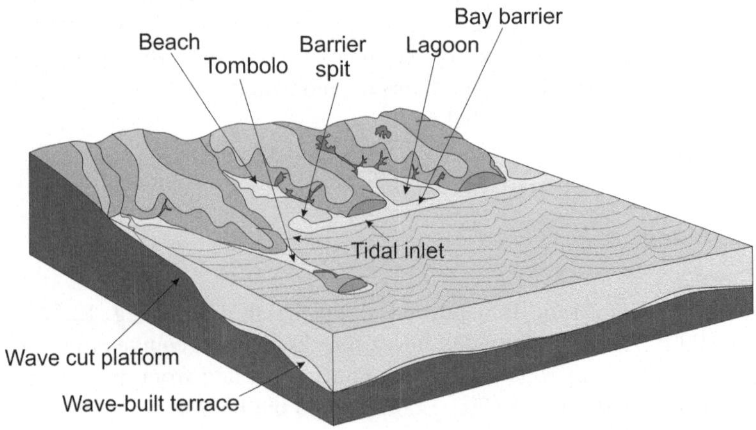

Fig. 12.10 Wave erosion, Spits, Bars, Bay barriers, Tombolos and Longshore drift currents

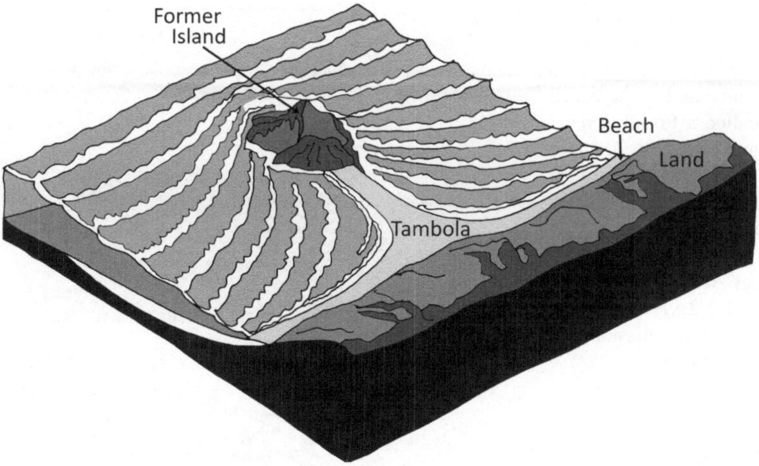

Fig. 12.11 A Tambola. It is formed when a spit continues to grow outwards (towards the Sea), thus, joining land to an offshore island

large quantities of clastic sediments (soils or portions of rocks that have been moved by water from where they formed), and (c) they are not near geologically active coastlines.

12.6.3 Spits

A spit is an area of sand or shingle which extends at a gentle angle out to sea or which grows across a river estuary. These are elongated deposits of sand or gravel that project from the land into open waters (Fig. 12.10). Many spits are characterized by a hooked or curved end (largely due to the refraction of waves around

Fig. 12.12 Barrier island. It is a narrow island of sand running along a shoreline that buffers the mainland from storms and large ocean waves. Redrawn from Earth's Dynamic Systems with permission of Eric H. Christiansen

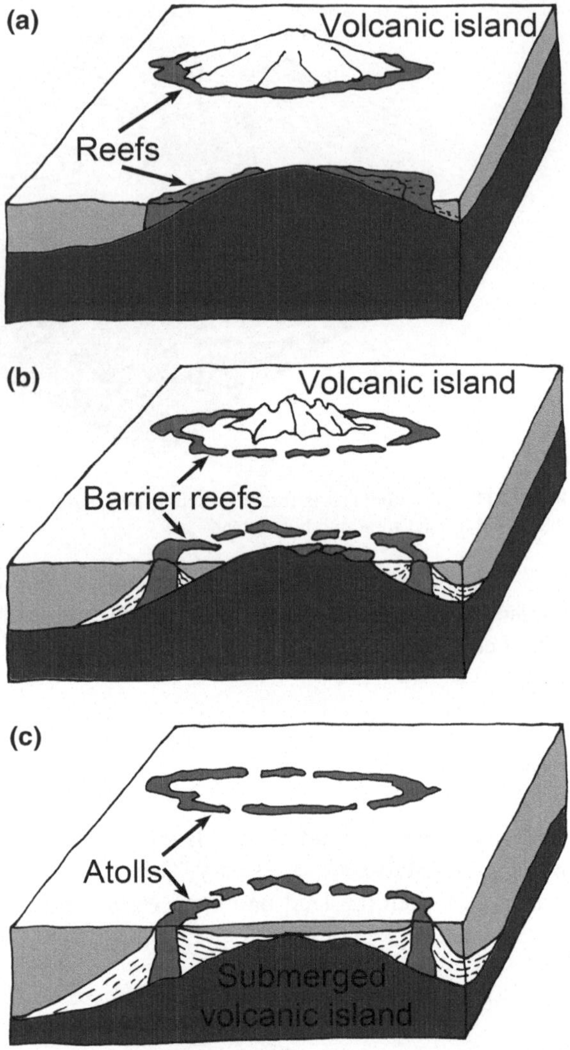

Fig. 12.13 Reefs and atoll. **a** When a volcanic island is rising or static, the reef remains attached to the beach and is called a Fringing reef. **b** As the island sinks, the reef continues to grow upward to form a Barrier reef. **c** Finally the island becomes submerged and the reef forms a circular Atoll. Redrawn from Earth's Dynamic Systems with permission of Eric H. Christiansen

the mouth of the bay). Spits usually form at the mouth of a bay due to long shore current and beach drift. However, spits develop particularly in places where (a) the Longshore drift moves large amounts of material along the beach, (b) there is a sudden change in the direction of the coastline, and (c) the sea is relatively shallow and becomes progressively more sheltered.

12.6.4 Bay Barriers

If a spit extends across a bay, it is called a Bay barrier (Fig. 12.10). These are elongated sand or pebble banks lying parallel to, but separate from the coastline. These form offshore bars, i.e., semi submerged sand deposits outside breaker zone. The exchange of water between the bay and the ocean is accomplished through the groundwater system.

12.6.5 Tombolos

A spit that connects the mainland to an offshore island is called a Tombolo (Fig. 12.11). Hence, it is a natural bridge formed when sand is deposited between the shoreline and an island. Tombolos are particularly vulnerable and depend on oceanic currents to continue their supply of sand on both sides of the landform. They can be breached or wiped out by a strong storm.

12.6.6 Barrier Islands

A Barrier island is a long narrow ridge of sand just offshore running parallel to the coast (Fig. 12.12) (see also Dolan and Lins 1987). Barrier islands are constantly changing. They grow parallel to the coast by beach drift and longshore drift, and they are eroded by storm surges that often cut them into smaller islands. Separating the island and coast is a narrow channel of water called a Lagoon.

12.6.7 Reefs and Atolls

Reefs consist of colonies of corals that secrete calcium carbonate. They form in shallow tropical seas as corals can only live in warm waters and need sunlight to survive. In the deep, reefs build upon the margins of volcanic islands, but they only do so after the volcanoes have become extinct. After volcanism ceases, volcanic island begins to erode and subside, due to the weight of the newly added material. As the island subsides, the reefs grow upwards. Eventually, the original volcanic island subsides and is eroded below sea level. However, the reefs trap sediment and a circular or annular island called the Atoll is formed (Fig. 12.13).

12.7 Summary

Ocean determines climate and plays a crucial role in earth's habitability. The Ocean basins are divided into Continental shelf, slope, and rise, Abyssal plains, Oceanic ridges, trenches, seamounts, Guyots, Transform faults, and Rift zones.

Rivers deposit large volume of sediments in the ocean. These are of four types—Terrigenous (often interchanged with Lithogenous sediments), Biogenous (Calcareous and Siliceous Oozes), Hydrogenous, and Cosmogenous sediment.

Deep-ocean waters are undersaturated in both calcium carbonate and opalline silica. Therefore, biogenic particles dissolve as they settle through the water column and as they sit on seafloor. The calcareous sediments are more prone to this effect. Hence, calcareous oozes are absent below a certain depth. This is called the Calcite Compensation Depth or CCD (often Carbonate is also used interchangeably for Calcite).

In regions where the input of terrigenous sediment is very high, such as along continental margins, the calcareous and siliceous components are diluted to less than 30 % of the total sediment volume (and hence, do not make the "biogenous ooze").

The ocean waves move through water in an oscillatory manner; floating objects move up and down rather than forward. Waves are generated by winds that blow over the surface of Oceans. The effectiveness of wind in generating a wave depends on three factors: (a) its average speed, which determines its force, (b) its duration, and (c) the extent of open water across which it blows (this is also called the Fetch).

Tsunamis are super destructive seismic sea waves triggered by massive changes to the sea floor. They are caused by earthquakes, volcanic eruptions, landslides, explosions, underwater nuclear tests and, more rarely, by the impact of cosmic bodies, such as meteorites or asteroids. Hence, any disturbance of appreciable magnitude, above or below the sea floor can trigger a Tsunami.

Tides cause the rise and fall of sea level to a meter or so, once or twice per day and are caused by the attraction of both Moon and Sun on the Earth. The highest high tide is called the Spring tide and occurs when the Sun, Moon and the Earth are in line on New and Full moon days. During the moon's quarter phases, the Sun and Moon work at right angles, causing the bulges to cancel each other. The result is a smaller difference between high and low tides. Such a tide is called a Neap tide. The rising tides produce flood currents, while falling tides produce ebb currents. Areas that are alternately submerged and exposed by the rising and falling of tides are called Tidal Flats.

Waves generally do not approach the shoreline parallel to the shore but bend as they enter shallow waters (primarily due to changes in their velocity). As the wave base touches the bottom, waves slow down. The part of a wave in shallow water moves slower than the part in deeper waters. Hence, when the depth under a wave crest varies with respect to the crest, waves bend.

The unidirectional flow of water is called a Current. Ocean currents are controlled by winds (like trade winds) or by solar heating of the earth. The currents are acted upon by the Coriolis Effect which, due to earth's rotation, has the tendency for currents in the Northern hemisphere to turn right, and left in the Southern hemisphere. There are several types of currents. Surface currents (less than 400 m deep) result from the drift of the upper 50–100 m of the ocean, caused by the drag of the wind. The deep water currents are also called Thermohaline circulation and are found below 400 m. These make up about 90 % of the Ocean. Like surface currents, gravity plays a role in the creation of deep water currents. However, besides gravity, density differences in water, is the principal cause for the formation of currents. Density difference is a function of changes in both water salinity and temperature. The Thermohaline circulation turns over all the deep water in the ocean every 600 years or so. Hence, this circulation is also known as the Global Conveyor Belt.

Rigorous erosion of sea floor occurs in the surf zone, i.e., between shoreline and breakers. A Coast is the boundary between the sea and land. A rocky coast is formed when a wave has had enough time to lower the coastline to the sea level. Due to resistance to erosion, a Wave cut bench and Wave cut cliff develops. If subsequent uplift of the Wave-cut bench occurs, a Marine terrace is formed. In areas of differential erosion, undercutting initially produces Sea caves. If Sea caves, from opposite sides of a rocky headland meet, then a Sea arch is formed. Gradual weakening of the Sea arch can lead to its collapse, thus, forming a Sea stack.

The depositional features formed by waves include—Beaches (pebble or a cobble beaches) and the Longshore drift currents form Spits, Bay barriers, Barrier Islands, and Tombolos. Other depositional landforms formed by waves include marine deltas that form due to the input of sediments from a river.

In the marine realm, after volcanism ceases, volcanic island begins to erode and subside, due to the weight of the newly added material. As the island subsides, the reefs grow upwards. Eventually, the original volcanic island subsides and is eroded below sea level. However, the reefs trap sediment and a circular or annular island called the Atoll is formed.

References

Bird E (2000) Coastal geomorphology, an introduction. Wiley, New York

Bruun AF (1957) Deep sea and abyssal depths. In: Hedgpeth JW (ed) Treatise of marine ecology and paleoecology, vol 1, Ecology, Geological Society of America, pp 641–672

Cartwright DE (1999) Tides, a scientific history. Oxford University Press, Oxford

Defant A (1961) Physical oceanography, vol I. MacMillan, New York

Dietrich G (1963) General oceanography, an introduction. John Wiley and Sons, New York

Dolan R, Lins HL (1987) Beaches and barrier islands. Sci Am 250(7):146

Fairbridge RW (ed) (1966) The encyclopedia of oceanography. Van Nostrand Reinhold Co, New York

Fred EC (1990) Tsunami. In: Herbich JB (ed) Handbook of coastal and ocean engineering, vol I. Wave phenomena and coastal structures. Gulf Publishing Co., pp 591–634

Gonzalez RT (1999) Tsunami! Sci Am 262(5):57

Hardisty J (1990) Beaches form and process. Harper-Collins, New York

Masselink G, Hughes M (2003) An introduction to coastal processes and geomorphology. Hodder Arnold, London

Snead RE (1982) Coastal landforms and surface features, a photographic atlas and glossary. Hutchinson Ross, Stroudsburg

Trenhaile AS (1997) Coastal dynamics and landforms. Oxford University Press, USA 384p

Wessel P, Sandwell DT, Kim SS (2010) The global seamount census. Oceanography 23:24–33

Wright LD (1978) River deltas. In: Davis RA Jr. (ed) Coastal sedimentary environments. Springer-Verlag, New York, pp 5–68

Chapter 13
Aeolian Processes

13.1 Introduction

Wind has the ability to transport, erode, and deposit sediments (Fig. 13.1). Generally, Aeolian (or Eolian or Æolian) processes remind us of a desert landscape of shifting sand dunes (Simonett 1968; Brookfield and Ahlbrandt 1983; Abrahams and Parsons 1994). However, there are many other environments where wind is a significant land forming force. These include coasts, braided rivers, glacial outwash plains, and dustbowls. Wind plays a key role in a geological process that starts with the transportation of sediments, followed by erosion and its eventual deposition. Deposition results from velocity decrease, and/or due to the presence of obstacles in its course. But, for these processes to occur, limited vegetation cover is a prerequisite. Vegetation protects the ground by binding the surface and keeping it moist, but more importantly by reducing wind velocity due to boundary layer friction.

13.2 Process of Wind Erosion

Wind erosion occurs by Deflation, Abrasion, and Attrition (see also Greeley and Iversen 1985). These are briefly defined below.

13.2.1 Deflation

It is the removal of loose fine-grained particles by wind (entrainment), and concentrating the coarser ones at the surface resulting in a surface that is largely composed of coarse grains that cannot be transported by the wind. A landscape thus formed is called a Desert Pavement (Fig. 13.2).

S. Jain, *Fundamentals of Physical Geology*, Springer Geology,
DOI: 10.1007/978-81-322-1539-4_13, © Springer India 2014

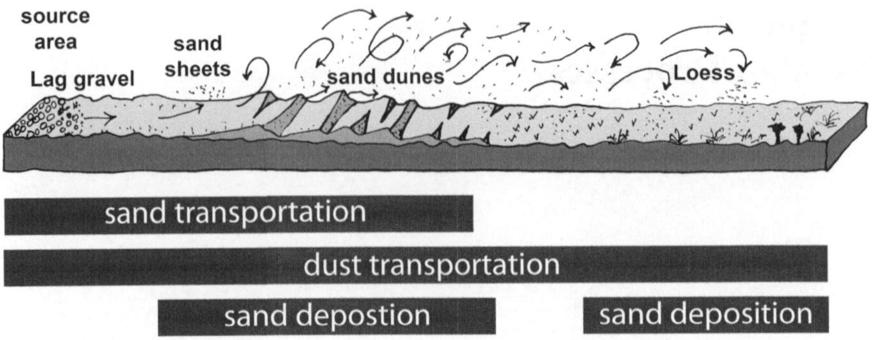

Fig. 13.1 Sediment transport by wind. Redrawn from Earth's Dynamic Systems with permission of Eric H. Christiansen

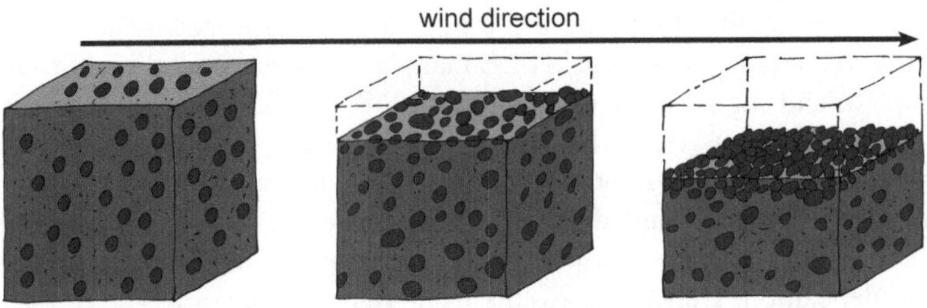

Fig. 13.2 Desert pavement. Fine-grained material is removed by wind, leaving a concentration of larger particles that form desert pavement. Redrawn from Earth's Dynamic Systems with permission of Eric H. Christiansen

13.2.2 Abrasion

It is the sand blasting of the rock surface by wind resulting in a polished but scratched and worn out surface. The degree of wind abrasion is a function of the character of the blow (the rock particles) and the bedrock that receives the blow. The resultant effect is maximum when the blow particles are hard, bedrock is soft, and the velocity of the wind is high. Those rocks made of hard and soft layers get abraded differentially, thus, producing a structure that resembles a Honeycomb (Fig. 13.3).

13.2.3 Attrition

Attrition occurs when windborne sediments roll against each other in collision, thereby wearing out each other so that their resultant grains are rounded like the Millet seed. Attrition is thus, a function of both wind velocity and the length (of time) of grain-to-grain interaction. Hence, the rock particles are not only abraded when exposed to the bedrock but also by colliding against each other.

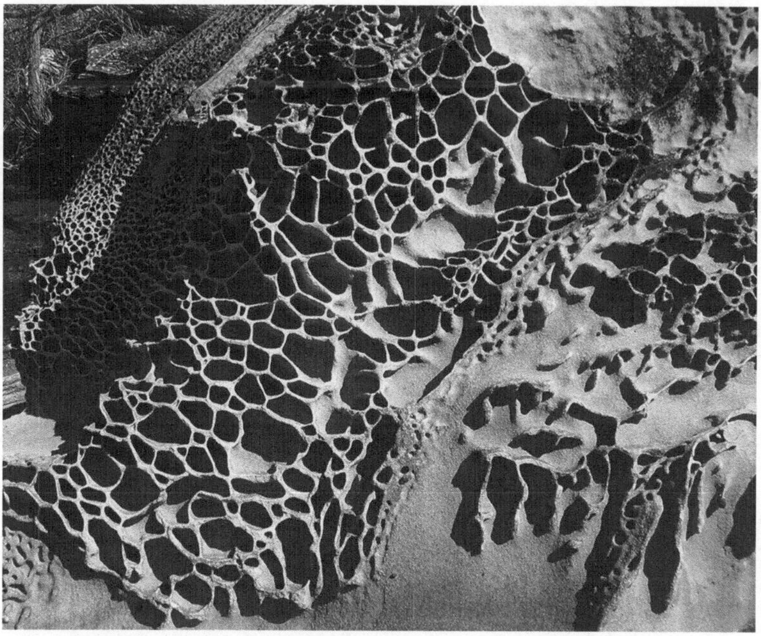

Fig. 13.3 Honeycomb structures. Wind abrades differentially hard and soft rocks thereby producing a structure that resembles a honeycomb

13.3 Sediment Transport by Wind

Wind transports and deposits sediment. Suspension, Saltation, and Creep are the three types of movement that occur (Fig. 13.4). Majority of the soil (>93 %) is blown away up to or below 1 m. Larger grains cannot move for a longer distances, hence, they accumulate in lines perpendicular to the wind direction and form sand ripples. The smaller particles, however, are suspended in wind and travel longer distances. These are windblown dust called Dust storms, which occur at lower levels in the atmosphere (<1 km).

Aeolian features form in areas where wind is the primary source of erosion (Brookfield and Ahlbrandt 1983; Greeley and Iversen 1985). The particles deposited are of sand, silt, and clay size (Table 13.1). These particles are entrained in by one of the four processes (Creep, Lift, Saltation, and Impact transport). Creep (Roiling) is when a particle rolls or slides across the surface (Fig. 13.4). Lift is when a particle rises off the surface due to the Bernoulli Effect (the same mechanism which causes aircraft to rise). If the flow is turbulent, larger particles are trajected by a process known as Saltation. As the grains fall back to the surface, they dislodge other grains and are then get carried further until they collide with the ground to dislodge other particles. Abrasion from the ballistic impact of

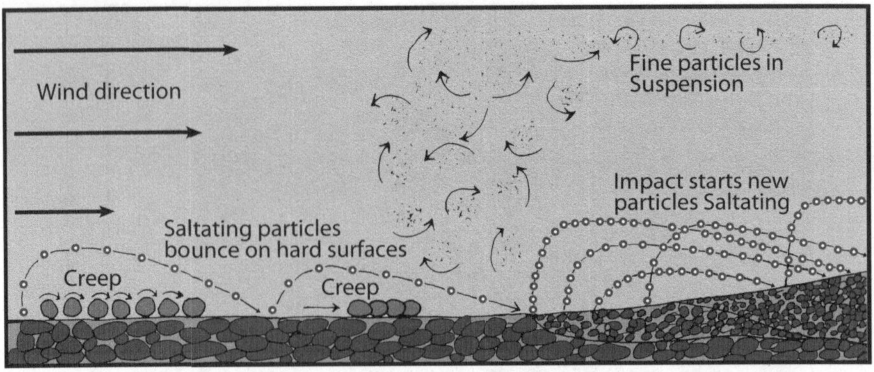

Fig. 13.4 The process of sediment transport. The deposited particles (of sand, silt, and clay size) are entrained in by one of the four processes (Creep, Lift, Saltation, and Impact transport). When a particle rolls or slides across the surface, it is called Creep (Roiling). Lift is when a particle rises off the surface due to the Bernoulli effect. If the flow is turbulent, larger particles are trajected by a process known as Saltation. Impact transport occurs when one particle strikes another causing the second one to move ahead

Table 13.1 The size of particles (or Grain size) by class

Size range (metric)	Size range (approx. inches)	φ scale	Aggregate name (Wentworth class)	Common names
<256 mm	<10.1 in	<−8	Boulder	
64–256 mm	2.5–10.1 in	−6 to −8	Cobble	
32–64 mm	1.26–2.5 in	−5 to −6	Very coarse gravel	Pebble
16–32 mm	0.63–1.26 in	−4 to −5	Coarse gravel	
8–16 mm	0.31–0.63 in	−3 to −4	Medium gravel	
4–8 mm	0.157–0.31 in	−2 to −3	Fine gravel	
2–4 mm	0.079–0.157 in	−1 to −2	Very fine gravel	Granule
1–2 mm	0.039–0.079 in	0 to −1	Very coarse sand	
½–1 mm	0.020–0.039 in	1 to 0	Coarse sand	
¼–½ mm	0.010–0.020 in	2 to 1	Medium sand	
125–250 μm	0.0049–0.010 in	3 to 2	Fine sand	
62.5–125 μm	0.0025–0.0049 in	4 to 3	Very fine sand	
3.90625–62.5 μm	0.00015–0.0025 in	8 to 4	Silt	Mud
<3.90625 μm	<0.00015 in	>8	Clay	
<1 μm	<0.000039 in	>10	Colloid	

Size ranges define limits of classes that are given names in the Wentworth scale (or Udden-Wentworth) used in the United States. The Krumbein phi (φ) scale, a modification of the Wentworth scale created by W. C. Krumbein, (Krumbein and Sloss 1963) is a logarithmic scale. Gravel is anything larger than sand (comprising granule, pebble, cobble, and boulder in the table above)

saltated grains can be a significant producer of small-scale polishing and sculpturing of the rock. Impact transport occurs when one particle strikes another causing the second one to move ahead.

13.4 Erosional Landforms

13.4.1 Desert Pavement

Desert pavements (Fig. 13.2) are flat areas that are free of vegetation. They are covered with tightly packed angular to subrounded gravels (gravel is any loose rock that is larger than 2 mm; Table 13.1). Desert pavements cover areas ranging from a few square meters to hundreds of square kilometres. They occur mostly in sand-poor regions, such as desert plains near bedrock outcrops, on plateaus, in dry wadis and terraces, and on alluvial fans. Thus, they generally characterize hot and arid regions (Abrahams and Parsons 1994).

13.4.2 Rock Pedestal

This is also called a Mushroom rock (Fig. 13.5), and is often found in hot and arid regions (generally deserts). They forms when erosion of an isolated rocky outcrop progresses at different rates at its bottom than at its top. Wind erodes the softer materials from the bottom leaving the top hard resistant rock as a pedestal, thus, resembling a Mushroom (Fig. 13.5). Good examples of Pedestal rock are seen near Vermilion Cliffs, Lees Ferry, Arizona (USA). The Pedestal Rocks Scenic Area in the Ozark National Forest (Arkansas, USA) is another good example to find Rock Pedestal.

13.4.3 Zeugen

These occur in areas that contain parallel alternating hard and soft rocks. The rate of wind erosion is different and the lower soft portions of rocks are eroded fast and

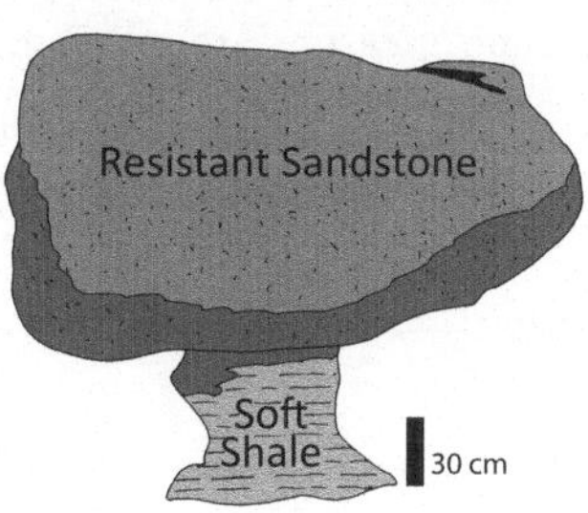

Fig. 13.5 Rock pedestal. This is also called a Mushroom rock. It forms when erosion of an isolated rocky outcrop progresses at different rates at its *bottom* than at its *top*. Wind erodes the softer materials from the *bottom* leaving the *top* hard resistant rock as a pedestal

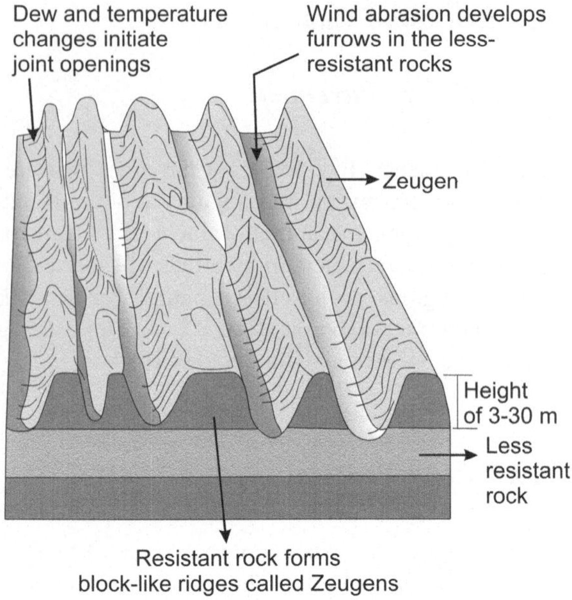

Fig. 13.6 Zeugen. They are parallel, flat-topped ridges of hard rock up to 30 m high formed as the rate of wind erosion progresses differently on the lower soft portions of rock (that are eroded fast and become *narrow*) as compared to hard resistant upper portions of the unit, giving it an appearance of a table on soft rocks. The hard rock stands out as a ridge. Modified from www.revisionworld.co.uk

become narrow but the upper portions of the hard rocks look like tables on soft rocks. These are known as Zeugen (Fig. 13.6). They are parallel, flat-topped ridges of hard rock up to 30 m high. The hard rock finally stands out as a ridge. A classic example of Zeugen is noted in the Monument Valley, located on the southern border of Utah with northern Arizona, USA.

13.4.4 Yardang

They resemble Zeugen (Fig. 13.7). Bands of softer rocks are eroded into long narrow corridors by the hard impacting winds. These rocks resemble ribs (being parallel to each other) with steep slopes. These streamlined wind-eroded ridges are commonly found in deserts, and measure around 7 m in height and 10–14 m in breadth.

13.4.5 Mesas and Buttes

Mesa in Spanish means "Table." Hence, Mesa is a flat, table-like landmass with a very resistant horizontal top layer and steep sides (Fig. 13.8). As erosion proceeds, columns of rock called Buttes remain that jut out of the landscape.

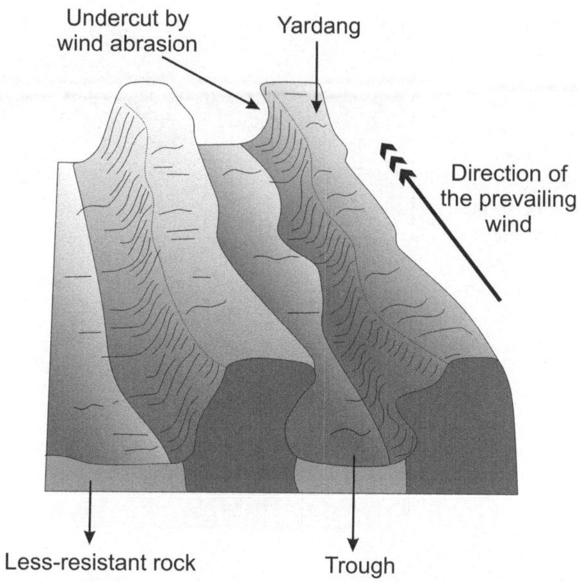

Fig. 13.7 Yardangs. These are bands of softer rocks are eroded into long *narrow* corridors by the hard impacting winds. They resemble Zeugen. Modified from www.revisionworld.co.uk

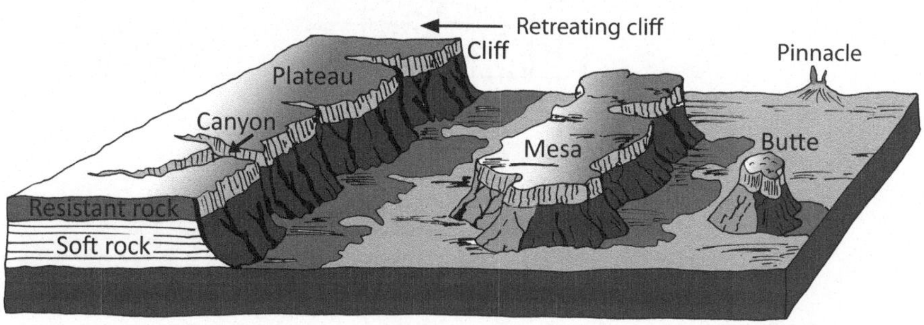

Fig. 13.8 Plateau, Mesa, Butte and Pinnacle. Redrawn from Earth's Dynamic Systems with permission of Eric H. Christiansen

13.4.6 Inselberg

In the German language it is "Island Mountain." Inselberg is an isolated residual hill rising abruptly from the ground level and consisting of steep slopes and rounded tops. It is often made of gneiss and granitic rocks. It is also called Bornhardt (Fig. 13.9), named after the South African scientist Wilhelm Bornhardt (1864–1946) who used this term while explaining its origin in 1900.

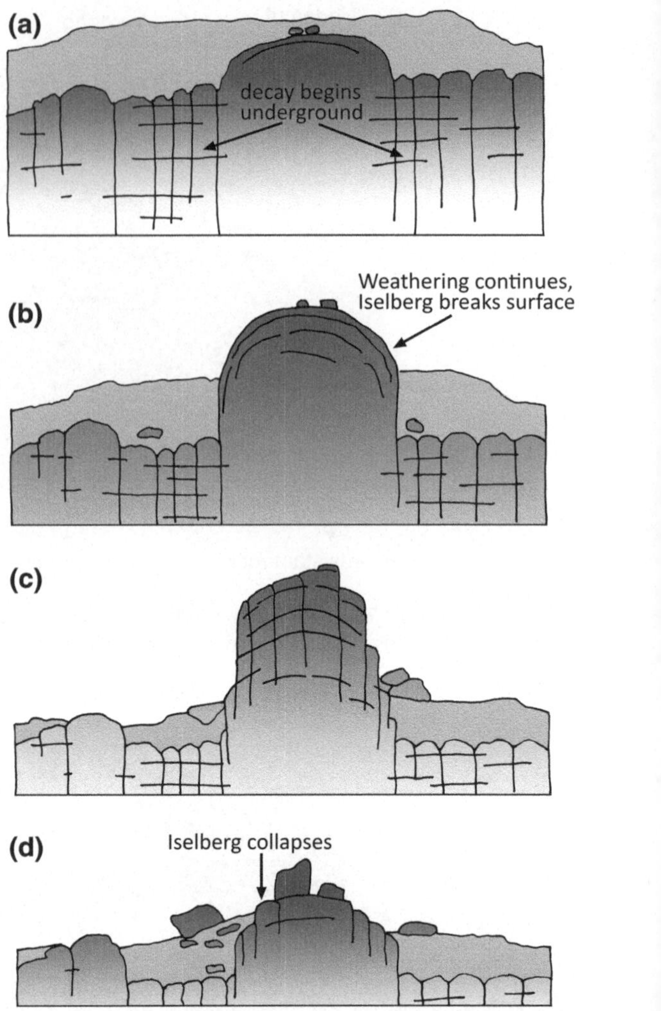

Fig. 13.9 Bornhardt. also called "Island Mountain" or Inselberg. These are bare surfaces, dome-like summits, and steep sides becoming steeper towards the base, marked by the absence of talus, alluvial cones or soils. The rounded tops are often made of hard rocks (a Gneiss or a Granite). Modified from www.revisionworld.co.uk

13.4.7 Ventifacts

Ventifacts are any bedrock surface or stone that has been abraded or shaped by windblown sediment in a process similar to sand blasting (Fig. 13.10). Ventifacts exhibit grooves and facets from the abrasion caused by windborne sand (Abrahams and Parsons 1994). Due to their dominantly triangular shape (facets), these are also called Brazil nuts. They typically form on valley floors and in open settings where wind transports sand and silt for great distances and literally sandblasts rocks.

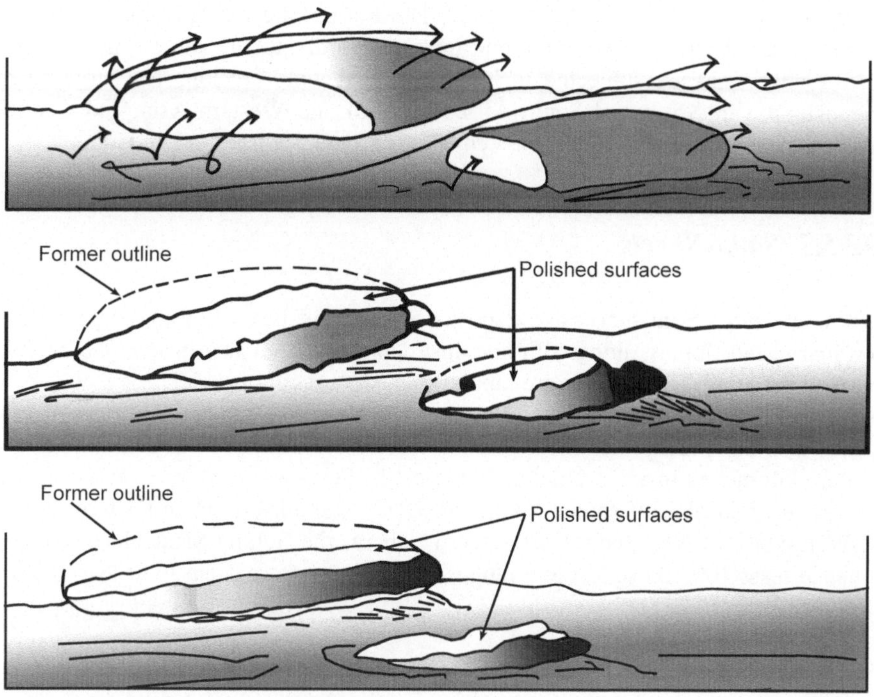

Fig. 13.10 Ventifacts. They are any bedrock surface or stone that has been abraded or shaped by windblown sediment in a process similar to sand blasting

13.5 Depositional Landform

Wind can deposit sediments when its velocity decreases to the point where the particles can no longer be transported. This happens when topographic barriers slow the wind speed on the downwind side of the barrier. As velocity decreases, some sediment in suspension can no longer be held, and thus, are dropped out to form deposits. Various landforms are formed by this process, such as Dunes, Barchans, Loess, etc., that occur at a range of differing spatial scales.

13.5.1 Sand Seas or Ergs

These cover vast areas (>100,000 sq km) and consist of large dune fields. Ergs are concentrated in two broad belts between 20–40° N and S latitudes. These are regions crossed by dry and subsiding air of the trade winds. Active Ergs are limited to regions that receive on an average and not more than 150 mm of annual rainfall. Additionally, sand seas and dune fields occur in regions downwind with ample source of dry and loose sand, such as in dry riverbeds, deltas, floodplains, glacial outwash plains, dry lakes, and beaches. These areas are not only located on the downwind side of the river beds but are also too dry to support extensive vegetative cover and are subjected to long-continued wind erosion. The largest Erg deposits are in North and South Africa, Central and Western Asia, and Central Australia. Ergs are also found on Venus, Mars, and Saturn's moon Titan.

13.5.2 Sand Sheets

These are built from successive deposits of sand left behind by the migration of ordinary sand ripples, along with fine dust deposited from suspension, and gravels moved by creep. Sand sheets are tabular deposits ranging in thickness from a few cm to a few meters. These cover similar large areas like sand seas but are of low relief, non dune areas with some grass cover. Most Sand sheets extend only a few square kilometers in and around the dune fields. Few others extend over thousands to tens of thousands of square kilometers. Best examples occur in Eastern Sahara (SW Egypt and NW Sudan). In Eastern Sahara, the Selima Sand Sheet extends over at least 100,000 sq km and ranges in thickness from 1 cm to at least 10 m.

13.5.3 Ripples

These are small scales (cm to m) bed forms that develop on sand surfaces.

13.5.4 Windblown Dust

Dust consists of silt and clay-sized particles (see Table 13.1 for particle size) that are often packed together with smooth surface. When dust is disturbed, dust storms develop, and are transported by wind over large distances. Most soil contains some silt and clay particles deposited by the wind.

13.5.5 Loess

Large deposits of wind dust are called Loess (Fig. 13.11). Much Loess was derived from the debris left by glacial erosion. Loess is formed mainly from silt-sized material (see Table 13.1 for particle size) and possesses high porosity, generally near 60 %. Usually Loess blankets the landscape in sheets from a few cm to several meters thick. The Loess Plateau (Fig. 13.12), also known as the Huangtu Plateau, is a plateau that covers an area of some 640,000 sq km in the upper and middle of China's Yellow River and China proper. The soil of this region is the most erodible soil on

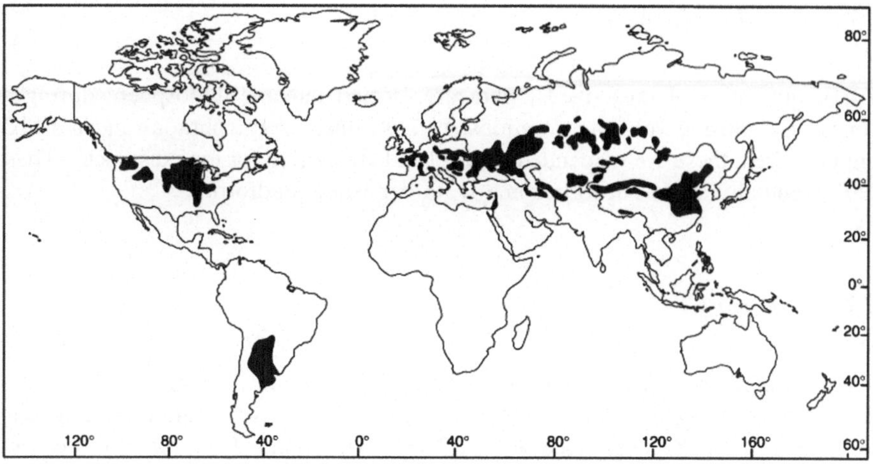

Fig. 13.11 Loess deposits around the world

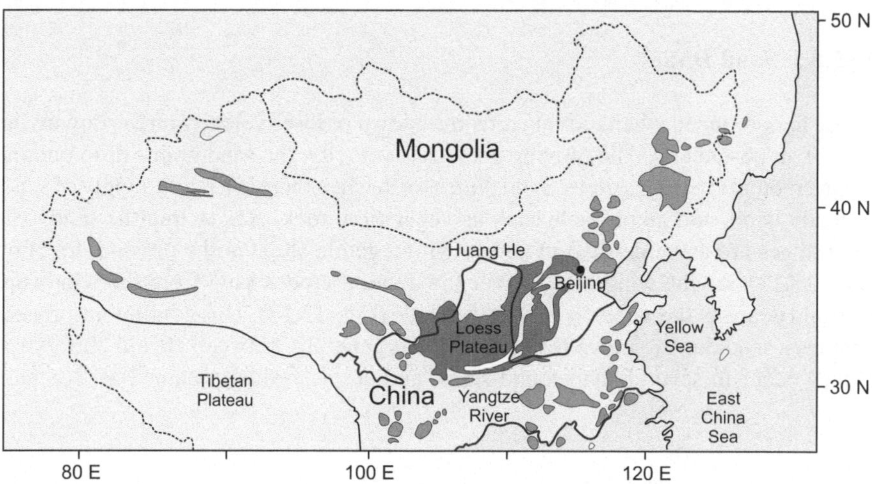

Fig. 13.12 Loess Plateau and the distribution of Loess within the Gobi Desert

Earth. In North America, the Palouse area in eastern Washington, the Great Plains and Midwest, Mississippi River valley, and Alaska exhibit large Loess deposits.

13.5.6 Dust in Ocean Sediments

Dust can be transported by wind onto oceans. Much of the fine-grained continent-derived sediment that reaches the abyssal plains of the oceans is due to this.

13.5.7 Volcanic Ash

Large quantities of dust-sized tephra are ejected into the atmosphere during at times of explosive volcanic eruptions. If these dust-sized tephra are ejected high enough, then, these are suspended in wind and are carried for long distances. These will eventually settle out to become wind-deposited landforms.

13.5.8 Dunes

These individual landforms can be as high as 500 m and as long as 10 s of km. They are often mobile, with an advancing slipface (downwind; high-angle) and an upwind low-angle stoss face (Fig. 13.13). The Sand Hills of north-central Nebraska (USA) have large dunes that were formed during the Pleistocene or Holocene Epochs.

13.5.8.1 Sand Dunes

These are produced when moving air slows down or loses velocity on the downwind side of an obstacle. As the moving air looses velocity, the sand grains drop out and form a mound creating a dune. Sand dunes form when there is a ready supply of sand, a steady wind, and an obstacle such as vegetation, rock, etc., to trap the sand. The sand dunes are asymmetrical in nature with a gentle slope in the upwind direction (Fig. 13.13). As the wind moves over the dune, it erodes sand from the windward side and deposits the same on the leeward side (Fig. 13.13). This continuous process enables a sand dune to move forward which is generally between 10 and 20 m/year. Dunes occur in several shapes and sizes, and are accordingly named as Barchan,

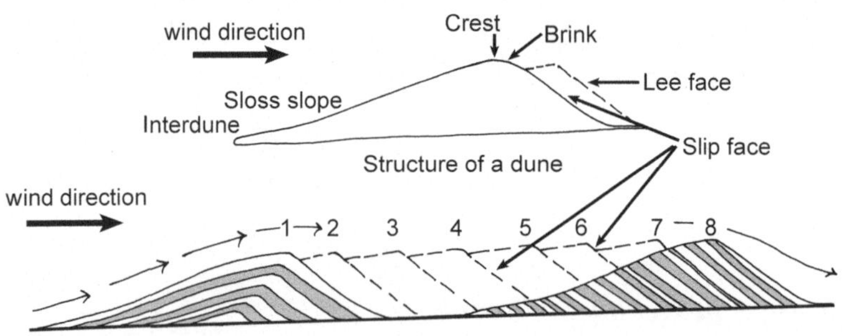

Fig. 13.13 Structure and migration of a Dune. They migrate when sand moves up the windward side and slides down the leeward slope. Such movement of the sand grains produces a series of crossbeds that slope in the direction of wind movement. Here, the dune migrates from 1 to 8. Redrawn from Earth's Dynamic Systems with permission of Eric H. Christiansen

Seif, Star, Parabolic, among others. Their shape is a product of several factor namely sand size, sand supply, wind speed, wind direction, and vegetation cover.

13.5.8.2 Sand Dune Types

Geomorphologists have classified major types of sand dunes (Simonett 1968; Pye and Tsoar 1990; Livingstone and Warren 1996; Brooks and Agate 1997) as (Fig. 13.14):

Barchan

These are also called Crescent dunes. A Barchan dune is an arc-shaped sand ridge, comprising of well-sorted sand grains. This type of dune possesses two "horns" that face downwind, with the slipface (the downwind slope; the long axis is transverse to the dominant wind direction) at the angle of repose of sand, or approximately 35° (Figs. 13.14a, b). Barchan dunes have a single slipface. They form in areas where the surface is hard and relatively flat coupled with a moderate supply of sand, and constant wind direction. Barchan dunes have been observed on Mars, where the thin atmosphere produces fierce winds, strong enough to move sand and dust.

Dome

These are modifications of stationary barchans. And are mounds of sand that are circular or elliptical in shape. There are no slipfaces.

Barchanoid

It is a long, asymmetrical dune that runs at right angles to the prevailing wind direction. A barchanoid ridge consists of several joined barchan dunes and looks like a row of connected crescents. Each of the Barchan dune produces a wave in the barchanoid ridge. Barchanoid dunes occur when the sand supply is higher than what is needed to form Barchan dunes.

Transverse

These long asymmetrical dunes form at right angles to the wind direction and resemble sand ripples on a large scale (Figs. 13.14c, d). Transverse dunes form in areas where there is abundant supply of sand, constant wind direction and relatively weak wind speed. If the supply of sand increases, then, the Barchan dunes merge into Transverse dunes. Transverse dunes have a single long slipface.

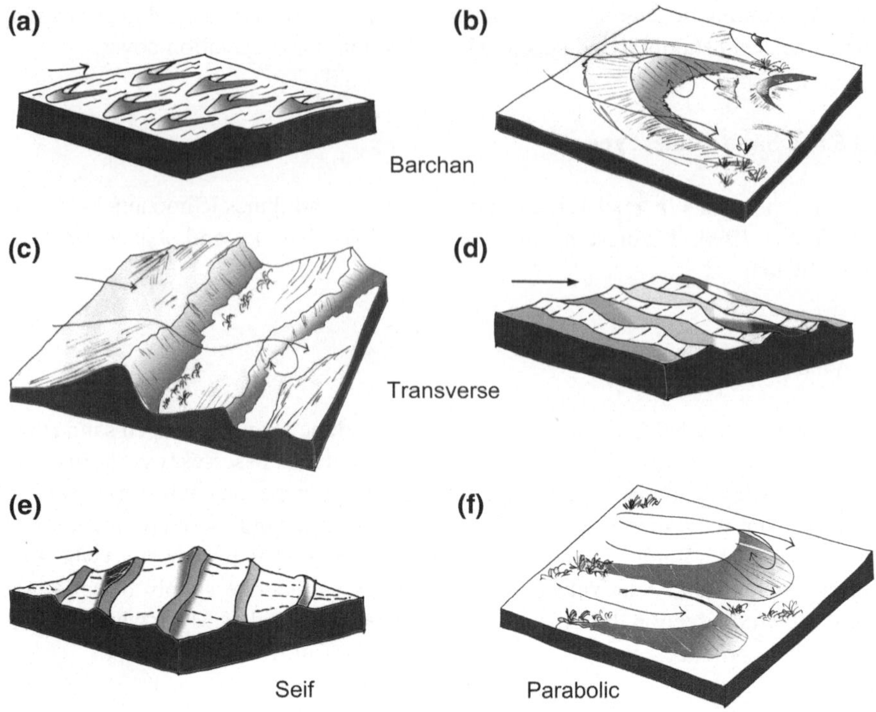

Fig. 13.14 Major types of sand dunes. **a–b** Barchan dunes form in areas that have a limited amount of sand, a nearly constant wind direction, and a generally flat, dry surface with little vegetation. The tips of barchan dunes point downwind. **c–d** Transverse dunes form long ridges of sand that are perpendicular to the prevailing wind direction in areas of little or no vegetation and abundant sand. **e** Seif dunes (also called Longitudinal dunes) form long, parallel ridges of sand aligned roughly parallel to the prevailing wind direction. They typically form where sand supplies are limited. **f** Parabolic dunes typically form in coastal areas that have a partial cover of vegetation, a strong onshore wind, and abundant sand

Seif

These are subtype of Longitudinal dunes (Fig. 13.14e). They are somewhat shorter with a more sinuous ridge.

Parabolic

These are also called Blowouts. These are crescent-shaped dunes (and more commonly U shaped) whose long axis is transverse to the dominant wind direction and an open end facing the upwind direction (Fig. 13.14f). They occur where there is abundant vegetation and sand supply, and a constant wind direction. They have multiple slipfaces. They are common in coastal areas.

Longitudinal

Sinuous dune can be more than 100 km long and 100 m high and are formed when there are strong winds from at least two directions (Fig. 13.14). The dune ridge is symmetrical, aligned parallel to the net direction of the wind. They have slipfaces on either side.

Star

These are large pyramidal or star-shaped dunes with three or more sinuous radiating ridges from a central peak of sand. These have several arms and variable slipface directions with three or more slipfaces. They form in areas where there is abundant sand with variable wind directions. They do not migrate along the ground, but grow vertically.

Reversing

These are intermediate dunes between Star and Transverse. Their ridge is asymmetrical with two slipfaces.

13.6 Deserts and Desert Types

Deserts are arid regions where average annual rainfall is less than 250 mm and where evaporation exceeds precipitation. Here more water is lost by evaporation and transpiration than gained by precipitation. Hence, drought characterizes such regions. The deserts are ever changing in size, distribution, and character. They are classified as extremely arid (i.e., 12 consecutive months without rain), arid or semiarid. Most deserts are located between latitudes of 30° N and 30° S (in the subtropical high belt), rainshadow and Polar Regions, or in the westside of continents (Fig. 13.15). These locations are controlled by the global circulation pattern. Temperature changes can be extreme in mid latitude deserts ranging from 57 °C in Algeria to −20 °C in Arizona and much colder in Antarctica. Deserts occur in areas close to the poles down to areas near the Equator (Fig. 13.11). The People's Republic of China has both the highest desert; the Qaidam Depression is 2,600 m above sea level, and one of the lowest ones, the Turpan Depression which is 150 m below sea level (Fig. 13.12). Deserts originate by several different mechanisms resulting in several different types such as subtropical deserts, continental deserts, rain shadow deserts, coastal deserts, and polar deserts. These are briefly discussed below.

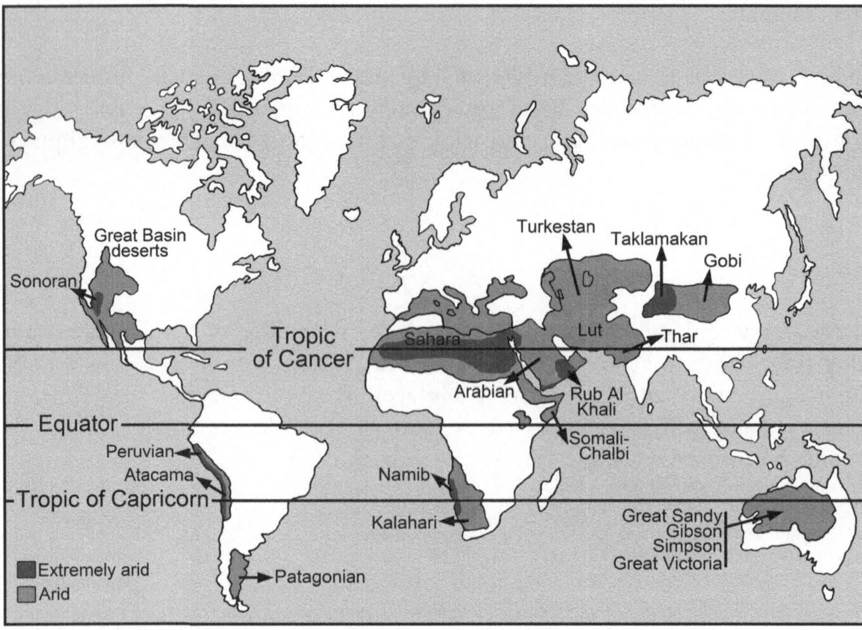

Fig. 13.15 Major deserts of the world. Note the global concentration of deserts at 30° N (Tropic of Cancer) and S (Tropic of Capricorn) latitudes

13.6.1 Subtropical Deserts (Trade Wind Deserts)

The trade winds in two belts on the equatorial sides of the Horse Latitudes heat up as they move toward the Equator. These dry winds dissipate cloud cover, allowing more sunlight to heat the land. Most of the major deserts of the world lie in areas crossed by the trade winds. The world's largest desert, the Sahara of North Africa, a trade wind desert, experiences temperatures as high as 57 °C. The Kalahari of Southern Africa and the Great Australian Desert are other examples (Fig. 13.15).

13.6.2 Continental Deserts

They are found in continental interiors, far from the source of moisture where hot summers and cold winters prevail. Best examples are Gobi and Taklamakan deserts (Fig. 13.15). The Gobi Desert is a large region in central Asia and is the fifth largest desert in the world and covers parts of north-, northwestern China, and southern Mongolia with an approximate size of 1,295,000 km^2 in area. The desert is located on a plateau roughly 910–1,520 m above sea level. Due to this unique location, temperatures in the Gobi desert fluctuate between −40 °C in winter and

+50 °C in summer. The Taklamakan Desert is a desert in Central Asia, in the Xinjiang Uyghur Autonomous Region of the People's Republic of China. It is one of the largest sandy deserts in the world, ranking 15th in size in a ranking of the world's largest nonpolar deserts. It covers an area of 270,000 km. Given its relative proximity with the cold to frigid air masses in Siberia, extreme low temperatures are recorded in wintertime; at times well below −20 °C.

13.6.3 Rainshadow Deserts

Mountainous area causes air to rise and condense, thereby dropping its moisture as it passes over. Classic examples are deserts formed in the east of the Sierra Nevada Mountains, California, and Nevada (the Sonora Desert) (Fig. 13.15), where moderately high mountain range prevents moisture-rich clouds from reaching areas on the leeward side, or the protected side, of the range. Hence, a desert is formed in the leeward side "shadow" of the range.

13.6.4 Coastal Deserts

These deserts are formed where cold upwelling seawater cools the air above and hence, decreases its ability to hold moisture. The coastal deserts are generally found on the western edges of continents near the Tropics of Cancer (23.5° N) and Capricorn (23.5° S). The Atacama Desert of coastal Peru and the Namib Desert of coastal South Africa are classic examples (Fig. 13.15). The Atacama of South America is the earth's driest desert. Here, rainfall of 1 mm or more may occur as rarely as once in every 5–20 years.

13.6.5 Polar Deserts

Polar deserts are areas with an annual precipitation of less than 250 mm and a mean temperature during the warmest month of less than 10 °C. Hence, here cold dry air prevails and the available moisture remains frozen throughout the year. Best examples are Northern Greenland, and the ice-free areas of Antarctica. Polar deserts cover nearly 5 million sq km of the earth's surface and are mostly bedrock or gravel plains. Snow dunes commonly occur. Temperature changes in Polar deserts frequently cross the freezing point of water. Thus, this "freeze–thaw" alternation forms patterned textures on the ground.

13.6.6 Monsoon Desert

The word "Monsoon" is an Arabic word for "Season". Monsoon refers to a wind system with a pronounced seasonal reversal. Monsoons develop in response to temperature variations between continents and oceans. The southeast trade winds of the Indian Ocean provide heavy summer rains in India as they move onshore. As the monsoon crosses India, it loses moisture on the eastern slopes of the Aravalli Range. The Thar Desert is a good example of a Monsoon desert (Fig. 13.15).

13.6.7 Paleodeserts

Climatic conditions have changed considerably in the recent geologic past. During the last 12,500 years, parts of deserts were more arid than they are today. About 10 % of the land between 30 °N and 30 °S was covered by sand seas. Nearly 18,000 years ago, sand seas in two vast belts occupied almost 50 % of this land area. Tropical rain forests and savannahs were present between these two belts. Fossil desert sediments, as old as 500 Ma, have been found in many parts of the world. Additionally, sand dune-like patterns have also been recognized in presently nonarid environments. Many such relict dunes now receive from 80 to 150 mm of rain each year and some ancient dunes are in areas now occupied by tropical rain forests. The Nebraska Sand Hills is one such an inactive 57,000 sq. km dune field in Central Nebraska, USA. It is also the largest sand sea in the Western Hemisphere, now stabilized by vegetation with $\sim$500 mm of rain annually. Here dunes are as high as 120 m.

13.6.8 Extraterrestrial Deserts

Mars is the only other planet that has wind-shaped (Aeolian) features. Although its surface atmospheric pressure is only about one-hundredth of earth, its global circulation patterns have formed a circumpolar sand sea of more than 5 million sq km, an area greater than the Empty Quarter of Saudi Arabia, the largest sand sea on our planet. Martian sand seas consist predominantly of crescent-shaped dunes on plains near the perennial ice cap of the North Polar area. Smaller dune fields occupy the floors of many large craters in the Polar region.

13.7 Surface Processes and Desert Landforms

Earth surface processes are influenced by lack of vegetation cover, wind, water, and physical weathering processes. Bedrock landscapes are the most common of arid landscapes, covering about 40 % of the total area. Mountains, Plateaus, Piedmonts, and Pediments occur. Physical weathering and erosion processes wear down the landscape based on the strength of the rocks. Fluvial forms cover 10–15 % of arid regions. Alluvial fans, Flood plains, River channels, and Badlands all occur in arid landscapes. The fluvial action of ephemeral Playas is a Salt lake, usually dry, but filled after heavy rain. They are quite common, but small, and cover only about 1 % of the arid zone. Sand landforms are quite common, but only account for about 30 % of the arid landscape. Desert flats are undifferentiated landscape types that cover about 15–20 % of the arid zone.

13.7.1 Weathering and Mass Wasting Processes

Deserts have little soil cover due to low moisture, slow rate of chemical weathering, and little vegetation cover. Mechanical weathering processes dominate here (Well and Haragan 1983; Atkinson 2004). In comparison to humid regions, the deserts are dominated by rock falls, rockslides, and the accumulation of coarse-grained sediments (for details see Chap. 8).

13.7.2 Streams and Fluvial Landforms

Surface water is rare in deserts. Streams that do flow, usually originate at higher elevations, and supply only sufficient water for them to pass through the desert region. Streams tend to be ephemeral as they come alive during the rainy season. For this reason, flash floods and braided streams are quite common (for details see Chap. 9).

13.7.3 Playa Lake

Due to scarce rainfall and intermittent input from streams, standing bodies of water like lakes are rare in a desert. Those that do occur, form during short periods of rainfall which later quickly evaporates, thus, leaving a dry lakebed behind. Such lakes are called Playa Lakes (also called dry lakes). These are mostly formed in basins of internal drainage. Their lakebeds consist of salts (called Evaporites) that were carried in by streams and precipitated during periods of intense evaporation.

These precipitated salts give the dry lakebed a white color resembling a beach (in Spanish playa means beach) (for more details on Playa lakes see Chap. 11).

13.7.4 Alluvial Fan and Bajada

An Alluvial fan is the washing of broken rocks and sediments from the mountain to the valley. It forms where a mountain stream enters a broad flat valley and deposits sediment. Velocity decreases as the stream enters a flatter valley. Typically, the smaller sediment washes to the bottom of the fan, and is called Silt. When a linear mountain range has several closely spaced valleys, the alluvial fans coalesce to form a gentle undulated slope on the sides of the bounding lowlands. Such coalesced alluvial fans are called Bajadas (for details on Alluvial Fans and Bajadas, see Chap. 11).

13.7.5 Pediment

A Pediment, usually formed in an arid region, is a broad gently inclined bedrock platform that extends outward from a mountain front. A Pediment is formed when running water erodes most of the mass of the mountain. Hence, it is a residual surface, formed by erosion and not by deposition. The highlands remain as residual hills as the pediment matures (for details on Pediment, see Chap. 11).

13.7.6 Inselberg

A prominent isolated residual knob, hill, or small mountain usually smoothed and rounded, rising abruptly and surrounded by an extensive lowland erosion surface in a hot, dry region (as in deserts of southern Africa or Arabia) is called an Inselberg. It is generally bare and rocky. Inselbergs are characteristic of an arid or semiarid landscape in the late stage of an erosion cycle. Inselbergs form because the rock making them up is more resistant to erosion than the rocks that once made up the surrounding plain. Although, Inselbergs are common in desert regions, they also occur in areas where differential erosion takes place.

13.7.7 Desertification

Climatic changes, such as changing positions of the continents, or changes in ocean and air circulation patterns facilitate Desertification. Other contributory factors include Human impacts, such as overgrazing, draining of land, and lowering of the groundwater table. As soil holds moisture, if it erodes (primarily due to lack of or dying out of vegetation), the area eventually becomes arid, resulting in the expansion of the desert.

13.8 Aridity

Arid and semiarid regions are distinguished on the basis of their annual precipitation. Deserts have an annual precipitation of less than 50 mm/year and are devoid of vegetation. Arid regions have 50–250 mm/year precipitation and with sparse vegetation whereas the Semiarid regions have a precipitation from 250 to 500 mm/year with a Steppe/Savannah/Prairie/Pampas vegetation. Most deserts and (semi) arid regions occur between 10 and 35° latitude (such as the Sahara and Kalahari deserts), or in the interior parts of the continents (such as the Australian and Gobi deserts) or in rain shadow areas of fold belts (such as in Peru and Nepal). A large part of the Arctic Tundra that receives precipitation less that 250 mm/year also qualifies as an arid region. A simple scheme of arid zone types, based on precipitation, is given in Table 13.2.

Based on changes in temperature, the zones are classified as hot, mild, cool, cold, and polar deserts (Table 13.3).

Taken together, these arid zones cover nearly half the global land area, occurring on all continents, and in all climatic zones. The general areas of the different arid zones are given in Table 13.4. Table 13.5 gives the distributions of arid zones on various continents.

13.8.1 Causes of Aridity

Arid zones are caused by climatic, topographic, location, and oceanographic factors. Over half of the arid zone is caused by atmospheric instability and is formed under the descending arm of the Hadley Cell circulation system (Fig. 13.16). Arid zones occur with distance away from the ocean due to the Continentality effect. Some arid zones are in the rain shadow areas, and downwind of major mountain ranges. Some arid zones are found inland of the cold ocean

Table 13.2 Types of arid zones based on precipitation

Zone Type	Annual Rainfall (mm)
Semiarid	250–500
Arid	50–200
Extremely arid	No seasonal rainfall regime
Polar	Very low snowfall precipitation regime.

Table 13.3 Types of Arid Zone based on temperature

Zone	In km^2	Percent of the arid zone
Hot	23×10^6 km^2	31
Mild	10×10^6 km^2	13
Cool	8×10^6 km^2	11
Cold	13×10^6 km^2	17
Polar	20×10^6 km^2	28

Table 13.4 Area of different arid zones

Zones	In km^2	Percent of the arid zone
Semiarid	24×10^6 km^2	32
Arid	24×10^6 km^2	32
Extremely (Hyper) arid	6×10^6 km^2	8
Polar	20×10^6 km^2	28

Table 13.5 Arid Zones on the continents

Continents	Area $\times 10^6$ km^2	Continent area (%)	Global arid zone area (%)
Africa	21	60	28
Asia	18.4	42	25
Australia	5.8	75	8
North America	4.4	18	6
South America	3.3	18	4
Europe	1.1	11	1
Polar	20	90	28

currents where there is low evaporation of the ocean. All these factors prevent moisture-bearing weather systems from reaching the land surface. Four factors, Atmospheric high pressure, Rainshadow, Cold ocean currents, and Continentality facilitate the formation of deserts. These interact together to produce arid conditions, but, in many cases only one is the dominant force.

13.8.1.1 Atmospheric High Pressure

Most deserts lie in the center or on the western coast of continents, between 15 and 30° N or S of the Equator (Fig. 13.15). Air which has been warmed at the equator rises, cools and eventually descends at 30° N and 30° S of the Equator. This descending air warms and, hence, its water-holding capacity increases, resulting in very low atmospheric humidity at these latitudes. This condition makes cloud formation or rain very unlikely.

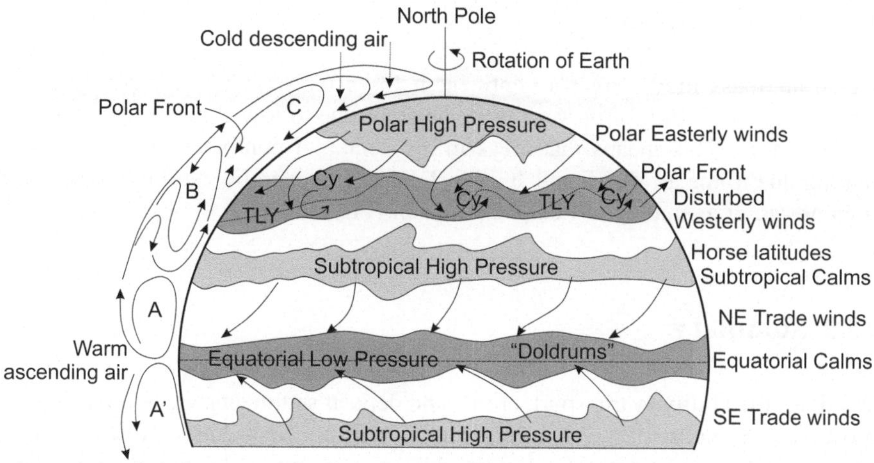

Fig. 13.16 General circulation in the lower atmosphere. The general circulation pattern of earth's atmosphere air flows from high-pressure zones to low-pressure zones, and the resulting winds are deflected to the *right* of their direction of movement (clockwise) in the Northern Hemisphere and to the *left* of their direction of movement (counterclockwise) in the Southern Hemisphere. A: Hadley Cell, B: Farrel Cell, and C: Polar Cell; Cy: Cyclones. TLP: temperate low pressure belts

13.8.1.2 Rainshadow

This effect is produced by tall mountain ranges (the Orographic belts). The windward slopes of such barriers are wetter than the leeward slopes, which experience an arid climate. An example of this Orographic effect is the western slopes of the Rockies and the Andes. As air rises on the windward side of the mountain, pressure decreases and it cools, resulting in condensation, cloud formation, and rain. As the air descends on the leeward sides of these mountains, pressure increases and the air is warmed, thus preventing condensation. It is suggested that the Gobi Desert of Central Asia (Fig. 13.15), which is largely north of the subtropics, has formed due to air descending from the Himalayas.

13.8.1.3 Cold Ocean Currents

Several deserts lie along the western coasts where, due to the action of circulating wind currents, there is upwelling of cold sea water. This cools the passing air, thereby reducing the amount of water that the air can hold, thereby limiting the amount of precipitation. The Atacama Desert on the western coast of South America and the Namib Desert on the western coast of Africa are classic examples (Fig. 13.15).

13.8.1.4 Continentality

As an air mass moves over a continent it loses moisture through precipitation. Equally, the air will take up very little moisture due to low evaporation rates over land surfaces. This means that areas in the center of continents have very little rainfall due to the air that is much drier. Deserts in Australia, North America, and Asia are good examples of this phenomenon (Fig. 13.15).

13.9 Summary

Wind has the ability to transport, erode, and deposit sediments. Generally, Aeolian processes remind us of a desert landscape of shifting sand dunes. However, there are many other environments where wind is a significant land forming force. These include coasts, braided rivers, glacial outwash plains, and dustbowls. Wind plays a key role in a geological process that starts with the transportation of sediments, followed by erosion and its eventual deposition. Deposition results from velocity decrease, and/or due to the presence of obstacles in its course. But, for these processes to occur, limited vegetation cover is a prerequisite.

Wind erosion occurs by Deflation, Abrasion, and Attrition.

Wind transports and deposits sediment. Movement occurs by three processes—Suspension, Saltation, and Creep. Majority of the soil (>93 %) is blown away up to or below 1 m. Larger grains cannot move for a longer distances, hence, they accumulate in lines perpendicular to the wind direction and form Sand ripples. The smaller particles, however, are suspended in wind and travel longer distances. These are windblown dust called Dust storms which occur at lower levels in the atmosphere (<1 km).

Aeolian features form in areas where wind is the primary source of erosion. The particles deposited are of sand, silt, and clay-sized. These particles are entrained in by one of the four processes (Creep, Lift, Saltation, and Impact transport).

Erosional landforms include–Desert Pavement, Rock Pedestal, Zeugen, Yardang, Mesas ,and Buttes, Inselberg, and Ventifacts.

Wind can deposit sediments when its velocity decreases to the point where the particles can no longer be transported. This happens when topographic barriers slow the wind speed on the downwind side of the barrier. As velocity decreases, some sediment in suspension can no longer be held, and thus, are dropped out to form deposits. Various landforms are formed by this process, such as Sand seas or Ergs, Sand sheets, Ripples, Windblown dust, Loess, Dust in ocean sediments, Volcanic ash, and Dunes, that occur at a range of differing spatial scales.

Sand dunes are produced when moving air slows down or loses velocity on the downwind side of an obstacle. Dunes occur in several shapes and sizes, and are accordingly named as Barchan, Dome, Barchanoid, Transverse, Parabolic, Longitudinal, Seif, Star, and Reversing dunes. Their shape is a product of several factor namely sand size, sand supply, wind speed, wind direction, and vegetation cover.

Deserts are arid regions where average annual rainfall is less than 250 mm and where evaporation exceeds precipitation. Here, more water is lost by evaporation and transpiration than gained by precipitated. Deserts originate by several different mechanisms resulting in several different types such as Subtropical Deserts (Trade wind Deserts), Continental Deserts, Rainshadow Deserts, Coastal Deserts, Polar Deserts, Monsoon Desert, Paleodeserts, and Extraterrestrial deserts.

Earth surface processes are influenced by lack of vegetation cover, wind, water, and physical weathering processes. Bedrock landscapes are the most common of arid landscapes, covering about 40 % of the total area. Mountains, Plateaus, Piedmonts, and Pediments occur. Physical weathering and erosion processes wear down the landscape based on the strength of the rocks. Fluvial forms cover 10–15 % of arid regions. Alluvial fans, Flood plains, River channels, and Badlands all occur in arid landscapes. The fluvial action of ephemeral Playas is a Salt lake, usually dry, but filled after heavy rain. They are quite common, but small, and cover only about 1 % of the arid zone. Sand landforms are quite common, but only account for about 30 % of the arid landscape. Desert flats are undifferentiated from other landscape types that cover about 15–20 % of the arid zone.

Fluvial Landforms include Playa Lake, Alluvial Fan, and Bajada, Pediment, and Inselberg.

Climatic changes, such as changing positions of the continents, or changes in ocean and air circulation patterns facilitate Desertification. Other contributory factors include Human impacts, such as overgrazing, draining of land, and lowering of the groundwater table. As soil holds moisture, if it erodes (primarily due to lack of or dying out of vegetation), the area eventually becomes arid, resulting in the expansion of the desert.

Arid and semiarid regions are distinguished on the basis of their annual precipitation. Arid regions have 50–250 mm/year precipitation and with sparse vegetation whereas the Semiarid regions have a precipitation from 250 to 500 mm/year with a Steppe/Savannah/Prairie/Pampas vegetation. Most deserts and (semi) arid regions occur between 10 and 35° latitude (such as the Sahara and Kalahari deserts), or in the interior parts of the continents (such as the Australian and Gobi deserts) or in rain shadow areas of fold belts (such as in Peru and Nepal). Large parts of the Arctic Tundra that receives precipitation less that 250 mm/year also qualifies as an arid region. Taken together these arid zones cover nearly half the global land area, occurring on all continents, and in all climatic zones.

Arid zones are caused by climatic, topographic, location, and oceanographic factors. Over half of the arid zone is caused by atmospheric instability and is formed under the descending arm of the Hadley Cell circulation system. Arid zones occur with distance away from the ocean due to the Continentality effect. Some arid zones are in the rain shadow areas, and downwind of major mountain ranges. Some arid zones are found inland of the cold ocean currents where there is low evaporation of the ocean. All these factors prevent moisture-bearing weather systems from reaching the land surface. Four factors, Atmospheric high pressure, Rainshadow, Cold ocean currents, and Continentality facilitate the formation of deserts. These interact together to produce arid conditions, but, in many cases only one is the dominant force.

References

Abrahams AD, Parsons AJ (eds) (1994) Geomorphology of desert environments. Chapman and Hall, London

Atkinson D (2004) Weathering slopes and landforms. Hodder Arnold, London

Brookfield ME, Ahlbrandt TS (1983) Eolian sediments and processes. Elsevier, New York

Brooks A, Agate E (1997) Sand Dunes: A Practical Handbook, 3rd edn. British Trust for Conservation Volunteers, Wallingford

Greeley R, Iversen J (1985) Wind as a geologic process. Cambridge University Press, Cambridge

Krumbein WC, Sloss LL (1963) Stratigraphy and sedimentation, 2nd edn. Freeman, San Francisco

Livingstone I, Warren AA (1996) Aeolian geomorphology, an introduction. Longman, Harlow

Pye K, Tsoar H (1990) Aeolian sand and sand dunes. Unwin Hyman, London

Simonett DS (1968) Wind action. In: Fairbridge RW (ed) Encyclopedia of geomorphology. Reinhold, New York, pp 1233–1236

Well SG, Haragan DR (eds) (1983) The origin and evolution of deserts. University of New Mexico Press, Albuquerque

Part IV
The Tectonic System

Chapter 14
Plate Tectonics

14.1 Introduction

Plate Tectonics provides a single unifying theory for understanding the dynamics of earth as everything about earth is mostly related either directly or indirectly to it (Condie 1998; Cox and Hart 1986; Hamblin and Christiansen 2003; Kearey et al. 2009). Few decades ago, it was believed that continents and ocean basins were fixed and permanent features and that the Theory of Continental Drift was just a radical idea (Hallam 1973).

14.2 Continental Drift

It was the predecessor of the theory of Plate Tectonics. Scientists noted that continents, particularly Africa and South America, fit together like a jigsaw puzzle (Fig. 14.1). In 1929, Alfred Wegener in his book "*The Origin of the Continents and Oceans*" ("*Die Entstehung der Kontinente und Ozeane*"), noted that the shapes of the continents, and their geology are similar (Fig. 14.1). He used geological, fossil, and glacial evidences gathered on the opposite sides of the Atlantic Ocean to support his theory of Continental Drift. He drew a series of maps showing the three stages in the drifting process; starting with an initial large super continental landmass called Pangea, whose breakup and the theory of continental drift were supported by several evidences from numerous regional geological studies (see also Kearey et al. 2009). These evidences are briefly enumerated below:

14.2.1 Paleontological Evidence

Although fossil records indicate that a new species appears at one place and then disperses outward. Additionally, floating and swimming organisms migrate in the

S. Jain, *Fundamentals of Physical Geology*, Springer Geology,
DOI: 10.1007/978-81-322-1539-4_14, © Springer India 2014

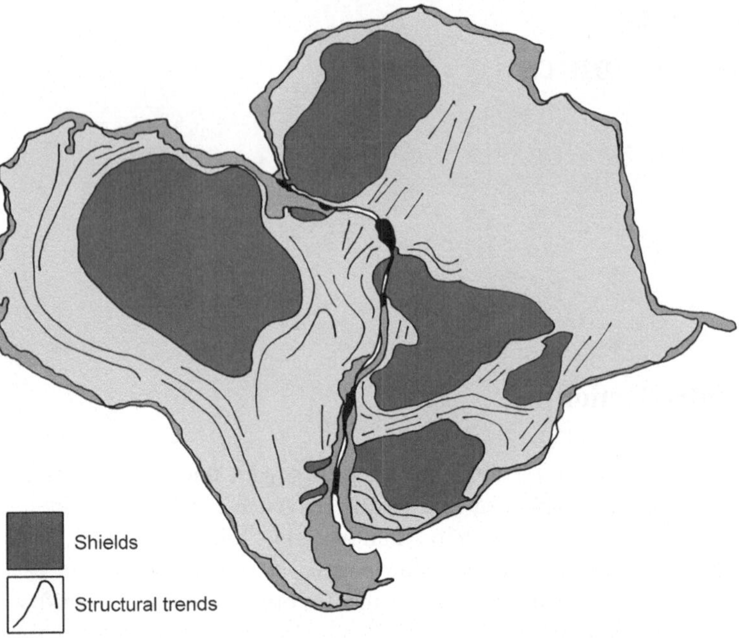

Fig. 14.1 The geological evidence of the past link between South America and Africa. Rock type, structure, sequence, fossils, ages, and degree of metamorphism are very similar between South America and South Africa up to about 200 million years ago and then the pattern diverges as continents parted. The isotopic ages of rocks also match between these two continents. Redrawn from Earth's Dynamic Systems with permission of Eric H. Christiansen

ocean, from the shore of one continent to another. However, finding striking similarity of certain fossils found on the continents on both sides of the Atlantic is difficult to explain due to the sheer vastness of the water in-between, unless, it is assumed that the continents were once connected to each other. This, that the landmasses were once connected, is borne out by collating paleontological evidences. These evidences are briefly discussed below.

14.2.1.1 *Glossopteris*

Fossils of *Glossopteris*, a fern-like plant (Fig. 14.2a, b), has been found in rocks of the same age from South America, South Africa, Australia, India, Madagascar, and Antarctica (Fig. 14.2c). The mature seeds of this plant are several millimeters in diameter, hence, too large to have been dispersed across the ocean by winds. However, the simultaneous presence of *Glossopteris* on all of the southern continents (Fig. 14.2b, c), strongly provides a supporting evidence that once these continents were connected to each other. If the continents were indeed joined, then the rocks and mountain ranges of the same age in adjoining locations on opposite

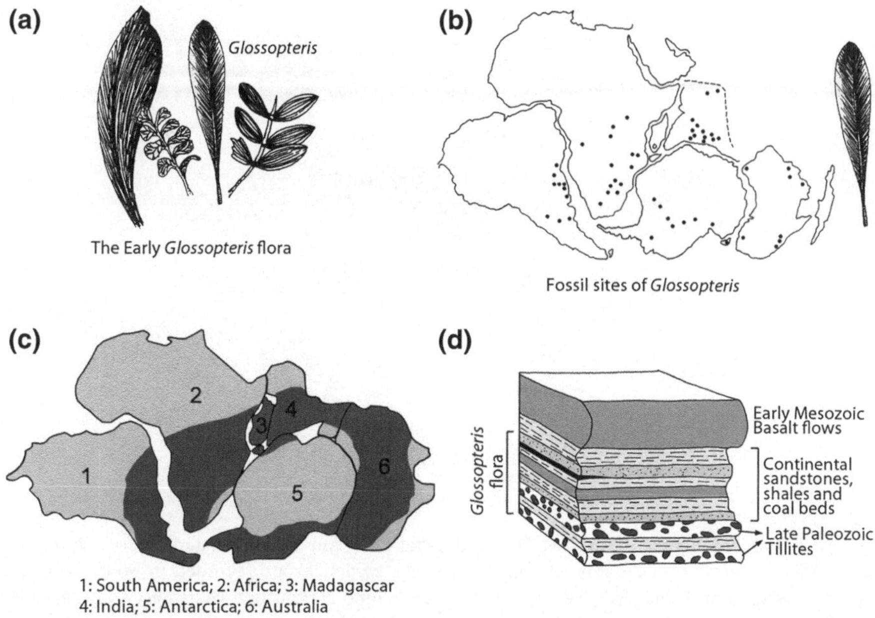

Fig. 14.2 **a** The Early *Glossopteris* flora constitutes a group of 27 species. *Glossopteris* is the genus name of one such common seed fern. *Glossopteris* is the one in the middle (**a**). **b** Recorded fossil sites of *Glossopteris* on all six Gondwana continents (South America, Africa, Madagascar, India, Australia, and Antarctica). **c** The global distribution of the Carboniferous-Permian *Glossopteris* flora (*dark gray shade* indicate the spread). **d** The common sequence of marine, nonmarine, and glacial rocks of Pennsylvanian (UC) to Jurassic (Jr) age recorded from all the six Gondwana continents. This similarity suggests that once all these continents were joined together

sides of the continents should also closely match. Such is the case for these five Gondwana continents (Fig. 14.2c) where marine, nonmarine, and glacial rock sequences of Pennsylvanian to Jurassic age are almost identical (Fig. 14.2d), strongly indicating that they were once joined.

14.2.1.2 *Lystrosaurus*

The distribution patterns of the Paleozoic and Mesozoic reptiles provide another strong evidence where fossils of several species are found in now-separated southern continents. Abundant fossil occurrence of Genus *Lystrosaurus*, a Permian-aged freshwater mammal-like reptile, in South Africa, South America, Asia, and in the Antarctica, is a good example indicating that the Gondwana continents were once joined (Fig. 14.3). *Lystrosaurus* being an exclusive land dweller (freshwater), clearly, suggests that these reptiles could not have swum thousands of kilometers of salty water across the Atlantic and Antarctic oceans. A former land bridge between the continents, thus, is the best possible explanation for their

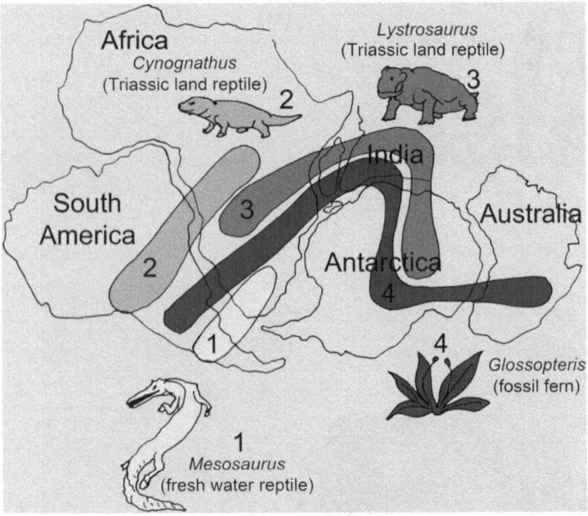

Fig. 14.3 Paleontological evidences for Continental Drift can be appreciated by considering the distribution of some fossil plants and animals. These evidences are recorded in South America, Africa, Madagascar, India, Antarctica, and Australia, some 200 million years ago, suggesting the presence of a once unified single supercontinent called the Pangea. *1 Mesosaurus*, a Permian freshwater reptile, is found in both Brazil and South Africa. *2 Cynognathus*, an older Triassic reptile, is found in Argentina and South Africa. *3 Lystrosaurus*, a Triassic land reptile, is found in South Africa, South America, India, and Antarctica and *4 Glossopteris*, a fossil fern, is found on all of the southern continents. Redrawn from Earth's Dynamic Systems with permission of Eric H. Christiansen

recorded distribution. Additionally, recent discoveries of dinosaur fossils in Gondwana continents further solidify the argument that these landmasses were close to each other during the Early Mesozoic Era.

14.2.2 Geologic Features

Across the Atlantic Ocean, several geologic features end abruptly at the coast of one continent and then suddenly reappear on the facing continent (Fig. 14.4). The folded mountain ranges at the Cape of Good Hope, at the southern tip of Africa, trend from east to west and terminate sharply at the coast. An equivalent structure, of the same age and style of deformation, appears near Buenos Aires, Argentina (Fig. 14.1). The folded Appalachian Mountains are another such example (Fig. 14.4). The deformed structures of the mountain belt extend northeastward across the eastern United States and through Newfoundland, terminating abruptly at the ocean. The mountain belt with a similar age, rock sequence, fossils, and structural style, reappears on the Scandinavian coasts in Ireland, Scotland, and

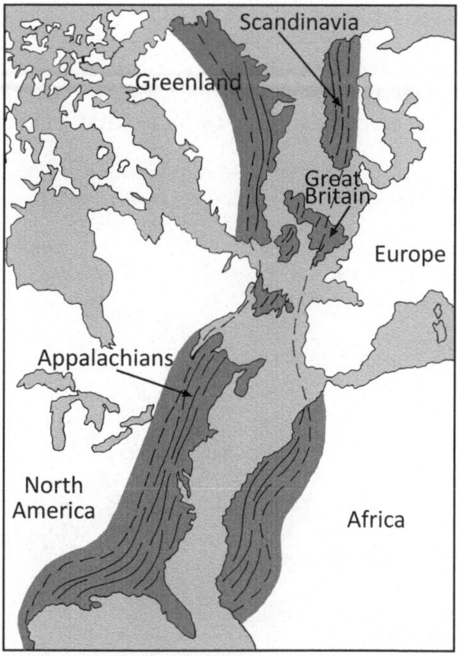

Fig. 14.4 The folded mountain ranges end abruptly at the coast of one continent and then suddenly reappear on the facing continent across the Atlantic Ocean such as the folded Appalachian Mountains. The deformed structures of the mountain belt extend northeastward across the eastern United States and through Newfoundland, terminating abruptly at the ocean. The mountain belt with a similar age, rock sequence, fossils, and structural style, reappears on the Scandinavian coasts in Ireland, Scotland, and Norway. Mountain ranges of the same age and deformational style are found in eastern Greenland, Ireland, Great Britain, and Norway indicating that once all these continents were joined

Norway (Fig. 14.4). Mountain ranges of the same age and deformational style are found in eastern Greenland, Ireland, Great Britain, and Norway. In fact, the same red sandstones used in the construction of many English and Scottish castles are used in various buildings throughout New York. So, even though the Appalachian Mountains and their equivalent-age mountain ranges in Great Britain are currently separated by the Atlantic Ocean, they form an essentially continuous mountain range when the continents are positioned next to each other as they were during the Paleozoic Era. But it must be noted that such geologic similarities on opposite sides of the Atlantic are found only in rocks older than the Cretaceous Period, which began 145 million years ago (Ma). It is now noted that about 200 Ma, the southern continents started splitting and drifted apart during the Jurassic time (from 145 to 199 Ma).

14.2.3 Glacial Evidence

During the latter part of the Paleozoic Era ($\sim$300 Ma), glaciers covered large portions of the Southern continents (Fig. 14.5). Distinct glacial deposits marked by

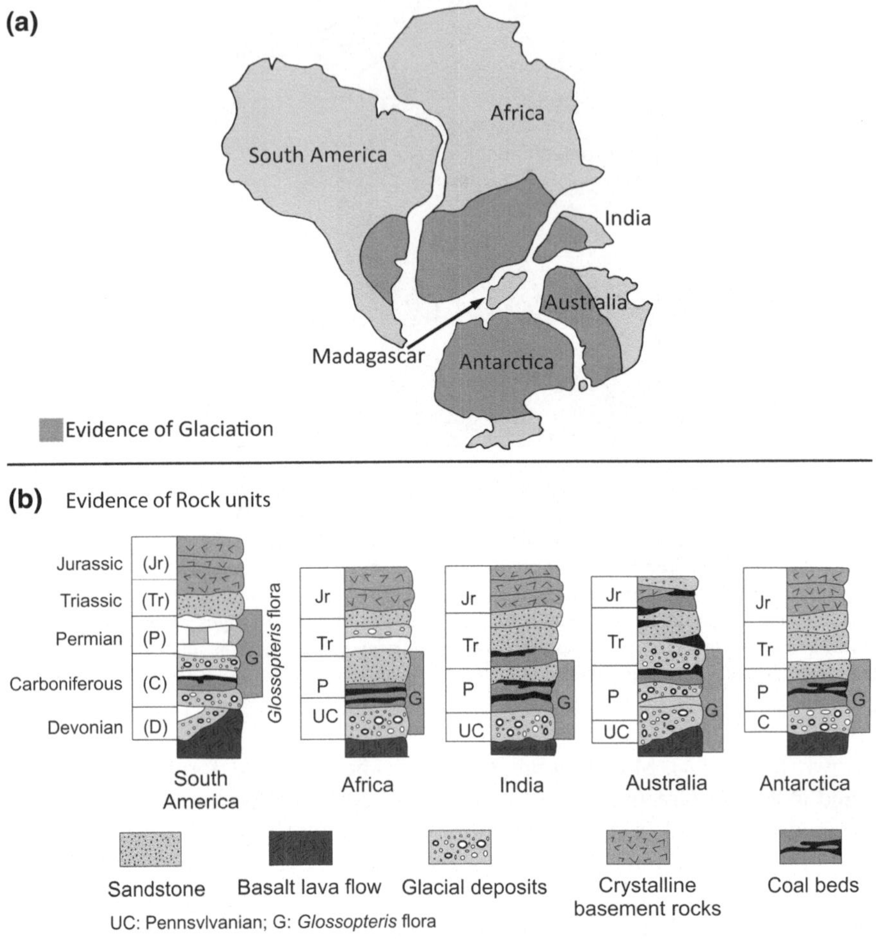

Fig. 14.5 **a** The glaciers (in *blue*) covered large portions of the Southern continents during the latter part of the Paleozoic Era ($\sim$300 Ma) suggesting that once all these continents were joined. This is also bore by the fact that the rock units in all the Gondwana continents show similar rock sequence (**b**). The common sequence of marine, nonmarine, and glacial rocks of Pennsylvanian (UC) to Jurassic (Jr) age are nearly the same on all six Gondwana continents indicating that once all these continents were joined together. Now, all these continents are widely separated and have different environments and climates ranging from tropical to polar. Hence, rocks formed on each continent are very different. But, in the past, when the continents were all joined together, the environments of adjacent continents were similar and the rocks formed in those areas were also similar (**b**)

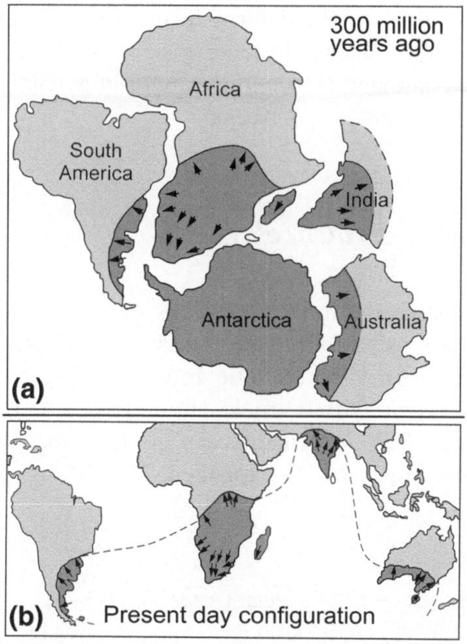

Fig. 14.6 Evidence for Continental Drift. **a** Presence of striations (*scratch marks*) and grooves (glacial deposits) etched on the underlying rocks indicate the direction in which ice moved (*black arrows*). When these Gondwana continents are placed together so that South Africa is located at the South Pole, the glacial movements indicated by striations (*black arrows*) found on rock outcrops on each continent suggest that once all these were joined together. In this situation, the glaciers (*light blue areas*) were located in a polar climate and appear to have moved radially outward from its thick central area toward its periphery. **b** All these continents, except for Antarctica, now lie close to the equator, and are quite far removed from a latitude that could have produced glaciation. Interestingly, the present-day Northern Hemisphere continents show no trace of glaciation during this time; fossil plants in North America and Europe indicate a tropical climate. This evidence is difficult to explain in the context of immovable continents as the climatic belts are determined by latitude. Blue color marks the extent of glaciation

striations (scratch marks) and grooves on the underlying rocks indicate the direction in which ice moved (Fig. 14.6). All these continents, except for Antarctica, now lie close to the equator, and are quite far removed from a latitude that could produce glaciation. Interestingly, the present-day Northern Hemisphere continents show no trace of glaciation during this time; fossil plants in North America and Europe indicate a tropical climate. This evidence is difficult to explain in the context of immovable continents as the climatic belts are determined by latitude. Even more difficult is to explain the direction in which the glaciers moved. Regional mapping of striations and grooves indicate that in South America, India, and Australia, ice accumulated in oceans and moved inland. Such movement of ice would be impossible unless there was a land mass where the oceans now exist. If glaciers could form in the sea, a large permanent glacier would exist in the Arctic Ocean. Instead, glaciers originate on land and move toward the edge of a continent. Hence, this pattern of glaciation was considered as a strong evidence of the drift of continents.

14.2.4 Paleoclimatic Evidence

Thick deposits of salt (Evaporites), formations of wind-blown sandstone (desert dune deposits), and extensive fossil coral reefs and coal provide additional clues that permit us to reconstruct the climatic zones of the past (Fig. 14.7). The paleoclimatic patterns shown by these rocks are baffling with the continents in their present positions, but if the continents are grouped together in their pre-drift positions, the patterns can be easily explained (Fig. 14.7).

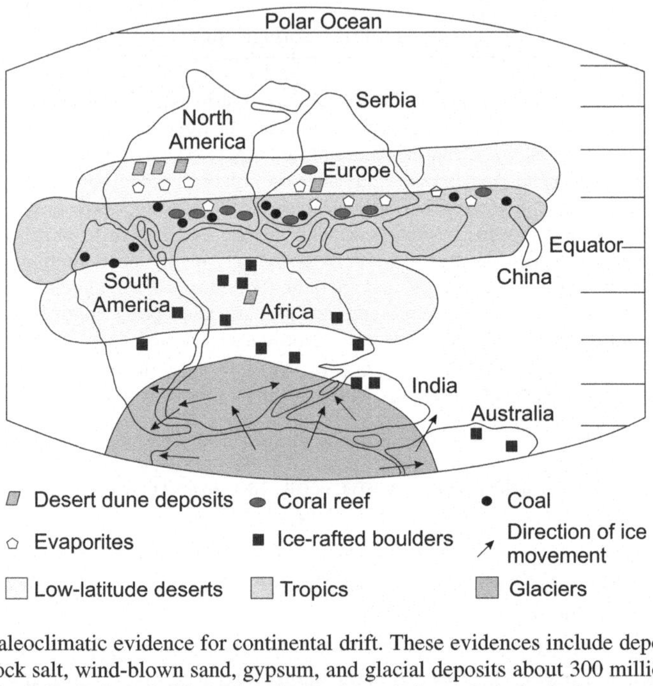

Fig. 14.7 Paleoclimatic evidence for continental drift. These evidences include deposits of coal, sandstone, rock salt, wind-blown sand, gypsum, and glacial deposits about 300 million years ago near the end of the Paleozoic Era. The distribution of these deposits is best explained if we assume that the continents were grouped together at the end of the Paleozoic Era, as shown here. Redrawn from Earth's Dynamic Systems with permission of Eric H. Christiansen

14.3 Plate Tectonics

Plate tectonics has enabled us to predict geologic events and explain almost all aspects of what the earth experiences. The theory explains why mountains, earthquakes, and volcanoes occur, and where they do, the ages of deformational events, the ages and shapes of continents and ocean basins, as well as other aspects of the earth. By combining the theories of Seafloor Spreading with Continental Drift along with the modern information on global seismicity, a new theory took birth called the Theory of Plate Tectonics. It is a holistic theory and explains crustal movements. According to this theory, the earth is made of Plates, composed of lithosphere (~ 100 km thick) that "floats" on the ductile Asthenosphere (Fig. 14.8). While the continents do indeed appear to drift, they do so only because they are part of a larger plate that floats and moves horizontally on the upper Mantle Asthenosphere. The plates behave as rigid bodies with some ability to flex, but deformation occurs mainly along plate boundaries (the margins; Fig. 14.9).

14.3.1 Plate Boundaries

The plates move in three ways—toward each other, away from each other, and sliding past each other (Fig. 14.8). These are accordingly classified into three types: Convergent, Divergent, and Transform (Fig. 14.9).

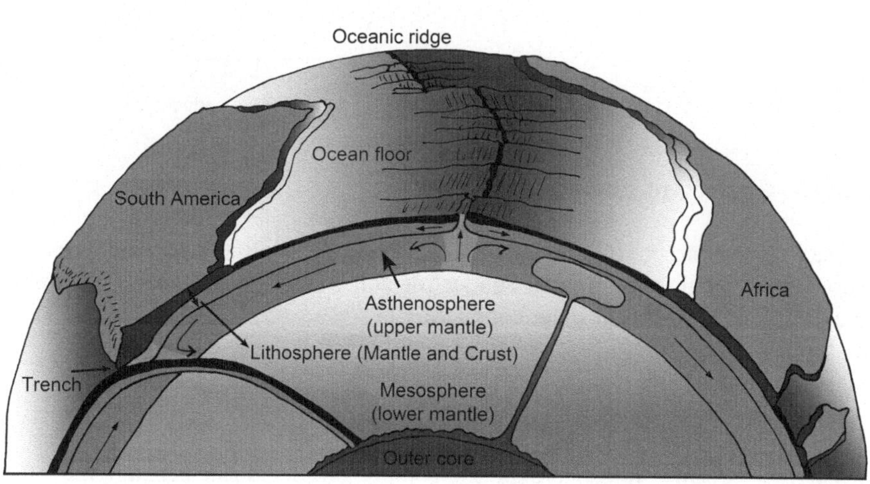

Fig. 14.8 The layers of the tectonic Earth. Redrawn from Earth's Dynamic Systems with permission of Eric H. Christiansen

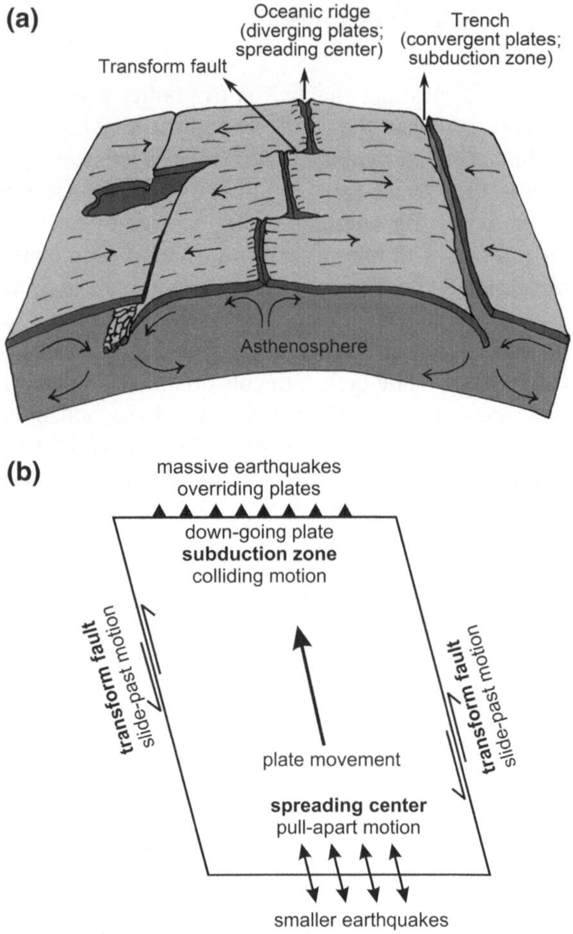

Fig. 14.9 The movement and position of tectonic plates (**a**). **b** These tectonic plates move in three different ways—away from each other in opposite directions, toward each other and sliding past each other and are accordingly classified into three types—divergent, convergent and transform. **a** Redrawn from Earth's Dynamic Systems with permission of Eric H. Christiansen

14.3.1.1 Convergent Plate Boundary

These occur where oceanic lithosphere is pushed back into the Mantle. These are marked by Oceanic trenches and Subduction zones (Fig. 14.10). Following are the three types of Convergent plate boundaries.

Ocean–Ocean Convergence

When two oceanic lithosphere plates collide, one subducts beneath the other resulting in the formation of an oceanic trench on the seafloor (Fig. 14.11). The sinking plate becomes a Subduction zone. As both the plates are oceanic, volcanism is therefore, basaltic. An Island arc is produced where subduction occurs (as in the western Pacific, including the islands of Tonga and Marianas and the

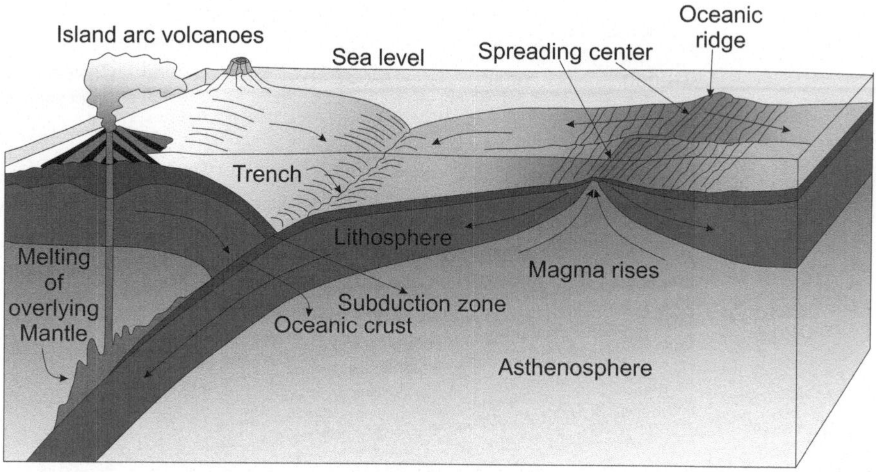

Fig. 14.10 The convergent plate boundary. Due to its higher density, the oceanic crust is pushed back into the mantle. Such subducted regions are marked by oceanic trenches and subduction zones

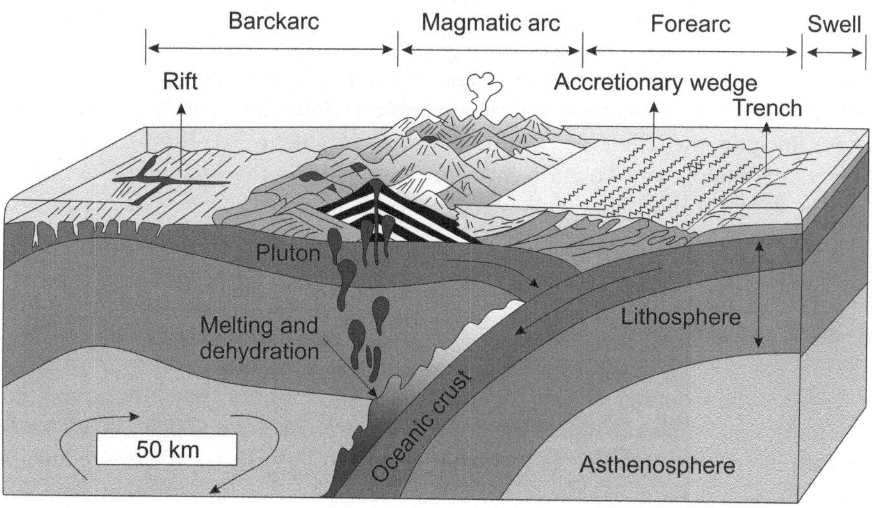

Fig. 14.11 The Ocean–Ocean convergence is dominated by volcanic activity and the construction of a Magmatic arc (Island arc). Redrawn from Earth's Dynamic Systems with permission of Eric H. Christiansen

Philippine Islands). Earthquakes extend down to depths of 660 km before the subducting plate heats up and loses its ability to deform in a brittle manner. These deep earthquakes indicate that the plates are at least brittle up to ~660 km. The Japanese Islands are also a good example of a volcanic island arc resulting from

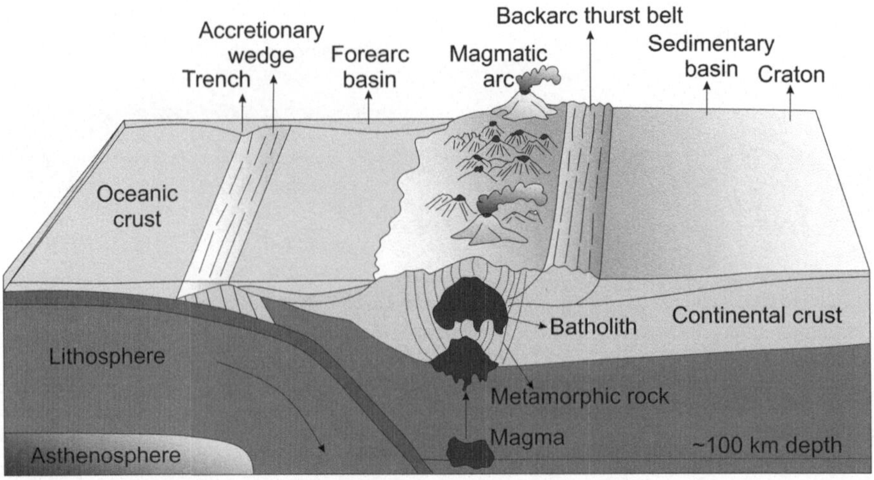

Fig. 14.12 The Ocean–Continent collision. It is also known as the Continental Convergent Margin. At ocean–continent convergent plate boundaries, major geologic processes include formation of an accretionary wedge, deformation of the continental margin into a folded mountain belt, metamorphism due to high pressures and temperatures in the mountain roots, and the partial melting of the mantle overlying the descending plate

the subduction of one oceanic plate beneath another. The western Pacific Ocean is full of examples of ocean–ocean convergence, stretching from north to south. These include the Aleutian, Kuril, Japanese, Ryukyu, Izu-Bonin, Philippine, Mariana, Solomon, and Tonga-Kermadec island arcs. In the Atlantic these include Caribbean and South Sandwich island arcs. In the Indian Ocean, is a tangle of arcs—the Indonesian archipelago.

Ocean–Continent Convergence

It is called the Continental Convergent margin (Fig. 14.12). When oceanic lithosphere runs into a plate with continental lithosphere, the oceanic plate is subducted beneath the continental, due to its higher density. As the denser plate subducts, it heats up and undergoes metamorphism. As it does so, dehydration reactions release water into the overlying mantle asthenosphere, causing a reduction in the melting temperature and facilitating the production of andesitic magma. These magma rise to the surface and form a volcanic arc (magmatic arc) parallel to the trench. The magma that is created by this convergence forms a magmatic arc. It is a broad term used both for island arcs at sea and for belts of igneous activity on the edges of the continents. On surface it manifests either as a line of andesitic islands (such as the Aleutian Islands) or a line of andesitic continental volcanoes (such as the Cascade volcanoes of the Pacific Northwest). Sediment deposited along the convergent margin, and particularly those in the trench are deformed by thrust faulting and break the rocks up into a chaotic mixture of broken, jumbled, and thrust faulted sediments called Mélange. The west coast of the Americas is a major example, with volcanic zones in Alaska, the Pacific Northwest states, and a continuous stretch from Central America to Tierra del Fuego ("Land of fire"), Italy, Greece, Kamchatka, and New Guinea.

Continent–Continent Convergence

It is also called as the Continental Collision margin (Fig. 14.13). Both plates have continental lithosphere; these collide with one another, and their margin becomes a continental collision margin (Fig. 14.13). They are characterized by fold-thrust mountain belts that develop along the zone of collision. Neither of the continental plates are subducted below as they are too buoyant. The Himalaya Mountain chain and the Tibetan Plateau are examples of this kind of margin (Fig. 14.13). Both were formed when the Indian plate collided with the Eurasian plate (Fig. 14.13). India drifted away from Africa and moved rapidly northwards until it collided with Asia, 71 Ma ago (Fig. 14.14). The collision started at about 25 Ma that eventually built the high Himalayan range and the Tibetan Plateau, to the North. Other examples include the Alps, Andes, and Rockies. Tibet on the Tibetan Plateau is a site of extreme tectonics.

14.3.2 Divergent Plate Boundary

At these boundaries, the plates move away from each other. This can occur in the middle of the ocean or in the middle of a continent. Those occurring at oceanic ridges, the plates move away from the ridge in opposite directions thereby creating new oceanic lithosphere by erupting basaltic magma (Fig. 14.15); the new oceanic crust pushes aside in opposite directions. Hence, the age of this diverging oceanic crust becomes progressively older in both directions away from the ridge. When occurring in the middle of a continent, the divergent boundary is marked by rifting, basaltic volcanism, and uplift. During rifting, the continental crust is stretched and thinned, thereby producing shallow-focus earthquakes on normal faults. A rift valley forms as a central graben. These faults act as a pathway for basaltic magma to raise from the mantle and erupt on the surface as cinder cones and basalt flows. Uplift at a divergent boundary is usually caused by the upwelling of hot mantle beneath the crust; the surface is elevated by the thermal expansion of the surface rock as it is warmed from below. This was noted for the opening of the Red Sea, where the crust was initially stretched and thinned (Fig. 14.16). Numerous normal faults broke the crust, and the surface subsided into a central graben. Shallow earthquakes and basalt eruptions occurred. Such a situation is also noted for the African Rift Valleys in eastern Africa. The valleys are grabens that formed; mark the site of the future breakup of Africa (Fig. 14.16).

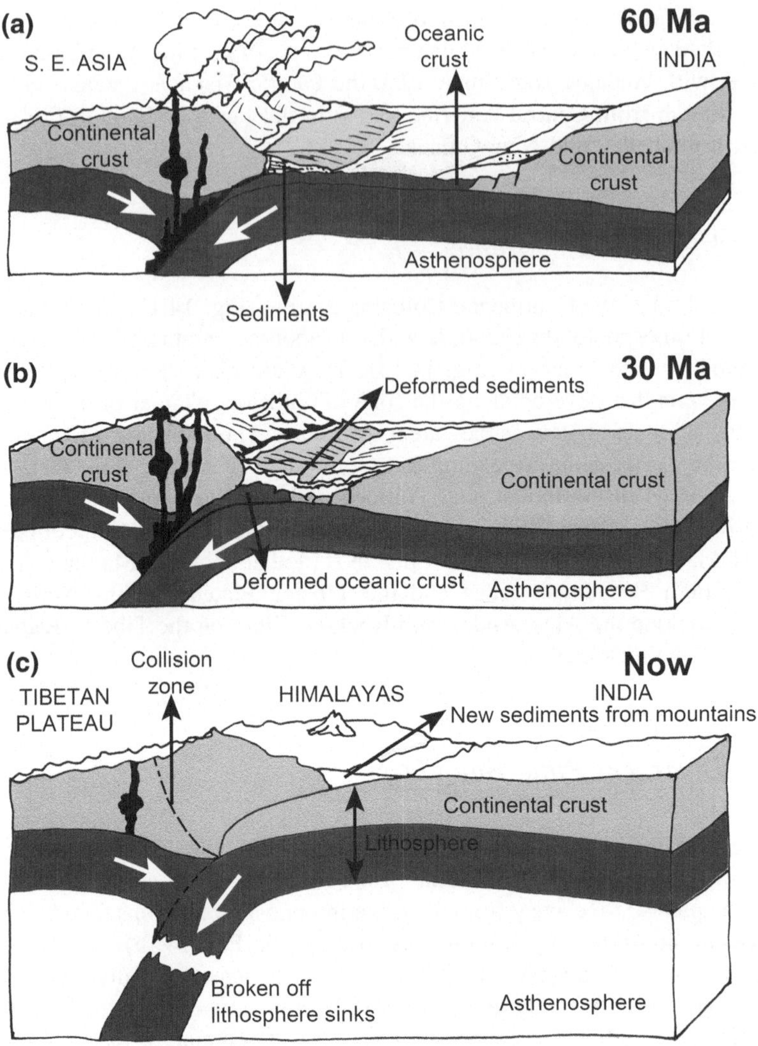

Fig. 14.13 The Continent–Continent Convergence: It is also called as the Continental Collision Margin and is marked by the complete subduction of the oceanic crust. A high mountain belt forms by folding, and thrust-faulting, resulting in the doubling of the crustal layer as one continent is thrust beneath the other

It must be noted that different plates have varied velocities of movement. These movements are a function of the amount of continental lithosphere within the diverging plate. Additionally, plates with continental lithosphere have lower relative velocities as compared to those with only oceanic lithosphere. The relative plate velocities are determined by the age of the crust and the distance from the ridge. Moving away from the ridge, the depth to the sea floor increases with

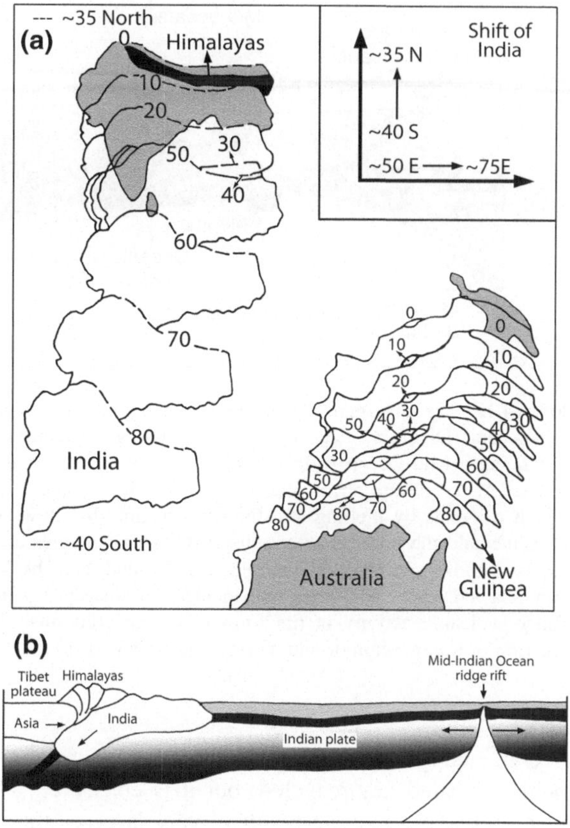

Fig. 14.14 The progressive evolution of India through time and space. **a** This movement resulted in the collision of India into the Eurasian plate and let to the formation of the Himalayan mountain belt. For reference, position of Australia and New Guinea are also shown. Inset a: This graph shows that the Indian plate moved north from ~40° South to ~35° North and ~50° East to ~75° East (tip to tip). **b** The crosssectional view of the collision and the involved tectonic plates

increasing age. There is very little sediment accumulation on the ridges and sediment thickness increases in both directions away from the ridge; it is thickest where the oceanic crust is the oldest. In the Atlantic, the oldest oceanic crust occurs next to the North American and African continents and is about 180 Ma old, i.e., of Jurassic age. In the Pacific, the oldest crust is also of Jurassic age, and occurs off the coast of Japan.

14.3.3 Transform Plate Boundary

Transform boundaries occur where two plates slide past one another horizontally (Fig. 14.17). Earthquakes are shallow like the ones noted in the famous San Andreas Fault (California, USA). Transforms come in two types. Most are short faults on the seafloor that run from one segment of spreading ridge to another. When diverging plates break apart, the shape of the break is preserved in the spreading ridge. The ridge quickly adopts the stair-step configuration, minimizing

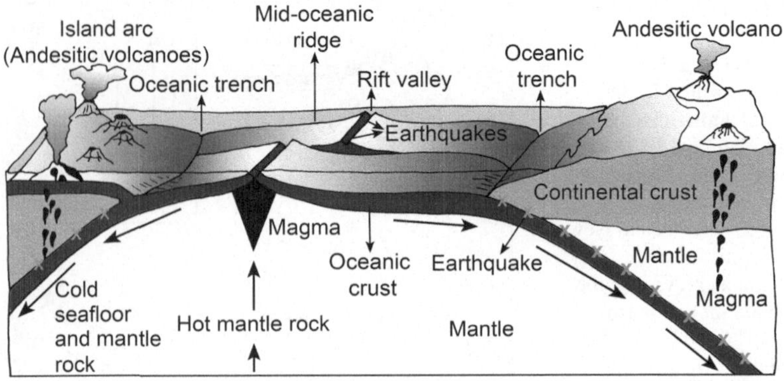

Fig. 14.15 The Divergent Plate Boundary. *Gray crosses* (x) mark the position of earthquakes. At this oceanic Divergent Plate Boundary the rising convection current below lifts the lithosphere producing a Mid-Ocean Ridge. Extensional forces stretch the lithosphere and produce a deep fissure. As the fissure opens, the pressure is reduced on the super-heated mantle material below which responds by melting and the new magma that flows into the fissure. This magma then solidifies and the process repeats itself. Classic examples are found at the Mid-Atlantic Ridge exposed above sea level on the island of Iceland, and the Mid-Atlantic Ridge between North America and Africa. Major features include a submarine mountain range such as the Mid-Atlantic Ridge, volcanic activity in the form of fissure eruptions, shallow focus earthquakes and the creation of a new seafloor and a widening ocean basin

overall energy. Here, one oceanic plate slips horizontally by another; lithosphere is neither created nor recycled, but it is conserved. Plate motion is parallel to the plate boundary and can result in very large earthquakes. With a few exceptions, transform plate boundaries do not exhibit volcanoes. The other type of transform occurs mostly on land. The San Andreas Fault of California is a prime example; others are the North Anatolian fault of northern Turkey, the Alpine fault crossing New Zealand, the Dead Sea rift in the Middle East, the Queen Charlotte Islands fault off western Canada, and the Magellanes-Fagnano fault system of southernmost South America. These continental transforms are more complex than their short oceanic counterparts.

14.3.4 Accreted Terrane Margins

A former convergent margin or transform fault margin that has been modified by numerous additions of small blocks of crust and that have been accreted to the continental margin is an Accreted Terrane margin (Accretionary belt; Fig. 14.18). Such a margin occurs close to a plate boundary as it once was a plate boundary of the Convergent or Transform type. Hence, a "terrane" is an area possessing unique tectonic assemblages (lithostratigraphic units representing a specific depositional

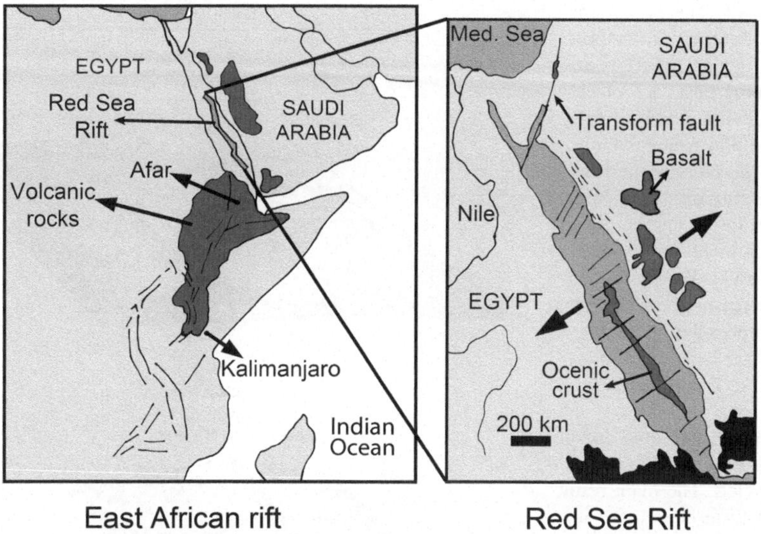

Fig. 14.16 The East Africa Rift Valley and the Red Sea rift. At the continental Divergent Plate Boundary, beneath a thick continental plate, the pull-apart is not forceful enough to create a clean, single break through the thick plate material. Here the thick continental plate is arched upwards from the convection current's lift, pulled thin by extensional forces, and fractured into a rift-shaped structure. As the two plates pull apart, normal faults develop on both sides of the rift and the central blocks slide downwards resulting in earthquakes. Early in the rift-forming process, streams and rivers will flow into the sinking rift valley to form a long linear lake. As the rift grows deeper it might drop below sea level allowing ocean waters to flow in. This will produce a narrow, shallow sea within the rift. This rift can then grow deeper and wider. If rifting continues a new ocean basin could be produced. The East Africa Rift Valley is a classic example of this type of plate boundary (**a**). The rift is in its nascent stage as the plate has not been completely rifted and the rift valley is still above the sea level, often occupied by lakes. On the other hand, the Red Sea (**b**) is an example of a more completely developed rift. Here, the plates have fully separated and the central rift valley has dropped below sea level

or volcanic setting responding to a tectonic event), which differs from adjacent terranes and is bounded by faults (Gabrielse et al. 1991). The Southern Cordilleran System of North America forms a ~500 km wide 'collage' of oceanic, arc, and continental margin tectono-stratigraphic terranes accreted during the Phanerozoic to the western margin of continental North America (Fig. 14.18). Each terrane is a fault bounded entity of regional extent characterized by an internally homogeneous stratigraphy and geologic history that is different from that of contiguous terranes.

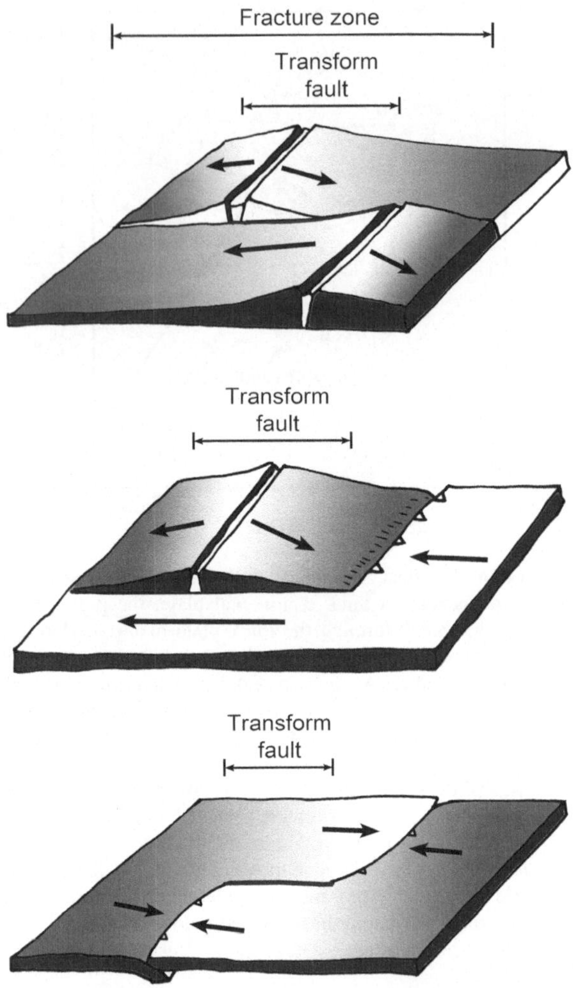

Fig. 14.17 The Transform Plate Boundary. Transform boundaries are where plates are sliding past each other (i.e., moving side by side) like the San Andreas fault where the North American plate is moving south east relative to the Pacific plate (the continental transform boundaries). Within the marine realm, the Mid Ocean Ridge spreading areas are also offset by transform faults. At Transform boundaries no new lithosphere is created and no old lithosphere is destroyed, or recycled. The main result of transform boundaries is horizontal plate movement, usually resulting in catastrophic earthquakes. Redrawn from Earth's Dynamic Systems with permission of Eric H. Christiansen

14.4 Continental Crust and Plate Tectonics

The continents can be divided into two kinds of structural units: Cratons and Orogens.

14.4.1 Cratons

Cratons form the core of a continent (Fig. 14.18); portions of the continental crust that have attained isostatic and tectonic stability. They were formed and deformed very early in the earth's history and hence, are the oldest constituents of a

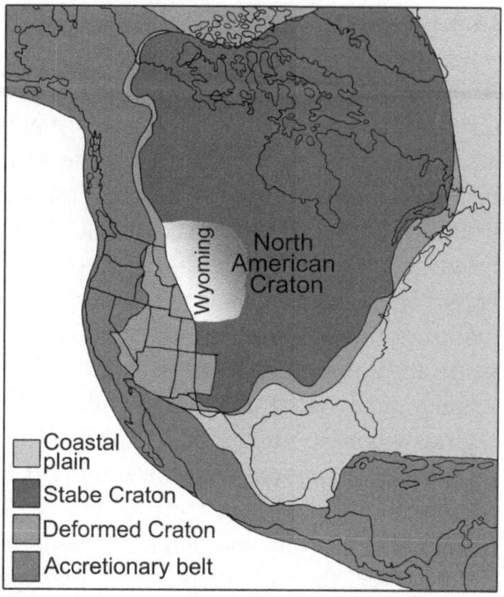

Fig. 14.18 The geologic regions of the North American craton (also called the Laurentina)

continent. The Wyoming craton (also called the Wyoming province) is a good example of a craton. It is a craton located in the west-central United States and western Canada (in Montana, Wyoming, southern Alberta, southern Saskatchewan, and parts of northern Utah) (Fig. 14.18). The Wyoming craton is the initial core of the continental crust of North America (Fig. 14.8; see also Mueller and Frost 2006).

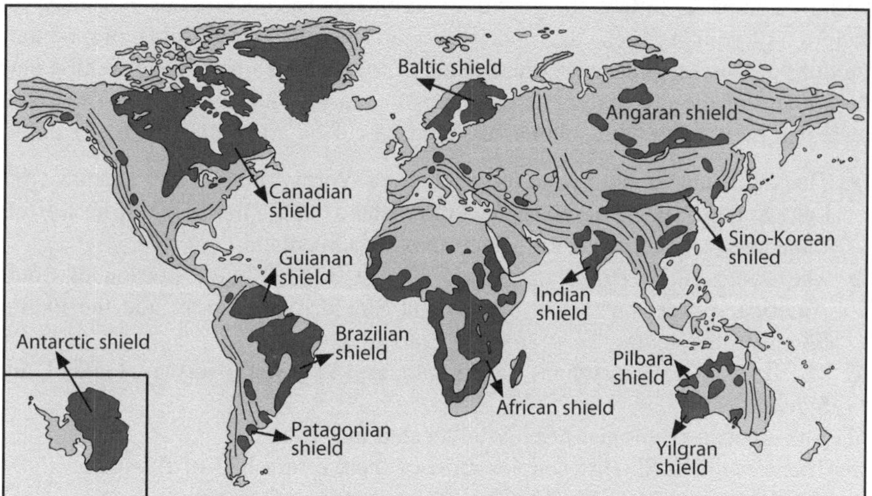

Fig. 14.19 The distribution of Precambrian shields around the world. *Solid lines* are Orogenic belts

14.4.1.1 Shields

A shield is that part of the continental crust in which Precambrian basement rocks (crystalline igneous and high-grade metamorphic rocks) form tectonically stable areas, cropping out extensively at the surface (Fig. 14.19). Shields themselves can be very complex; they consist of vast areas of granitic or granodioritic gneisses, usually of tonalite composition, and they also contain belts of sedimentary rocks, often surrounded by low-grade volcano-sedimentary sequences, or greenstone belts. These rocks are frequently metamorphosed greenschist, amphibolite, and granulite facies. In all cases, the age of these rocks is greater than 570 million years and sometimes dates back 2–3.5 billion years. They have been little affected by tectonic events following the end of the Precambrian Era, and are relatively flat regions where mountain building, faulting, and other tectonic processes are greatly diminished compared with the activity that occurs at the margins of the shields and the boundaries between tectonic plates. Shields are normally the nucleus of continents and most are bordered by belts of folded Cambrian rocks.

Together, the shield, platform and basement are the parts that comprise the stable interior portion of the continental crust known as the "Craton."

The margins surrounding a shield generally constitute relatively mobile zones of intense tectonic or plate-like dynamic mechanisms. In these areas, complex sequences of mountain building (orogeny) events have been documented over the past few hundred million years. For example, the Ural Mountains to the west of the Angaran Shield are at the top of the mobile zone that separates the shield from the Baltic Shield. Similarly, the Himalayas are on the mobile boundary between the Angaran and Indian shields. Shield margins have been subject to geotectonic forces that have both destroyed and rebuilt the margins and the cratons that they partially comprise. In fact, the growth of continents has occurred as a result of the accretion of younger rocks that underwent deformations during series of mountain building processes. In a sense, these belts of folded rocks have been welded onto the borders of the preexisting shields, thus increasing the size of the proto-continents that they make up. Continental shields occur on all continents, for example:

a. The Canadian Shield forms the nucleus of North America and extends from Lake Superior in the south to the Arctic Islands in the north, and from western Canada eastward across to include most of Greenland.
b. The Amazonian (Brazilian) Shield on the eastern bulge portion of South America. Bordering this is the Guiana Shield to the north, and the Platian Shield to the south.
c. The Baltic (Fennoscandian) Shield is located in eastern Norway, Finland, and Sweden.
d. The African (Ethiopian) Shield is located in Africa.
e. The Australian Shield occupies most of the western half of Australia.
f. The Arabian-Nubian Shield on the western edge of Arabia.
g. The Antarctic Shield.

h. In Asia, an area in China and North Korea is sometimes referred to as the China-Korean Shield.
i. The Angaran Shield, as it is sometimes called, is bounded by the Yenisey River on the west, the Lena River on the east, the Arctic Ocean on the north, and Lake Baikal on the south.
j. The Indian Shield occupies two-thirds of the southern Indian peninsula.

14.4.1.2 Stable Platforms

The shields are eroded down to within a few tens of meters of sea level, and any rise of sea level will lead to flooding of vast areas of the shield. At present only 18 % of the continental crust is flooded, but there were times in the past where vast portions of the continents were covered by a shallow sea (example: the interior of North America).

14.4.2 Orogens and Folded Mountain Belts

These are broad elongated belts of deformed rocks draped around a Craton. They appear to be the eroded roots of former mountain belts formed by continent–continent collision. Only the youngest of these orogens still form mountain ranges. The orogens are generally younger toward the outside of any continent, thereby suggesting that the continents were built by collisions of plates that added younger material to the outside edges of the continents. This is also evidence that plate tectonics has been in operation for at least the last 2 billion years.

The Folded Mountain Belts are usually found along the margins of continents, and the folding and thrusting indicates that as much as 30 % of crustal shortening has taken place during their formation. We know now that this shortening is a direct reflection of the compressive stress regime and subduction of oceanic crust along convergent plate margins. The location of these fold belts along continental margins implies that by convergence of plates material is piled up along the continents, and finally becomes part of the continental crust. Fold belts that are terminated abruptly at the continental margin, such as the Appalachians and the Caledonides (Fig. 14.20), suggest that the fold belts were once much longer, and have been separated when continents broke up by continental rifting.

14.5 Cause of Plate Tectonics

Asthenosphere can flow like a liquid under stress. As the deeper level material is heated to the point where it expands and becomes less dense than the material above it, the hot, less dense material raises, cools, and becomes denser than its

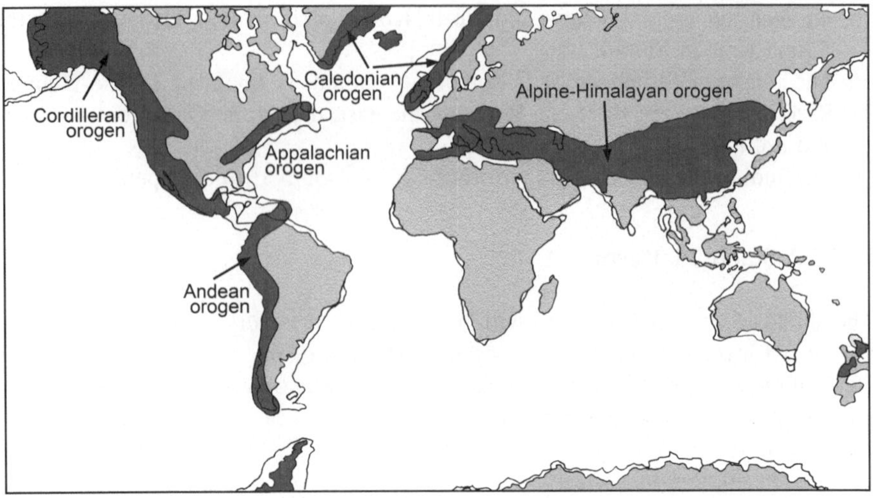

Fig. 14.20 The Folded Mountain Belts. These are broad elongated belts of deformed rocks formed by continent–continent collision. Only the youngest of these orogens still form mountain ranges. The orogens are generally younger toward the outside of any continent, thereby suggesting that the continents were built by collisions of plates that added younger material to the outside edges of the continents. This is also evidence that plate tectonics has been in operation for at least the last 2 billion years

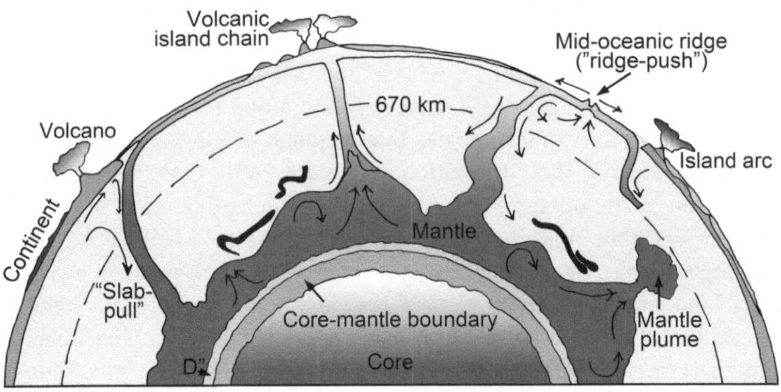

Fig. 14.21 The mechanism of plate tectonics. A combination of dragging of the lithosphere along the top of the convection cell, ridge push, sliding, and slab pull all contribute to the cause of plate tectonics and the movement of plates. Modified from Kellogg et al. (1999)

surroundings and eventually sinks (Fig. 14.21). This mechanism gives rise to convection cells, which are characterized by hot rising and cool descending currents. This convection is also responsible for moving of the asthenosphere and driving plate tectonics. Hot rising currents occur beneath oceanic ridges. Magma pushes the lithosphere apart at the ridge. This newly formed lithosphere eventually becomes cool and dense and pulls the rest of the lithosphere downwards. Thus, a combination of dragging of the lithosphere along the top of the convection cell, ridge push, sliding, and slab pull all contribute to the cause of plate tectonics (Fig. 14.19). Thus, the plate motion that was once thought to be caused by mantle convection, is now attributed to the cold, dense, leading edge of a subducting plate pulling the rest of the plate along with it (the slab pull). The plates near mid-oceanic ridges also slide down the sloping lithosphere–asthenosphere boundary at the ridge (the ridge push) and the Trench suction has been noted to help continents diverge. The Trench suction is thought to result from small-scale convection within the mantle wedge, driven by the subducting lithosphere.

14.6 Summary

Plate Tectonics provides a single unifying theory for understanding the dynamics of earth. Few decades ago, it was believed that continents and ocean basins were fixed, permanent features, and that the Theory of Continental Drift was just a radical idea. The theory of Continental drift was the predecessor of the theory of Plate Tectonics. Alfred Wegener in his book "*The Origin of the Continents and Oceans*" noted that the shapes of the continents and their geology are similar. He used geological, fossil, and glacial evidences gathered on the opposite sides of the Atlantic Ocean to support his theory of Continental Drift. He drew a series of maps showing the three stages in the drifting process; starting with an initial large super continental landmass called Pangea, whose breakup and the theory of continental drift were supported by several evidences from numerous regional geological studies. These were—the paleontological evidence (*Glossopteris* flora and the *Lystrosaurus*), the geologic features (folded mountain ranges), the glacial evidence and the paleoclimatic evidence [thick deposits of salt (Evaporites), formations of wind-blown sandstone (desert dune deposits), and extensive fossil coral reefs, and coal deposits].

By combining the theories of Seafloor Spreading with Continental Drift along with the modern information on global seismicity, a new theory took birth called the Theory of Plate Tectonics. It has enabled to predict geologic events and explain almost all aspects of what the earth experiences. According to this theory, the earth is made of Plates, composed of lithosphere ($\sim$100 km thick) that "floats" on the ductile Asthenosphere. While the continents do indeed appear to drift, they do so only because they are part of a larger plate that floats and moves horizontally on the upper Mantle Asthenosphere. The plates move in three ways—toward each other, away from each other, and sliding past each other. These are accordingly classified into three types of plate boundaries: Convergent, Divergent, and Transform. The Convergent plate boundary is marked by Oceanic trenches and Subduction zones and is of three types Ocean–Ocean Convergence, Ocean–Continent Convergence, and Continent–Continent Convergence.

The Accreted Terrane Margins are former convergent margin or transform fault margin that has been modified by numerous additions of small blocks of crust that have been accreted to the continental margin is an Accreted Terrane margin.

The continents can be divided into two kinds of structural units: Cratons and Orogens.

Asthenosphere can flow like a liquid under stress. As the deeper level material is heated to the point where it expands and becomes less dense than the material above it, the hot, less dense material raises, cools, and becomes denser than its surroundings and eventually sinks. This mechanism gives rise to convection cells, which are characterized by hot rising and cool descending currents. This convection is also responsible for moving of the asthenosphere and driving plate tectonics. Hot rising currents occur beneath oceanic ridges. Magma pushes the lithosphere apart at the ridge. This newly formed lithosphere eventually becomes cool and dense and pulls the rest of the lithosphere downwards. Thus, a combination of dragging of the lithosphere along the top of the convection cell, ridge push, sliding, and slab pull all contribute to the cause of plate tectonics.

References

Condie KC (1998) Plate tectonics and crustal evolution, 4th edn. Butterworth-Heinemann, New York

Cox A, Hart RB (1986) Plate tectonics, how it works. Blackwell, Palo Alto

Gabrielse H, Monger JWH, Wheeler JO and Yorath CJ (1991) Part A. Morphogeological belts, tectonic assemblages and terranes. In: Geology of the Cordilleran Orogen in Canada (eds. Gabrielse H. and Yorath CJ) Geology of Canada, No. 4, pp. 15–28. Geological Survey of Canada

Hallam A (1973) A revolution in the sciences, from continental drift to plate tectonics. Oxford University Press, New York

Hamblin WK, Christiansen EH (2003) Earth's Dynamic Systems, 10th edn. Prentice-Hall Inc., New Jersey, 816 p

Kearey P, Klepeis KA, Vine FJ (2009) Global tectonics, 3rd edn. Wiley-Blackwell, Chichester

Kellogg LH, Hager BH, van der Hilst RD (1999) Compositional stratification in the deep mantel. Science 283:1881–1884

Mueller PA, Frost CD (2006) The Wyoming Province: a distinctive Archean craton in Laurentian North America. Can J Earth Sci 43(10):1391–1397

Chapter 15
Earthquakes

15.1 Introduction

Any trembling of earth's surface that follows a release of energy in the earth's crust, generated by a sudden dislocation of a segment of crust, or by a volcanic eruption, or at times by man-made explosions, is defined as an Earthquake (Pakiser 1991; Abbot 1998; Bolt 1999). However, most destructive earthquakes are caused by the sudden crustal movement and often along plate boundaries (Fig. 15.1). As plates move (over, under, and past each other), these boundaries (ocean ridges, continental rifts, subduction zones, and transform faults) are sites of most intense earthquake activity (Fig. 15.1). Although this movement is gradual, though at times, it is violent when plates are locked and are unable to release this accumulating energy. As this energy reaches a threshold, plates break free thereby generating seismic waves. These waves travel outward from the source of the earthquake along the surface and through earth at varying speeds depending on the material through which they traverse. Most natural earthquakes are caused by a sudden slippage along a fault (Fig. 15.2). If this slippage is obstructed, elastic strain energy builds up in the deforming rocks on both sides of the fault, which is eventually released when slippage occurs, resulting in an earthquake. This is called the Elastic Rebound theory (Fig. 15.3). This theory was proposed by H. F. Reid of the Johns Hopkins University (San Francisco, USA) in 1906 by studying the measurements taken across a portion of the San Andreas Fault that had broken during the 1906 earthquake. It was noted that rocks adjacent to the fault were bending, prior to an earthquake. These bends disappeared after the earthquake, suggesting that the energy stored in bending the rocks was released during the earthquake (Fig. 15.3). The opposite sides of the fault had moved 3.2 m during the 50-year period prior to the breakage in 1906, with the west side moving northward.

Faults that occur close to the surface of a plate usually generate the most significant quakes as these are zones of weakness within the earth's crust. The earth is made up of several layers (Fig. 15.4; see also Fig. 4.3) marked by very different physical and chemical properties. The outer layer consists of about a dozen irregularly shaped large plates (Fig. 15.5), about 80 km in thickness

S. Jain, *Fundamentals of Physical Geology*, Springer Geology,
DOI: 10.1007/978-81-322-1539-4_15, © Springer India 2014

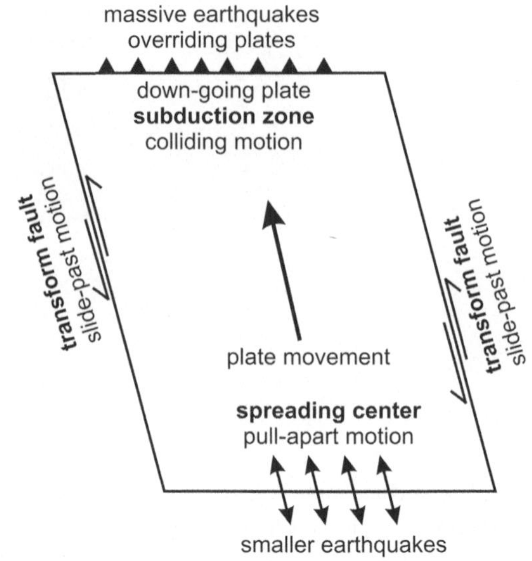

Fig. 15.1 Schematic diagram of the three tectonic plate boundaries

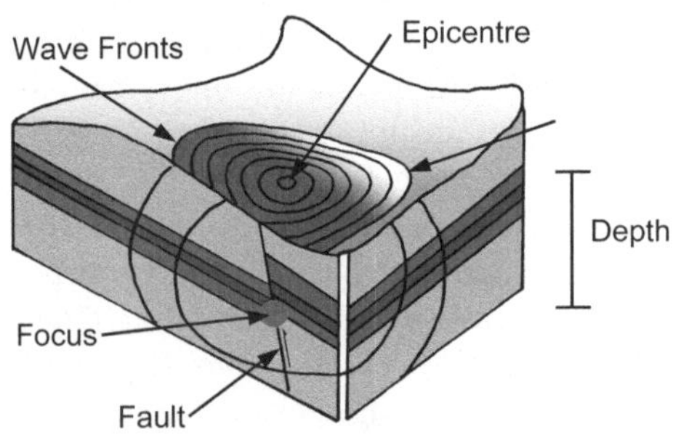

Fig. 15.2 Earthquake terminology

(ranging from 50 to 100 km) moves on top of a partly molten inner layer. The plates slide over under and past each other (Fig. 15.1). Most earthquakes occur at boundaries where the plates meet. In fact, the locations of earthquakes and the kinds of rupture they produce help scientists define a plate boundary (Fig. 15.6). At times, the plates are locked together, unable to release the accumulating energy. When this accumulated energy grows strong enough, the plates break free. When two pieces that are next to each other get pushed in different directions, they will stick together for a long time, but eventually the forces pushing on them will compel them to break apart and move. This sudden shift in the rock shakes all of the ground around it. These aftershocks (shocks after the main earthquake) are common and result from the realignment of the crust around a major fault movement. Aftershocks usually occur within a few days of the main earthquake, becoming subdued over time.

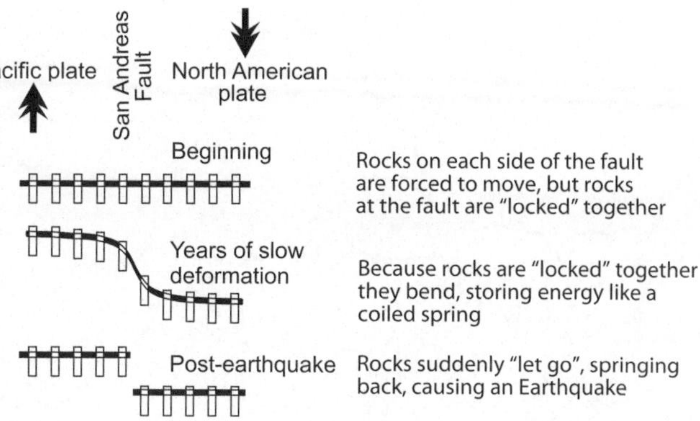

Fig. 15.3 The elastic rebound theory. The theory states that if slippage along a fault is obstructed, the elastic strain energy builds up in the deforming rocks on both sides of the fault. This energy is eventually released when slippage occurs suddenly, causing an earthquake

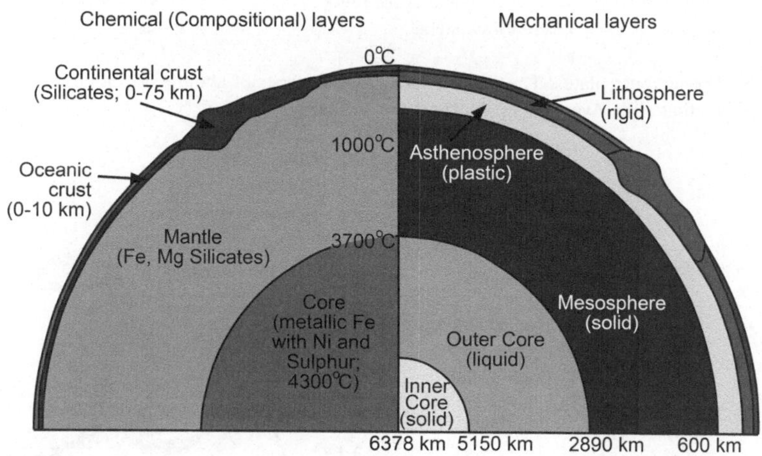

Fig. 15.4 The chemical (compositional) and mechanical layers of the earth

15.2 Locating an Earthquake (and some Terminologies)

Although, a general location for an earthquake can be inferred from the records of a single station, but to determine the exact location of an earthquake, the distance of the earthquake must be determined from at least three seismic recording stations (Fig. 15.7). Circles with the appropriate radii are then drawn around each station (here A, B, and C; Fig. 15.7). The intersection of these three circles uniquely identifies the epicenter of the earthquake. This is called the Triangulation method.

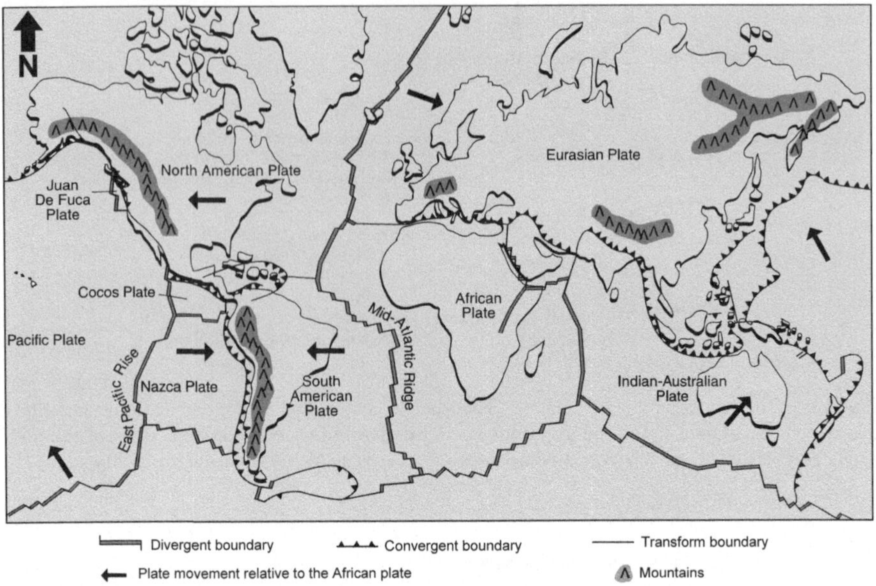

Fig. 15.5 The tectonic plates. The earth's outer layer consists of about a dozen large, irregularly shaped plates that slide over, under and past each other

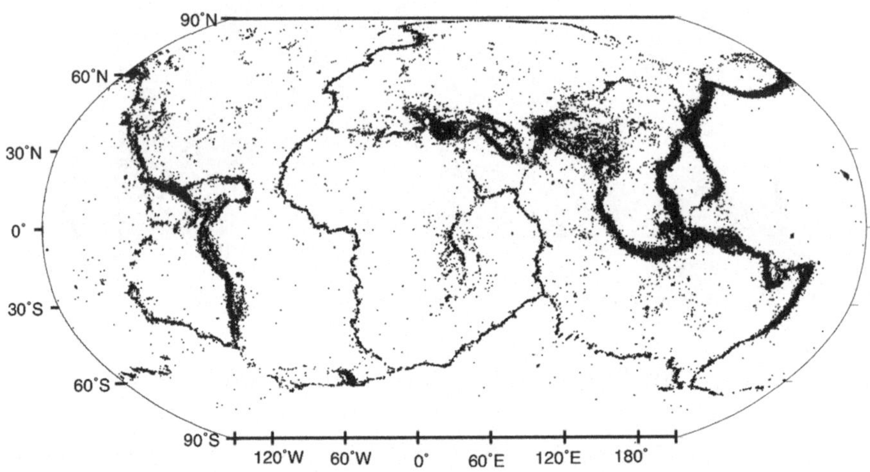

Fig. 15.6 Earthquake distribution around the globe. Note that distribution closely parallels plate boundaries (compare with Fig. 15.5)

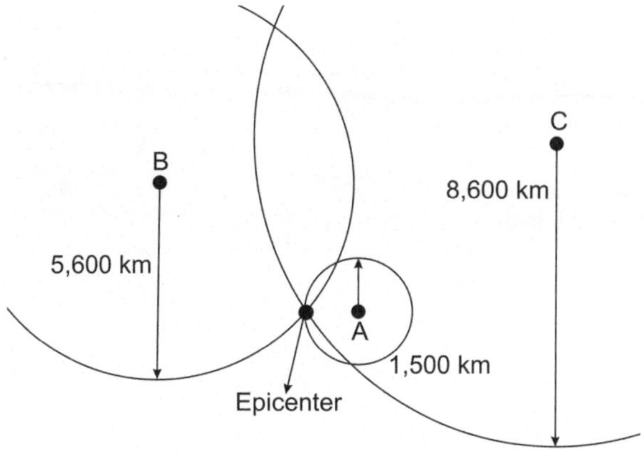

Fig. 15.7 Locating an earthquake. To determine the exact location of an earthquake, the distance of the earthquake must be determined from at least three seismic recording stations. *Circles* with the appropriate radii are then drawn around each station (here *A*, *B*, and *C*). The intersection of these three circles uniquely identifies the epicenter of the earthquake. This is called the Triangulation method

15.2.1 Seismology

As earthquake happens, elastic energy is released by vibrations that travel throughout the earth. These vibrations are called Seismic waves and their study is called Seismology.

15.2.2 Seismographs

Seismic waves travel as vibrations throughout the earth which are recorded graphically by an instrument (a simple pendulum) called a Seismograph (Fig. 15.8). When ground shakes, the base and frame of the instrument moves with it, but the inertia keeps the pendulum bob in place. It will then appear to move, relative to the shaking ground. As it moves, it records the pendulum displacements as they change with time, tracing out a record called a Seismogram. The first seismograph was developed by the Chinese almost 2,000 years ago. One seismograph station, having three different pendulums sensitive to the north–south, east–west, and vertical motions of the ground, will record seismograms that allow scientists to estimate the distance, direction, Richter Magnitude, and type of faulting of the earthquake.

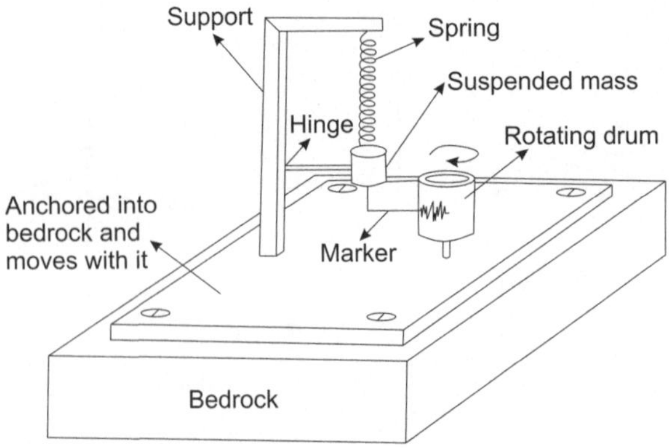

Fig. 15.8 The seismograph. Seismic waves travel as vibrations throughout the earth which are recorded graphically by an instrument (a simple pendulum) called a Seismograph

15.2.3 Seismogram

The zig-zag line made by a Seismograph is called a Seismogram. It reflects the changing intensity of the vibrations by responding to the motion of the ground surface beneath the instrument.

15.2.4 Focus

This marks the source of an earthquake and is the exact location within the earth where seismic waves are generated by a sudden release of the stored elastic energy (Fig. 15.2).

15.2.5 Focal Depth

It is the depth from the earth's surface to the region where an earthquake's energy originates (the focus; Fig. 15.2).

15.2.6 Epicenter

It is the point on the surface directly above the Focus (Fig. 15.2). It is the location of an earthquake commonly described by the geographic position of its Epicenter and by its Focal depth (Fig. 15.2)

15.2.7 Seismic Waves

Enormous sea waves (Tsunamis) are formed when earthquakes occur beneath the ocean floor. These waves travel across the Ocean at speeds as much as 960 km/h and can be as high as 15 m or higher as they reach ashore. Earthquakes produce two types of waves—Surface and Body waves.

15.2.7.1 Surface Waves

They are also called as Long waves or L-waves. They are the slowest-moving seismic waves but are responsible for maximum damage. They travel along paths nearly parallel to the surface of the earth and not inside the earth. They have a complex motion with both up-and-down and side-to-side. They behave like S-waves (with similar up-and-down and side-to-side movements; Fig. 15.9), however, they travel slower and do not travel through the body of the earth.

There are two types of Surface waves—Love waves (L-waves) and Rayleigh waves. L-waves shake the ground side-to-side like an S wave and the Rayleigh waves displace the ground like rolling ocean waves. The ground rolls forward up and then down and backwards. This motion is similar to the P-wave (Primary waves) motion but with an extra up-down segment (Fig. 15.9).

15.2.7.2 Body Waves

These originate from the Focus (Fig. 15.2) and travel in all directions through the body of the earth. They are of two types—Compressional or Primary waves and Shear or Secondary waves (Fig. 15.10).

P-Waves

They are the fastest waves to travel (at speeds of 4–7 km/s) and are the first to reach the surface. They are called Primary waves or P-waves (Fig. 15.10). They are compressional (or longitudinal) waves where the rock vibrates back and forth parallel to the direction of wave propagation. P-waves are sound waves, and hence, are able to travel through all mediums (solids, liquids, and gasses; Fig. 15.10). However, their velocity varies and depends on rigidity and density of the material and the elastic properties of the rock through which they traverse. Hence, they traverse inside earth at different speeds, depending of the density of the material through which they are moving. As they move deeper, their speed increases, as the inside material is much more dense (Fig. 15.10).

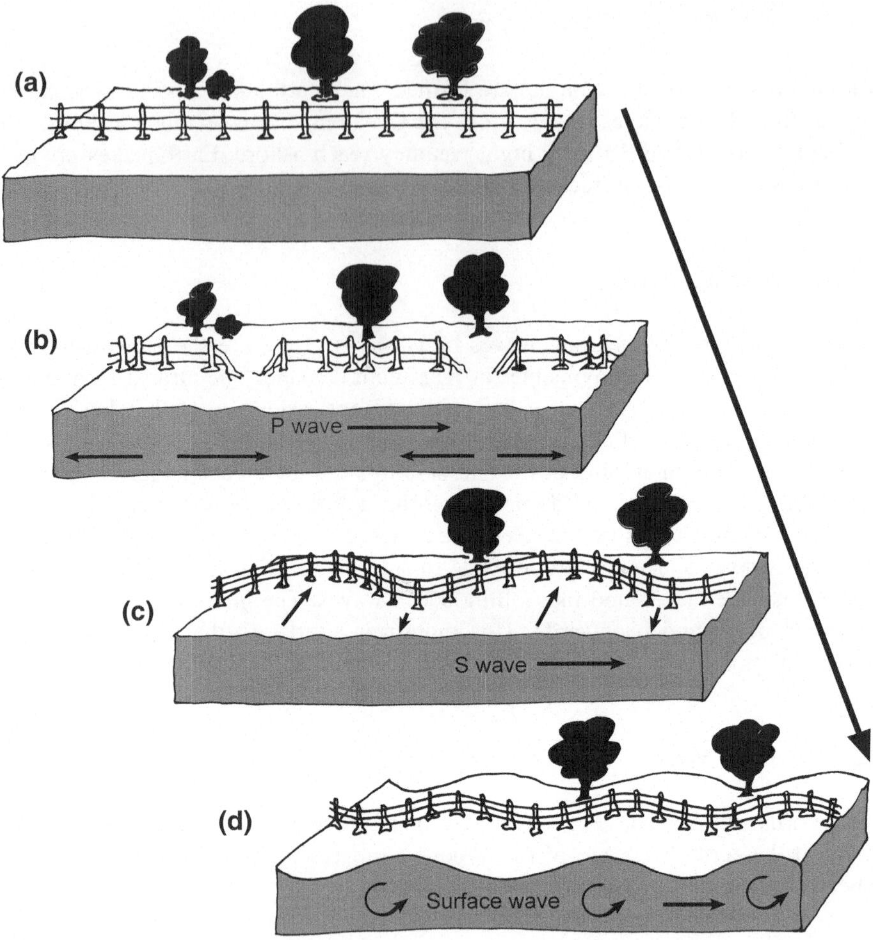

Fig. 15.9 Surface waves unlike body waves do not travel through the earth, but instead travel along paths nearly parallel to the surface. They have complex motion, up-and-down and side-to-side. Redrawn from Earth's Dynamic Systems with permission of Eric H. Christiansen

S-Waves

These are also called Shear or Secondary waves or S-waves. They travel through material by shearing it or changing its shape in the direction perpendicular to the direction of travel (Fig. 15.10) at speeds of 2–5 km/s. The resistance to shearing of a material is called Rigidity. Liquids and gasses have no rigidity; hence, the velocity of an S-wave in them is zero. S-waves arrive at a given point after P-waves. S-waves travel at 60–70 % less speed as of P-waves. The first indication of an earthquake is often a sharp thud, signaling the arrival of compressional waves. This is followed by the shear waves and then the ground rolls due to the surface waves (Fig. 15.9).

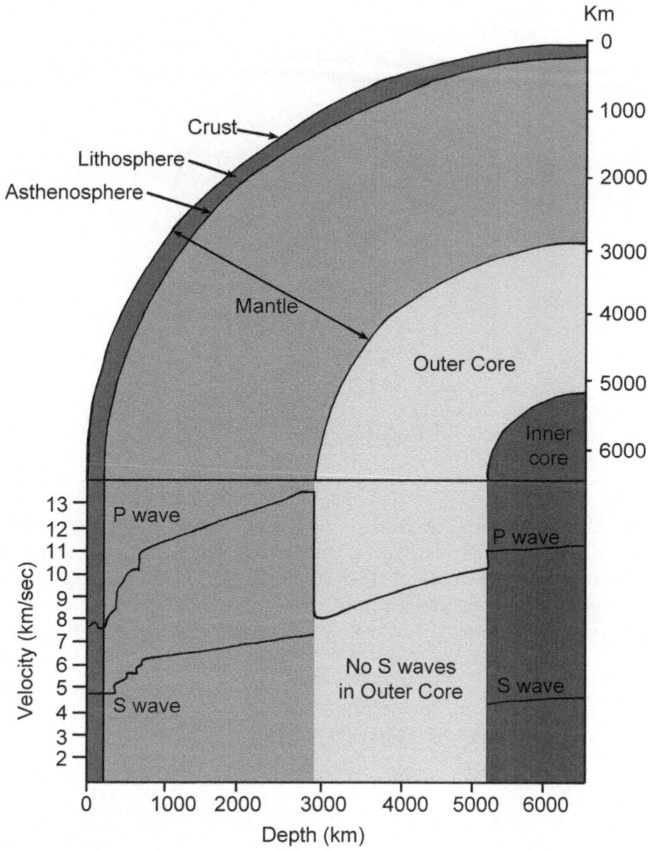

Fig. 15.10 The internal structure of the earth is revealed by following the velocity changes in P- and S-waves. These are also known as the compression P-, or push and pull, waves and the shear S, or shake, waves. The P-waves are almost twice as fast as the S-waves. The P-waves pass through the fluid Outer Core whereas the *S*-waves cannot. Hence, the boundary between the Mantle and Core is marked be a precipitous drop in the velocity of the P-waves. The S-waves do not propagate beyond this boundary. P-waves increase in velocity at the boundary between the liquid Outer Core and the solid Inner Core

15.2.8 Shadow Zone

The shadow zone is the area of the earth from angular distances of 103–142° that, for a given earthquake, does not experience seismic waves (Fig. 15.11). The shadow zone results from S-waves that are stopped by the liquid core and the P waves that get bent (refracted). By measuring how the P- and S-waves travel through earth and out on the other side, a seismic wave shadow zone was discovered in about 1910. From the lack of S-waves and a great slowing of the P wave velocity (by about 40 %), it was deduced that the outer Core is made of liquid. Thus, the shadow zone also defines the diameter of the Core.

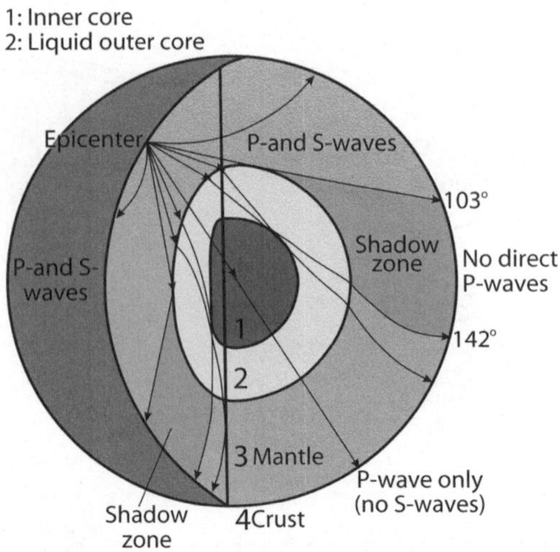

Fig. 15.11 The Shadow Zone. The shadow zone is the area of the earth from the angular distances between 103 and 142° that, for a given earthquake, do not experience seismic waves. The shadow zone results from S-waves that are stopped by the liquid Outer Core and the *P*-waves that get bent (refracted). Seismic waves refract (*bend*) inside the earth because of the change in speed of the waves as they move through material of variable density, composition, and temperature

15.3 How to Measure an Earthquake

Magnitude and Intensity are the most commonly used terms to measure the severity of an earthquake (see also Stein and Wysession 2002). Magnitude is a measure of the amplitude of the seismic waves, whereas, Intensity, as expressed by the Modified Mercalli Scale (MMS), is a subjective measure that describes how strong a shock was felt at a particular location.

15.3.1 Magnitude

Magnitude is expressed by the Richter scale and is a measure of the amplitude (height) of a seismic wave, determined by measuring the amplitude of the largest waves on the Seismogram. The Richter scale, named after Dr. Charles F. Richter of the California Institute of Technology (USA), is the best known scale for measuring the magnitude of an earthquake. He showed that larger the intrinsic energy of the earthquake, the larger the amplitude of ground motion at a given distance. He calibrated his scale of magnitudes using measured maximum amplitudes of shear waves on seismometers particularly sensitive to shear waves with periods of about one second. The scale is logarithmic. Each increase of 1 on

Table 15.1 Relationship of earthquake magnitude and energy released

Richter magnitude (M)	Energy (ergs)	Factor
1	2.0×10^{13}	31 ×
2	6.3×10^{14}	
3	2.0×10^{16}	
4	6.3×10^{17}	
5	2.0×10^{19}	
5.5 (Hiroshima bomb)		
6	6.3×10^{20}	
7	2.0×10^{22}	
8	6.3×10^{23}	
8.6: Largest earthquake—Alaska, 1964		

Table 15.2 Earthquakes magnitude and effects

Richter magnitude	Noted effects of the earthquake
<3.5	Generally not felt, but electronically recorded.
3.5–5.4	Often felt, but rarely causes damage
Under 6.0	Slight damage to well-designed buildings. Major damage likely to poorly constructed ones; effects are limited to a small region
6.1–6.9	Destructive in areas up to ~100 km across where people live
7.0–7.9	A major earthquake; serious damage over large areas
8 and>	A great earthquake; serious damage in areas several 100 km across

the Richter Magnitude is a tenfold increase in the size of the earthquake. The size of the earthquake is dependent on the amount of energy released. Hence, both the amount of energy release and Richter scale are related by the following equation: $\text{Log10E}_{\text{(Energy in ergs)}} = 11.8 + 1.5_{\text{Magnitude}}$. Thus, an increase by each 1 M involves a release of energy by a factor of 31 times (Table 15.1).

Seismologists use a Magnitude scale to express the amount of seismic energy released by each earthquake. Typical effects of earthquakes in various magnitude ranges are given in Table 15.2.

Although each earthquake has a unique magnitude, its effects vary greatly according to distance, ground conditions, construction standards, and other factors. Seismologists use a different Mercalli Intensity Scale to express the variable effects of an earthquake.

15.3.2 Intensity

Intensity is the estimate of the effects of an earthquake. Intensity should not be confused with magnitude. Although each earthquake has a single magnitude value, its effects vary from place to place, and hence, there are many different intensity estimates. The intensity is expressed by the Mercalli Scale. It is a subjective measure that describes how strong a shock was felt at a particular location. The Modified Mercalli Scale (MMS; Table 15.3) expresses the intensity of an earthquake's effects in a given locality in values ranging from I to XII. The most commonly used adaptation covers the range of intensity from the condition of I (not felt except by few) to XII (state of total damage). An earthquake's destructiveness depends on many factors. In addition to magnitude and the local geologic conditions, focal depth, distance from the epicenter, and the design of buildings is some other contributing factors. Add to that the density of population and the quality and amount of construction in the area of the earthquake. An area underlain by unstable ground (sand, clay, or other unconsolidated materials) will experience

Table 15.3 Modified Mercalli intensity (MMI) scale (modified from Wood and Neumann 1931, Bulletin of the Seismological Society of America)

MMI Scale	Noted effects of the earthquake
I	People do not feel any earth movement
II	A few people might notice movement if they are at rest and/or on the upper floors of tall buildings
III	Many people indoors feel movement. Hanging objects swing back and forth. People outdoors might not realize that an earthquake is occurring
IV	Most people indoors feel the movement. Hanging objects swing. Dishes, windows, and doors rattle. A few people outdoors may feel the movement. Parked cars rock
V	Almost everyone feels movement. Sleeping people are awakened. Doors swing open or close. Dishes are broken. Pictures on the wall move. Small objects move or are turned over. Trees might shake. Liquids might spill out of open containers
VI	Everyone feels movement. People have trouble walking. Objects fall from shelves. Pictures fall off walls. Furniture moves. Plaster in walls might crack. Trees and bushes shake. Damage is slight in poorly built buildings. No structural damage
VII	People have difficulty standing. Drivers feel their cars shaking. Some furniture breaks. Loose bricks fall from buildings. Damage is slight to moderate in well-built buildings; considerable in poorly built ones
VIII	Drivers have trouble steering. Houses that are not bolted down might shift on their foundations. Tall structures such as towers and chimneys might twist and fall. Well-built buildings suffer slight damage. Poorly built structures suffer severe damage. Tree branches break. Hillsides might crack if the ground is wet. Water levels in wells might change
IX	Well-built buildings suffer considerable damage. Houses not bolted down move off their foundations. Some underground pipes are broken. The ground cracks. Reservoirs suffer serious damage
X	Most buildings and their foundations are destroyed. Some bridges are destroyed. Dams are seriously damaged. Large landslides occur. Water is thrown on the banks of canals, rivers, lakes. The ground cracks in large areas. Railroad tracks are bent slightly
XI	Most buildings collapse. Some bridges are destroyed. Large cracks appear in the ground. Underground pipelines are destroyed. Railroad tracks are badly bent
XII	Almost everything is destroyed. Objects are thrown into the air. The ground moves in waves or ripples. Large amounts of rock may move

more noticeable effects than an area equally distant from an earthquake's epicenter but underlain by firm ground such as granite or a basalt.

Rating the Intensity of an earthquake's effects does not require any instrumental measurements. Seismologists use newspaper accounts, diaries, and other historical records to make intensity ratings of past earthquakes. Such research helps estimate future hazards. The relationship between Modified Mercalli Scale (MMS) and Modified Mercalli Intensity (MMI) Scale is given in Table 15.4.

15.4 Classification of Earthquakes

Earthquake classification is based on depth of focus, cause of their origin, and their intensity and magnitude.

15.4.1 Depth of Focus

Earthquake foci are shallow beneath oceanic tranches and become progressively deeper under tectonically active continental margins (Fig. 15.12). They are grouped into three broad categories: (a): Shallow-foci (0–70 km deep); (b) Intermediate-foci (70–300 km deep); and (c) Deep-foci (300–700 km deep). No earthquakes deeper than 670 km has so far been recorded.

15.4.2 The Cause of Their Origin

Earthquakes can be both Tectonic and Nontectonic in origin. The tectonic ones are produced when rocks break suddenly in response to global forces. Such earthquakes are scientifically important to the study of the earth's interior. But, these also pose the greatest hazard and destruction (see also Reiter 1990). They are dealt in detail in this chapter under the section earthquakes and plate boundaries below.

The nontectonic ones can be grouped into four types:

15.4.2.1 Denudation Earthquakes

These earthquakes break up rocks and thrust sediments upward, exposing sections of the Crust and are thus, susceptible to erosion and weathering. Denuded areas may take thousands of years to recreate the subsurface and the topsoil that supports life.

Table 15.4 Relationship between Modified Mercalli Scale (MMS) and Modified Mercalli Intensity (MMI) Scale

MMI scale	Major effects		MMS scale	Richter scale	Description of shaking severity	Full description
I	Detected	Felt only on a seismograph	<10		.	Felt by very few people
II	Feeble	Some people feel it	10 to <25			Felt by persons at rest, on the upper floors, or favorably placed
III	Slight	Felt by people at rest Like the rumbling of a large truck	25 to <50	4.2		Felt indoors. Hanging objects swing. Vibrations feel like passing of light trucks. May not be recognized as an earthquake
IV	Moderate	Felt by people walking. Loose objects on shelf rattle	50 to <100			Hanging objects swing. Vibration like passing of heavy trucks; or sensation of a jolt like a heavy ball striking the walls. Windows, dishes, doors rattle. Glasses clink. Crockery clashes. In the upper range of IV, wooden walls and frames creak
V	Slightly strong	Sleepers wake up Ringing of church bells	100 to <250	4.8	Light	Felt outdoors; Sleepers are wakened. Liquids disturbed, some is spilled. Small unstable objects displaced or upset. Doors swing, close, open. Shutters, pictures move
VI	Strong	Swaying of trees, Suspended objects swing and/or fall from shelves	250 to <500	5.4	Moderate	Felt by one and all. Many frightened and run outdoors. People walk unsteadily Windows, dishes, glassware break, Knickknacks, books, etc., fall off shelves. Pictures come off the wall. Furniture is moved or overturned. Weak plaster and masonry work cracks. Trees, bushes shaken (visibly, or heard to rustle)

(continued)

Table 15.4 (continued)

MMI scale	Major effects		MMS scale	Richter scale	Description of shaking severity	Full description
VII	Very strong	Mild alarm, cracks in walls, the wall plaster falls off	500 to <1,000	6.1	Strong	Difficult to stand. Noticed by drivers of cars. Hanging objects quiver. Furniture is broken. Damage to masonry D, including cracks. Weak chimneys break at the roof line. Plaster, loose bricks, stones, tiles, cornices (also unbraced parapets and architectural ornaments) fall off. Some cracks in masonry C. Waves on ponds; water turbid with mud. Small slides and caving in along sand or gravel banks. Large bells ring. Concrete irrigation ditches damaged
VIII	Destructive	Moving cars uncontrollable, Chimneys fall, Fracture of masonry and poorly constructed buildings fail	1,000 to <2,500		Very strong	Steering of motor cars affected. Damage to masonry C; partial collapse. Some damage to masonry B; none to masonry A. Fall of stucco and some masonry walls. Twisting, fall of chimneys, factory stacks, monuments, towers, elevated tanks. Frame houses moved on foundations if not bolted down; loose panel walls thrown out. Decayed piling broken off. Branches broken from trees. Changes in flow or temperature of springs and wells. Cracks in wet ground and on steep slopes

(continued)

Table 15.4 (continued)

MMI scale	Major effects		MMS scale	Richter scale	Description of shaking severity	Full description
IX	Ruinous	Some houses collapse, pipes break, ground cracks	2,500 to <5,000	6.9	Violent	This causes general panic. Masonry D destroyed; Masonry C is heavily damaged, sometimes with complete collapse; Masonry B seriously damaged (a general damage to foundations) Frame structures, if not bolted, are shifted off the foundations. Frames racked. Serious damage to reservoirs. Underground pipes broken. Conspicuous cracks in ground. In alluvial areas sand and mud ejected, earthquake fountains, sand craters noted
X	Disastrous	Ground cracks profoundly, buildings destroyed, liquefaction and landslides widespread	5,000 to <7,500	7.3	Very violent	Most masonry and frame structures destroyed with their foundations. Some well-built wooden structures and bridges destroyed. Serious damage to dams, dikes, and embankments. Large landslides occur. Water thrown on banks of canals, rivers, lakes, etc. Sand and mud shifted horizontally on beaches and flat land. Rails bent slightly

(continued)

Table 15.4 (continued)

MMI scale	Major effects		MMS scale	Richter scale	Description of shaking severity	Full description
XI	Very disastrous	Most buildings destroyed bridges and other communication network destroyed, general triggering of other hazards	7,500 to <9,800	8.1	.	Rails bent greatly. Underground pipelines completely out of service
XII	Catastrophic	Total destruction; Trees totally uprooted, rise and fall of ground	>9,800			Damage nearly total. Large rock masses displaced. Lines of sight and level distorted. Objects thrown into the air

Masonry A: shows good workmanship, mortar and design; reinforced, especially laterally, and bound together using steel, concrete, etc.; designed to reduce lateral forces. Masonry B: Good workmanship and mortar; reinforced, but not designed in detail to resist lateral forces. Masonry C: Ordinary workmanship and mortar; no extreme weaknesses like failing to tie in at corners, but neither reinforced nor designed against horizontal forces. Masonry D: Weak materials, such as adobe; poor mortar; low standards of workmanship; weak horizontally

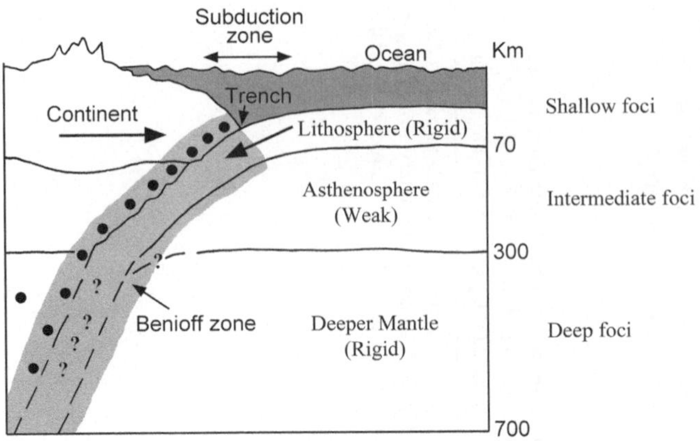

Fig. 15.12 The Benioff Zone. It is the broad region that is marked by earthquakes produced by the interaction of a downgoing oceanic crustal plate with a continental plate. It is also known as the Wadati-Benioff zone. The Benioff Zones are only found near trenches and nowhere else on earth. *Solid circles* mark the position of earthquakes

15.4.2.2 Volcanic Earthquakes

These occur in conjunction with volcanic activity. However, it is believed that eruptions and earthquakes both result from tectonic forces in the rocks and need not occur together.

15.4.2.3 Collapse Earthquakes

These are small earthquakes occurring in regions of underground caverns and mines. The immediate cause of ground shaking is the collapse of the roof of the mine or cavern. An often-observed variation of this phenomenon is the so-called "mine burst." This happens when the induced stress around the working mine causes large masses of rock to fly off the mine face explosively, producing seismic waves. Collapse earthquakes are also produced by massive land sliding.

15.4.2.4 Man-Made Earthquakes

These are produced by the detonation of chemicals or nuclear devices. When a nuclear device is detonated in an underground borehole, enormous nuclear energy is released. In millionths of a second, the pressure jumps thousands of times the pressure of the earth's atmosphere and the temperature increases by millions of degrees. The surrounding rock is vaporized, creating a spherical cavity many meters in diameter.

Table 15.5 World's major earthquakes in the last 100 years

Richter magnitude	Date	Human deaths	Location
9.5	22nd May, 1960	61	Chile
9.3	26th December, 2004	230,000	West coast of Sumatra, Indonesia
9.2	27th March, 1964	131	Prince William Sound, Alaska
9.1	9th March, 1957	0	Andreanof Islands, Alaska
9.1	26th December, 2004	227,898	Of west coast of northern Sumatra
9	4th November, 1952	10,000	Kamchatka, Russia
9	3rd November, 2011	20,896	Near the east coast of Honshu, Japan
8.9	11th March, 2011	20,000	Japan
8.8	27th February, 2010	700	North East Concepcion, Central Chile
8.8	31st January, 2006	1,500	Ecuador, Colombia
8.7	28th March, 2005	1,300	Nias, Indonesia
8.7	3rd February, 1965	15,000	Rat Islands, Alaska
8.6	15th August, 1950	780	Assam, India and Tibet
8.5	4th February, 1923	3	Kamchatka, Russia
8.5	2nd February, 1938	0	Banda Sea, Indonesia
8.5	13th October, 1963	0	Kuril Islands, Russia
8.3	28th July, 1976	250,000	Tangshan, China
8.0	12th May, 2008	87,000	Chengdu, China's SW Sichuan province
8.0	19th September, 1985	10,000	Mexico City
8.0	31st May, 1970	66,000	Yungay, Peruvian Andes
8.0	15th August, 2007	514	Near the coast of Central Peru
7.9	30th September, 2009	1,000	Indonesian island of Sumatra
7.9	15th August, 2007	519	Ica, Peru
7.9	6th April, 2009	103	Southern Sumatra, Indonesia
7.9	26th January, 2001	20,000	Gujarat, North Western India
7.9	1st September, 1923	142,800	Tokyo, Japan
7.9	18th April, 1906	3,000	San Francisco, USA
7	5th December, 2008	87,587	Eastern Sichuan, China
7.7	26th January, 2001	20,023	Kachchh, India
7.7	17th July, 2006	650	Southern coast of Java, Indonesia
7.6	8th October, 2005	73,000	Pakistan and Kashmir region, India
7.6	21st September, 1999	2,500	Taiwan
7.5	27th May, 1995	1,989	Eastern island of Sakhalin, Russia
7.5	30th September, 2009	1,117	Southern Sumatra, Indonesia
7.4	17th August, 1999	17,000	Turkish cities of Izmit and Istanbul
7.4	30th September, 1993	10,000	Western and Southern India
7.4	21st June, 1990	50,000	Gilan, Northern Iran

(continued)

Table 15.5 (continued)

Richter magnitude	Date	Human deaths	Location
7.2	23rd October, 2011	200	Ercis, South Eastern Turkey
7.2	12th November, 1999	400	Ducze, in North West Turkey
7.2	17th January, 1995	6,430	Kobe in Japan
7.2	4th March, 1977	1,500	Bucharest, Romania
7.1	2nd May, 1997	1,600	Birjand, Eastern Iran
7	12th January, 2010	230,000	Haitian capital Port au Prince
7	1st Dec 2010	316,000	Haitian capital Port au Prince
6.9	14th April, 2010	400	Qinghai province, Western China
6.9	7th December, 1988	25,000	North West Armenia
6.9	26th July, 1963	1,000	Skopje, Macedonia
6.9	30th May, 1998	5000	Takhar Province, Northern Afghanistan
6.8	21st May, 2003	2,000	Algeria and Spain
6.7	2nd January, 2012	113	Negros-Cuebo region, Philippines
6.6	26th December, 2003	30,000	Bam, Southern Iran
6.5	23rd December, 1972	10,000	Managua, Nicaraguan
6.4	29th October, 2008	300	Balochistan, Pakistan
6.4	22nd February, 2005	3000	Zarand, Iran's Kerman province
6.4	1st May, 2003	160	Bingöl, South Eastern Turkey
6.4	24th February, 2003	260	Xinjiang region, in western China
6.3	11th August, 2012	250	Tabriz and Ahar, North West Iran
6.3	22nd February, 2011	160	Christchurch, New Zealand
6.3	26th May, 2003	5,749	Java, Indonesia
6.3	6th April, 2009	300	L'Aquila, Italy
6.2	27th May, 2006	5,700	Yogyakarta, Java, Indonesia
6.1	25th March 2002	1,000	Hindu Kush region, Afghanistan
6	1st April, 2006	1,200	Western Iran
5.9	31st October, 2002	27	San Giuliano di Puglia, Italy

Table 15.6 Worldwide earthquakes per year

Magnitude (Richter)	Average number of earthquakes
8	2
7	20
6	100
5	3,000
4	15,000
3	>100,000

15.4.3 Intensity and Magnitude

This is explained in detail under the sections Intensity and Magnitude later in the chapter.

15.5 History of Earthquakes

15.5.1 World

Aristotle was one first to theorize about the origin of earthquakes. He thought that they were the result of heavy winds. In European history, the earliest recorded earthquake occurred in 580 BC; better described ones start from the mid-sixteenth century. However, the earliest well recorded earthquake occurred in China in 1177 BC. Table 15.5 gives a list of world's major earthquakes in the last 100 years and their human casualties.

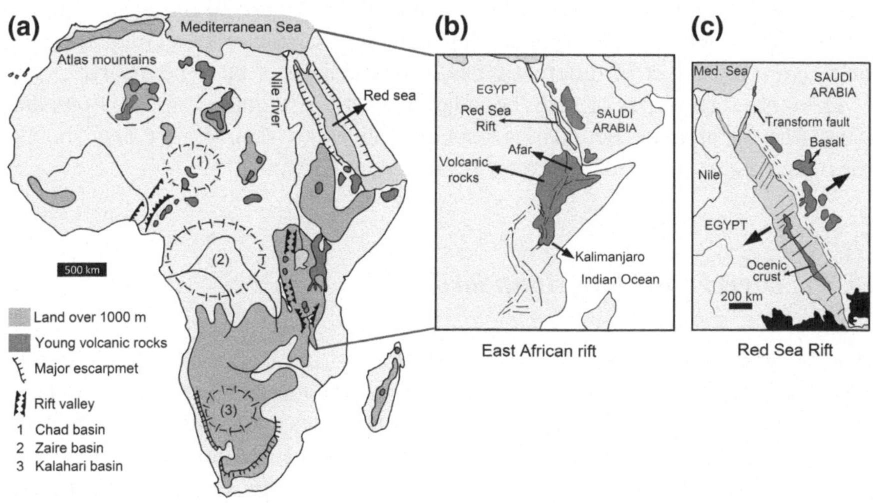

Fig. 15.13 The rifts of the Red Sea and East Africa. **a** The Red Sea Rift, a spreading center, lies between the two tectonic plates (African and Arabian) and extends down the length of the Red Sea. It stretches from the southern end of the Dead Sea Transform fault to the Afar Triple Junction in the Afar Depression of eastern Africa. **b** The East African Rift (in Kenya and Ethiopia) runs from the Afar Triple Junction in in the Afar Depression southward through eastern Africa and includes a number of active as well as dormant volcanoes, among them Mount Kilimanjaro is the most famous. **b-c** Redrawn from Earth's Dynamic Systems with permission of Eric H. Christiansen. **a** Redrawn from Marsh and Kaufman (2012) with permission of Cambridge University Press, UK

15.6 World Distribution of Earthquakes

The plates consist of an outer layer of earth, the lithosphere, which is cool enough to behave as a more or less rigid shell. Occasionally, the lower hot asthenosphere finds a weak place in the lithosphere to rise buoyantly as a plume, or a hotspot. Most earthquakes occur along relatively narrow belts, coinciding with plate boundaries, but few also occur away from plate boundary margins. (Figs. 15.6 and 15.12). Other major earthquake regions include the Mediterranean Sea area, through Iran and on to the Himalayas, Asia (Indonesia, Himalayan region), and the Mid-Ocean Ridges. Table 15.6 lists the worldwide earthquakes recorded per year and their corresponding magnitude on the Richter scale.

15.7 Earthquakes and Plate Boundaries

Almost 95 % of earthquakes take place in seismic belts corresponding to plate boundaries where plates converge, diverge, and slide past each other (see also Scholz 1990). There are seven major plates which are subdivided into a number of smaller ones (Fig. 15.5). These plates are thick ($\sim$70 km) and in constant motion relative to one another (at varying rates ranging from 10 to 130 mm/year). Most of the geological action such as mountains, rift valleys, volcanoes, earthquakes, and faulting are a product of different types of interactions at plate boundaries.

There are three types of plate boundaries where earthquakes occur—spreading zones (divergent plate boundaries), subduction zones (convergent plate boundaries), and transform faults (transform fault boundaries).

15.7.1 Divergent Plate Boundaries

Plate separation or where two plates move away from each other (the Divergent plate boundaries) is a very slow process. These boundaries are mostly found in oceans where molten rock rises, pushing the two plates apart, adding new material at plate edges. For example, the North American and Eurasian plates are moving away along the Mid-Atlantic Ridge (Fig. 15.1) at the rate of 2 cm/year. In such extensional areas, the lithosphere is in a state of tensional stress, resulting in the formation of normal faults and rift valleys. Here, the lithosphere is very thin and weak; hence, the strain cannot build up enough to cause large earthquakes. Hence, most shallow focus and lower magnitude earthquakes with focal depths less than 30 km occur here. Such shallow earthquakes indicate that not only the lithosphere is brittle but that it is relatively thin along these boundaries.

When a divergent boundary crosses land, a Rift valley is formed. These are typically 30–50 km wide. Classic examples of such include the East Africa rift in

Kenya and Ethiopia (Fig. 15.13), and the Rio Grande rift in New Mexico. However, when a divergent boundary crosses the ocean floor, the rift valley is much narrower, only about a km or less across, and it runs along the top of a Mid-Oceanic Ridge. Oceanic ridges rise to a km or so above the ocean floor and form a global network of tens of thousands of miles (Fig. 15.1). Examples of these include the Mid-Atlantic ridge and the East Pacific Rise.

15.7.2 Convergent Plate Boundaries

Here two plates run into each other and are thus, zones where compressional stresses are active. There are two types of converging plate boundaries: subduction (where oceanic lithosphere is pushed beneath either oceanic or continental lithosphere) and Collision (where two plates with continental lithosphere collide) (Fig. 15.1).

15.7.2.1 Subduction Boundaries

The most numerous and most severe earthquakes happen within the Subduction zones. These are found where one plate overrides, or subducts, another (usually an ocean plate), pushing it downward into the Mantle where it melts (Fig. 15.12). The cold oceanic lithosphere is pushed back down into the Mantle where two plates converge in an oceanic trench. As the subducted lithosphere is cold it remains brittle as it descends and thus, fractures under compressional stress. As it fractures, it generates earthquakes that define a zone with increasing focal depths beneath the overriding plate (Fig. 15.12). This zone of earthquakes is called the Benioff Zone (Fig. 15.12). Focal depths of earthquakes in the Benioff Zone can reach down to 700 km (Fig. 15.12). Some earthquakes on the subduction zones as in Alaska and Chile have exceeded magnitude 9. A recent earthquake in Japan (11th March 2011 of magnitude 9) near the east coast of Honshu, resulted from thrust faulting on or near the subduction zone plate boundary between the Pacific and the North America plates. The Pacific plate moved westwards with respect to the North America one at a rate of 83 mm/year, descending beneath Japan at the Japan Trench. The location, depth, and focal mechanism of the 11th March earthquake are consistent with the event having occurred on the subduction zone plate boundary. Modeling of the rupture of this earthquake indicate that the fault moved upwards of 30–40 m, and slipped over an area approximately 300 km long (along-strike) by 150 km wide (in the down-dip direction). This earthquake was the largest plate boundary thrust-fault earthquake in the southern Japan Trench.

15.7.2.2 Collision Boundaries

Here two plates of continental lithosphere collide to form fold-thrust mountain belts. Earthquakes occur due to thrust faulting and range in depth from shallow to about 200 km deep. The broad swath of seismicity from Burma to the Mediterranean, crossing the Himalayas, Iran, and Turkey, to Gibraltar are typical examples. Within this zone, shallow earthquakes are associated with high mountain ranges where intense compression takes place. Intermediate- and deep-focus earthquakes occur in the Himalayas.

15.7.3 Transform Fault Boundaries

Here, the lithospheric plates slide past one another (Fig. 15.1), producing less striking features as compared to extensional or compressional environments discussed above. At Transform fault boundaries, earthquakes show strike-slip motion of faults. At these boundaries, the earthquakes tend to be shallow focus occurring at depths less than ~50 km with magnitudes smaller than 8.5 and forming a fairly linear pattern down the earth. A classic example of a transform-fault plate boundary is the San Andreas Fault, along the coast of California and NW Mexico, separating the Pacific from the North American plate. It is one of the longest transform fault boundary known. However, besides the above settings, rare earthquakes (less than 10 %) do occur within plates, although plate boundary earthquakes are much more common. As plates continue to move and plate boundaries change over geologic time, weakened boundary regions become part of the interiors of the plates. These zones of weakness within the continents can cause earthquakes in response to stresses that originate at the edges of the plate or in the deeper crust. Though, these earthquakes may be as severe as the earthquakes in the subduction zones, they tend to be far less frequent in space and time. Table 15.7 lists the number of earthquakes worldwide between 2000–2012.

15.8 Earthquake Hazards

The effects of any earthquake depends on three factors—(a) intrinsic to the earthquake (including its magnitude, type, location or depth), (b) geologic conditions where effects are felt (including the distance from the earthquake, path of the seismic waves, types of soil, and the water saturation of soil), and (c) societal (including the quality of construction, preparedness of populace, and time of the day) (see also Reiter 1990).

Earthquakes don't kill people, but buildings do. This is actually true as most deaths from earthquakes are caused by buildings or other human construction falling down during an earthquake and not by the earthquakes itself. Earthquakes

Table 15.7 Number of earthquakes Worldwide from 2000 to 2012

Magnitude	2000	2001	2002	2003	2004	2005	2006	2007	2008	2009	2010	2011	2012
8.0–9.9	1	1	0	1	2	1	2	4	0	1	1	1	2
7.0–7.9	14	15	13	14	14	10	9	14	12	16	23	19	12
6.0–6.9	146	121	127	140	141	140	142	178	168	144	150	185	108
5.0–5.9	1,344	1,224	1,201	1,203	1,515	1,693	1,712	2,074	1,768	1,896	2,209	2,276	1,401
4.0–4.9	8,008	7,991	8,541	8,462	1,0888	1,3917	12,838	12,078	12,291	6,805	1,0164	13,315	9,534
3.0–3.9	4,827	6,266	7,068	7,624	7,932	9,191	9,990	9,889	11,735	2,905	4,341	2,791	2,453
2.0–2.9	3,765	4,164	6,419	7,727	6,316	4,636	4,027	3,597	3,860	3,014	4,626	3,643	3,111
1.0–1.9	1,026	944	1,137	2,506	1,344	26	18	42	21	26	39	47	43
0.1–0.9	5	1	10	134	103	0	2	2	0	1	0	1	0
No magnitude	3,120	2,807	2,938	3,608	2,939	864	828	1,807	1,922	17	24	11	3
Total	22,256	23,534	27,454	31,419	31,194	30,478	29,568	29,685	31,777	14,825	21,577	22,289	16,667
Estimated Deaths	231	21,357	1,685	33,819	228,802	88,003	6,605	712	88,011	1,790	320,120	21,953	768

Data from the US Geological Survey's (USGS) National Earthquake Information Center. Note that since January 2009, the USGS National Earthquake Information Center no longer locates earthquakes smaller than magnitude 4.5 outside the United States, unless the earthquake was felt or had caused damage located in isolated areas and far from human population rarely cause any deaths. Based on their impact, there are three classes of earthquake effects—Direct, Secondary, and Transient.

15.8.1 Direct Effects

This includes the region of the exposed fault rupture, the area where the earthquake happened. Elastic rebound (the permanent deformation of the ground due to fault rupture), extends to several km from the fault, and is often measured even where the rupture itself remains buried.

15.8.2 Secondary Effects

The secondary effects cause maximum damage primarily due to the propagation of seismic waves. Damage is wide spread and encompasses a large region.

Damages are grouped into four categories:

15.8.2.1 Fire

Both power and natural gas lines are affected causing widespread fires. The effects are compounded when water lines are also broken.

15.8.2.2 Rapid Mass wasting

This occurs in mountainous areas where earthquakes trigger rock and debris falls, slides, slumps, and debris avalanches.

15.8.2.3 Landslides

These are triggered by earthquakes causing more destruction than the earthquake itself.

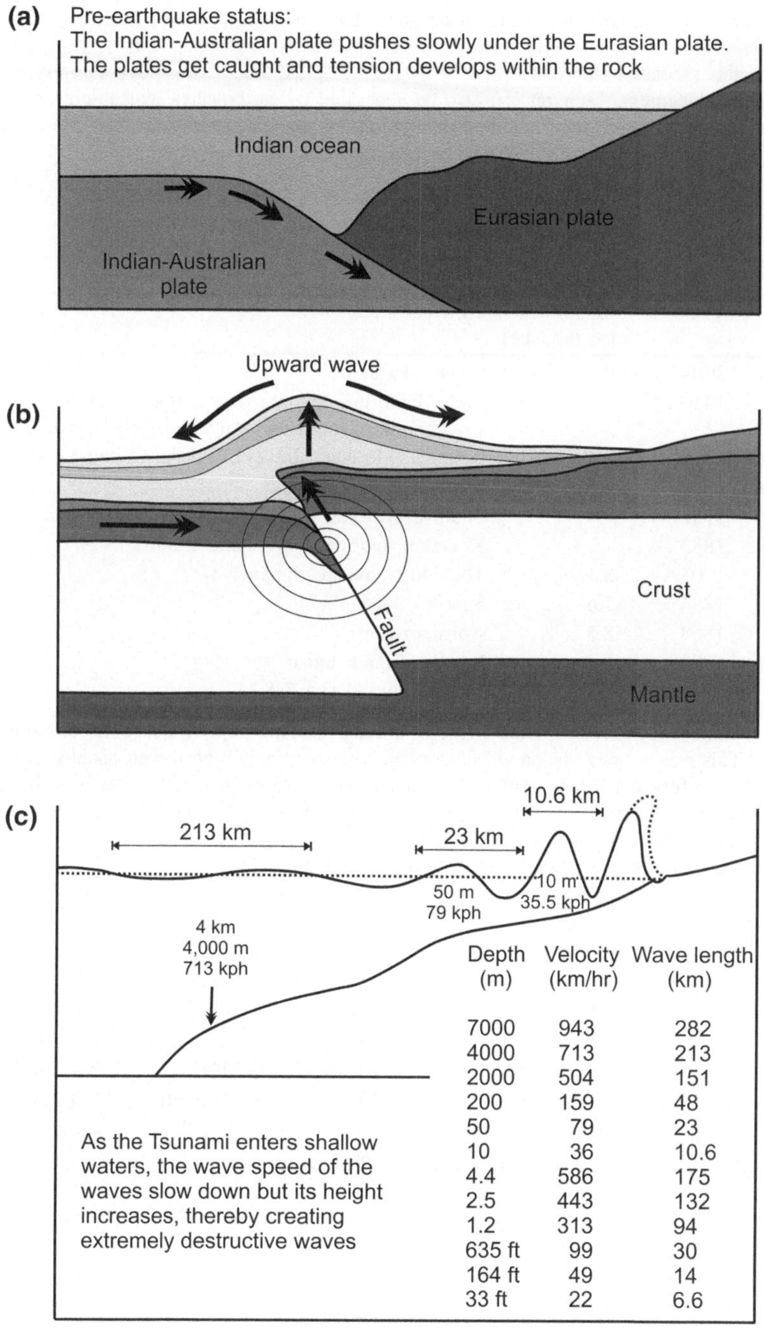
(a)
Pre-earthquake status:
The Indian-Australian plate pushes slowly under the Eurasian plate.
The plates get caught and tension develops within the rock
Indian ocean
Eurasian plate
Indian-Australian plate
(b)
Upward wave
Crust
Fault
Mantle
(c)
213 km
23 km
10.6 km
50 m
79 kph
10 m
35.5 kph
4 km
4,000 m
713 kph
Depth (m) Velocity (km/hr) Wave length (km)
7000 943 282
4000 713 213
2000 504 151
200 159 48
50 79 23
10 36 10.6
4.4 586 175
2.5 443 132
1.2 313 94
635 ft 99 30
164 ft 49 14
33 ft 22 6.6
As the Tsunami enters shallow waters, the wave speed of the waves slow down but its height increases, thereby creating extremely destructive waves

◀ **Fig. 15.14** The Tsunami. Tsunami can be generated when the sea floor abruptly deforms and vertically displaces the overlying water. More specifically, a tsunami can be generated when thrust faults associated with convergent or destructive plate boundaries move abruptly, resulting in water displacement. Tsunami can also be generated by earthquakes, volcanic eruptions, and other underwater explosions (including detonations of underwater nuclear devices), landslides, glacier calvings, and meteorite impacts, among others

Table 15.8 Deadliest Tsunamis

Fatalities	Years	Magnitude (earthquake)	Principal areas
350,000	2004	9	Indian Ocean
100,000	1410 BC	–	Crete-Santorini, Ancient Greece (caused by the massive eruption of the Thera volcano in the Aegean Sea)
100,000	1755	8.5	Portugal, Morocco, Ireland, and the United Kingdom
100,000	1908	7.1	Messina, Italy
40,000	1782	7	South China Sea, Taiwan
36,500	1883	–	Krakatau, Indonesia (volcanic eruption)
30,000	1707	8.4	Tokaido-Nankaido, Japan
26,360	1896	7.6	Sanriku, Japan
25,674	1868	8.5	Northern Chile
15,030	1792	6.4	Kyushu Island, Japan
2	1958	8.3	Lituya Bay, Alaska, USA[a]

Source National Geophysical Data Center, National Oceanic and Atmospheric Administration (USA).[a] This was a "mega-tsunami" (a tsunami with extremely high waves, usually caused by a landslide). On July 9, 1958, an earthquake caused a landslide in the Crillon Inlet at the head of the bay, generating a massive megatsunami measuring 524 m (1,719 ft). For comparison, the Empire State Building (New York, USA) is only 448 m (1,470 ft.) high (including its antenna spire). Miraculously, only two people died in this mega-tsunami

15.8.2.4 Liquefaction

It is a process that occurs due to shaking in water-saturated unconsolidated sediments. Loosely packed, water-logged sediments lose their strength in response to strong ground shaking during an earthquake. Areas underlain by such sediments loose grain-to-grain contact, resulting in their flow.

15.8.3 Transient Effects

This includes Tsunamis, ground shaking, sand blows, mud volcanoes, brontides (loud sounds), and earthquake lights (refraction due to atmospheric oscillation). Of these, Tsunamis and ground shaking are most devastating.

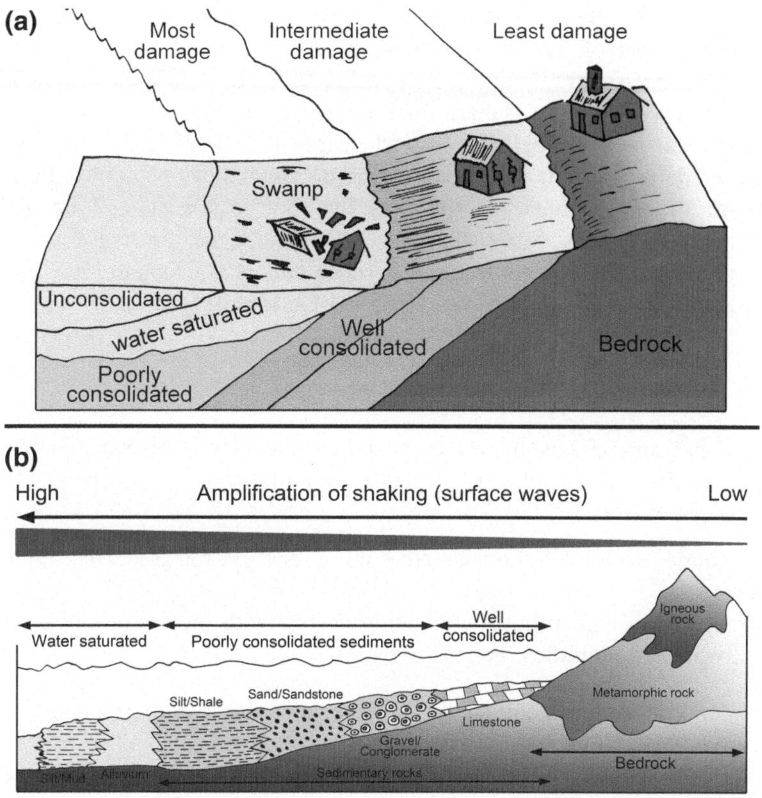

Fig. 15.15 The passage of seismic waves near the epicenter is responsible for much damage and the collapse of structures. **a** Loose, saturated and unconsolidated sediments amplify ground shaking, resulting in greater damage. **b** The type of soil is also a determinant factor. Loose sediments such as silt, mud and alluvium greatly amplify seismic waves, resulting in severe ground shaking

15.8.3.1 Tsunamis

Strong earthquakes along coastal areas generate giant ocean waves called Tsunamis (Fig. 15.14), that travels fast causing extensive damage even thousands of km away from its origin. These waves can travel between 500 and 960 km/h and are hence, are fast, high energy waves. In the deep sea, these are not tall (<1 m), but when they reach shallow waters, they slow down and the water stacks up. Then, the wave height may even reach up to 15 m or higher as they approach the shore. The withdrawal of water from the coast is the first sign of a Tsunami. Then in a matter of 10–30 min, the water recedes as a huge wave (5–30 m high) resulting in flooding of coastal cities, washing ashore boats into far inland and killing scores of people.

During the 2004 Sumatra earthquake (Table 15.8), Tsunamis engulfing coastal areas caused most of the destruction in India (Andaman and Nicobar Islands, Tamil Nadu, and Andhra Pradesh), Indonesia and Sri Lanka. Interestingly, historic descriptions reveal a large tsunami had an impact on the fleet of Alexander the Great in October–November 325 BC, when the ships were in the vicinity of the Indus river delta. In fact, the Greek historian Thucydides (460–395 BC) in his History of the Peloponnesian War was the first to associate Tsunamis with underwater earthquakes. Since, the past 3,423 years, the top five countries that have experienced the maximum number of Tsunamis are Japan (250), Indonesia (67), Russia (63), the United States (56), and Chile (39).

15.8.3.2 Ground Shaking

The brief passage of seismic waves near the epicenter is responsible for much damage and the collapse of most structures. Ground shaking is dependent on the following parameters:

(a) magnitude of the earthquake (the more energy released, the higher the destruction)
(b) distance from the earthquake (shaking decays with increasing distance) and
(c) local soil that amplifies the shaking

In general, loose unconsolidated sediments are subject to more intense shaking than solid bedrock (Fig. 15.15a). The construction type is also a factor to count in when shaking happens. Concrete and masonry structures, due to their brittleness are more susceptible to damage than more flexible wood and steel ones. It is important to note that the intensity of ground shaking also depends on distance from the epicenter and on the type of bedrock underlying the area (Fig. 15.15b).

15.9 Summary

Any trembling of earth's surface that follows a release of energy in earth's crust, generated by a sudden dislocation of a segment of crust, or by a volcanic eruption, or at times by man-made explosions is defined as an Earthquake. However, most destructive earthquakes are caused by the sudden crustal movement and often along plate boundaries. As plates move (move over, under, and past each other), these boundaries (ocean ridges, continental rifts, subduction zones, and transform faults) are sites of most intense earthquake activity. Most natural earthquakes are caused by a sudden slippage along a fault. If this slippage is obstructed, elastic strain energy builds up in the deforming rocks on both sides of the fault, which is eventually released when slippage occurs, resulting in an earthquake. This is called the Elastic Rebound theory.

Although, a general location for an earthquake can be inferred from the records of a single station, but to determine the exact location of an earthquake, the distance of the earthquake must be determined from at least three seismic recording stations. Circles with the appropriate radii are then drawn around each station and the intersection of these three circles uniquely identifies the epicenter of the earthquake. This is called the Triangulation method.

As earthquake happens, elastic energy is released by vibrations that travel throughout the earth. These vibrations are called Seismic waves and their study is called Seismology. These vibrations are recorded graphically by an instrument (a simple pendulum) called a Seismograph that traces out a record called a Seismogram. The first seismograph was developed by the Chinese almost 2,000 years ago. This record allows scientists to estimate the distance, direction, Richter Magnitude, and type of faulting of the earthquake.

Focus marks the source of an earthquake and is the exact location within the earth were seismic waves are generated. Focal depth is the depth from the earth's surface to the region where an earthquake's energy originates. Epicenter is the point on the surface directly above the Focus. It is the location of an earthquake commonly described by the geographic position of its epicenter and by its focal depth.

Earthquakes produce two types of waves—Surface and Body waves. Surface waves are also called as Long waves or L-waves. They are the slowest moving seismic waves but are responsible for maximum damage. They travel along paths nearly parallel to the surface of the earth and not inside the earth. There are two types of Surface waves—Love waves (L-waves) and Rayleigh waves. L-waves shake the ground side-to-side like an S-wave and the Rayleigh waves displace the ground like rolling ocean waves. The ground rolls forward and up and then down and backwards. Body waves originate from the Focus and travel in all directions through the body of the earth. They are of two types—Compressional or Primary waves and Shear or Secondary waves. P-waves are compressional (or longitudinal) waves where the rock vibrates back and forth parallel to the direction of wave propagation. They are the fastest and are the first to reach the surface. They are sound waves, and hence, are able to travel through all mediums (solids, liquids, and gasses). However, their velocity varies and depends on rigidity and density of the material and the elastic properties of the rock through which they traverse. Hence, as they move deeper, their speed increases, as the inside material is much more dense. The S-waves are also called Shear or Secondary waves. They travel through material by shearing it or changing its shape in the direction perpendicular to the direction of travel. The resistance to shearing of a material is called Rigidity. Liquids and gasses have no rigidity; hence, the velocity of an S-wave in them is zero. S-waves arrive at a given point after P-waves. S-waves travel at 60–70 % less speed as of P-waves. The first indication of an earthquake is often a sharp thud, signaling the arrival of compressional waves. This is followed by the shear waves and then the ground rolls due to the surface waves.

The shadow zone is the area of the earth from angular distances of 103–142° that, for a given earthquake, does not experience seismic waves. The shadow zone results from S-waves that are stopped by the liquid core and the P-waves that get bent (refracted).

Magnitude and Intensity are the most commonly used terms to measure the severity of an earthquake. Magnitude is a measure of the amplitude of the seismic waves, whereas, Intensity, as expressed by the Modified Mercalli Scale (MMS), is a subjective measure that describes how strong a shock was felt at a particular location.

Earthquake classification is based on depth of focus (Shallow-foci: 0–70 km deep; Intermediate-foci: 70–300 km deep; Deep-foci: 300–700 km deep), cause of their origin (Tectonic and Nontectonic; the latter has four types: Denudation; Volcanic; Collapse, and Man-made earthquakes) and their intensity and magnitude.

Aristotle was one first to theorize about the origin of earthquakes. He thought that they were the result of heavy winds. In European history, the earliest recorded earthquake occurred in 580 BC; better described ones start from the mid-sixteenth century. However, the earliest well recorded earthquake occurred in China in 1177 BC.

Most (95 %) earthquakes occur along relatively narrow belts, coinciding with plate boundaries, but few also occur away from plate boundary margins. There are three types of plate boundaries where earthquakes occur—spreading zones (divergent plate boundaries), subduction zones (convergent plate boundaries), and transform faults (transform fault boundaries).

The effects of any earthquake depends on three factors—(a) intrinsic to the earthquake (including its magnitude, type, location, or depth), (b) geologic conditions where effects are felt (including the distance from the earthquake, path of the seismic waves, types of soil, and the water saturation of soil), and (c) societal (including the quality of construction, preparedness of populace, and time of the day). Based on their impact, there are three classes of earthquake effects—Direct, Secondary, and Transient.

Strong earthquakes along coastal areas generate giant ocean waves called Tsunamis. Tsunamis travel fast causing extensive damage, even thousands of km away from its origin. These waves can travel between 500 and 960 km/h and are hence, fast, high energy waves. In the deep sea, these are not tall (<1 m), but when they reach shallow waters, they slow down and the water stacks up. Then, the wave height may even reach up to 15 m or higher as they approach the shore. The withdrawal of water from the coast is the first sign of a tsunami. Then in a matter of 10–30 min, the water recedes as a huge wave (5–30 m high) resulting in flooding of coastal cities, washing ashore boats into far inland and killing scores of people.

The brief passage of seismic waves near the epicenter is responsible for much damage and the collapse of most structures. Ground shaking is dependent on three parameters: magnitude of the earthquake; distance from the earthquake and local soil that amplifies the shaking. In general, loose unconsolidated sediments are subject to more intense shaking than solid bedrock. The construction type is also a factor to count in when shaking happens. Concrete and masonry structures, due to their brittleness are more susceptible to damage than more flexible wood and steel ones.

References

Abbot PL (1998) Natural disasters. McGraw-Hill, New York, 400 p

Bolt BA (1999) Earthquakes. W. H. Freeman, New York

Marsh WM, Kaufman MM (2012) Physical Geography: Great Systems and Global Environments. Cambridge University Press, United Kingdom, 720 p

Pakiser LC (1991) Earthquakes. U.S. Government Printing Office, Washington, D.C.

Reiter L (1990) Earthquake hazard analysis, issues and insights. Colombia University Press, New York

Richter CF (1958) Elementary seismology. W. H. Freeman and Company, San Francisco, pp 135–149

Scholz CH (1990) The mechanics of earthquakes and faulting. Cambridge University Press, New York

Shearer P (1999) Introduction to seismology. Cambridge University Press, New York, 260 p

Stein S, Wysession M (2002) Introduction to seismology, earthquakes and earth structure. Blackwell Publishing, Oxford, 498 pp

United States Geological Survey's (USGS) National Earthquake Information Center (NEIC). http://earthquake.usgs.gov/earthquakes/eqarchives/year/eqstats.php

Wood HO, Neumann F (1931) Modified Mercalli intensity scale of 1931. Bull Seismol Soc Am 21:277–283

Chapter 16
Volcanoes

16.1 Introduction

A volcano is an opening in the earth's surface from which molten lava, rock fragments, and other gases erupt. When hot material (Lava) comes out from a volcano, it is called a volcanic eruption (Fig. 16.1). In such eruptions, temperatures can reach as much as 1,250 °C. Under some conditions, magma temperature can reach as much as 1,400 °C (Bullard 1984; Fisher et al. 1997; Schmincke 2004; Decker and Decker 2005; Claybourne, 2007). Magma is an extremely hot liquid that is generated inside the earth, usually at depths from 10 to 200 kms. Often magma starts building pressure and moves up to the surface. This pressure buckles the surface of the earth and eventually forms a mountain. If this liquid comes out to the surface of the earth, then, a Volcano is born. Around 80 % of our earth's surface is a product of volcanoes, as also the surface of the sea and mountains that are formed by countless eruptions (Camp 2010). About 20 % of all volcanoes are under water (Harris and Armstrong 2003). The atmosphere was also formed by gases produced during such eruptions. In fact, many scientists are of the view that that all water on earth was originally vented into the atmosphere by volcanoes. Areas with high volcanic activity have some of the world's most fertile farmlands due to release of nutrients such as potassium and phosphorus.

The word volcano comes from the little island of Vulcano in the Mediterranean Sea off the coast of Sicily (Fig. 16.2). Centuries ago, people living in this area believed that the hot lava fragments and clouds of dust erupting from Vulcano came from Vulcan's forge as he beat out thunderbolts for Jupiter, the King of Gods, and weapons for Mars, the God of War (Claybourne 2007).

In August 24, 79 AD, two letters from Pliny the younger to the Roman historian Tacitus recorded the world's first volcanic eruption that suddenly exploded and destroyed the Roman cities of Pompeii, Herculaneum and Stabiae (modern Italy; Fig. 16.12). Pompeii was buried in 3 m of pyroclastic material whereas, Herculaneum by 20 m. Thus, the science of volcanology originated with the accurate descriptions of the eruption of Mount Vesuvius (Fig. 16.2). In 1847, an observatory was established on the flanks of Vesuvius, upslope from the site of

S. Jain, *Fundamentals of Physical Geology*, Springer Geology,
DOI: 10.1007/978-81-322-1539-4_16, © Springer India 2014

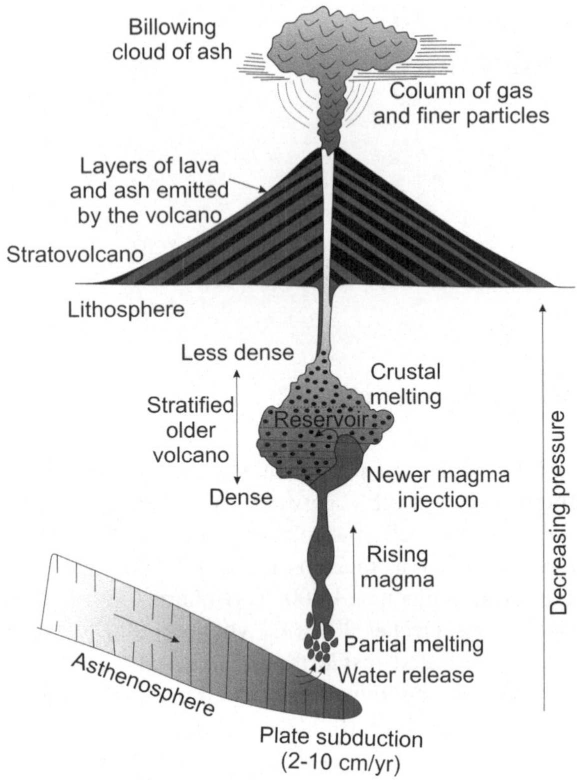

Fig. 16.1 The process of formation and eruption of a volcano in a Subduction zone setting

Herculaneum, which led to a more or less continuous recording of the activity of this volcano. Mount Vesuvius has continued its activity intermittently ever since 79 A.D. with numerous minor eruptions and several major ones occurring in 1631, 1794, 1872, 1906, and 1944. However, modern study of volcanology started in 1912, when Thomas A. Jaggar, Head of the Geology Department of the Massachusetts Institute of Technology (USA), founded the Hawaiian Volcano Observatory (HVO; see also Babb et al. 2011) (Fig. 16.3a), located on the rim of Kilauea's caldera (Fig. 16.3b).

16.2 Volcanic Features

The volcanic material that is emitted on the earth's surface through an opening is called a Volcanic vent. But the main conduit through which magma moves upward to the surface is called a Central vent (Fig. 16.4). It is connected to a magma chamber (a reservoir) at depth, the main storage area for the eruptive material. The volcano's cone-shaped structure is built by more-or-less symmetrical accumulation of lava and/or pyroclastic material around this central vent system. Lateral

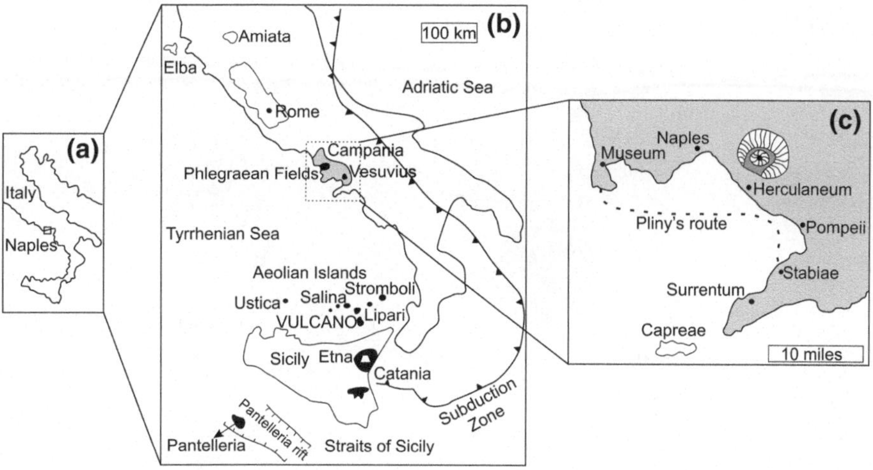

Fig. 16.2 Italy **a**. **b** A detailed map showing the Island of Vulcano in the Mediterranean Sea off Sicily and of Mount Vesuvius. **c** Detailed map of Mount Vesuvius. In August 24, 79 AD, Mt. Vesuvius erupted and destroyed the Roman cities of Pompeii and Herculaneum. The eruption remained active for more than 20 h. Dotted lines show the tentative route taken by Gaius Plinius Caecilius Secundus, better known as Pliny the Younger. He witnesses the eruption of Vesuvius and wrote letters, many of which still survive and are regarded as a historical source for the time period

vents (also called side vents) are found on the sides of some volcanoes where lava is extruded. In some volcanoes, the flanks contain fractures that descend downward toward the central vent or to a shallow level magma chamber. Such fractures may occasionally tap the magma source and act as conduits for flank eruptions along the sides of the volcanic edifice. These eruptions generate cone shaped accumulations of volcanic material called Parasitic cones or Secondary cones (Fig. 16.4). A radial arrangement of such cones is quite common and on most large composite volcanoes; they number from 10 to 12.

16.3 Magma

Magma is a naturally occurring liquid. In any given magma, 99 % of it is made up of 10 elements, including Silicon (Si), Titanium (Ti), Aluminum (Al), Iron (Fe), Magnesium (Mg), Calcium (Ca), Sodium (Na), Potassium (K), Hydrogen (H), and Oxygen (O). Magmas also contain many other chemical elements (generally in trace quantities), molten rock fragments, some crystals, fragments of surrounding (unmelted) rocks, and dissolved gases. More than 90 % of the gas emitted from hot magma is water and CO_2 (Diaz et al. 2002). Upon cooling, magma precipitates crystals of various minerals to form an igneous rock, leading to a large range of

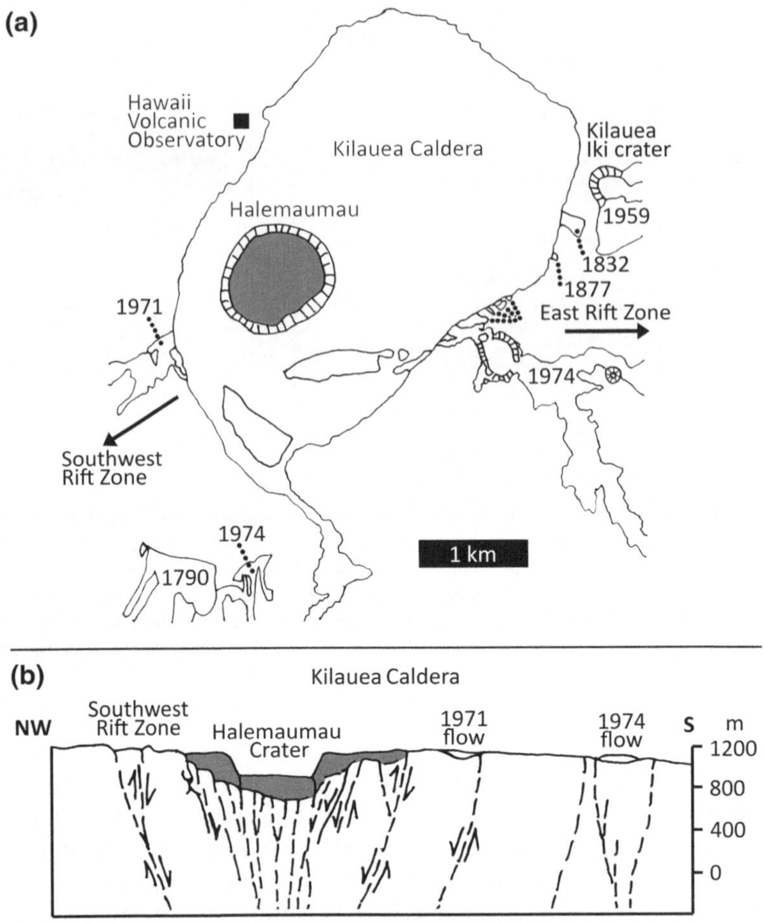

Fig. 16.3 **a** Location of the Hawaii Volcano Observatory adjacent to the Kilauea caldera, Hawaii (USA). **b** A cross section (North West to South) of the Kilauea caldera

different compositions of magma associated with different types of volcanoes under different settings (Fig. 16.5).

Lava (the magma that extrudes out) can occur either as a fluid or as fiery clouds of ash and fragments. The ability of lava to flow depends on the viscosity of the parent magma (Fig. 16.6). Viscosity is a function of temperature, silica content, and incorporated gases (Table 16.1; see also Fig. 16.6). However, the primary compositional difference in magma is in the amount of silica that makes-up the melt. Based on this, four distinct types of magma are commonly found; the silica content varies between 48 and 77 % (Table 16.1). The principle structural difference between these magmas is due to their viscosity (Fig. 16.6). The complex silica ion likes to bond to others of its kind to form complex three dimensional

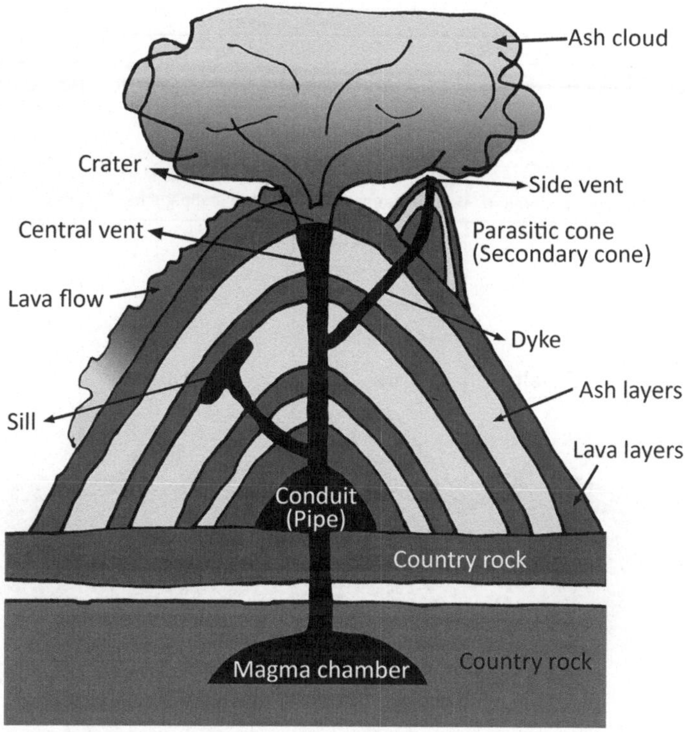

Fig. 16.4 Components of a volcano

networks. The greater the amount of silica in the melt, the more bonding there is and the more viscous the magma is (Table 16.1; Fig. 16.6).

Another important factor that differentiates magmas is in the amount of dissolved gas they contain. Gas content typically varies from 0.2 to 3 %. Volcanic gases are mostly water vapor and CO_2, with smaller amounts of more noxious gasses such as H_2S (Hydrogen Sulfide). Under pressure, deep in the crust, magma can hold more dissolved gas. As the magma rises to the surface, pressure decreases and gases come out of solution, sometimes explosively. Magma can be divided into four types based primarily on their composition (Table 16.1).

16.3.1 Basaltic Magma

Basaltic magma has about 48–52 % SiO_2 and the smallest amount of incorporated gas. Hence, it readily flows. Basaltic magma is derived from the dry melting of the mantle rock and so it is also called the Mantle magma. Basaltic magma often forms shield volcanoes (Fig. 16.6). Basalt, the rock, has a density of 2.9 and it is

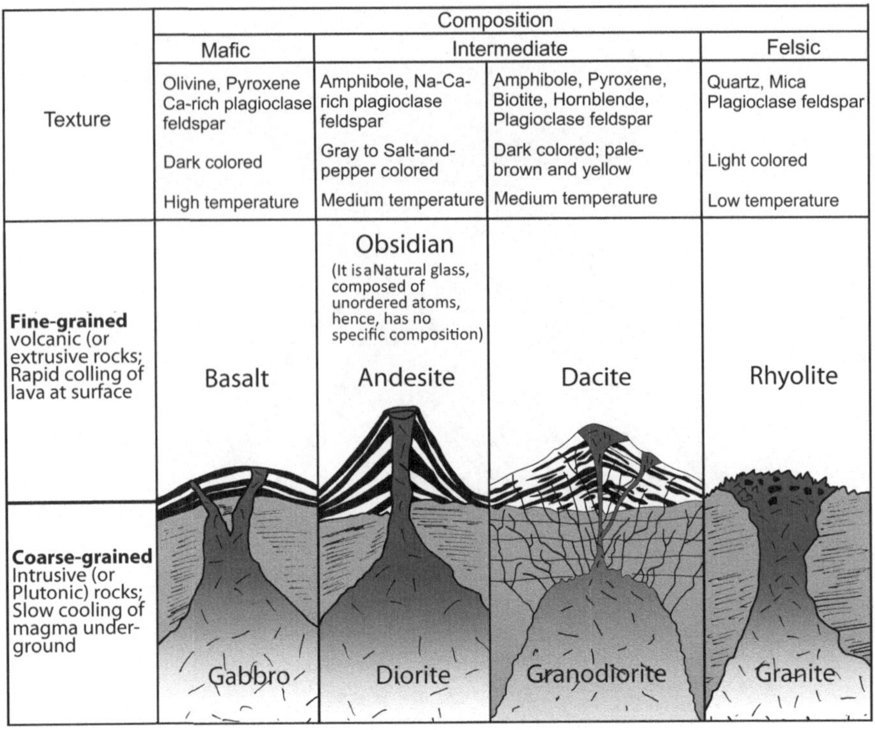

Fig. 16.5 Igneous rock texture, composition, and placement of the major magma types discussed in the text. Texture is largely determined by its rate of cooling and composition by its constituent minerals. Modified from B Perry, CSU, Long Beach (http://geology.campus.ad.csulb.edu)

dark and dense containing olivine, pyroxene and Ca-feldspar in solid solution (Fig. 16.7). Basaltic rocks (like basalt and gabbro) are largely made up of feldspars, pyroxenes and other minerals common in planetary crusts. They have also been noted as major surface rocks on the dark lunar planes and in much of Mars and Venus.

16.3.2 Andesitic Magma

Andesitic magma contains 52–63 % SiO_2 with a lot of incorporated gases. It has intermediate viscosity between the Basaltic and the most mobile Rhyolytic magmas (Figs. 16.5, 16.6 and 16.7). The Andesitic magma is derived from the melting of mantle rocks in the presence of water and so it is also called the Continental margin magma. Andesitic rocks are associated with steep-sided cone volcanoes (the Strato or Composite volcanoes) (Fig. 16.6).

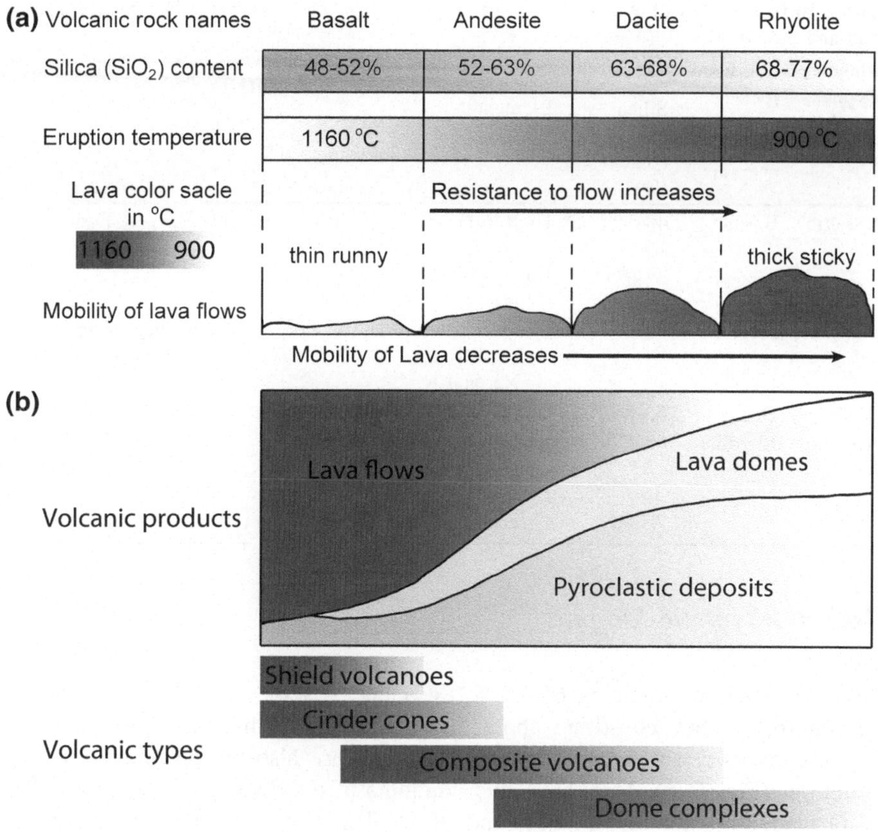

Fig. 16.6 Flow characteristics (**a**), volcanic products and volcanic types (**b**). The flow of lava is a function of its viscosity (resistance to flow), slope of the ground over which it travels, and the rate of lava eruption **a**: From USGS Glossary

16.3.3 Dacite Magma

The Dacite magma occurs in very small quantities. It has 63–68 % SiO_2 content and contains moderately high amounts of gas. It consists primarily of plagioclase feldspar with biotite, hornblende and pyroxene. Because of the moderate silica content, these magmas are quite viscous (Fig. 16.6) and therefore, prone to explosive eruptions. A classic example of this is Mount St. Helens in which Dacite domes formed from previous eruptions (Decker and Decker 1981). The rock Dacite usually forms as an intrusive rock such as a dike or sill.

Table 16.1 Summary of magma types. The 63–68 % SiO_2 range belongs to Dacite magmas (see text and Fig. 16.4)

Magma Type	Solidified rock	Chemical composition		Temperature (°C)	Viscosity	Gas content
		Silica (SiO_2) content (%)	Elemental			
Basaltic	Basalt	48–52	High in Fe, Mg, and Ca Low in K, and Na	1,250–1,100	Low	Low
Andesitic	Andesite	52–63	Intermediate in Fe, Mg, Ca, Na, and K	1,100–900	Intermediate	Intermediate
Dacitic	Dacite	63–68		1,000–800	Intermediate	Intermediate
Rhyolitic	Rhyolite	68–77	High in K, and Na Low in Fe, Mg, and Ca	800–650	High	High

16.3.4 Rhyolytic Magma

Rhyolytic magma contains 68–77 % SiO_2 and has the largest amount of gas (Table 16.1). They constitute about 10 % of the total magma. They have the highest viscosity among all the lava types and are also responsible for violent eruptions (Fig. 16.6). They have high amounts of dissolved gasses such as CO_2, H_2O, and SO_2, but low in FeO and MgO. The composition is more like the composition of a continental crust and is considered to be derived from the melting of continental crust, hence, also called Continental magmas. Magmas erupted from volcanoes that once were active at Yellowstone National Park (USA) were mostly rhyolitic in composition.

16.3.5 Carbonatite Magma

In addition to the above four basic types, this one is rare and is composed of 75 % carbonate with lesser amounts of clinopyroxene, phlogopite, alkali amphibole, apatite, magnetite, olivine, etc., with less than 10 % SiO_2. However, Carbonatites are highly variable in composition. They belong to alkaline igneous provinces and generally found in stable cratonic regions sometimes with major rift faulting such as the East African Rift Valley. However, not all alkalic rock provinces and complexes have associated carbonatites. Carbonatitic activity is episodic and seems to have been related temporally and spatially to orogenic events. They often form clusters or provinces within which there may have been several episodes of

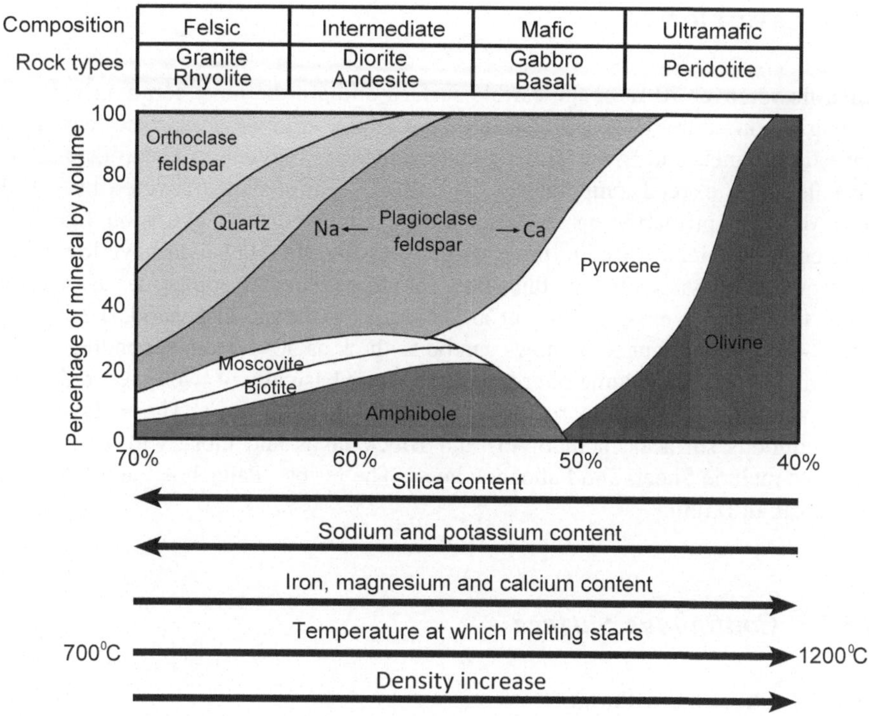

Fig. 16.7 The genetic relationship between rock type and rock composition. Igneous rocks are classified on the basis of texture and composition; major types of igneous rocks include granite, diorite, gabbro, rhyolite, andesite, and basalt. Rocks crystallizing slowly grow large crystals; rapidly cooled ones have a fine-grained or glassy texture. Rocks that cool below the earth's surface are called Intrusive and those above are called Extrusive igneous rocks. The composition of a rock provides information about the nature and origin of the magma. Mafic magmas high in iron and magnesium but poor in silica generally originate from the partial melting of the mantle, erupted in continental rift systems and along Mid-Oceanic Ridges (MOR). Rocks richer in silica (like andesite and rhyolite, or their intrusive equivalents, diorite and granite), form at convergent plate margins and in other settings, such as rifts or above hotspots, where continental is partially melted by the hot basalt. The mafic lava is around 10,000 times as viscous as water, whereas silicic magma is about 100 million times. Hence, lavas rich in silica are associated with most violent eruptions

activity. The Carbonatite lavas generally have low eruption temperatures, between 500 and 600 °C. There are only 350 known carbonatite localities on earth, mostly occurring in association with larger intrusions of alkali-rich silicate igneous rocks. The Natrocarbonatites are carbonatites enriched in alkalies (Na and K). A nephelinite/phonolite volcano in Tanzania (the Oldoinyo Lengai volcano) is currently erupting Na–Ca–K carbonate magma (around 600 °C). This is unlike all other intrusive and effusive carbonatites that are dominantly composed of Ca, Mg, Fe carbonates, and have negligible alkali contents. Carbonatites are found in alkaline provinces, restricted largely to continental regions (see also Bailey 1993).

16.4 Lava Flows

Lava flows cover 70 % of the earth's surface and are the most common volcanic formation on earth. They are also the most common geologic surface feature on terrestrial planets and cover 50 % of Mars, and 90 % of Venus. On earth, an active lava flow can exceed temperatures of 1100 °C. The distance covered by a lava flow depends primarily on its temperature, silica content, extrusion rate, and gradient of the land. Basaltic flows (Fig. 16.6) like the Hawaiian have low silica content and low viscosity, and thus, flow for large distances, sometimes as much as 4 km from their source (Tilling et al. 1989). Lava flows, in general, are usually only 1–10 m thick, but some flows can be as thick as 50–100 m, depending on the type of lava and the volume of eruption. They are basically of two types, reflecting their internal movement in relation to their congealing crust. Those flows with discontinuous surfaces include Aa- and Block lavas and those with continuous surfaces include Sheet- and Pahoehoe lavas. The terms "Pahoehoe" and "Aa" are of Hawaiian origin.

16.4.1 Continuous Surface

16.4.1.1 Sheet Lava Flows

Sheet lava flows emerge from fissure systems forming broad, laterally extensive blankets of lava flows commonly ranging between 10 and 30 m in thickness. Because of their relatively high effusion rate, the eruptions are very fluid, and tend to fill-in and allow individual lobes of lava to quickly coalesce back together into a sheet. This then grows into a solid crust on its upper surface while in the flow's interior, the lava remains molten for some time. But, the solid crust may later collapse if the lava in the interior drains away; areas of collapse are quite common in the middle of sheet flows. Subject to local flow conditions, a variety of surface textures are exhibited by Sheet lavas such as Lobate, Ropy, Lineated, and Jumbled.

16.4.1.2 Pahoehoe Lava (Pronounced as Pah- hoe-ee-hoe-ee)

This basaltic lava contains higher amounts of gas and hence, is thinner and flows quickly, often as fast as 50 m/h. The temperature typically reaches 1100–1200 °C when they flow out of a volcanic vent. Pahoehoe lava progresses forward in tongues or lobes and is characterized by a glassy, plastic skin. Its surface cools quickly and retains a smooth wrinkled or ropy-looking surface. The classic "ropy" texture of a pahoehoe lava flow is best seen in the lava eruptions at Hawaii. Cooling Pahoehoe flows sometimes turn into Aa lava flows. However, the Aa lava flows never become Pahoehoe flows. The Aa flows are cooler and move at a much slower speed than Pahoehoe (Peterson and Tilling 1980)

16.4.2 Discontinuous Surface

16.4.2.1 Aa Lava Flows (pronounced as Ah-ah)

Aa is a basaltic lava containing low amounts of gas and hence, moves slowly (and is thus, more viscous). The surface of this flow is marked by rough, jumbled mass of broken, angular blocks, and clinkers. Aa is a Hawaiian term meaning "stony with rough lava." In Hawaii and typical to it, the transition between Aa and Pahoehoe lava flow is controlled by two factors, viscosity and strain rate. But, each of these factors are further controlled by various parameters such as crystallinity, dissolved gas content, temperature, bubble content, slope, eruption rate, and lava composition. In Hawaii, all eruptions >5–10 m^3/sec form Aa and those <5–10 m^3/sec form Pahoehoe (Peterson and Tilling 1980; Tilling et al. 1989).

If lava cools slowly and does not move too fast, it forms smooth ropy lava called Pahoehoe. However, if it cools quickly and moves fast it can tear into clinkery pieces called Aa (Peterson and Tilling 1980; Tilling et al. 1989).

16.4.2.2 Block Lava Flows

Block-lavas are fairly viscous, stronger, and thicker than Aa lava flows. The more silicic the magma, the shorter and stubbier the flow is (Fig. 16.6). Block lavas move slowly at a rate ranging from 1 to 5 m/day. When solidified, they are characterized by cubic masses with relatively smooth faces. In comparison with Aa lava flows, the surfaces of Block lavas are much less rough and pitted.

16.5 Lava Tubes or Lava Caves

Sometimes basaltic lava flowing down a moderate slope becomes confined into a channel. If that channel cools and becomes roofed over, the resultant lava tube may continue to transmit molten lava and extend downhill for miles. These drained, cooled tubes are called Lava tubes or Lava caves. Because the walls and roofs of such tubes are good thermal insulators, lava flowing through them can remain hot and flowing, much longer than the surface flows. Hence, the lava can be transported to great distances from the eruption site. While flowing through the tube, the lava produces large amounts of gas and steam. This produces a certain air pressure inside, thus, keeping the tube open. However, increased pressure may also press a hole into the covering crust, which is still hot and rather soft. Such venting holes are called Hornitos (Spanish for "little ovens") (Fig. 16.8). The reason for their origin is simple, the gas pressure from below throws small pieces of lava out of the venting holes, which form cones, up to 15 m high, on top of the vent. The most famous place for lava tubes on earth is Hawaii and the Big Island has the biggest tubes; the Kazumura Cave is 65,500 m long with a height difference of 1,102 m from end to end.

16.6 Pressure Ridges

Pressure ridges form as the semi-rigid lava surface is squeezed with its interior remaining more fluid and flexible. Cracks often adorn the peaks of these ridge formations; both lava and steam issue secondarily from them. Explosive eruptions blast large quantities of volcanic rock out of the volcano. Some of this rock is excavated from the volcano itself, the rest consists of fragments of lava that has cooled in the air. This material accumulates around the volcano as pyroclastic (from the Greek *pyro*, "fire," and *clast*, "broken") deposits (also called Tephra). Within the pyroclastic deposits, include large fragments Blocks (no rounded edges or corners) and Bombs (spindle or lens-shaped pyroclast) (both >64 mm; Spatter bombs), and the gravel sized ones include Cinder and Lapilli (2–64 mm size).

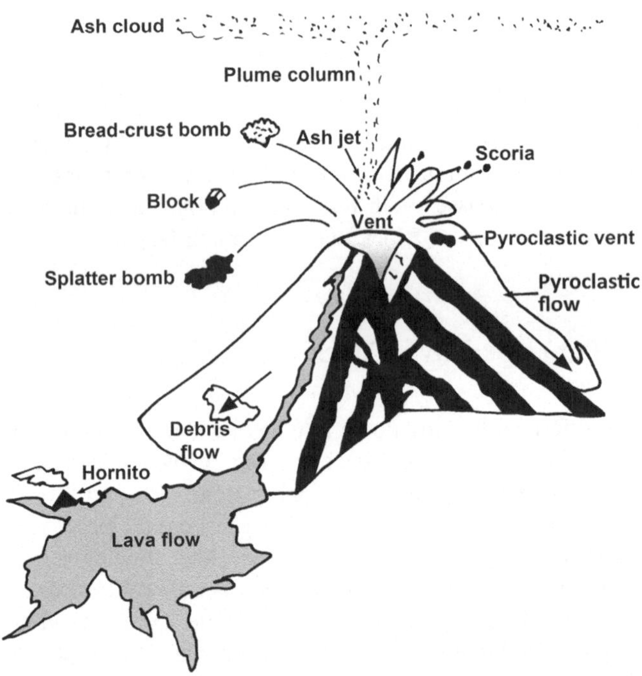

Fig. 16.8 Major products of a volcano

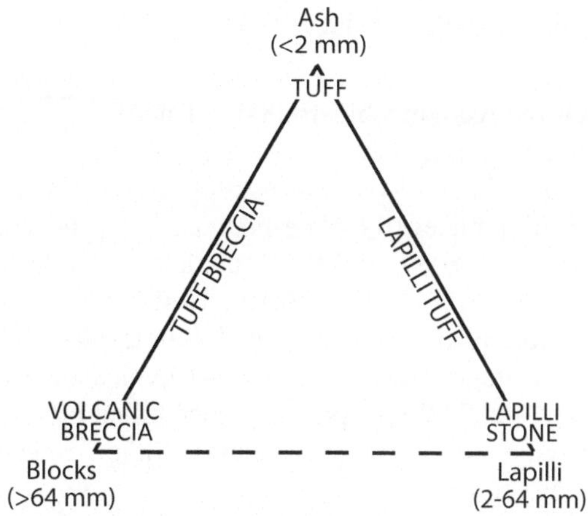

Fig. 16.9 The nomenclature of volcanic rocks based on the size of fragments

Cinder is often used as a less-restricted, general term for smaller pyroclasts. The fine-grained tephra is called Ash and Dust (Fig. 16.9). Ash is silt to sand sized (1/8–2 mm; Fig. 16.8) and Dust measures <1/8 mm.

16.7 Nuées Ardentes

It is a French word meaning "glowing cloud." Nuées Ardentes is used to describe a fast moving gaseous cloud of hot ashes and other material thrown out from an erupting volcano that are often incandescent. These pyroclastic flows glowed red in the dark. Nuées Ardentes was first used to describe the disastrous 1902 eruption of Mount Pelée over the village of St. Pierre on the French Caribbean Island of Martinique.

16.8 Pillow Lava

These basaltic lava flows may form by the discharge of lavas into rivers, lakes, ponds or under glaciers, and into oceans. The pillow structures are formed as a result of the protrusion of elongate lava lobes that detach and fall down the moving flow front. Pillow lava flows can be many hundreds of m to km long. These are very common on the Mid-Atlantic Ridge, East Pacific Rise, and the Juan de Fuca Ridge, which are spreading at slow to intermediate rates (about 2–5 cm/year). Pillow lavas can also form on continents when lava from an on-land volcano flows into a lake or a river. Lobate flows resemble on-land Pahoehoe flows, but are a bit more inflated-looking.

16.9 Global Distribution of Volcanoes

Of the recognizable 10,000 volcanoes, around 539 (53 of these are in the United States and of them most are in Alaska; Simkin et al. 1989) are active volcanoes and ~62 % of them are strung like beads along, or near the boundaries between shifting plates (the plate-boundary volcanoes; mostly at Convergent and Divergent plate margins), and 90 % of them encircle the Pacific Ocean as a Ring of Fire (Bullard 1976) (Fig. 16.10). However, volcanic eruptions can also occur in the middle of plates and are called Intra-plate volcanoes or Hotspot (Fig. 16.11). On an average, there are about 5–10 volcanoes erupting every month (~50/year). So far, ~262,000 people have been killed by volcanoes since 1600 A.D. (around one person per year per volcano). Major volcanic killers are given in Table 16.2 (Fig. 16.12).

16.9.1 Convergent Plate Margin

These boundaries are often associated with intense volcanism (Fig. 16.11). As plates collide, heavier plates slide under lighter ones and melt as they move down into the hot mantle (Fig. 16.11). All around the Pacific Ocean is a zone called the

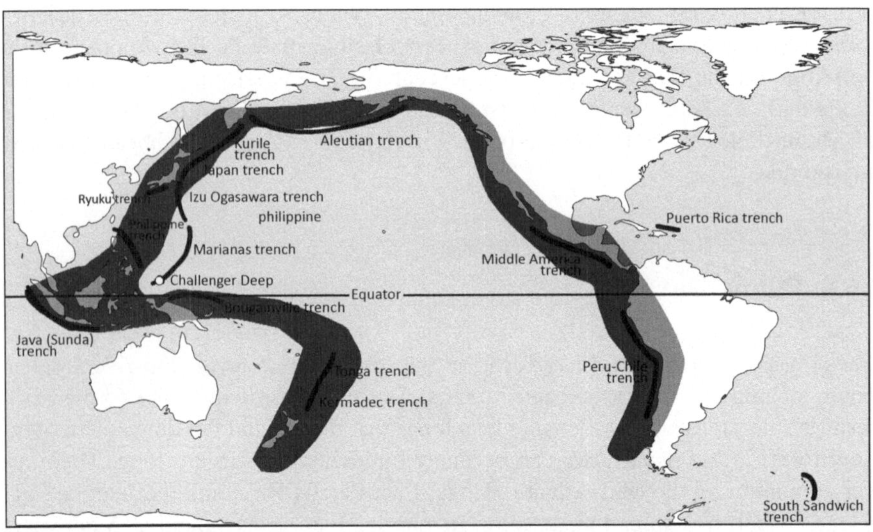

Fig. 16.10 The Ring of Fire. Volcanic arcs and oceanic trenches encircle the Pacific Basin and form the Ring of Fire, a zone of frequent volcanic eruptions (and earthquakes). The trenches are shown in solid black lines. The volcanic island arcs (not labeled) are parallel to, and always landward of, trenches. For example, the island arc associated with the Aleutian Trench is represented by the long chain of volcanoes that make up the Aleutian Islands

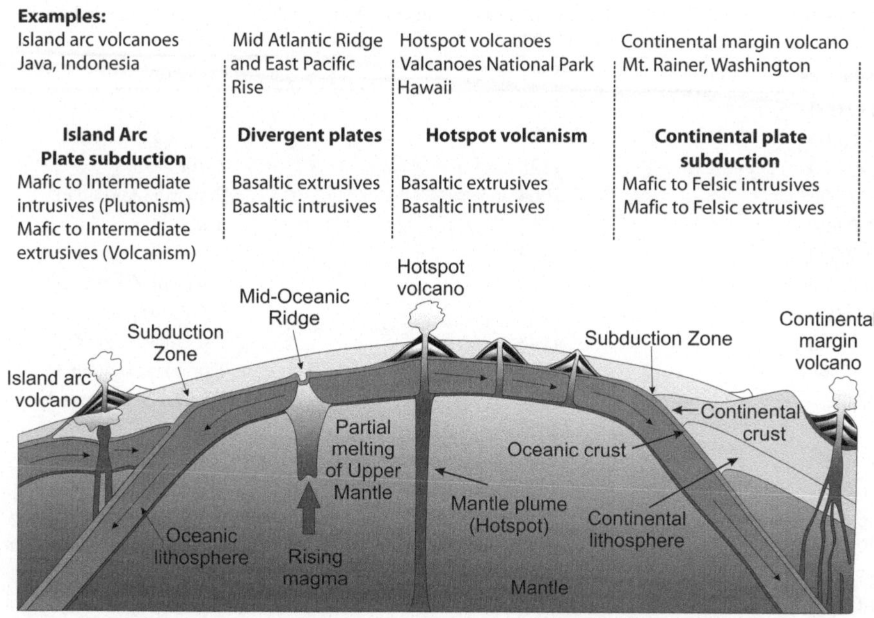

Fig. 16.11 Volcanism and tectonic plate boundary settings. Volcanism occurs along, or near boundaries between shifting plates (the plate-boundary volcanoes; at Convergent and Divergent plate margins) or at the interior of plates (Hotspots)

Pacific Ring of Fire (Fig. 16.10), where a large part of the world's most active and dangerous volcanoes are located. This ring occurs as most of the margins of the Pacific Ocean coincides with converging margins along which subduction continuously occurs. Thus, the subduction-related volcanism is of two types:

16.9.1.1 Ocean–Ocean Convergence

Here oceanic lithosphere is subducted beneath oceanic lithosphere. The subsequent volcanism is expressed on the surface as chains of islands referred to as Island arcs (Fig. 16.11; the Island Arc plate subduction). Classic examples include the Caribbean Arc, Aleutian Arc, Kurile Kamachatka Arc, Japan, Philippines, South Sandwich Arc, Indonesian Arc, Marianas, Fiji, and Solomon Islands.

16.9.1.2 Ocean–Continent Convergence

Here the oceanic lithosphere is subducted beneath continental lithosphere and volcanism occurs as chains of volcanoes near the continental margin, called the Continental margin arc (Fig. 16.11; the Continental plate subduction). Classic

Table 16.2 Major fatalities due to volcanic eruptions (After catalog of significant earthquakes, USGS; see also Ganse and Nelson 1982)

Fatalities	Year (AD)	Location
830,000	1556	China
650,000	1976	Tangshan, China
230,000	1138	Syria
227,000	2004	Sumatra–Andaman
200,000	856	Iran
200,000	1927	Nan-Shan, China
200,000	1920	Haiyuan, China
142,000	1923	Tokyo, Japan
110,000	1948	Ashgabat, USSR
92,000	1815	Tambora, Indonesia
87,000	2008	Eastern Sichuan, China
86,000	2005	Pakistan
72,000	1908	Messina, Italy
66,000	1970	Lima, Peru
36,417	1883	Krakatau, Indonesia
29,025	1902	Mount Pelee, Martinique
25,000	1985	Ruiz, Colombia
14,300	1792	Unzen, Japan
9,350	1783	Laki, Iceland
5,110	1919	Kelut, Indonesia
4,011	1882	Galunggung, Indonesia
3,500	1631	Vesuvius, Italy
3,360	79	Vesuvius, Italy

examples include the Andes Mountains, Central American Volcanic Belt, Mexican Volcanic Belt, the Cascades, the part of the Aleutian arc on Continental crust, and the North Island of New Zealand.

16.9.2 Divergent Plate Margins

The Oceanic Ridges or spreading centers (Fig. 16.11) are sites of active volcanism along diverging plate margin (Plate Divergence; Fig. 16.11). However, most of this volcanism is submarine and often does not pose any threat to humans. One of the only places where an oceanic ridge reaches above sea level is at Iceland, along the Mid-Atlantic Ridge. Here, most eruptions are basaltic and are explosive in nature belonging to the Strombolian, Phreatic, or Phreatomagmatic types (discussed later in the chapter).

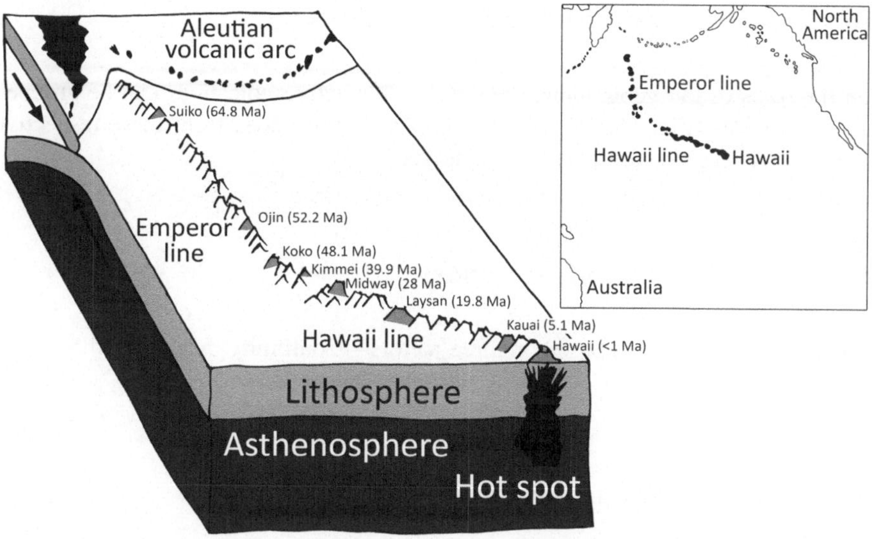

Fig. 16.12 The Hawaii Hotspot. A hot spot is the location where a stationary mantle plume has risen to the surface and formed a volcano. The Hawaii Hotspot is the archetype hotspot volcano sitting in the middle of the Pacific plate, far away from any plate boundary. Over the past 80 Ma basaltic volcano eruptions and the continued movement of the Pacific plate over the stationary Hawaiian hotspot have *left* a long trail of (volcanic) seamounts and volcanoes across the Pacific Ocean floor. This formed the Hawaiian Ridge-Emperor Seamounts chain that consists of 129 volcanoes extending ~6,000 km from the Aleutian Trench off Alaska to the "Big Island" of Hawaii. The *chain bend* occured ~40 Ma ago when the motion of the Pacific plate changed from nearly due north to northwesterly. The volcanoes are still active on the large southern island of Hawaii, but not on the more eroded Maui, and other islands to the northeast. The Hawaiian hot spot currently underlies the southern half of the island of Hawaii and adjoining offshore area

16.9.3 Hot Spots

Hot spots are volcanic eruptions in the interior of plates (Fig. 16.11). They result from plumes of hot mantle material upwelled to the surface, independent of the convection cells thought to cause plate motion. Hot spots are fixed in position, and the plates move over them (Fig. 16.12). As the rising plume of hot mantle moves upward it begins to melt the plate to produce magmas. These magmas then rise producing a volcano at the surface. However, as the plate carrying the volcano moves away from the position over the hot spot, volcanism ceases and new volcanoes are formed in the position that is now over the hot spot. This produces chains of volcanoes or seamounts (former volcanic islands that have eroded below the sea level). Volcanism resulting from hotspots occurs in both Atlantic and Pacific Oceans, but is more evident on the sea floor of the Pacific Ocean, as the plates move at higher velocity than those under the Atlantic. Hence, a hot spot trace shows up as a linear chain of islands and seamounts, many of which can be seen in the Pacific Ocean (as the Hawaii; Fig. 16.11).

16.10 Geographic Distribution of Volcanoes

On the basis of the geographical region, earth's active volcanoes can be grouped into 12 regions (Fig. 16.13). Major examples of volcanoes from these respective geographical regions are listed in Table 16.3.

16.11 Classification of Volcanoes

Volcano classification is based on three factors—Continuity, Nature, and Mode of eruption.

16.11.1 Continuity of Eruption

Volcanoes are either active or not active. However, depending partly on the average repose interval between eruptions, inactive volcanoes are at times considered as extinct, or just dormant. Hence, this classification is quite subjective.

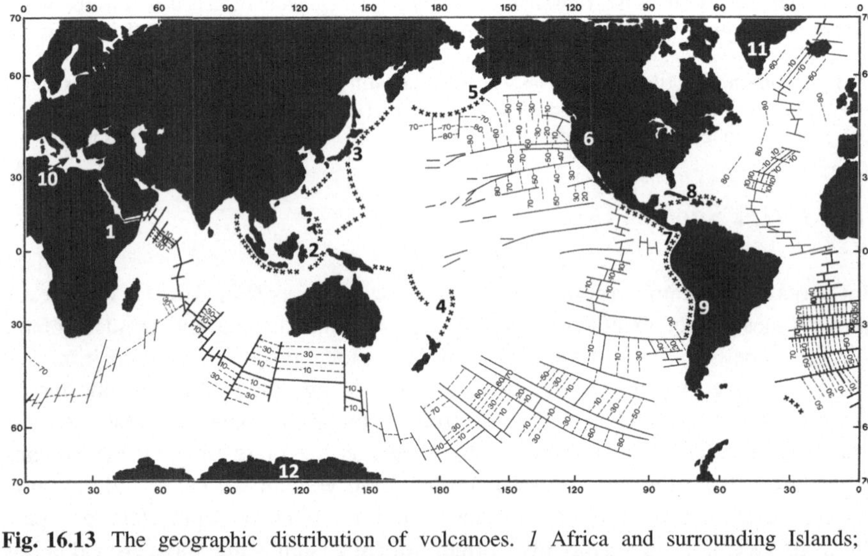

Fig. 16.13 The geographic distribution of volcanoes. *1* Africa and surrounding Islands; *2* Southwest Pacific, Southeast Asia, and India; *3* East Asia (including Japan and Kamchatka); *4* Central Pacific and South Pacific, New Zealand; *5* Alaska and the Northern Pacific Region; *6* North America; *7* Central America; *8* West Indies; *9* South America and surrounding Islands; *10* Mediterranean, *11* North Atlantic, Iceland; *12* Antarctica

Table 16.3 Major example of volcanoes from respective geographical regions

Regions	Volcano names
1. Africa and nearby Islands	Fogo Caldera, SW Cape Verde Is. Atlantic Ocean
2. SW Pacific and SE Asia	Aoba (Ambae Island), Vanuatu Islands
	Batur Voclano, Bali, Indonesia
	Bulusan, Luzon, Philippines
	Gemini Seamount, New Hebrides Island Arc, Vanuatu Islands
	Merapi Volcano, Java, Indonesia
	Mt Canlaon, Negros Islands, Philippines
	Parker, Southern Mindanao, Philippines
	Pinatubo Volcano, Central Luzon, Philippines
	Rabaul Caldera, Papua New Guinea
India	Barren Island, Andaman Islands, Indian Ocean
	Baratang, an island in the Andaman Islands, Indian Ocean
	Narcondam or Narcondum a small volcanic island in the Andaman Sea
3. East Asia (including Japan and Kamchatka, Russia)	Avachinsky Volcano, Kamchatka, Russia
	Bezymianny Volcano, Kamchatka, Russia
	Karymsky Volcano, Kamchatka, Russia
	Klyuchevskoi Volcano, Kamchatka, Russia
	Mt Unzen, Japan
4. Central Pacific and South Pacific, New Zealand	Kilauea Volcano, Hawaii
	Loihi Seamount, Hawaii
	Marianis Islands
	Metis Shoal, Tonga
	Ruapehu, New Zealand
	Taupo Volcanic Zone, New Zealand
5. Alaska and Northern Pacific Region	Akutan Volcano, Aleutian Islands
	Gorda Ridge, Northeast Pacific Ocean
	Mt Spurr, Alaska
	Pavlof Volcano, Alaska Peninsula
	Shishaldin Volcano, Aleutian Islands
6. North America	Lake Superior Ice Volcanoes, Michigan
	Mount Lassen, California
	Mount St. Helens
	Popocatepetl, Mexico Archived Info
	Popocatepetl, Mexico more recent

(continued)

Table 16.3 (continued)

Regions	Volcano names
7. Central America	Arenal Volcano, Costa Rica
	Cerro Negro, Nicaragua
	Cerro Quemado Volcano, Guatemala
	Coatepeque, El Salvador
	Fuego Volcano, Guatemala
	Ilopango, El Salvador
	Izalco, El Salvador
	Pacaya Volcano, Guatemala
	San Miguel, El Salvador
	San Salvador, El Salvador
	San Vicente, El Salvador
	Santa Ana, El Salvador
	Santa María Volcano, Guatemala
	Tacaná Volcano, Guatemala
	Volcano Rincon de la Vieja, Costa Rica
8. West Indies	Soufriere Hills, Montserrat, West Indies
9. South America and surrounding Islands	Cerro Negro de Mayasquer, Azufral
	Galapagos, Fernandina
	Galeras, Nevado Cumbal, Dona Juana
10. Mediterranean	Etna Volcano, Italy
	Stromboli Volcano, Italy
11. North Atlantic and Iceland	Askja Volcano
	Bardarbunga/Grimsvotn Volcanoes
	Hekla Volcano
	Katla Volcano
	Krafla Volcano
	Vestmannaeyjar Volcano
12. Antarctica	Mount Erebus

(After Bullard 1984; Decker, and Decker 1991; Etienee 1992; Fisher, et al. 1997; Francis 1993; Kraft 1993; Schmidt, and Shar 1971; Simkin, and Siebert 1994; Tilling 1982; Wright, and Pierson 1992)

16.11.1.1 Active

An active volcano is one that erupts either continually or periodically. In precise terms, an active volcano is defined as having erupted within the last 10,000 years. There are approximately 1,500 active volcanoes in the world today and 75 % of them are located in the “Pacific Ring of Fire” (Fig. 16.9). On average, 50–70 volcanoes erupt every year. Japan has 10 % of the world’s active volcanoes (Simkin and Siebert 1994; Tilling 1991). Table 16.4 gives the approximate number of active volcanoes in the world and Table 16.5 lists the top 10 active volcanoes of the world.

Table 16.4 Note that these figures exclude submarine eruptions

Active volcanoes	Approximate numbers
Erupting now	20
Each year	50–70
Each decade	160
Historical eruptions	550

16.11.1.2 Dormant

A volcano that has not been known to erupt within modern times and is now inactive is classified as a dormant, or a sleeping volcano.

16.11.1.3 Extinct

A volcano not known to have erupted within recent history is classified as an extinct volcano ("ancient" is also used interchangeably). Volcanologists consider truly extinct volcanoes to be those that have been worn away almost to the level of their magma chamber. But even extinct volcanoes can prove unpredictable. There are about ~ 1,500 volcanoes in the world. In addition, there are 539 volcanoes that have not erupted in historic times, but which exhibit clear evidence of eruption in the past 10,000 years. These latter volcanoes are probably best considered dormant, since they have the potential to erupt again.

16.12 Nature of Eruption

All magmas contain dissolved gases, and as they rise to the surface to erupt, the confining pressures are reduced and the dissolved gases are liberated either quietly or explosively. Based on the nature of volcanic eruption, the volcanoes are classified into three types: Quite, Intermediate, and Violent. During an episode of activity, a volcano commonly displays a distinctive pattern of behavior. Some mild eruptions (Quite type) merely discharge steam and other gases, whereas other eruptions quietly extrude quantities of lava (Intermediate type). The most spectacular eruptions consist of violent explosions that blast great clouds of gas-laden debris into the atmosphere (Violent type).

16.13 Mode of Eruption

If the lava is not viscous, the gases escape easily. But if the lava is thick and highly viscous, the gases will not move freely but will build up enormous pressure, and ultimately escape with a sudden expansion, causing violent explosions that throw

Table 16.5 Top ten active volcanoes of the world

No.	Name	Country	Major characteristics		Last eruption
1.	Nyamuragira	Democratic	Republic of Congo	This is Africa's most active volcano, located in the Virunga National Park. The main crater is 2 km wide and since 1882, it has erupted at least 34 times	The volcano emits more lava than almost any other volcano in the world with flows that race downhill at speeds of up to 60 mph (97 km/h)
	The last eruption was in January of 2010				
2.	Mount Etna	Italy	The ancient Greeks believed Mount Etna to be the home of Vulcan, the god of fire. To them, Mount Etna's erupting meant Vulcan was forging weapons for Mars, the god of war. Etna is a composite volcano on the east coast of Sicily, has been continuously erupting for over 3,500 years	It has the longest period of documented eruptions and is the largest active volcano in Europe. It has an altitude of 3,329 m that is also in an almost constant state of eruption. Mt Etna consists of a series of nested stratovolcanoes, characterized by summit calderas—the most important one being the Ellittico Caldera, which formed about 14,000–15,000 years ago	Mount Etna has erupted at least 190 times. It spewed lava as recently as January of 2011
3.	Kilauea Volcano	Hawaii (U.S.A.)	Kilauea is earth's most active volcano. It is the most recent of a series of volcanoes that created the Hawaiian archipelago of islands and has been in a state of almost constant basaltic eruption since 1983	The volcano only rises 1,247 m above sea level, but it's still growing. Since 1983, it has been spewing out an average of 325,000 m^3 of lava per day (equivalent to 40,000 dump truck-loads). In 30 years, almost 3.5 billion m^3 of lava has been produced -enough to build a highway that circles the world over seven times	There were 45 eruptions of the volcano in the twentieth century alone. It ejected lava as recently as March of 2011

(continued)

Table 16.5 (continued)

No.	Name	Country	Major characteristics		Last eruption
4.	Santa Maria	Guatemala	At Guatemala's Pacific coastal plain, Santa Maria is a 3,772 m tall stratovolcano consisting of alternating layers of hardened ash, lava and rock	Santa María is part of the Sierra Madre range of volcanoes, which extends along the western edge of Guatemala, separated from the Pacific Ocean by a broad plain. The volcanoes are formed by the subduction of the Cocos Plate under the Caribbean Plate, which led to the formation of the Central America Volcanic Arc	One of the largest eruptions of the twentieth century occurred in 1902. Since 1922, a lava-dome complex, Santiaguito, has formed in the 1902 crater. It's most recent eruption occurred in March of 2011
5.	Piton de la Fournaise	La Reunion Island	French for the "Peak of the Furnace". It is a shield volcano on the eastern side of Renunion Island in the Indian Ocean	It is referred locally as "le Volcan," and is 2,631 m tall. There are many craters and cinder cones inside the volcano's caldera and around its outer flanks. It is considered to be one of the three most active volcanoes in the world	It has had more than 150 eruptions since the seventeenth century, erupting most recently in October of 2010
6.	Stromboli	Italy	The island of Stromboli is the tip of a massive underwater volcano off the west coast of southern Italy and the north coast of Sicily	It has erupted nearly continuously for over 2,000 years, earning it the nickname "Lighthouse of the Mediterranean." Eruptions typically result in mild energetic bursts that last for few seconds and emit ash, incandescent lava fragments and lithic blocks up to a few hundred meters in height. This explosive style of volcanic eruption is called "Strombolian"	The volcano was particularly active during April of 2011

(continued)

Table 16.5 (continued)

No.	Name	Country	Major characteristics		Last eruption
7.	Mount Yasur	Vanautu	Mount Yasur is an active volcano on Tanna Island, part of the archipelago of the nation Vanuatu in the South Pacific	It is called the "Lighthouse of the Pacific". Its glow is rumored to have attracted Captain James Cook to the island in 1774. It is 361 m above sea level	Yasur has been erupting nearly continuously for over eight centuries, and its eruptions, which often occur several times an hour, are classified as Strombolian or Vulcanian (a relatively low-level type of eruption). The volcano is hurling volcanic bombs since June of 2011
8.	Láscar	Chile	It is located in northern Chile and is currently the most active volcano of the Central Volcanic Zone of the Andes. Láscar has two cones—the Western Extinct Cone and the Eastern or Active Cone	Volcán Aguas Calientes is an older higher stratovolcano located 5 km east of Láscar	The last eruption occurred here in May of 2007
9.	Sangay	Ecuador	The present-day volcano was built within the horseshoe-shaped calderas of two previous edifices that were destroyed by a collapse to the east, producing large debris avalanches that reached the Amazonian lowlands	The modern edifice dates back to at least 14,000 years ago. Access to the volcano is very difficult, and approaching the summit is dangerous due to constant ejection of stones and material into the air from the crater	Sangay has had frequent eruptions in historic times. Due to its remote location, the level of surveillance is less than other more dangerous volcanoes in Ecuador

(continued)

Table 16.5 (continued)

No.	Name	Country	Major characteristics		Last eruption
10.	Mt. St. Helens	USA	It is situated near major metropolitan centers of the Pacific Northwest and its eruptions are highly explosive	Thirty years after it blew its top, the peak is still the second most dangerous volcano in the U.S. after Kilauea on Hawaii	The famous May 18, 1980 eruption killed 57 people and flattened more than 200 mi^2 (518 km^2) of forest. This eruption had 500 times the power of an atomic bomb. Geologists considered this a moderate eruption. It erupted as recently as March of 2005

out great masses of solid rocks, lava, dust and ashes. Such types of volcanic eruptions are often labeled with the name of a well-known volcano where their characteristic behavior is noted. These include Icelandic, Hawaiian, Strombolian, Vulcanian, Vesuvian, Phreatic, Peléan (Nuées Ardentes), and Plinian (Fig. 16.14). Some volcanoes may exhibit only one characteristic type of eruption but some others even display an entire sequence of types.

16.13.1 Icelandic

This type of volcanic eruption is classified in order of increasing violence. They involve the quiet outpouring of lava from long fissures (Fig. 16.14a). More than 2 1/2 mi^3 of lava flowed from Iceland's Lake Fissure in 1783, the largest emission of lava in historic times. Eruptions of the Icelandic type during the Tertiary Period built up vast basalt plateaus in many parts of the World. One of the largest is the Columbia Plateau, extending more than 200,000 km^2 in the northwestern United States (eastern Washington, northern Oregon, and western Idaho) containing 350,000 km^3 of basalt with lava thickness of more that 3,000 m erupted 15 million years ago (Fig. 16.15).

16.13.2 Hawaiian

The 1950 eruption of Mauna Loa volcano (Hawaii) characterizes Hawaiian eruption (see also Tilling et al. 1989). Hawaiian eruptions occur along fissures or fractures that serve as linear vents or they may also occur through the central vent (as in the case of the 1959 Kilauea volcanic eruption, Hawaii). Hawaiian eruptions are the calmest types of volcanic events, consisting of basaltic, highly fluid lavas of low gas content that produce effusive lava flows with some pyroclastic debris. Thin, fluid lava flows (growing no more than 3 m at a time) eventually build up large broad shield volcanoes (Fig. 16.14b). Flows can be divided into two types by their structural characteristics. Pahoehoe is a relatively smooth lava flow that is ropey in nature. Aa lava flows are denser and more viscous and tend to move slower than Pahoehoe. The Hawaiian style eruptions can be massive with extrusion rates as high as 1,000 m^3/sec and have also been observed to reach heights of 1600 m (like the 1986–1987 Izu-Oshima eruption in Japan). Hawaiian eruptions are generally continuous, but exhibit sizeable pulses every 1–5 s.

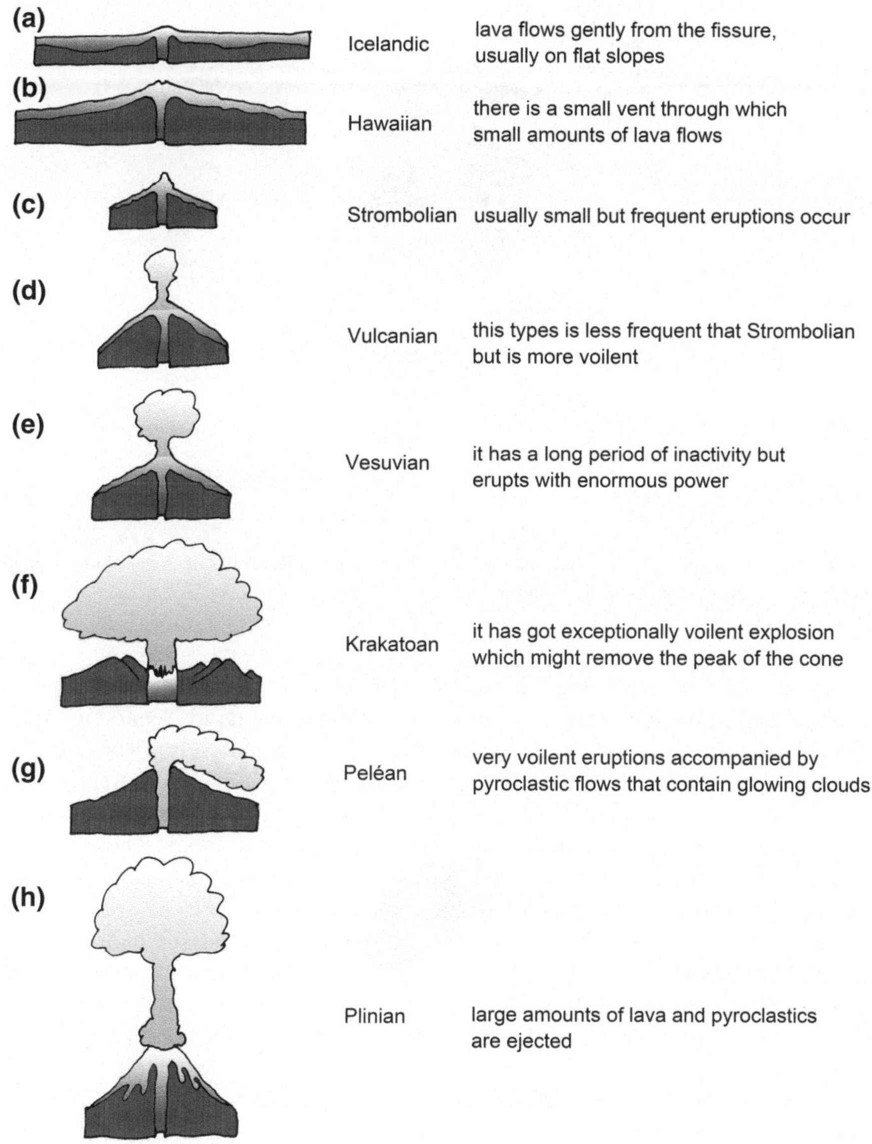

Fig. 16.14 The classification of volcanoes according to their nature of eruption

16.13.3 Strombolian

They are the most beautiful to watch and least destructive in nature. This type of eruption was observed during the 1965 activity of Irazú Volcano in Costa Rica. This style of eruption is named after Stromboli volcano in Italy (Fig. 16.2) and is

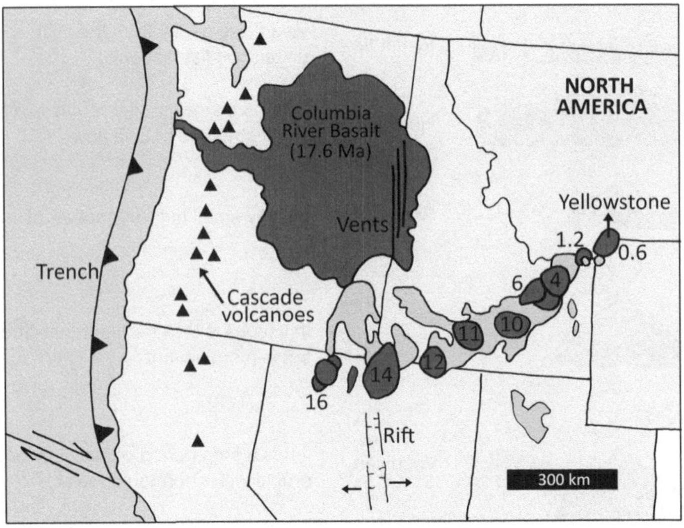

Fig. 16.15 The eruption history and the extent of the Columbia River basalt in the Northwestern United States. Numbers are in Millions of years

markedly different from the Hawaiian style (Fig. 16.14c). The Strombolian eruption begins with a Hawaiian fissure eruption (i.e., rapid degassing) or a Vulcanian eruption (where the over-pressurized gas explosively clears out the blocked vent). However, the Strombolian eruptions are distinctly different in that they do not form fountains, but rather sizeable explosion-like bursts of magma from a vent. Hence, Strombolian eruptions are generally more subdued, with extrusion rates no higher than 100 m^3/explosion and the ejecta heights of less than 100 m. The eruptions occur at somewhat regular intervals, but can accelerate in frequency by one or two orders of magnitude during periods of high activity. Hence, in Strombolian eruptions, huge clots of molten lava burst from the summit crater to form luminous arcs producing bombs and scoria (Fig. 16.9). They also eject volcanic bombs and Lapilli fragments that travel in parabolic paths before landing around their source vent. The steady accumulation of small fragments builds cinder cones composed completely of basaltic pyroclasts. This form of accumulation tends to result in the fragmentation of well-ordered rings of Tephra. Eruption of Mount Etna, Sicily (Italy; Fig. 16.2) is a good example of this. Like the Hawaiian eruptions, gas plays an integral role, but viscosity is what distinguishes between these two types. Hawaiian style eruptions are low viscosity while Strombolian eruptions are the result of higher viscosity magmas. Strombolian eruptions are usually basaltic or andesitic in nature.

16.13.4 Vulcanian

This type of volcanic eruption was recorded by studying the activity of Fossa cone at Vulcano in the Aeolian Islands (Italy; Fig. 16.2) between 2nd August 1888 and 22nd March 1890. Dense cloud of ash-laden gas, cinders, pumice, breadcrust bombs, and blocks explode from the crater (Fig. 16.14d). Vulcanian eruptions can last, with long intervals of repose, for several millennia. These eruptions burst into life suddenly after a dormant period and are similar to hydrovolcanic activity where fragments of magma are expelled with loud bangs as highly viscous magma within the volcano makes it difficult for vesiculate gases to escape. This leads to the buildup of high gas pressure, eventually popping the cap holding the magma down and resulting in an explosive eruption, similar to the Strombolian. However, unlike it, the ejected lava fragments are not aerodynamic; this is due to the higher viscosity of the Vulcanian magma and the greater incorporation of crystalline material broken off from the cap. Vulcanian eruptions are also more explosive than their Strombolian counterparts. The Vulcanian eruptions are also similar to the Plinian (discussed below), however, are characterized by more explosive activity that produces a mushroom-shaped eruption cloud. The Vulcanian deposits are andesitic to dacitic rather than basaltic. Volcanoes with such activity include—the Sakurajima in Japan, a site of near-continuous Vulcanian activity since 1955; the Tavurvur in Papua New Guinea, and volcanoes in the Rabaul Caldera and Irazu Volcano in Costa Rica. The latter is exhibiting vulcanian activity since its 1965 eruption.

16.13.5 Vesuvian

On August 24, 79 A.D. eruption of Mount Vesuvius (that destroyed Pompeii, South of Napoli, Italy; Fig. 16.2) violently discharged large quantities of ash-laden gas and formed cauliflower shaped clouds (Fig. 16.14e). Such clouds reached great heights (around 45 km into the stratosphere) and deposited tephra. Vesuvian eruptions occur after long interval of quiescence of mild activity. The vent tends to be emptied to considerable depth and lava ejects in an explosive spray. Mount Vesuvius has continued its activity intermittently ever since 79 A.D. with numerous minor eruptions and several major ones in 1631, 1794, 1872, 1906, and in 1944. It is also the only active volcano on the European mainland. Some consider Vesuvian as a subset of the Plinian volcanic eruption; the latter being a violent form of the former.

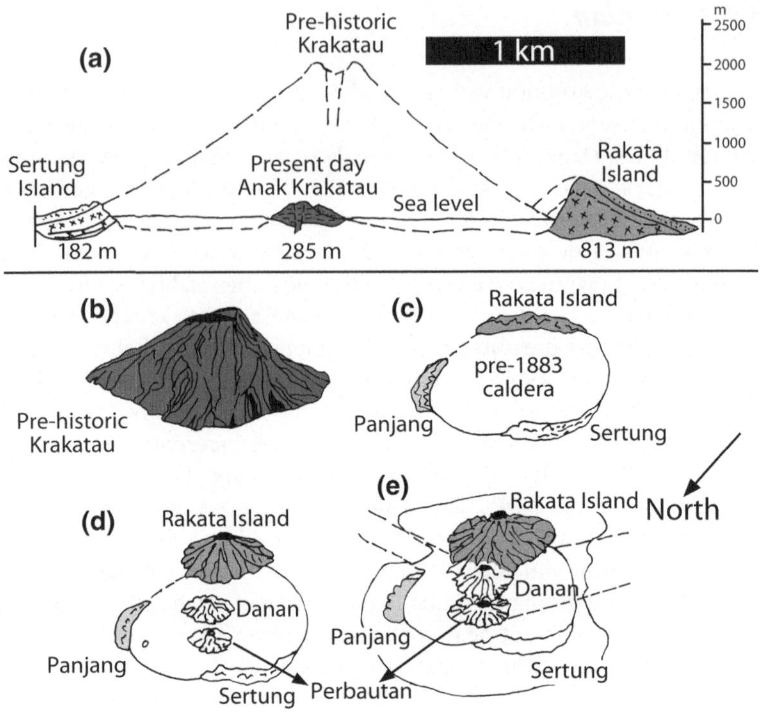

Fig. 16.16 The Krakatau volcano. **a** Cross section (pre-1183 and post-eruption), **b–e**: Eruption history of the Krakatau volcano. The volcano lies in the Sunda Strait between Java and Sumatra. Collapse of the ancestral Krakatau edifice, sometime in 416 A.D., formed a 7-km-wide caldera. Remnants of this ancestral volcano are preserved in the Verlaten and Lang Islands. Since 1927, small eruptions have been frequent inis region resulting in the formation of a new island, Anak Krakatau (meaning the Child of Krakatau)

16.13.6 Krakatoan

The renowned Krakatau volcano (Fig. 16.14a) lies in the Sunda Strait between Java and Sumatra. Collapse of the ancestral Krakatau edifice, perhaps in 416 A.D., formed a 7-km-wide caldera (Fig. 16.16). Remnants of this ancestral volcano are preserved in Verlaten and Lang Islands; subsequently Rakata, Danan and Perbuwatan volcanoes were formed, coalescing to create the pre-1883 Krakatau Island. Caldera collapse during the catastrophic 1883 eruption destroyed Danan and Perbautan volcanoes, and left only a remnant of the Rakata volcano (Fig. 16.16). Pyroclastic surges traveled 40 km across the Sunda Strait and reached the Sumatra coast. After a quiescence of less than a half century, the post-collapse cone of Anak Krakatau (Child of Krakatau) was constructed within the 1883 caldera at a point between the former cones of Danan and Perbautan. Anak Krakatau has been the site of frequent eruptions since 1927. Recent eruptions of Krakatau have been at Anak Krakatau, an island that emerged in 1927 (Fig. 16.16). The eruption and collapse of the caldera in 1883 produced one of the largest explosions on earth in recorded time (VEI = 6) (see also Simkin and Siebert 1994; Breining 2007) and destroyed much of Krakatau island, leaving only a remnant. Since 1927, small eruptions have been frequent and have constructed a new island, Anak Krakatau (Child of Krakatau).

16.13.7 Peléan

When Mount Pelée destroyed St. Pierre (Martinique) in 1902, it heralded a new type of volcanic eruption, marked by a high degree of fragmentation and thick accumulation, although, the dispersed area was comparatively small ($\sim$50 km^2) (Fig. 16.14g). Peléan eruptions develop chiefly on stratovolcanoes, mainly from rhyolitic, dacitic, trachytic and andesitic magmas. Interestingly, Peléan eruptions can remain inactive for decades, until they become active by first ejecting ash for few weeks, culminating into spasms of decreasing intensity for several months afterwards. The climax develops very suddenly when Nuées Ardentes (or glowing avalanches that moves down slope with velocities as much as 160 km/h at 700 °C or more) blast from the volcano to form directed deposits. Later, the viscous domes that often rise out of the vents may be partly destroyed by further eruptions.

16.13.8 Plinian

The 18th May 1980 eruption at Mount St. Helens (Fig. 16.17) and more recently the 15th June 1991 eruption at Pinatubo in the Philippines are classic example of Plinian eruption. The fast moving deadly pyroclastic flows (Nuées Ardentes) are commonly associated with them, though, they also occur during Pelean eruptions (discussed below). The Plinian eruptions are the largest, most violent, and most destructive of all eruptions (Fig. 16.14h). They are named after Pliny the Younger, who provided a remarkably accurate description of the Italian Mount Vesuvius eruption in 79 A.D. (Fig. 16.2). This eruption buried the Roman towns of Pompeii and Herculaneum. Plinian eruptions are a violent form of the Vesuvian eruption and involve explosive ejection of relatively viscous lava.

The process powering Plinian eruptions starts in the magma chamber, where dissolved volatile gases are stored. The gases vesiculate and accumulate as they rise through the magma conduit. These bubbles come together and once they reach a certain size (about 75 % of the total volume of the magma conduit) they explode. The narrow confine of the conduit forces the gases and associated magma up, forming a massive eruptive column. These eruptive columns are a distinctive feature of the Plinian eruption. The highly explosive eruptions are associated with volatile-rich dacitic to rhyolitic lavas, mostly associated with Stratovolcanoes.

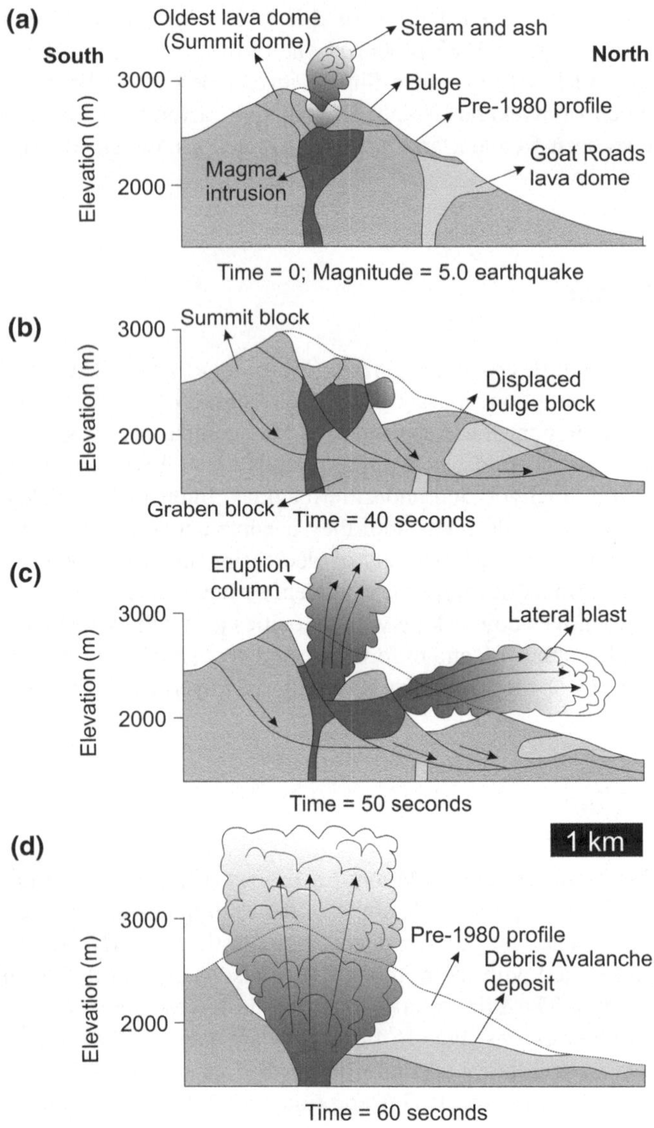

Fig. 16.17 The time line and eruptive mode of Mount St. Helens on 18th May 1980

Eruptions can last anywhere from hours to days, with longer eruptions being associated with more felsic volcanoes. Although Plinian eruptions are associated with felsic magma, they can just as well occur at basaltic volcanoes, provided that the magma chamber differentiates and has material rich in silicon dioxide. These types of eruptions can send ash and volcanic gas tens of miles into the air and across hundreds of miles.

Plinian eruptions are similar to both Vulcanian and Strombolian eruptions, except that rather than creating discrete explosive events, Plinian eruptions form sustained eruptive columns. They are also similar to Hawaiian lava fountains in that both eruptive types produce sustained eruption columns maintained by the growth of bubbles that move up at about the same speed as the magma surrounding them.

In some classifications, the term Krakatoan (Fig. 16.14) is used instead of Plinian which is based on the Krakatau eruption of 1883 in Indonesia. However, Krakatoan is extremely violent and generally removes the peak of the volcanic cone. Ultra-Plinian is the most extreme type of Plinian eruption, in which the column of ejecta reaches a height of more than 45 km. Such an eruption occurred in New Zealand near Taupo (North Island) in 186 AD. This had a 50 km high column.

Interestingly, the effects of the Pinatubo eruption extended beyond Philippines. The fine volcanic dust and gasses that blasted into the atmosphere took time to settle out and for a while, worldwide, the sunsets were more colorful. For a couple of years, due to the filtering effect of the solar radiation, the global average temperature also dropped by almost 0.5 °C. The estimated volume of magma erupted from the Pinatubo eruption was 5 km^3, the world's largest eruption since 1917.

16.13.9 Other Eruptions

16.13.9.1 Hydrovolcanic Eruptions (the Surtseyan Eruption)

It is a hydrovolcanic eruption caused by the shallow-water interactions between water and lava. Surtseyan eruptions take place mainly in shallow seas and lakes. The eruption is named after its most famous example, the eruption and formation of the island of Surtsey off the coast of Iceland in 1963. These eruptions are the "wet" equivalent of the ground-based Strombolian eruptions, but because of where they occur, they are much more explosive. As water is heated by lava, it flashes in steam and expands violently; fragmenting the magma it is in contact with, into fine-grained ash. Hence, these eruptions are characteristic of shallow-water volcanic oceanic islands. But these can also occur on land, caused by rising magma (basaltic or andesitic) that comes into contact with an aquifer (a water-bearing rock formation) at shallow levels under the volcano. Hence, these are also referred to as Phreatomagmatic eruptions. The Surtseyan eruption is similar to Strombolian but is generally continuous or otherwise rhythmic.

Surtseyan eruptions tend to form Maars (a broad low-relief volcanic craters dug into the ground), and tuff rings (circular structures built of a rapidly quenched lava) (Fig. 16.18). These structures are characteristic of single vent eruptions. Littoral cones are another hydrovolcanic feature, generated by the explosive deposition of basaltic tephra. They form when lava accumulates within cracks in lava,

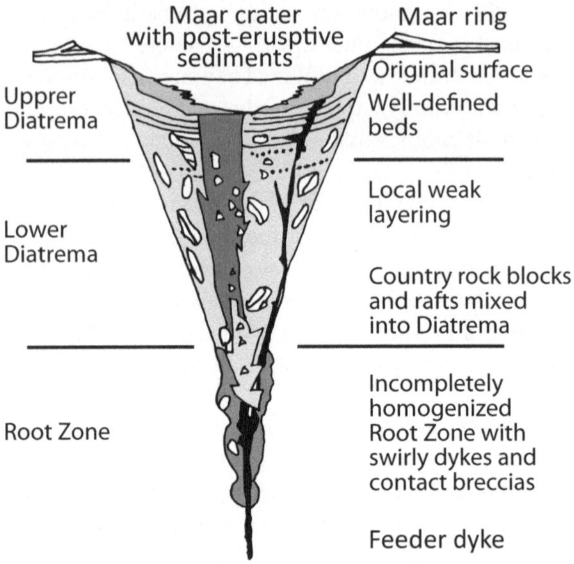

Fig. 16.18 A schematic diagram of the Coombs Hills diatreme complex. The Maar-Diatreme volcanoes are produced by explosive eruptions that cut deeply into the country rock. A maar is the crater cut into the ground and surrounded by an ejecta ring, while the diatreme structure continues downward and encloses diatreme and root zone deposits. Redrawn with permission from Dr James D. L. White (New Zealand)

superheats and explodes in a steam explosion, breaking the rock apart and depositing it on the volcano's flank. Volcanoes known to have Surtseyan activity include the Surtsey in Iceland. This volcano was built from depth and emerged above the Atlantic Ocean off the coast of Iceland in 1963. Initial hydrovolcanics were highly explosive, but as the volcano grew out, the rising lava started to interact less with the water and more with the air, until finally the Surtseyan activity reduced and became more Strombolian in character. The underwater volcano Hunga Tonga in Tonga, 30 km SSW of the Falcon Island, breached sea level in 2009. Both of its vents exhibited Surtseyan activity for much of the time. It also erupted earlier in May 1988.

16.13.9.2 Submarine Eruptions

These volcanic eruptions occur underwater. Approximately 75 % of the total volcanic eruptive volume on the earth is generated by submarine eruptions near the Mid-Ocean Ridge (MOR) alone. Submarine eruptions are generated by seamounts (the underwater volcanoes). MOR volcanoes are mostly basaltic, whereas subduction flows are generally calc-alkaline; more explosive and viscous due to higher silica content. There are about 100,000 deepwater volcanoes in the world, although most are beyond the active stage of their life. The largest of the oceanic plateaus is the Ontang Java Plateau in the western Pacific Ocean. It is larger in area than Alaska (USA). A thick sequence of sedimentary rocks covers the huge volume of basalt that formed the plateau around 90 Ma.

16.13.9.3 Phreatomagmatic Eruptions

These eruptions are a product of the interactions between water and magma. This temperature difference between the two states causes violent water-lava interactions that make up the eruption. They are driven from thermal contraction (as opposed to magmatic eruptions, which are driven by thermal expansion) of magma when it comes in contact with water. In Phreatomagmatic eruptions, the products are more regular in shape and fine grained than the products of magmatic eruptions due to differences in the eruptive mechanisms. The Phreatomagmatic eruptions only blast out fragments of preexisting solid rock from the volcanic conduit, and no new magma is erupted. Phreatic activity is generally weak, but can be quite violent as noted in the 1965 eruption of Taal Volcano, Philippines, and the 1975–1976 activity at La Soufrière, Guadeloupe (Lesser Antilles).

16.13.9.4 Phreatic Eruptions (Steam-Blast Eruptions)

These eruptions are driven by the expansion of steam. When cold ground or surface water comes into contact with hot rock or magma it superheats and explodes, fracturing the surrounding rock and thrusting out a mixture of steam, water, ash, volcanic bombs, and volcanic blocks. The characteristic feature of such explosions is that they only blast out fragments of pre-existing solid rock from the volcanic conduit and no new magma is erupted. However, as these eruptions are driven by the cracking of rock strata under pressure, phreatic activity does not always result in an eruption. Phreatic eruptions are generally weak, although there have been exceptions. Some Phreatic events may be triggered by earthquake activity, and they may travel along dike lines. Phreatic eruptions form Lahars, and Avalanches. They may also release deadly toxic gas (such as CO_2) capable of suffocating anyone within range of the eruption. The Taal Volcano (Philippines) in 1965 exhibited phreatic activity. Interestingly, Mount St. Helens, prior to its catastrophic 1980 eruption, also exhibited similar phreatic activity (Decker and Decker 1981).

16.13.9.5 Subglacial Eruptions

This type of eruption is characterized by the interaction between lava and ice, often under a glacier in areas of high latitude and high altitude. It has been suggested that subglacial volcanoes that are not actively erupting, often dump heat into the ice covering them, thereby producing meltwater. This meltwater mix means that subglacial eruptions often generate dangerous floods (Jokulhlaups) and Lahars (Myers et al. 2008). Lahars are volcanic mudflows. Glaciovolcanic products have been identified in Iceland, British Columbia, Hawaii and Alaska, the Cascade Range, South America, and even on planet Mars. Volcanoes known to have subglacial activity also include the Mauna Kea, Hawaii.

Table 16.6 The volcanic explosivity index (after Newhall and Self 1982; Cas and Wright 1988)

VEI scale	Description	Plume height	Volume	Classification	How often	Example
0	Non-explosive	<100 m	1,000 s m^3	Hawaiian	Daily	Kilauea (Hawaii, USA)
1	Gentle	100–1,000 m	10,000 s m^3	Hawaiian/Strombolian	Daily	Stromboli (Italy)
2	Explosive	1–5 km	1,000,000 s m^3	Strombolian/Vulcanian	Weekly	Galeras 1992 (Colombia)
3	Severe	3–15 km	10,000,000 s m3	Vulcanian	Yearly	Nevado del Ruiz 1985 (Colombia)
4	Cataclysmic	10–25 km	100,000,000 s m^3	Vulcanian/Plinian	10s of years	Galunggung 1982 (West Java, Indonesia)
5	Paroxysmal	>25 km	1 km^3	Plinian	100s of years	St. Helens 1981 (USA)
6	Colossal	>25 km	10 s km^3	Plinian/ Krakatoan	100s of years	Krakatau 1883 (Indonesia)
7	Super-colossal	>25 km	100 s km^3	Ultra-Plinian	1,000s of years	Mount Tambora 1815 (Indonesia)
8	Mega-colossal	>25 km	1,000 s km^3	Ultra-Plinian	10,000s of years	Yellowstone 2 Ma (USA)

VEI is logarithmic, meaning that each interval on the scale represents a tenfold increase in the size of the eruption. An eruption of VEI 3 is ten times bigger than a 2 and one hundred times smaller than a 5. The Tambora eruption of 1815 in Indonesia was the largest, single eruption in a millennium where almost 40 km^3 of material were blasted out of a volcanic island, leaving a 6-km-wide depression. The following year, 1816, was known as "the year without summer"

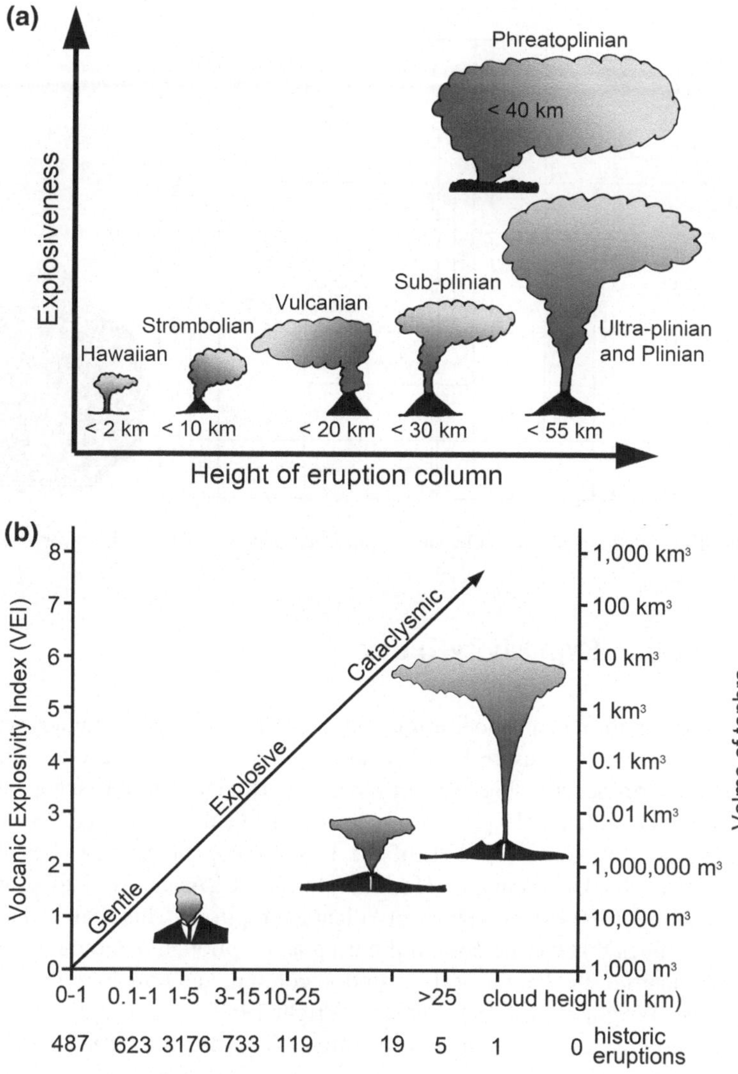

Fig. 16.19 The Volcanic Explosivity Index (VEI). **a** The relationship between explosiveness and the height of the eruption column associated with different styles of eruption. **b** The relationship of VEI to the volume of tephra, the height of eruptive cloud and the number of eruptions along with the nomenclature of destruction (i.e., gentle, explosive and cataclysmic). An eruption with a VEI value of 5 has an eruption cloud height up to 25 km high and ejects about 1 km^3 of tephra. Mount Saint Helens is a good example of category 5 VEI. VEI 5 eruptions normally occur once in a decade

VEI	Classi-fication	Description	Ejecta volme	Plume	Frequency	Duration	Tropospheric injection	Stratospheric injection	Examples	Erup.	
0	Hawaiian	effusive	<10,000 m³	<100 m	constant		negligible	none	Kilauea, Piton de la Fournaise	755	non-explosive
1	Hawaiian/ Strombolian	gentle	>10,000 m³	100-1000 m (above crater)	daily	continuous blast	minor	none	Stromboli, Nyiragongo (2002)	963	small
2	Strombolian / Vulcanian	explosive	>100,000 m³	1-5 km	weekly	< 1 hr	moderate	none	Galeras (1993), Mt. Sinabung (2010)	3631	moderate
3	Vulcanian / Pelean	severe	>10,000,000 m³	3-15 km	few months	1-6 hrs	substantial	possible	Nevado del Ruiz (1985), Siufriere Hills (1995)	924	
4	Pelean/ Plinian	cata-clysmic	> 0.1 km³	10-25 km	> 1 yr.	6-12 hrs	substantial	definite	Mt. Pelee (1902), Eyjafallajokull, (2010)	307	large
5	Plinian	paro-xysmal	> 1 km³	20-35 km	> 10 yrs.		substantial	significant	Mt. Vesuvius (79 CE), Mt. St. Helens (1980)	106	very large
6	Plinian/ Ultra-Plinian	colossal	> 10 km³	> 30 km (above sea level)	> 100 yrs.	> 12 hrs	substantial	substantial	Krakatoa (1882), Mt. Pinatubo (1991)	46	
7	Ultra-Plinian	super-colossal	> 100 km³	> 40 km	> 1,000 yrs.	> 12 hrs	substantial	substantial	Thera (~1600 BC), Tambora (1815)	4	
8	Super-volcanic	mega-colossal	> 1,000 km³	> 50 km	> 10,000 yrs.		substantial	substantial	Yellowstone (640,000 BP), Toba (74,000 BP)	0	

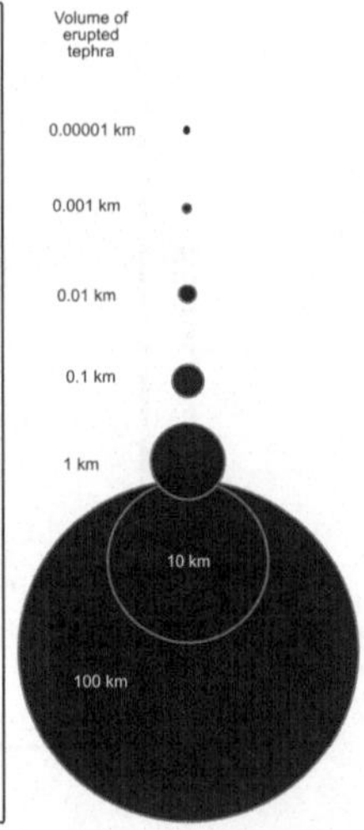

Fig. 16.20 The relationship of Volcanic Explosivity Index (VEI) with several eruptive parameters

16.14 Volcanic Explosivity Index

Every year about 50–70 volcanoes erupt, but most of the activity is rather weak. So how do volcanologists measure how big an eruption is? There is not any singular feature that determines the "bigness" of a volcano. But a magnitude scale has been created called the Volcanic Explosivity Index or VEI (Table 16.6; Fig. 16.19). The VEI was devised by Chris Newhall of the U.S. Geological Survey (USGS) and Stephen Self at the University of Hawaii (Hawaii, USA) in 1982 to provide a relative measure of the explosiveness of volcanic eruptions. This scale is based on a number of things that can be observed during an eruption, the volume of erupted pyroclastic material, the height of the eruption column, and how long the eruption lasts (see also Wright et al. 1980; Cas and Wright 1988). According to this scale, really huge eruptions don't happen very often. The 18th May 1980 eruption of Mount St. Helens, which destroyed 632 km^2 of land, expelled 1.4 km^3 of magma, and produced an eruption column that rose to 24 km, was assigned a VEI value of 5. The last large eruption from the Yellowstone caldera, which occurred 600,000 years ago and expelled over 1,000 km^3 of magma, is assigned a 7 and that which occurred 2 Ma is assigned 8 (Christiansen 1984; Wood and Kienle 1990; Smith and Braille 1994; Smith and Siegel 2000). However, most volcanic eruptions have a value ranging from 0 to 2 (Fig. 16.20). The Lake Taupo's Oruanui eruption is the most recent one that occurred some 26,500 years ago. This means that there have not been any eruptions within the last 10,000 years (i.e., in the Holocene) with a VEI of 8 (Courtillot 1999; Mason et al. 2004).

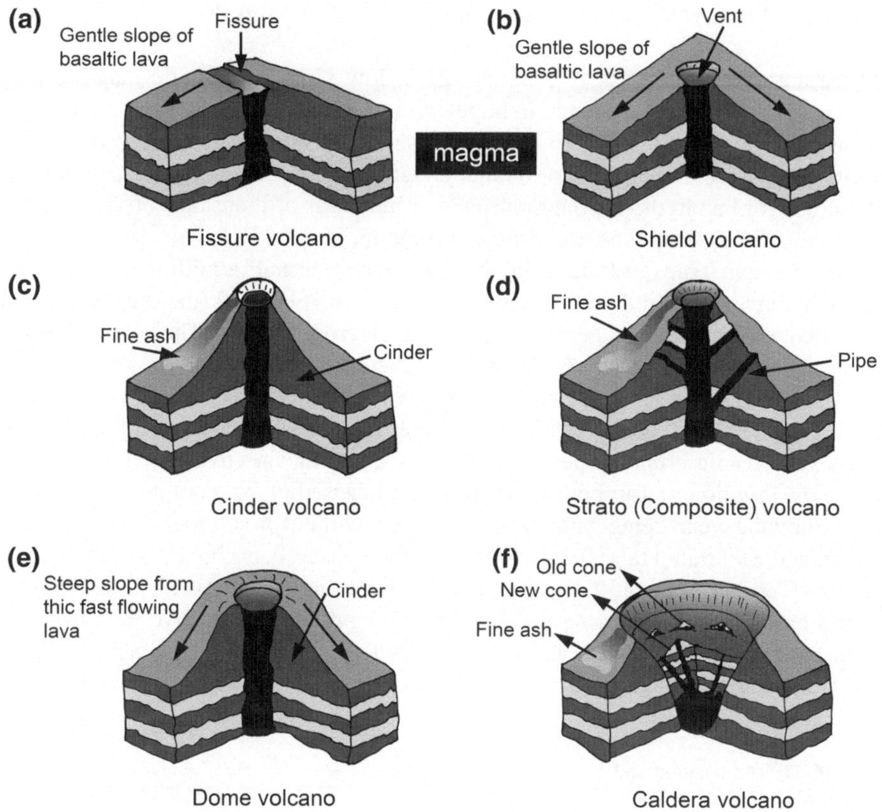

Fig. 16.21 The volcano types discussed in the text

16.15 Volcanic Landforms

Volcanic landscapes contain extremely diverse landforms varying markedly in size, shape, composition, and eruptive history. Hence, only the major ones are discussed below:

16.15.1 Type of Volcanoes

The form of a volcano is determined by the ingredients of the erupting magma. Their shapes are determined by the exclusivity of the eruptions and to the amount of water in the magma. Although every volcano has a unique eruptive history, but most can be grouped based largely on their eruptive patterns and their general form (Fig. 16.21) (see also Wright et al. 1992; Wright and Pierson 1992).

16.15.1.1 Fissure Volcano (Plateau Basalts or Flood Basalts)

These are extremely massive outpourings of low viscosity basaltic magma from fissure vents. The basalts spread in huge areas of relatively low slope and builds up plateaus. Flood basalts (Fig. 16.21a) are products of enormous eruptions from long vents, or fissures, rather than individual craters. Unlike present volcanic eruptions, these last for hundreds or even thousands or hundreds of thousands of years rather than days. Their eruptions are often intermittent.

The Deccan Traps in India (Fig. 16.22), cover over half a million square miles and may have covered three times that area when first erupted (the remainder of it either covered by younger sediments or has been eroded away). These flood basalts have an average thickness of a kilometer. The most recent flood basalt province, the Columbia River Province in the north west of the USA (Fig. 16.15) covers an area of 165,000 km^2 and is estimated to have produced 170,000 km^3 of lava. Such massive volcanic eruptions are known to have significant environmental impacts also. The eruption of Tambora volcano in 1815 caused global cooling which led to the following year being known as "the year without a summer." The Deccan Traps and Siberian Traps flood basalt provinces have been linked to major Cretaceous/Tertiary and end-Permian faunal extinctions, respectively. Interestingly, Flood basalt eruptions during the Tertiary have produced extensive lava fields in both continental areas (such as the Columbia River Plateau in the Pacific NW of the U.S.) and oceanic areas (such as the Ontong-Java Plateau in the SW Pacific).

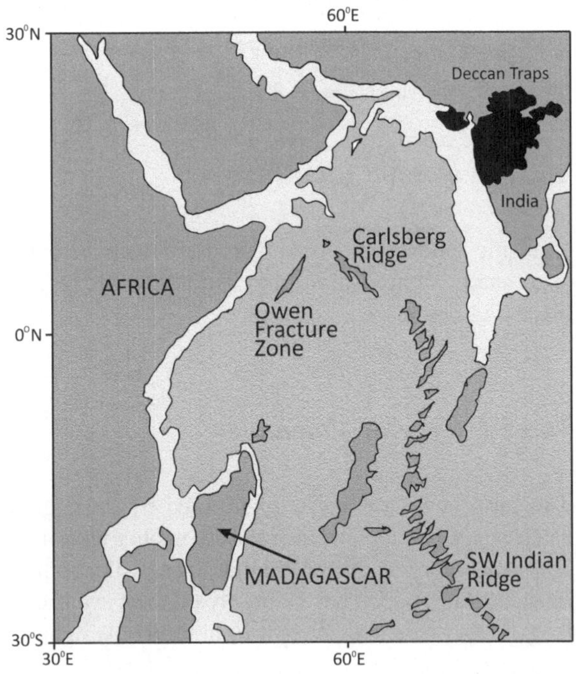

Fig. 16.22 The location and distribution of Deccan Traps in India. The light yellow shade marks the 3,000 m (below mean sea level) depth contour

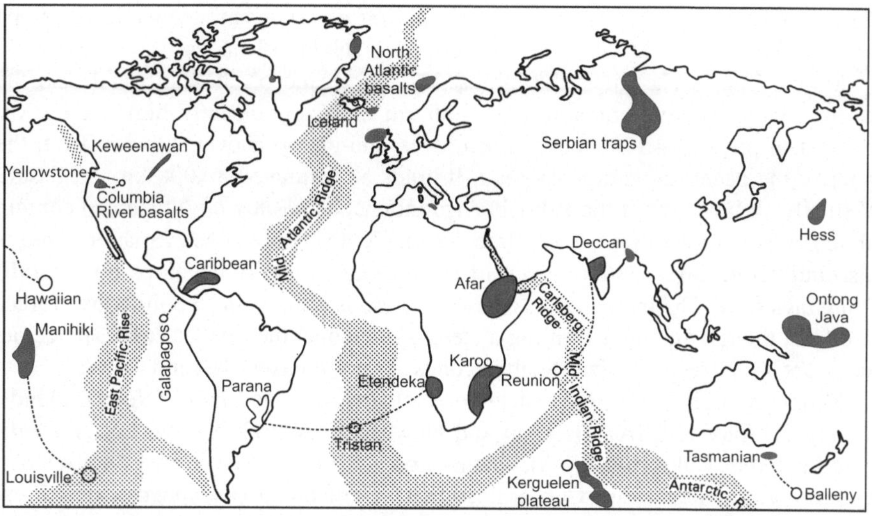

Fig. 16.23 The global distribution of Flood basalts (also called Large Igneous Provinces, LIPS)

The Columbia Plateau area of Washington (Fig. 16.15), Idaho, and Oregon is constructed of layer upon layer of basalt at places as thick as 3,000 m. The area covered is over 400,000 km^2. Each individual flood of lava added a layer usually between 15 and 100 m thick and sometimes thousands of square kilometers in extent. The outpourings of lava that built the Columbia Plateau took place from 17.5 to 6 million years ago but 95 % erupted between 17 and 15.5 million years ago. Similar huge, lava plateau-building events have not occurred since then.

Major examples of Flood basalt eruptions (see Fig. 16.23) include: Banda Api (Indonesia), Pagan (Mariana Islands), Kuchinoerabu-jima and Komaga-take (Japan), Tolbachik (Kamchatka, Russia), and Krafla (NE Iceland).

16.15.1.2 Shield Volcanoes

These are large volcanic forms with broad summit areas and low, sloping sides (Fig. 16.21b). They have a low-angle profile (ranging between 2 and 10° from the horizontal), which resembles the broad shields used by Hawaiian warriors. They are built almost entirely of fluid lava flows (non explosive eruptions of low viscosity basaltic magma). Multiple flows pour out in all directions from a central summit vent, or a group of vents, forming a broad, gently sloping cone of flat, domical shape, hence, their name. The shield volcanoes are composed of thousands of fluid lava flows that spread great distances, sometimes greater than 50 km. The erupted low viscosity basaltic magma permits gas to escape and so the lava flows easily down the slope and away from the summit vent. However, this slow movement also allows the lava to cool, thereby increasing its viscosity; its thickness builds up on the lower slopes giving a somewhat steeper lower slope (<10°). Some of world's largest volcanoes are shield volcanoes.

Mauna Loa in Hawaii is the largest shield volcano on the earth and also the world's largest active volcano. It is 8.85 km high, significantly higher than Mt. Everest. It projects 4,170 m above sea level, and its top is over 9,144 m from the floor of the ocean to its highest peak. It holds an estimated 80,000 km^3 of basalt. Virtually all lavas are of the Pahoehoe (pronounced *pah-hoy-hoy*) type and contain abundant small lava-tubes. Sometimes, a small, steep-sided cone (Spatter cone) is also built from lava sputtering out of a vent that generally develops on a solidifying lava flow. The increased gas concentration is trapped in a cooling lava forces itself out though an outlet forming a steep-sided cone, the Spatter cone. Hence, its sides are very steep. In size, Spatter cones do not measure beyond 10 m.

Major examples of Shield volcanoes include: Auckland Field (New Zealand), Wrangell (Alaska, USA), Belknap and Newberry (Central Oregon, USA), Medicine Lake (NE California, USA), Mauna Loa and Mauna Kea (Hawaii, USA), San Quintín Volc Field (Baja California, USA), Fernandina (Galápagos), and Prestahnukur (Central Iceland).

16.15.1.3 Cinder Volcano (Cinder, Tephra or Pyroclastic Cones)

Cinder cones are the simplest of the volcano types (Fig. 16.21c). They expel ash and cinders and have a cone-shaped peak that is associated with a lava-spewing Strombolian type eruption. They are small, hill-sized volcanoes built from pyroclastic fragments (Tuff) and blobs of congealed lava ejected from a single vent (central vent). As the gas-charged lava is blown violently into the air, it breaks into small fragments that solidify and fall as cinders around the vent to form a circular or oval cone, hence, their name. They are also called Pyroclastic cones. Most cinder cones have a bowl-shaped crater at the summit and rarely rise more than a 500 m or so above their surroundings and usually occur around summit vents and flank vents of Stratovolcanoes and Shield volcanoes. For example, the Newberry volcano (Oregon, USA) has more than 400 cinder cones on its flanks, and Wizard Island is a small cinder cone that formed following the origin of the Crater Lake caldera. The Cinder cones are characterized by their steeply angled sides (usually between 25° and 33°; higher angles are close to the vent) and conical shapes. Most Cinder cones are less than 300 m high, although a large one can be up to 700 m high. Tuff cone is a variety of the pyroclastic cone formed by more energetic explosions involving groundwater that extensively spew out ejecta, producing finer-grained ash deposits. Tuff cones typically have steeper slopes and wider craters than Cinder cones. Silicic cinder cones, which are made of fragments of Pumice are called Pumice cones. It must be noted that most cinder cones are short-lived features in terms of geologic time as the unconsolidated pyroclastic materials are eroded quite easily.

Major examples of Cinder volcanoes include: Barren Island (India), Aso (Kyushu, Japan), Tolbachik (Kamchatka, Russia), Okmok (Alaska, USA),

Bachelor, Newberry and Crater Lake (Oregon, USA), Pinacate, San Quintín Volc Field, Bárcena, Isla Isabel and Michoacán-Guanajuato (Mexico), Cerro Negro (Nicaragua) and Darwinand Santiago (Galápagos Islands).

16.15.1.4 Stratovolcanoes (Composite Volcanoes)

Some of earth's grandest mountains (Mount Fuji in Japan; Mount Cotopaxi in Ecuador; Mount Shasta (in California), Mount Hood (in Oregon), and Mount St. Helens and Mount Rainier (in Washington) in USA; Orizaba (third highest peak in North America) and Popocatépetl in Mexico (North America's highest mountain); Mount Erebus in Antarctica, the southernmost active volcano in the world; Mount Etna, on the island of Sicily, is Europe's largest and one of world's most active volcano, etc.) are all Composite volcanoes (Fig. 16.21d). These are also called Stratovolcanoes. Many of the highest mountains of the Andes and some of the most spectacular mountains of western North America (Fig. 16.24) are composite

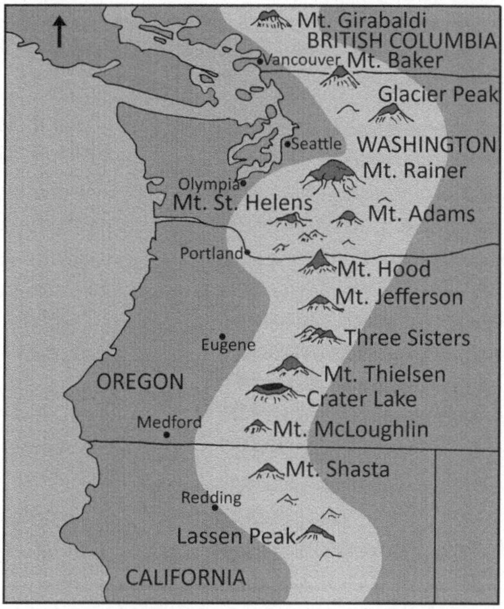

Fig. 16.24 The distribution of Cascade Range volcanoes in Western North America. Their distribution stretches from the Lassen Peak in northern California, north through Oregon and Washington (USA) to the Meager Mountain in British Columbia (Canada; Mount Meager is part of the Garibaldi Volcanic Belt, the northernmost segment of the Cascade Range). Most of the large volcanoes in the range are composite volcanoes, such as the Mount Shasta in California, Mount Hood in Oregon, and Mount St. Helens in Washington. The Lassen Peak in California is the world's largest lava dome. Cinder cones are common throughout the Cascade Range, such as the Wizard Island in Crater Lake, Oregon and the Cinder Cone in Lassen Volcanic National Park (Lessen Peak), California (USA)

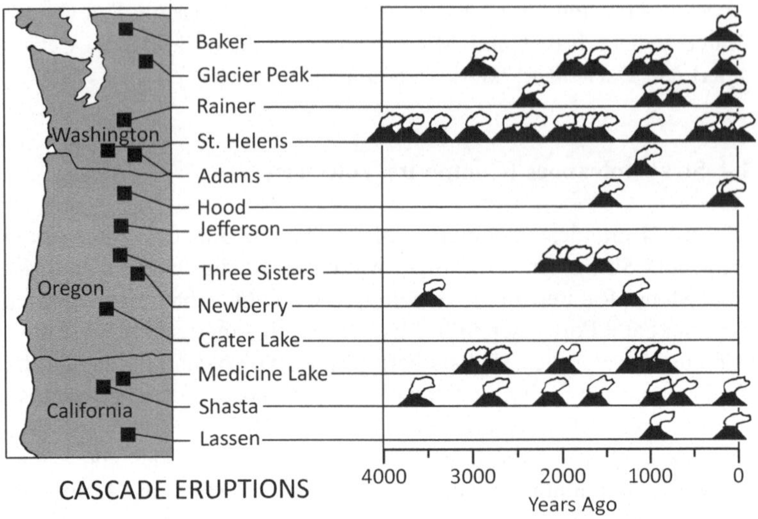

Fig. 16.25 The last 4,000 years eruption history of the 13 major Cascade Range volcanoes in Western North America. Eruptions in the last 4,000 years have become more frequent to about 2 per century (modified from USGS data and USGS Open-File Report 95-585)

cones. Within the Cascade Range, those erupting in the last 4,000 years are shown in Fig. 16.25.

Stratovolcanoes reach great heights (as much as 2,440 m above from their base) through a combination of explosive eruptions of alternating layers of viscous lava flows, volcanic ash, cinders, blocks, and bombs (pyroclastic fragments). Due to this, inter-layering of lava flows and pyroclastic material (mostly tephra), these are also called Composite volcanoes (Myers et al. 2008). Sometimes, the pyroclastic material can make up as much as 50 % of their total volume. However, for some other such as the Mount Rainier (Fig. 16.24), are composed of 90 % lava flows and only 10 % pyroclastic layers. On the other hand, Mount St. Helens is built mostly from pyroclastic eruptions. Andesite is the rock most associated with Stratovolcanoes.

The Stratovolcanoes have a crater at the summit that contains a central vent or a clustered group of vents. However, the most essential feature of a Stratovolcano is a conduit system through which magma from a deep seated reservoir rises to the surface. The lava either flows through breaks in the crater wall or issues from fissures on the flanks of the cone. It solidifies within the fissures and forms dikes that act as ribs which further strengthens the cone. These eruptions from multiple vents often result in a greatly variable morphology. The slopes of stratovolcano rarely exceeds the angle of repose for fragmental material (between 30 and 33°) and is intermediate in steepness between cinder and shield volcanoes; steeper than a shield volcano and broader than a cinder volcano (Fig. 16.21c).

Nearly all the larger and better known volcanoes of the world are composite volcanoes. They tend to align along three major belts. The circum-Pacific belt, or "Ring of Fire," is the largest where more than 60 % of all active volcanoes are found. The Cascade Range volcanoes (Fig. 16.24) described earlier make up a small segment of this circum-Pacific belt. The second major volcanic belt is the Mediterranean belt (about 20 % of all active volcanoes), which includes Mount Vesuvius. Mount Etna, on the island of Sicily, is Europe's largest volcano and one of the world's most active volcanoes. Its largest eruption in 300 years began in 1991 and lasted for 473 days. Some 250 million m^3 of lava covered 7 km^2 of land. The third is along the mid-oceanic ridges which has the remaining 20 % of active volcanoes.

Major examples of Stratovolcano include: Majon (Philippines), Haku-san and Iwate (Japan), Kliuchevskoi (Russian), Vsevidof (Alaska, USA), Baker and Rainier (Washington, USA), Jefferson (Oregon, USA), Colima and Popocatépetl (Mexico), Santa María and Fuego (Guatemala) and Pilas, Las (Nicaragua).

16.15.1.5 Lava Dome

They are also called volcanic domes. They are steep-sided, dome- or spine-shaped masses of volcanic rock formed from viscous lavas that solidifies in or immediately above a volcanic vent (Fig. 16.21e). Lava Dome are formed by the extrusion of highly viscous, relatively small, bulbous masses of lava. As the lava viscosity is high, it does not flow away from the vent, but piles over and around it. Hence, the dome largely grows by expansion from within. Such explosive and slowly rising lava domes can grow for months or for several years in the aftermath of explosive eruptions. However, younger domes whose lava is not entirely degassed may occasionally explode. Additionally, they can be extremely dangerous, as they form unstable slopes that may collapse to expose gas-rich viscous magma to atmospheric pressure resulting in lateral blasts or Plinian type pyroclastic flow eruptions. The lava composition ranges from basalt andesite to rhyolite, but mostly are crystal-rich dacites. Most of the viscous lavas that form volcanic domes are high in silica. Commonly, they solidify as Obsidian, the chemical equivalent of Rhyolite.

There are several types of lava dome depending upon their morphology. Although, lava domes are typically thick, steep extrusions, but their shapes can vary from circular, low-profile domes (Tortas) to cylindrical spines with thick talus slopes (Peléean domes). Less sluggish types are gradational to lava flows and are sometimes referred to as Dome flows or Coulées. A coulee is a large mass of viscous magma that erupts and flows down slope for short distances, forming a feature that is a hybrid between a lava dome and a lava flow. Rhyolitic volcanoes often produce layered deposits of volcanic ash called Tuff (as do stratovolcanoes also). When fine-grained pyroclastic material (such as dust and ash) accumulates and is cemented or otherwise consolidated, the rock is called Tuff. Additionally, the silica-rich rhyolitic magmas commonly produce solid volcanic glass (obsidian; a volcanic glass that is usually silicic, and is one of the few rocks that is not composed of minerals), and frothy volcanic glass (pumice) (see also Myers et al. 2008). Compound lava domes, such as currently forming at Santiaguito at the base of Santa María volcano in Guatemala, are common. Some volcanoes, such as Augustine in Alaska, consist entirely of a complex of overlapping summit lava domes surrounded by fragmental material produced during growth and collapse of the domes.

Major examples of Lava domes include: Merapi (Central Java), Unzen (Japan), Bezymianny (Kamchatka, Russia), Redoubt (Alaska, USA), St. Helens and Lassen Volcanic Center (Washington, USA), Ceboruco and El Chichón (Mexico), Santa María and Pululagua (Guatemala), Soufrière Hills (Montserrat, West Indies), and Soufrière St. Vincent (Windward Islands of the Caribbean). Some of the most destructive volcanic explosions known have been associated with volcanic domes.

16.15.2 Calderas

Calderas are formed when an erupting volcano empties a shallow-level magma chamber; the edifice of the volcano collapses into the voided reservoir, thus, forming a steep, bowl-shaped depression (Figs. 16.21f; 16.26). Caldera is a Spanish word for kettle or cauldron. Calderas are a volcanic depression much larger than the original crater, having a diameter of at least 1 km. However, Calderas are highly variable in size, ranging from 1 to 100 km in diameter with wall as high as 1 km. They resemble summit craters but can be easily differentiated as summit craters are mostly much smaller and form by explosive erosion of the central vent. The felsic calderas are surrounded by thick blankets of pumice derived from the eruption of voluminous pyroclastic sheet flows. It is important to

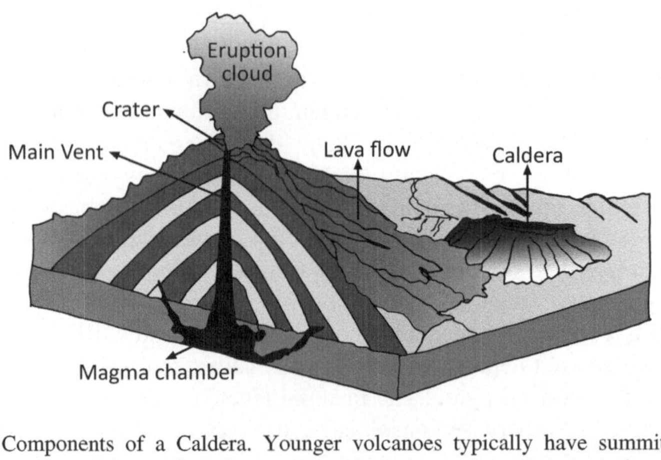

Fig. 16.26 Components of a Caldera. Younger volcanoes typically have summit craters, or calderas, as much as 5 km wide and several hundred meters deep that result from subsidence following the eruption of magma from below

mention the difference between a Crater and a Caldera here. If the circular or elliptical depressions are less than a km in diameter, they are called Craters whereas those with diameters ranging from 1 to 100 km are called Calderas. Both are formed by explosion, collapse, or either. The caldera can sometimes be filled with groundwater, precipitation, or by melt water to forms a lake within a crater, then it is called a Crater Lake type caldera.

Major examples of Caldera include: Toba (Sumatra), Banda Api (Indonesia), Taal and Pinatubo (Philippines), Mashu (Hokkaido, Japan), Akademia Nauk (Central Kamchatka), Crater Lake (Oregon, USA), and Masaya (Nicaragua).

Based on changes in calderas form and origin, they are classified into three types:

16.15.2.1 Crater Lake Type Calderas

This caldera type is named after the deepest (600 m) fresh water lake in North America. Crater Lake type calderas are smaller features typically defined as being less than 1 km in diameter. They are formed after the main phase of a Plinian eruption, during the collapse of a Stratovolcano, into the void of the underlying depleted magma chamber. Although the waning phase of a Plinian eruption is often associated with the generation of pyroclastic flows, the piston-like collapse of the volcanic edifice can generate additional eruption of voluminous, pumice-dominated sheet flows along ring fractures surrounding the collapsing mass. These sheet flows form thick deposits of Ignimbrite, the hallmark of both Crater-Lake and Resurgent calderas. Ignimbrite is an igneous rock formed by the lithification of ash flow or pyroclastic flow deposits. Hence, Calderas are huge structures that form following voluminous eruptions during which part of a magma chamber drains and the mountain's summit collapses into the vacated space below. An excellent example from Oregon (USA) is Crater Lake which is misnamed (Fig. 16.24). It is actually a steep rimmed caldera that formed 7700 years ago. It is more than 1200 m deep and measures 9.7 km long and is 6.5 km wide.

Major examples of Crater Lake type caldera include—West Eifel Volc Field (Germany), Ol Doinyo Lengai (Tanzania), Kelimutu (Indonesia), Maly Semiachik (Kamchatka), Douglas (Alaska, USA), Pinacate and Chichinautzin (Mexico), Telica and Las Pilas (Nicaragua), Cotopaxi (Ecuador), and Erebus (Antarctica).

16.15.2.2 Resurgent Calderas (Also Called Yellowstone Type)

They are the largest volcanic structures with diameters ranging between 15 and 100 km. Calderas become resurgent if the magma refills the chamber beneath the caldera. This refilling results in the uplift of the caldera floor with the eventual eruption of the lava within the caldera. In them, uplift is commonly more than one kilometer. Resurgent calderas are associated with massive eruptions of voluminous pyroclastic sheet flows. The Toba caldera (on the Indonesian Island of Sumatra) is one of the youngest of these (about 74,000-year-old). It generated 2,800 times more pyroclastic material than the 1980 Plinian eruption of Mt. St. Helens. The most distinctive feature of Resurgent calderas is their broad topographic depression with a central elevated mass resulting from the post-collapse upheaval of the caldera floor. Additionally, Resurgent calderas lack a single centralized vent. The caldera floor is typically filled with rhyolitic lavas, obsidian flows, and domes.

The most famous Resurgent calderas in the United States are - the Valles Caldera (Jemez volcanic field in New Mexico; erupted <1.5 million years ago), the Long Valley Caldera (California; formed around 760,000 years ago and erupted almost 600 km^3 of ash and pyroclastic flows) and the Yellowstone Caldera (Wyoming and Montana). The oldest eruption in the Yellowstone took place 1.9 million years ago and ejected around 2,500 km^3 of pyroclastic material. The next major eruption occurred 1.3 million years ago (280 km^3 of pyroclastic material) and the most recent one occurred around 0.6 million years ago (~630,000 years ago). It ejected 1,000 km^3 of ash and other debris and produced the Yellowstone caldera in the center of the Yellowstone National park. Intervals of 0.6–0.7 million years separate the three Yellowstone eruptions; 0.6 million years have already passed since the most recent one (see also Fig. 16.15).

16.15.2.3 Basaltic Shield Volcano Calderas (Also Called Hawaiian Type)

Basaltic calderas subside in increments to produce a nested structure of pits and terraces. They gradually enlarge by episodic collapse, due to the extraction of lava from shallow-level magma chambers. A good example is the Kilauea caldera on Kilauea, the Erta Al caldera in Ethiopia, the Summit caldera of Piton del la Fournaise on Reunion Island (Indian Ocean), and the shield volcanoes of the Galapagos Islands. Most basaltic shield volcano calderas range from 1 to 5 km in diameter.

16.16 Submarine Volcanoes

Submarine volcanoes occur in the ocean floor. Currently there are over 5,000 active volcanoes underwater marked by blasting steam and rock-debris above the sea surface. However, some others lie at great depths. In these, the weight of the water results in high confining pressure and prevents the formation and explosive release of steam and gases. Some are violent steam-blasting eruptions when sea water pours into an active shallow submarine vent. During such an explosive submarine eruption enormous piles of debris are built up around the active volcanic vent, in a shallow open ocean. Here at shallow depths, ocean currents rework the debris, while in others, the debris slumps from the upper part of the cone and flows into deep water along the sea floor. Submarine volcanism produces Pumice, a light igneous volcanic rock that floats on the water and drifts with the ocean currents over large areas. Pumice is considered a glass because it has no crystal structure.

16.17 Other Types of Landforms

16.17.1 Volcanic Necks and Plugs (or Lava Neck)

They are formed when the volcano becomes inactive and deeply eroded; the exhumed plug stands out in bold relief as an irregular, columnar structure (Fig. 16.27). This columnar structure is more resistant to erosion than the enclosing rock formation. Plugs are commonly funnel shaped that taper downwards. They are increasingly elliptical in plan or elongated to have dike-like forms. Compositionally, the igneous material in a plug is similar to the associated lava or ash, but may also include fragments and blocks of denser, coarser grained rocks that are higher in iron and magnesium, and lower in silicon. Good examples of volcanic plugs in the United States include Morro Rock (California), Lizard Head (Colorado), Laurel Hill (New Jersey), and Little Devils Postpile (in the

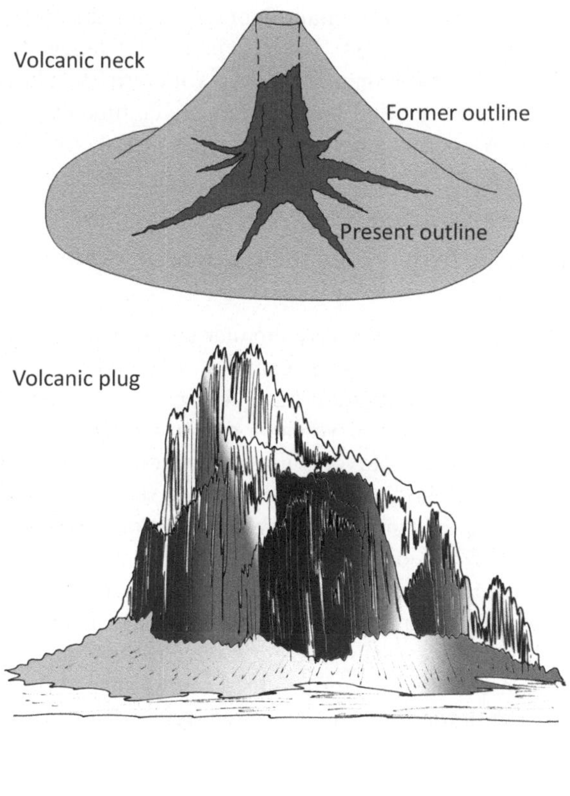

Fig. 16.27 Volcanic Neck (**a**) and Plug (**b**). The resistant rock remains while the softer rock units weather away. The upper panel (in the shade of light gray) shows the former un-weathered outline of a volcanic neck

Yosemite National Park). Plugs that bear a particularly strong imprint of explosive eruption of highly gas-charged magma are called Diatremes or Tuff-breccia (Fig. 16.18). The vents are filled with pyroclastic material that has collapsed back into the pipe. These include breccia pipes, tuff pipes, and Kimberlites. Kimberlites are the only known source of diamonds. Most known pipes formed between 70 and 140 million years ago, and most occur in continental crust older than 2.5 billion years. The most famous diamond-rich kimberlite pipes are located in South Africa, although others are known in Canada's Northwest Territories, Arkansas, and Russia. Diatremes are cylindrical features that converge downward like steep-walled funnels (Fig. 16.18). They are usually associated with composite volcanoes.

16.17.2 Maars (Tuff Cones)

These are flat-floored craters produced when the rising basaltic magma comes into contact with the groundwater close to the earth's surface (Fig. 16.18). Hence, it is an explosive interaction of hot magma (about 1200 °C) and cold water. A typical Maar volcano is marked by a wide funnel-shaped crater, ringed by a very low cone which slopes gently away from the crater. The cone is built of layers of ash and large basalt lumps, known as bombs, along with local rock material ripped from the crater walls. The crater floor lies below the pre-eruption land surface, evidence of a collapse process. The volcanic crater ranges from about 100–3,000 m wide, about 10 to more than 50 m deep with a rim height of from a few meters to nearly 100 m above ground level. Maars are commonly filled with water to form natural lakes. A good example is the Zuni Salt Lake in New Mexico (USA).

16.17.3 Non-Volcanic Craters (Cryptovolcanic)

The occurrence of nearly circular areas of intensely deformed sedimentary rocks with a central vent-like feature surrounded by a ring-shaped depression; resembling volcanic structures have been recorded at several places across the world. As no clear evidence of volcanic origin is noted in or near these structures, they are described as Cryptovolcanic. Impact craters, formed by collisions with the earth of large meteorites, asteroids, or comets, share with volcanoes the imprints of violent origin, as evidenced by severe disruption, and local melting of the rocks. Fragments of meteorites or chemically detectable traces of extraterrestrial materials and evidences of strong forces acting from above, rather than from below, distinguish such impact craters from volcanic features. A classic example of a Non-volcanic crater is the Meteor Crater in Arizona (USA), the largest known crater with associated meteorites. It is a bowl-shaped depression 180 m deep and about 1.2 km in diameter encompassed by a ridge or rim that rises 30–60 m above the surrounding plain. The kinetic energy of the projectile required to form such a crater is estimated from computer simulation to be equivalent to about 15 megatons of TNT. This energy is equal to that of a spherical body of meteoritic iron about 40 m in diameter traveling at a speed of 20 km/sec.

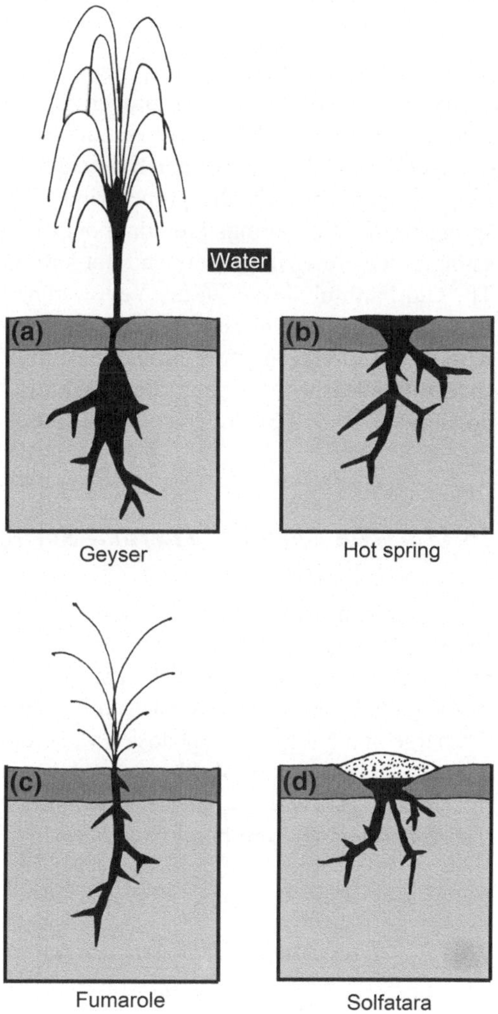

Fig. 16.28 Geysers, Fumaroles, Hot Springs and Solfataras

16.17.4 Geysers, Fumaroles, Hot Springs and Solfataras

Geysers, fumaroles, hot springs and Solfataras are often found in areas of young volcanic activity (Fig. 16.28). The surface water percolates downward through the rocks below to high-temperature areas bordering a magma reservoir. There, the water is heated, becomes less dense, and rises back to the surface along fissures and cracks.

16.17.5 Geysers

A geyser (Fig. 16.28a) requires three critical elements in order to form: a water supply, a heat source, and the proper kind of underground water circulation system. If only two of these are present, different features form. For example, if plenty of water is available but not much heat, a hot pool is formed. If plenty of water and heat are there, but not the proper kind of plumbing system, then a boiling hot spring forms. Hot springs are commonly found, but geysers are quite rare. Interestingly, the water in geysers and hot springs is often several hundred years old. The Old Faithful Geyser in the Yellowstone National Park (Wyoming, USA) is a classic example of a geyser. It erupts on an average of about once in every 65 min. Although, erupting geysers provide spectacular displays of underground energy suddenly unleashed, but their mechanism is not properly understood, so far. It is possibly controlled by subcrustal heat source from the mantle plume.

16.17.6 Hot Springs (Thermal Springs)

They occur in areas where hot water is naturally heated underground and then surfaces up (Fig. 16.28b). The cooler groundwater moves downward and is heated by a body of magma or hot rock. This hot water is usually seen along a fault zone that eventually finds its way back to the surface. The temperature and the rate of discharge of a hot spring depends on water circulation, the amount of heat supplied at depth and the extent of dilution of the heated water by cool ground water near the surface. Hot springs reach temperatures between 37.8 and 65.6 °C, whereas, the ground temperature barely averages 15 °C.

16.17.7 Fumaroles and Solfataras

Fumaroles are vents through which volcanic gases are released at the surface (Fig. 16.28c). Since most magmatic gas is water vapor, fumaroles emit mixtures of steam and other gases. Hydrogen sulfide (H_2S) is a typical gas issuing from fumaroles. Generally, water in fumaroles comes mainly from recycled atmospheric precipitation and very rarely from the rising magma. Some fumaroles are as hot as 900 °C (intense fumarolic activity was recorded between August–October 1992 at the Kudriavy volcano, Kuril Island, Far-Eastern Russia, where maximum fumarole temperature measured was 940 °C; Korzhinsky et al. 2002). Relatively cool fumaroles have temperatures between 100 and 300 °C. They often give off hydrogen sulfide that reacts with oxygen to produce sulphur deposits called by their Italian name—Solfataras (Fig. 16.28d). Iceland has the most number of Solfataras.

16.18 Summary

A volcano is an opening in the earth's surface from which molten lava, rock fragments, and other gases erupt. When hot material (Lava) comes out from a volcano, it is called a volcanic eruption. Often, Magma starts building pressure inside the earth's surface, and moves up to the surface. This pressure buckles the surface of the earth and eventually forms a mountain. If this liquid comes out to the surface of the earth, then, a Volcano is born. Around 80 % of our earth's surface is a product of volcanoes, as also the surface of the sea and mountains that are formed by countless eruptions. The word volcano comes from the little island of Vulcano in the Mediterranean Sea off the coast of Sicily.

The volcanic material is emitted on earth's surface through an opening called a Volcanic vent. But the main conduit through which magma moves upward to the surface is called a Central vent. Lateral vents (also called side vents) are found on the sides of some volcanoes where lava is extruded. These eruptions generate cone shaped accumulations of volcanic material called Parasitic cones or Secondary cones.

Magma is a naturally occurring liquid. Upon cooling, magma precipitates crystals of various minerals to form an igneous rock, leading to a large range of different compositions of magma associated with different types of volcanoes under different settings.

The ability of Lava to flow depends on the viscosity of the parent magma. Viscosity is a function of temperature, silica content, and incorporated gases. However, the primary compositional difference in magma is in the amount of silica that makes-up the melt. Based on this, three distinct types of magma are commonly found; silica content varies between 48 and 77 %. The greater the amount of silica in the melt, the more bonding there is and the more viscous the magma is! Another important factor that differentiates magmas is in the amount of dissolved gas they contain. Magma can be divided into three types based primarily on their composition—Basaltic, Andesitic, and Rhyolytic. However, there is another magma (Dacite) that occurs, but, in very small quantities—the Dacitic magma; it contains moderately high amounts of gas.

Lava flows cover 70 % of the earth's surface and are the most common volcanic formation on the earth. There are two types of flows, reflecting their internal movement in relation to their congealing crust. Those flows with discontinuous surfaces include Aa and Block lavas and those with continuous surfaces include Sheet and Pahoehoe lavas. The terms "Pahoehoe" and "Aa" are of Hawaiian origin.

If the basaltic lava flowing down a moderate slope is confined into a channel and if that channel cools and becomes roofed over, the resultant lava tube may continue to transmit molten lava and extend downhill for miles. These drained, cooled tubes are called Lava tubes or Lava caves.

Pressure ridges form as the semi-rigid lava surface is squeezed with its interior remaining more fluid and flexible. Cracks often adorn the peaks of these ridge formations; both lava and steam issue secondarily from them. Explosive eruptions blast large quantities of volcanic rock out of the volcano. Some of this rock is excavated from the volcano itself, the rest consists of fragments of lava that has cooled in the air. This material accumulates around the volcano as pyroclastic (from the Greek *pyro*, "fire," and *clast*, "broken") deposits (also called Tephra). Within the pyroclastic deposits, include large fragments—Blocks, Bombs (Spatter bombs), Cinder and Lapilli. Fine-grained tephra are called Ash and Dust; Ash is silt to sand sized and Dust measures $<1/8$ mm.

Nuées Ardentes is used to describe a fast moving gaseous cloud of hot ashes and other material thrown out from an erupting volcano that are often incandescent.

Pillow lavas are basaltic lava flows that form by the discharge of lavas into rivers, lakes, ponds or under glaciers, and into oceans. Lobate flows resemble on-land Pahoehoe flows, but are a bit more inflated-looking.

Of the recognizable 10,000 volcanoes, around 539 (53 of these are in USA, Alaska) are active volcanoes and $\sim$62 % of them are strung like beads along, or near the boundaries between shifting plates (the plate-boundary volcanoes; mostly at Convergent and Divergent plate margins), and 90 % of them encircle the Pacific Ocean as a Ring of Fire. However, volcanic eruptions can also occur in the middle of plates and are called Intra-plate volcanoes or Hotspot. On an average, there are about 5–10 volcanoes erupting every month ($\sim$50/year).

On the basis of the geographical region, earth's active volcanoes can be grouped into 12 regions.

Volcano classification is based on three factors - Continuity (Active, Dormant, and Extinct volcanoes), Nature (Quite, Intermediate, and Violent) and Mode of eruption Strombolian, Vulcanian, Vesuvian, Hawaiian, Phreatic, Peléan (Nuées Ardentes), Plinian, Phreatomagmatic eruptions, Hydrovolcanic eruptions (the Surtseyan eruption), Submarine eruptions, Subglacial eruptions, and Phreatic eruptions (Steam-blast eruptions).

Volcanic Explosivity Index is a measure of how big an eruption is? It is a magnitude scale as there is not any singular feature that determines the "bigness" of a volcano.

Volcanic landscapes contain extremely diverse landforms varying markedly in size, shape, composition, and eruptive history.

The form of a volcano is determined by the ingredients of the erupting magma. Their shapes are determined by the exclusivity of the eruptions and to the amount of water in the magma. Although every volcano has a unique eruptive history, but most can be grouped based largely on their eruptive patterns and their general form. Major ones are—Fissure volcano (Plateau Basalts or Flood Basalts), Shield Volcanoes, Cinder volcano (Cinder, Tephra or Pyroclastic Cones), Stratovolcanoes (Composite volcanoes), and Lava Dome.

The Calderas are classified into three types based on changes in their form and origin—Crater Lake type calderas, Resurgent calderas (also called Yellowstone Type) and Basaltic shield volcano calderas (also called Hawaiian Type).

Other Types of landforms include—Volcanic Necks and Plugs (or lava neck), Maars (Tuff cones), Non-volcanic Craters (Cryptovolcanic), Geysers, Fumaroles, Hot Springs, and Solfataras.

References

Babb JL, Kauahikaua JP, Tilling RI (2011) The story of the Hawaiian Volcano Observatory—A remarkable first 100 years of tracking eruptions and Earthquakes: U.S. Geological Survey General Information Product 135, p 60. Available at http://pubs.usgs.gov/ gip/135/

Bailey JC (1993) Geochemical history of sediments in the northwestern Pacific Ocean. Geochem. J. 27, 71–91.

Brantley SR (1994) Volcanoes in the United States. US. Geological Survey General Interest Publication Available at http://vulcan.wr.usgs.gov/Volcanoes/Publications/Brantley VolcanoesUS/windows_into_the_past.html.

Breining G (2007) "The Deadliest Volcanoes". Super Volcano: The Ticking Time Bomb Beneath Yellowstone National Park. Voyageur Press. p 256

Bullard F (1976) Volcanoes of the Earth. Austin, Texas, University of Texas Press, p 579

Bullard FM (1984) Volcanoes of the earth, 2nd edn. University of Texas Press, Austin

Cas RAF, Wright JV (1988), Volcanic Successions—Modern and Ancient, Unwin Hyman, London, p 528

Camp V (2010), How volcanoes work. Available at http//www.geology.sdsu.edu/ how_volcanoes_work/

Christiansen RL (1984) Yellowstone magmatic evolution: its bearing on understanding large-volume explosive volcanism In: Boyd FR et al. (eds), Explosive volcanism: inception, evolution and hazards. National Research Council Studies in Geophysics, National Academy Press, Washington DC, 84–95

Claybourne A (2007) Volcanoes. Kingfisher Publications, Boston, Massachusetts, USA

Courtillot V (1999) Evolutionary catastrophes: the science of mass extinction. Cambridge University Press, New York, p 173

Decker R, Decker B (1981) The eruption of Mount St. Helens, Scientific American 244(3), 68–80.

Decker R, Decker B (2005) Volcanoes, 4th Ed., W. H. Freeman and Company, New York, p 285

Diaz JA, Giese CF, Gentry WR (2002) Mass spectrometry for in-situ volcanic gas monitoring. Trends in Analytical Chemistry 21(8), 498-514.

Etienee R (1992) Pompeii, The Day a City Died. Harry N. Abrams, New York

Fischer RV, Schminke H-U (1984) Pyroclastic rocks. Springer-Verlag, Berlin

Fisher RV, Heiken G, Hulen JB (1997) Volcanoes: Crucibles of Change. Princeton University Press, Princeton, NJ

Fisher RV, Schmincke H-U (1984) Pyroclastic Rocks. Springer-Verlag, Berlin, pp 472

Francis P (1993) Volcanoes: a planetary perspective. Oxford University Press, New York

Ganse RA, Nelson JB (1982) Catalog of significant earthquakes 2000 B.C. to 1979, including quantitative casualties and damage. Bulletin of the Seismological Society of America 72, 873–877

Harris N, Armstrong L (eds.) (2003) Volcanoes. Greenhaven Press, San Diego, California, USA, p. 122

Kraft M (1993) Volcanoes: Fire from the Earth. Harry N. Abrams, New York

Mason BG, Pyle DM, Oppenheimer C (2004) The size and frequency of the largest explosive eruptions on Earth. Bulletin of Volcanology 66(8), 735–748.

Myers B, Brantley SR, Stauffer P, Hendley-II JW (2008) Reducing The Risk From Volcano Hazards, U.S. Geological Survey Factsheet available at http://pubs.usgs.gov/fs/fs002-97/.

Newhall CG, Self S (1982) The volcanic explosivity index (VEI): An estimate of explosive magnitude for historical volcanism. Journal of Geophysical Research 87 (C2), 1231–1238

Perry B California State University, Long Beach, USA. Available at http://geology.campus.ad.csulb.edu

Peterson DW, Tilling RI (1980) Transition of basaltic lava from pahoehoe to aa, Kilauea Volcano, Hawaii, field observations and key factors, Journal of Volcanology and Geothermal Research 7, 271–293

Schmidt RG, Shar HR (1971) Atlas of volcanic phenomena: U.S. Geological Survey

Schmincke H-U (2004) Volcanism. Springer-Verlag, Heidelberg. pp 324 ISBN 3-540-43650-2.

Simkin T, Siebert L (1994) Volcanoes of the World, 2nd edn., Geoscience Press, Tucson, AZ, p 349

Simkin T, Tilling Rl, Taggart JN, Jones WJ, Spall H (1989) This dynamic planet. World Map of volcanoes, earthquakes, and plate tectonics. U.S. Geological Survey, Reston, Virginia, prepared in cooperation with the Smithsonian Institution, Washington, D.C. (scale 1,30,000,000 at equator)

Smith RL, Braille LW (1994) The Yellowstone hotspot. Journal of Volcanology and Geothermal Research 61, 121–127.

Smith RB Siegel LJ (2000) Windows into the Earth, The Geologic Story of Yellowstone and Grand Teton National Park, Oxford University Press, New York, p 256

Tilling RI (1982) Volcanoes: U.S. Geological Survey Publication

Tilling Rl, Heliker C, Wright TL (1989) Eruptions of Hawaiian Volcanoes. Past, Present, and Future. Reston, Virginia, U.S. Geological Survey, p 54

Tilling Rl (1991) Monitoring Active Volcanoes, Reston, Virginia, U.S. Geological Survey, p.13 (Revised edition)

USGS Glossary: Classification and Flow Characteristics of Volcanic Rocks Available at http://volcanoes.usgs.gov/about/pglossary/VolRocks.php. Retrieved on 1/03/2013.

White JDL, McClintock MK (2001) Immense vent complex marks flood-basalt basalt eruption in a wet, failed rift: Coombs Hills. Antarctica Geol 29(10):935–938

White JDL, Ross P-S (2011) Maar-diatreme volcanoes: a review. J Volcanol Geoth Res 201(1–4):1–29 doi: 10.1016/j.jvolgeores.2011.01.010

Wood CA, Kienle J (1990) Volcanoes of North America: United States and Canada: Cambridge University Press, p 263–267.

Wright JV, Smith AL, Self S (1980) A Working Terminology of Pyroclastic Deposits, Journal Volcanology and Geothermal Research 8, p 315-336.

Wright TL, Pierson TC (1992) Living with volcanoes. U.S. Geological Survey Circular 1073

Wright TL, Takahashi TJ, Griggs JD (1992) Hawaii Volcano Watch: a Pictorial History, 1779-1991. Honolulu, University of Hawaii Press and Hawaii Natural History Association, p 162.

About the Author

Dr. Sreepat Jain has 12 years of teaching and over two decades of research experience. Dr. Jain has two doctorates—one from India (on Middle Jurassic Ammonites) and the other from United States (on Neogene Benthic Foraminifers). He completed his second Ph.D. from the Department of Earth Sciences, Florida International University, USA in 2006 where he was also awarded the "TA Excellence in Teaching" for meritorious teaching. During 2007–2008, he joined Smithsonian Institution, Washington D.C., USA for his Postdoctoral Fellowship. In 2013, he was awarded the "Prof. S. K. Singh Memorial Gold Medal" of The Paleontological Society of India for the best paper in PSI Journal for 2012. His areas of research include Micropaleontology—Jurassic Foraminifers and Nannofossils; Macropaleontology—Jurassic Ammonites; Paleoecology—Trace Fossils and Paleoenvironment. He has published several articles in International peer-reviewed journals and conference proceedings. Dr. Jain has also authored a book with Lap Lambert Academic Publishing Gmbh & Co. KG, Germany (Changes in Late Neogene Caribbean Benthic Foraminifers: Paleoproductivity, Diversity and Test Size). He is also a reviewer for several paleontology journals.

S. Jain, *Fundamentals of Physical Geology*, Springer Geology
DOI: 10.1007/978-81-322-1539-4, © Springer India 2014

Glossary

This glossary provides a brief explanation of the major terms used in this book. Whereever possible (almost 60 %), figure numbers are mentioned as an illustrative supplement to individual terms.

A

A horizona (top solid) It is the second layer of the soil that contains organic material and leached minerals and is the most fertile part of the soil profile. The topmost layer is called the O layer and contains plant litter (in various stages of decomposition) and humus (Fig. 7.2)

AA lava flow A blocky and fragmented form of basaltic lava that occurs in flows (usually ~3–20 m thick) and whose surface is marked with angular and jagged blocks. It has a characteristic rough, spiny, and "cindery" appearance, with sometimes being "rubbly" in nature. AA is a Hawaiian word and is pronounced as "ah-ah." It was introduced as a geological term by American geophysicists, Clarence Edward Dutton (1841–1912) in 1883

Ablation (glacial) The annual loss of ice and snow from a glacier through melting, evaporation, iceberg calving, deflation, and wind erosion. Ablation reduces a glacier. Hence, glaciers change, accumulating or losing area annually (Fig. 11.5)

Abrasion A common mechanical process in deserts that involves the wearing away of a rock by friction, rubbing, scraping, or grinding

Absolute age Also called the Numerical age. Absolute age is the age measured in years of a particular geologic event obtained through radiometric dating techniques. The first radiometric age dates were calculated in 1907 by a study of how long it takes for Uranium (^{238}U) to decay into Lead (^{206}Pb). The half-life of ^{238}U is 4.468 billion years. In contrast, Relative age, involves only the placement of chronologic order of events (Figs. 6.3 and 6.12)

Abyssal fan Also called deep sea fans. They are fan-shaped deposits formed by turbidity currents on the deep-sea floor at the base of many submarine canyons. A good example of an abyssal fan is the Bengal Fan (also known as the Ganges

S. Jain, *Fundamentals of Physical Geology*, Springer Geology
DOI: 10.1007/978-81-322-1539-4, © Springer India 2014

Fan) of the East coast of India. It is the largest submarine fan on earth and is about 3,000 km long, 1,000 km wide with a maximum thickness of 16.5 km

Abyssal hills Ocean floor consisting of submarine hills that rise as much as 1,000 m above the surrounding floor (at water depths of 3,000–6,000 m). These are found seaward of most abyssal plains and occur in basins isolated from continents by trench, ridges, or rises. Abyssal hills underlie most of the ocean floor, locally buried by accumulations of marine sediments. In the Atlantic, long abyssal-hill provinces parallel both flanks of the Mid-Atlantic Ridge (Fig. 12.1)

Abyssal plains Flat areas of the ocean floor where abyssal hills are completely covered with sediment. Most abyssal plains usually occur at the base of the continental rise with a slope <11° (Fig. 12.1)

Accreted terrane A Terrane that did not form at its present site on a continent. It is a fragment of crustal material that formed on, or was broken off from, one tectonic plate and accreted or "sutured" to crust lying on another plate. Hence, this crustal block maintains its own distinct geologic history (Fig. 14.18)

Accretionary wedge accretionary prism A wedge-shaped body of faulted and folded material scraped off from the subducting oceanic crust and added (accreted) to an island arc or continental margin at a subduction zone in a subaqueous thrust zone. This is often also called the accreted Mélange. They form on the inner wall of ocean trench, but not all trenches have them. They are also often called accretionary prisms (Fig. 14.11)

Aftershock A smaller earthquake that follows a larger one. After a major earthquake, several aftershocks occur over a period of days or months

Alluvial fan A fan-shaped or cone-shaped deposit of terrigenous sediments (largely sand and gravel) is formed by a fast flowing stream when it emerges from an upland or a mountain range into a broad valley or plain (i.e., in a relatively flat surface) or undergoes an abrupt widening as it leaves a mountain front for an open valley. Alluvial fans are common in arid and semiarid regions (Fig. 8.28)

Amphibole In the nineteenth century, Rene Haüy (a French mineralogist) named "amphibole" meaning "ambiguous" a group of minerals that show the same appearance but very different chemical composition. Amphiboles are important mineral constituents (group of rock-forming minerals of mafic silicates) of the metamorphic and igneous rocks and some ore deposits. They were found also in meteorites. Amphibole minerals generally contain iron, magnesium, calcium, and aluminum in varying amounts along with silicon, oxygen, and water. All members of the Amphibole mineral group are double-chain silicates

Amphibolite A mostly metamorphic rock often foliated, sometimes banded, medium to coarse grained, dark green to black rock, consisting mainly of amphibole minerals (often hornblende) and plagioclase feldspar without quartz (or with very low quartz content). It is a product of regional metamorphism of either sedimentary or magmamagmatic protoliths

Amphibolite facies A rock assemblage made mostly of amphibole minerals (such as hornblende and plagioclase feldspar) formed under conditions of moderate to high temperatures (~500 °C) and pressures. Amphibole, diopside, epidote, plagioclase, certain types of garnet, and wollastonite are minerals typically found in amphibolite facies rocks. Such rocks are widely distributed in the Precambrian gneisses and were probably formed in the deeper parts of folded mountain belts

Andesite A fine-grained volcanic igneous rock composed mainly of plagioclase feldspar containing 25–40 % pyroxene, amphibole, or biotite with no quartz or potassium feldspar. Andesite is intermediate in composition between a Rhyolite and Basalt and is an extrusive equivalent of Diorite. It is abundant in mountains bordering the Pacific Ocean, such as the Andes of South America, from which the name was derived. Andesite rocks are found near the subduction zones of ocean tectonic plates, along continental margins (Fig. 16.5)

Angle of repose It is the maximum slope at which grains are stable without sliding downslope

Angular unconformity An unconformity in which the older strata dips at a different angle (mostly steeper) than the younger strata, i.e., the bedding planes of the rocks above and below are not parallel (Fig. 6.5)

Anion A negatively charged atom or group of atoms produced by the gain of electrons

Anorthosite A coarse-grained intrusive igneous rock composed mainly of calcium-rich plagioclase feldspar, almost exclusively of Precambrian in age. When the earth was first forming, Anorthosite was probably produced the same way as it was on the Moon. They are also the oldest rocks on the Moon

Antecedent stream A stream that was there before the present topography was created. Thus, it maintains its original course despite changes in both the structure of the underlying rock(s) and in topography

Anthracite A very hard, and shiny black coal with a slight golden shine. It has the highest carbon content, low volatile content (<10 % on an air dry basis), high fixed carbon content (>80 % on an air dry basis) high heating value, and very low density. It burns dust-free and is smokeless. Most Anthracite deposits were formed in the Carboniferous Period, 340–280 million years ago. It is part of the three main ranks of coal–Lignite, Bituminous coal, and Anthracite. It burns with the cleanest flame and hence, is a preferred coal for use as a domestic fuel

Aphanitic texture A texture in which individual crystals are too small to be identified without the aid of a microscope. They are usually <0.5 mm in size. The Aphanitic texture results from rapid cooling in a volcanic or hypabyssal (shallow subsurface) environment. In hand specimens, such rocks appear dense and structureless

Aquiclude A stratum with low permeability that acts as a barrier to the flow of groundwater. It is also called a "confining layer." An Aquiclude does not yield useful quantities of water (Figs. 9.8, 9.9)

Aquifer A permeable formation, stratum or zone below earth's surface that stores and transmits groundwater (an underground reservoir). Here, the water is much cleaner with almost no bacteria. Around 97 % of the planet's fresh water is stored in aquifers with (Figs. 9.8, 9.9)

Arête A French word for fishbone. They are formed when two glaciers work on opposite sides of the same wall, leaving a long narrow ridge (Fig. 11.12)

Artesian flow Flow in a confined aquifer, in which the groundwater is at a greater pressure than in an unconfined aquifer at similar depths, thereby causing water in a well that penetrates a confined aquifer (an artesian well) to rise above the level of the aquiclude. The word artesian comes from the town of Artois in France, the old Roman city of Artesium, where the best known flowing artesian wells were drilled in the Middle Ages (Fig. 9.8)

Ash (volcanic) Fine particles (the size of dust; <4 mm) of volcanic rock and glass blown into the atmosphere by a volcanic eruption (Fig. 16.9)

Ash flow A turbulent blend of unsorted and fine grained pyroclastic material that is mixed with high-temperature gases and ejected explosively from a fissure or crater

Ash-flow tuff A rock composed of volcanic ash and dust, formed by the deposition and consolidation of ash flows

Asteroid A small, rocky planetary body that orbits the Sun. Majority of the asteroids are located between the orbits of Mars and Jupiter, having diameters less that 1000 km. They number in tens of thousands. Ceres in 1801 was the first asteroid discovered (Fig. 3.2)

Asthenosphere A weak layer below the lithosphere. The asthenosphere extends from 100 km depth to 660 km beneath the earth's surface. This layer is marked by low seismic wave velocities. Movement within the Asthenosphere occurs by plastic deformation (Fig. 5.4)

Astronomical unit (AU) A distance unit based on the average distance of the Earth from the Sun. 1 AU = 149,597,871 km

Atoll A continuous or broken circle of coral reefs and low coral islands surrounding a central lagoon. An atoll is formed from the upward growth of corals combined with the sinking of the island inside the reef (Fig. 12.13)

Atomic mass number The total number of neutrons and protons in an atom (Fig. 6.11)

Atomic number The total number of protons in an atom (Fig. 6.11)

Atomic weight The sum of the weight of the subatomic particles in an average atom of an element, given in atomic mass units (Fig. 6.11)

B

B horizon Also called the subsoil. It is characterized by the enrichment of organic matter and clay. It is the zone of accumulation of minerals and clay. It underlies the A horizon of the soil profile, formed by the accumulations leached from the A horizon (Fig. 7.2)

Back-arc basin Extensional basins formed behind subduction zones by rifting volcanic arcs and accreting new volcanic seafloor. These basins are associated with tensional forces caused by asymmetric seafloor spreading and oceanic trench rollback at some convergent plate boundaries (Fig. 14.11)

Back-arc spreading A type of seafloor spreading that moves an island arc away from a continent, or tears an island arc in to two, or splits the edge of a continent, in each case forming new sea floor. Spreading rates vary from very slow spreading, a few cm per year (as in the Mariana Trough in the western Pacific), to very fast, 15 cm/year (as for the Lau Basin; close to the Tonga Islands in the southwest Pacific) (Fig. 14.11)

Backshore The upper, mostly dry, zone of the shore, extending landward from the upper limit of wave wash at high tide to the upper limit of shore-zone processes. Or in other words, it is the upper part of the active beach above the normal reach of the tides (high water), but affected by large waves occurring during a high (Fig. 12.6)

Badlands A topography nearly devoid of vegetation and heavily dissected by an intricate patterns of stream erosion over unconsolidated or poorly cemented clays, silts, or sands. A dry terrain where softer sedimentary rocks and clay-rich soils have been extensively eroded by wind and water. Badlands form in semi-arid or arid regions with infrequent but intense rain-showers, sparse vegetation, and soft sediments; excellent recipe for substantial erosion

Bajada The surface of a system of coalesced alluvial fans. A bajada commonly occurs in semiarid and desert regions. It is a broad, gently sloping, depositional surface formed at the base of a mountain ranges (Fig. 8.28)

Bar An offshore, submerged (at least during the time of a high tide), elongate offshore ridge of sand, gravel or other unconsolidated material built on the seafloor by waves and currents especially at the mouth of a river or estuary, or lying parallel to and a short distance away from the beach. It is deposited due to decrease in stream velocity (Fig. 12.6)

Barchan dune Crescent-shaped eolian sand dune that moves across a flat surface with its convex face upwind (angle $<15°$) and its concave slip face downwind (angle of repose of $\sim 32°$). They form in deserts where sand is scarce. Sizes of individual dunes range from a meter to hundreds of meters, from horn to horn.

They are called by several other names such as Crescent dunes, Sand hills, Barkans, Demkhas, Giant crescent, Bourrelets, Draas (for the large forms) and, Megadunes (Fig. 13.14)

Barrier island A long, narrow island of sand or gravel parallel to the shore, built by wave action and separated from the mainland by open water in the form of estuaries, bays, or lagoons. Barrier islands occur in chains, anything from a few islands to more than a dozen. There are a total of 2,149 barrier islands worldwide measuring 20,783 km in length. About 74 % of the islands are found in the northern hemisphere. They are found along all continents (except Antarctica) and in all oceans and make up ~10 % of earth's continental shorelines (Fig. 12.12)

Barrier reef An elongate coral reef that trends parallel to the shore of an island or a continent, separated from it by a lagoon (deep waters where corals cannot grow). Australia's Great Barrier Reef is the largest living coral reef on the planet, stretching ~2,600 kms. It is home to some 400 coral species

Basal sliding Movement in which the entire glacier slides along as a single body on its base over the underlying rock (Fig. 11.7)

Basalt A dark, fine-grained, extrusive (volcanic) igneous rock with low silica content (40–50 %), but rich in iron, magnesium and calcium containing plagioclase feldspar (>50 %) and pyroxene. Olivine may or may not be present. It is the extrusive equivalent of Gabbro and generally occurs in lava flows or as dikes. Most ocean floor is made of basalt and is also the most abundant volcanic rock in the earth's crust (Figs. 16.5, 16.6)

Basement Oldest rocks recognized in a given area, usually Precambrian or Paleozoic in age. A complex of metamorphic and igneous rocks that underlies all sedimentary formations

Basement complex A series of undifferentiated group of igneous and metamorphic rocks lying beneath the oldest stratified rocks of a region

Batholith Also called granite domes. They are complex intrusive bodies composed of plutonic igneous rocks, mostly of felsic or intermediate rock-types, such as granite, quartz monzonite, or diorite. Batholiths are defined as having a surface area of more than 100 sq km. Smaller ones are called Stocks. Batholiths form when magma solidifies at depth. Sometimes a Batholith is also derived from the country rock through metamorphism at very high temperature and pressure (Fig. 7.3)

Bathymetry The measurement of ocean depths and mapping of the topography of the ocean floor

Bay A wide, curving inlet between two headlands (Fig. 12.10)

Beach A deposit of wave-washed sediment along a coast between the landward limit of wave action and the outermost breakers (Fig. 12.10)

Bed A layer of sediment generally 1 cm or more in thickness. One of several parallel layers of rock that are arranged on top of the other

Bed load The sediment (from sand to boulders) that a stream moves along the bottom of its channel by sliding, rolling, and saltating (skipping) along the bed of a stream (Fig. 8.20)

Bedding A characteristic of sedimentary rocks in which parallel planar surfaces separate layers of different grain sizes or compositions deposited at different times

Bedrock The solid rock underlying unconsolidated surface materials, such as soil or regolith (Fig. 7.2)

Benioff zone It is also called the Wadati-Banioff Zone; the Japanese seismologist Kiyoo Wadati and the American seismologist Hugo Benioff. It is a zone of earthquakes that dips away from a deep-sea trench and slopes (between 33–60°) beneath the adjacent continent or island arc a seismically active zone inclined from a deep sea trench. The earthquakes occurring can be as deep as about 700 km (Fig. 15.12)

B horizon It is also called the Zone of accumulation. It is a soil layer characterized by the accumulation of material leached downward from the A horizon above (Fig. 7.2)

Biochemical sediment When shelled organisms (mostly microorganisms) die, their shells fall to the sea floor forming biochemical sediments (Foraminifera and Coccoliths secrete $CaCO_3$: Carbonate, and Diatoms and Radiolaria produce SiO_2: Siliceous). The primary biochemical rock is the carbonate Limestone but if the shells are not finely grounded, the material is a Bioclast

Biosphere Parts of the atmosphere, hydrosphere, and lithosphere occupied by living organisms (Fig. 4.1)

Biotite Rock-forming mineral of the mica family. It is an important mafic silicate with silicon-oxygen tetrahedral arranged in sheets. It is a black or dark brown silicate rich in iron, magnesium, potassium, aluminum, and silica. Hence, it is also called Black Mica. Like other micas, it forms flat book-like crystals that peel apart into individual sheets on cleavage planes

Bird-foot delta A delta with distributaries extending seaward. In map view, it resembles the claws of a bird like the Mississippi Delta (USA)(Fig. 8.24)

Blowout 1. A parabola-shaped eolian sand dune that has its convex slip face oriented downwind; commonly formed along shorelines (same as a parabolic dune). 2. A shallow circular or elliptical depression in sand or dry soil formed by wind erosion

Boulder A rock fragment with a diameter of more than 256 mm (same size of a volleyball); one size larger than a cobble

Bowen reaction series A simple schematic description of the order in which different minerals crystallize during the cooling and progressive crystallization of magma (Fig. 7.12)

Braided stream A stream so choked with sediments that it divides and recombines numerous times, forming many small and meandering channels, separated by bars or islands. Hence, braided streams form when more sediment is available than that can be removed by the discharge of the stream (Fig. 8.13)

Breaker A wave that has become so steep that the crest of the wave topples forward, moving faster than the main body of the wave. In other words, a collapsing water wave (Fig. 12.6)

Breccia Sediment consisting of angular fragments interspersed in a matrix of finer particles such as a sedimentary breccias, volcanic breccias, fault breccias, or impact breccias

Butte An isolated hill, usually capped with a resistant layer of rock and bordered by talus. It is an erosional remnant of an old extensive slope (Fig. 13.8)

C

C-horizon Underlying the B horizon, it is the lowest zone of soil consisting of partly decomposed bedrock. It grades downward into fresh, un-weathered bedrock (Fig. 7.2)

Caldera A large basin-shaped (generally circular) fault-bounded volcanic depression or basin associated with a volcanic vent. It is formed after an eruption if the volcano collapses through the roof of the emptied magma chamber caused by the withdrawal of magma from below a volcano or volcanoes. Its diameter is many times greater than that of the included vents. Calderas are believed to result from subsidence or collapse and may or may not be related to explosive eruptions (Fig. 16.26)

Calving The breaking off of large blocks of ice from a glacier that terminates in a body of water

Carbon-14 A radioactive isotope of Carbon with a half-life of 5,730 years. Carbon is a nonmetallic element found native (in the form of Diamond and Graphite) or as a component of organic compounds (Fig. 6.15)

Carbonate compensation depth The ocean depth below which the solution rate of Calcite (Calcium Carbonate; $CaCO_3$) is so great that no carbonate organisms or sediments are preserved on the seafloor. It is also called the Calcite Compensation Depth (CCD) (Fig. 12.2)

Carbonates They largely consist of two types of rocks. Limestones: composed mostly of Calcite ($CaCO_3$) or high Mg Calcite [$(Ca, Mg)CO_3$], and Dolostones: composed mostly of Dolomite [$Ca\ Mg(CO_3)_2$]. Carbonates include the radical

"CO_3" which reacts with acids to produce CO_2. Carbonate rocks make up 10–15 % of the total sedimentary rocks

Cement Minerals precipitated from groundwater in pore spaces of a sedimentary rock and that which binds the rock particles together

Cementation A lithification process in which minerals are precipitated in pore spaces of sediments, often binding the grains. It is the chemical, physical, and biological change where sediments are converted to rocks through the deposition or precipitation of minerals in spaces between grains

Central ventvent The largest vent of a volcano, situated at the center of its cone (Fig. 16.4)

Chalk A variety of Limestone composed of shells of microscopic oceanic organisms

Channel The trough through which water flows in a stream valley. The term is sometimes reserved for the deepest part of the streambed, in which the main current flows

Chemical weathering The process that changes the chemical makeup of a rock or mineral at or near the earth's surface, thereby altering the internal structure of minerals by removing and/or adding elements. It acts on rocks exposed to water and atmosphere and dissolves the minerals and/or changes them into more stable forms

Cinder (volcanic) A bubbly (vesicular) volcanic rock fragment that forms when molten, gas-filled lava is thrown into the air, then solidifies as it falls. It is a fragment of the volcanic ejecta and ranges in size from 0.5 to 2.5 cm in diameter

Cinder cone A steep, conical hill built up around a volcanic vent. It is composed of coarse pyroclastic (loose volcanics) rock fragments expelled from the central vent by escaping gases

Circum-Pacific belt A major belt around the edge of the Pacific Ocean on which most composite volcanoes are located and where many earthquakes occur (Fig. 16.10)

Cirque An amphitheater-shaped depression at the head of a glacial valley, excavated mainly by ice plucking and frost wedging. The upper edges have the steepest slopes, approaching nearly vertical; the base may be flat and commonly occupied by a small lake or pond after deglaciation (Figs. 11.8, 11.12)

Clastic sediment (rock) A sediment or sedimentary rock formed from particles (clasts such as mud, sand, and gravel or even pieces of preexisting rock) derived from the erosion of preexisting rocks and mechanically transported. Its formation involves mechanical breakdown of rocks

Clastic texture The texture of sedimentary rocks consisting of fragments of minerals, rocks, and organic skeletal remains (Fig. 16.5)

Clay Any hydrous aluminosilicate mineral with sheet-like crystal structure, formed by the weathering and hydration of other silicates. It is also any mineral fragment smaller than 0.0039 mm or with a diameter less than 1/256 mm. Clay is smaller than silt

Clay minerals A group of hydrous silicates formed by weathering of minerals such as Feldspar, Pyroxene, or Amphibole

Cleavage The tendency of a mineral to split along planes as determined by their crystal structure

Cleavage (crystal) The tendency of a crystal to break along certain preferred planes in the crystal lattice

Coal Metamorphic product of stratified terrestrial plant remains. It contains more than 50 percent carbon compounds and burns readily; it is a common fuel mineral

Coast The strip of land adjacent to an ocean or sea and extending from low tide landward

Cobble A rock fragment with a diameter between 6.4 cm (size of a tennis ball) and 25.67 cm (size of a volleyball). Cobbles are larger than pebbles but smaller than boulders

Columnar jointing A system of fractures that splits a rock body into long prisms, or columns, characteristic of lava flows and shallow intrusive igneous flows

Comet A small icy object in orbit around the Sun. The orbits of many comets are elliptical and when they near the Sun, the ice sublimes to make a fuzzy head and long tail of gas and dust (as noted for Hailey's comet) (Fig. 3.4)

Compaction Burial-induced decrease in volume and porosity of a sediment

Composite volcano (stratovolcano) A volcanic cone containing layers of both lava flows and pyroclastic rocks that are built by the extrusion of ash, lava, and shallow intrusions (Fig. 16.21)

Condensation The process by which a vapor becomes a liquid or a solid

Cone of depression A conical depression of the water table surrounding a well after heavy pumping. It is shaped like an inverted cone

Confined aquifer (artesian aquifer) An aquifer overlain by relatively impermeable strata (aquiclude), thereby causing the water to be contained under pressure (Fig. 9.8)

Conglomerate A coarse-grained sedimentary rock whose major proportion is composed of rounded pebbles, cobbles, and boulders; a lithified equivalent of gravel. All grains are coarser than 2 mm

Continent A large landmass composed mostly of granitic rock. Continents rise abruptly above the deep-ocean floor and include marginal areas submerged beneath sea level

Continental accretion Growth of continents by incorporation of deformed sediments, arc magmas, and accreted terranes along their margins (Fig. 14.18)

Continental crust The type of crust underlies continents, including the continental shelves. It is commonly about 35–70 km thick with an average density of 2.7 g/cm^3 (Fig. 5.1)

Continental drift The theory that the continents move (horizontal displacement or rotation) with respect to each other

Continental glacier A continuous thick ice sheet covering large parts of a continent, covering more than 50,000 km^2 and moving independently of minor topographic features such as in Greenland and Antarctica (Fig. 11.10)

Continental margin The zone of transition from the continent to the adjacent ocean basin. In general, the portion of the ocean floor extending from the shoreline to the landward edge of the abyssal plain and including the continental shelf, slope, and rise

Continental rise A broad and gently sloping ramp that rises from an abyssal plain to the continental slope (Fig. 12.1)

Continental shelf A gently sloping submerged margin of a continent, extending to a depth of about 200 m (the first prominent break in the slope) to the edge of the continental slope (Fig. 12.1)

Continental shelf deposits Sediments laid down in a tectonically quiet syncline at a passive continental margin (Fig. 12.1)

Continental slope The region of steep slopes between continental shelf and continental rise. The slope extends from the continental shelf down to the ocean deep (Fig. 12.1)

Continuous reaction series A reaction series in which the same mineral crystallizes throughout the range of temperatures in question, but in which there is gradual change in the chemical composition of the mineral with changing temperature (Fig. 7.12)

Convection A mechanism of heat transfer in a flowing material in which hot material from the bottom rises because of its lesser density, while the cool surface material sinks (Fig. 14.21)

Convergent plate boundary A boundary at which earth's plates collide and area is lost either by shortening and crustal thickening or by subduction of one plate beneath the other. These are sites of massive geologic activity and are marked by volcanism, earthquakes, trenches, and mountain building (Fig. 14.10)

Core The central part of the earth below a depth of 2890 km. It is the innermost layer of the earth and is made mostly of iron and nickel. It is divided into a liquid outer core and a solid inner core. The core is the densest of the earth's layers. The overlying Mantle is made of silicate rocks (Fig. 5.1)

Country rock Rocks that were pre-existing. The rock into which an igneous rock intrudes or a mineral deposit is emplaced. It is defined as the part of earth above the Mohorovičić discontinuity. It represents <1 % of earth's total volume

Crater (volcano) A bowl-shaped circular depression of volcanic material at the summit of most volcanoes, around the central vent formed by collapse, or by the impact of a meteorite

Craton A portion of a continent (a stable continental crust including the shield and stable platform areas) that has not been subjected to major deformation or tectonic activity for a prolonged span of time, typically since Precambrian or early Paleozoic time (Fig. 14.18)

Creep The imperceptibly slow, downhill mass movement of material (soil and regolith) under the gravitational force

Crevasse A deep large vertical crack in the upper surface of a glacier or snowfield (Fig. 11.7)

Cross-bedding A sedimentary layer deposited at an angle to an underlying set of beds. These are inclined beds in a sedimentary rock (stratification inclined to the original horizontal surface) that were formed at the time of deposition by currents of wind or water in the direction in which the bed slopes downward. Often produced by deposition on the slope of a dune or sand wave

Crosscutting relations, principle of The principle that a rock body is younger than any rock across which it cuts (Fig. 6.4)

Crust The outermost compositional layer of the lithosphere, consisting of relatively low-density, low-melting temperature materials compared to the underlying Mantle. Continental crust consists largely of granite and granodiorite. In contrast, the Oceanic crust is mostly of basalt (Fig. 5.1)

Crustal rebound The rise of earth's crust after the removal of glacial ice

Crystallization Process of crystal growth occurring as a result of condensation from a gaseous state, precipitation from a solution, or cooling of a melt

Cuesta An asymmetrical elongate ridge formed where gently dipping beds of erosion-resistant rocks are undercut by erosion of a weaker bed underneath (Fig. 8.26)

Current The portion of a stream or body of water, which is moving with a velocity much greater than the average of the rest of the water

Cycle of erosion A sequence of changes within a landscape that progresses from high, rugged, tectonically formed mountains to low, rounded hills and finally to worn-down, tectonically stable plains

D

Dacite A volcanic equivalent of a Granodiorite (Fig. 16.5)

Daughter isotope An isotope produced by the radioactive decay of its parent isotope. The quantity of a daughter isotope continually increases with time (Fig. 6.12)

Debris A type of mass movement (landslide) made up of a mixture of water-saturated rock debris and soil. It has a consistency similar to that of a wet cement. Debris flows moves rapidly downslope under the influence of gravity (Fig. 7.14)

Debris avalanche A fast downhill mass movement of soil and rock debris (Fig. 7.14)

Debris flow A fluid rapid downslope mass movement of rock fragments supported by a muddy matrix. Debris flows differ from Earthflows in that the former generally contains coarser material and moves faster (Fig. 7.14)

Debris slide A mass movement of rock material and soil along planes of weakness at the base of or within the rock material (Fig. 7.14)

Deflation Removal of clay and dust from dry soil by strong winds that scoop out shallow depressions within the ground

Delta A body of sediment deposited at the mouth of a river or in an ocean or lake or a stream. It happens when the strong current of the river slows down and deposits the sediments it has been transporting. In the United States, the best-known delta is that of the Mississippi River (in Louisiana). Other well-known North American deltas include the Rio Grande (United States and Mexico) and the Mackenzie River (Alaska and Canada). The largest in the world are the Amazon River (Brazil) and the Ganges-Brahmaputra (India) (Fig. 8.24)

Dendritic drainage pattern The term dendritic is taken after the Greek word 'dendron' meaning a tree. This is the most common type of drainage pattern and is characterized by irregular branching of tributary streams flowing in many directions and at almost any angles, although usually at less than a right angle. This drainage pattern occurs mainly where the rocks below have a uniform resistance to erosion. Georgia's (USA) Coastal Plain are a good example for such a drainage pattern (Fig. 8.18)

Density current A current that flows as a result of differences in density. In oceans, density currents are produced by differences in temperature, salinity, and turbidity (the concentration of material held in suspension)

Depth of focus Distance between the focus and the epicenter of an earthquake (Fig. 15.2)

Deranged drainage A distinctively disordered (highly irregular) drainage pattern formed in a recently glaciated area or in areas that have experienced geological disturbance (disruption). The drainage pattern is characterized by irregular direction of stream flow, few short tributaries, swampy areas, and many lakes (Fig. 8.18)

Desert pavement A residual deposit left when continued deflation removes the fine grains of a soil (water and / or wind removes these finer materials) and leaves a surface covered with a veneer of closely packed pebbles and cobbles. Processes like wetting-drying cycles, freeze-thaw cycles and bioturbation accelerate this process. Desert pavement have also been used as relative-age indicators on arid-region alluvial surfaces. Best examples are seen in the Mojave Desert, USA. All Mojave Desert pavements above 400 m elevation are suggested to be of Holocene age (Fig. 13.2)

Desert varnish It is also called Weathering rinds. It is a dark coating (commonly 0.005–0.5 mm thick) found on the surface of rocks in the desert. The coating consists of clays, iron oxides, and magnesium oxides produced during the process of weathering. Desert varnish is common on gravel plains, especially on alluvial fans with desert pavement. Tone, intensity, and color of varnish depend on climate variations (like precipitation) and on the relative age of the exposure of the rock surface (Fig. 7.11)

Diatreme A volcanic vent filled with volcanic breccia by the explosive escape of gases (Fig. 16.18)

Differential erosion Variation in the rate of erosion. Resistant rocks such as Limestones would form steep cliffs and nonresistant ones (like Shale) would form gentle slopes

Differentiated planet A planet that has layers composed of elements and minerals of different densities. Earth is differentiated with heavy metals (iron and nickel) concentrated in the corecore; lighter minerals in the mantle; and still lighter ones within the crust hydrosphere, and atmosphere

Dike A sheet-like or tabular igneous intrusion that cuts across sedimentary layering, metamorphic foliation, or other texture of a pre-existing rock

Diorite A gray to dark gray colored intrusive plutonic rock with composition intermediate between granite and gabbro. It is the intrusive equivalent of Andesite and mostly consists of intermediate plagioclase feldspar and pyroxene, with some amphibole and biotite (Fig. 16.5)

Disappearing stream A surface stream that drains rapidly and completely into a sinkhole and follow channels through caves. It does not reappear in the same, or in an adjacent, drainage basin. Very common in Karst regions (Fig. 10.1)

Discharge It is the rate of flow. The rate of water movement through a stream, measured in units of volume per unit time (Fig. 9.10)

Disconformity A specific type of hiatus (time of nondeposition) wherein the layers above and below an erosional boundary have the same orientation (often horizontal) (Fig. 6.5)

Discontinuity A sudden or rapid change in physical properties of rocks within the earth. Seismic data enable the recognition of discontinuities (Fig. 5.4)

Discontinuous reaction series A reaction series in which the end members have different crystal structures (and hence, possess distinct mineral phases) (Fig. 7.12)

Dissolution The process by which materials/minerals are dissolved

Dissolved load Part of a stream's load that is carried in solution (Fig. 8.20)

Distributary A smaller branch of a large stream that receives water from the main channel; opposite of a tributary.

Divergent plate boundary A boundary or plate margin where earth's plates drift apart relative to one another and a new lithosphere is created. These are areas subjected to tension and are sites of Mid-Ocean Ridges (MOR), shallow-focus earthquakes, and volcanism (Fig. 14.15)

Divide A ridge of high ground separating two drainage basins and that are emptied by different streams (Fig. 8.16)

Dolomite A sedimentary rock composed mainly of mineral Dolomite [$CaMg(CO_3)_2$]

Dome An uplift that is circular or elliptical in map view, with beds dipping away in all directions from a central high point

Drainage basin A region of land surrounded by divides and crossed by streams that channel all its water into the network of streams draining a single drainage system (an area), usually to converge eventually to one river or lake (Fig. 8.16)

Drainage network The pattern of tributaries, large and small, of a stream system (Fig. 8.18)

Drift A collective term for all the rock, sand, and clay that are deposited by a glacier either as till or as outwash on land by glacial ice or deposited in lakes, oceans, or streams as a result of glaciation (Fig. 11.10)

Drip curtain (Drapery) A thin sheet of dripstone hanging from the wall of a cave (Fig. 9.11)

Dripstone A cave deposit formed by the precipitation of $CaCO_3$ from groundwater entering an underground cavern

Drumlin A smooth, glacially streamlined hill that is elongate in the direction of ice movement and composed of till and, in many cases, even the bedrock (Fig. 11.14)

Dune An elongated mound of fine-grained sand formed by wind or water that accumulates as a result of sediment transport in a current system. Dunes have characteristic geometric forms that are maintained as they migrate (Fig. 13.13)

Dwarf planet An object in our solar system that is in orbit around the Sun has sufficient mass for its self-gravity to overcome rigid body forces so that it assumes a hydrostatic equilibrium (nearly round) shape, has not cleared the neighborhood around its orbit, and is not a satellite

E

Earthflow A fluid mass movement, at slow or moderate speeds, of mainly fine-grained material, along with some broken rock (Fig. 7.14)

Earthquake The violent motion of the ground caused by the passage of seismic waves radiating from a fault along which sudden movement has taken place giving rise to a series of elastic waves that are propagated within the earth. These waves are initiated when the stress along a fault exceeds the elastic limit of the rock resulting in a sudden movement along the fault (Fig. 15.2)

Effluent stream A stream or portion of a stream that receives some water from groundwater discharge because the stream's elevation is below the groundwater water table

Ejecta Rock fragments, glass, and other material thrown out of an impact crater or a volcano

Ejecta blanket Rock material (crushed rock, large blocks, breccia, and dust) ejected from an impact crater or explosion crater and deposited over the surrounding area

Elastic-rebound theory A theory of fault movement and earthquake generation holding that faults remain locked while strain energy accumulates in the rock formations on both sides, temporarily deforming them until a sudden slip along the fault releases the energy (Fig. 15.3)

End moraine (terminal moraine) A ridge of till that accumulates at the margin of a glacier (Fig. 11.14)

Entrenched meander A meander cut into the underlying rock as a result of a change in the volume of water or the capacity of ocean basinsbasins

Eolian Pertaining to or deposited by wind

Eon The largest unit of geological time

Epicenter The point on earth's surface directly above the focus of an earthquake (Fig. 15.2)

Epoch Each period of the standard geologic time scale is divided into epochs (e.g., Pleistocene Epoch of the Quaternary Period)

Equilibrium line It is also called the Snow line It is an irregular line marking the highest level to which the winter snow cover on a glacier is lost during a melt season (Fig. 11.5)

Era Major subdivision of the standard geologic time scale (e.g., Mesozoic Era)

Erg Extensive region, or "sea," of sand dunes formed by wind-transported sand. These are found in major deserts

Erosion The set of all processes by which soil and rock are loosened and moved downhill or downwind

Erratic Rock fragment (especially large boulder-sized) carried by a glacier away from the outcrop from which it was derived, often into an area underlain by a rock type different from that of the rock fragment

Escarpment A cliff or very steep slope (Fig. 8.25)

Esker A glacial deposit of sand and gravel (stratified glacial drift) in the form of a long, continuous, winding ridge formed from the deposits of a stream flowing beneath a glacier in a tunnel or in a subglacial stream bed (Fig. 11.14)

Estuary A body of water along a coastline that opens to the ocean but is diluted by fresh water

Evaporite A rock made of minerals precipitated from solutions concentrated by the evaporation of solvents such as Rock Salt, Gypsum and Anhydrite

Exfoliation A physical weathering process in which concentric sheets, shells, or slabs of rock are successively broken loose (or fractured) and stripped away from a larger rock otherwise massive unit (Fig. 7.3)

Exfoliation dome A large, rounded landform developed in a massive rock, such as granite, by the process of exfoliation (Fig. 7.3)

Extrusive igneous rock These are also called Volcanic rocks. An igneous rock formed from lava or from other products of volcanic material spewed out onto the surface of the earth. They cool and solidify rapidly at or very near the earth's surface such as Basalt

F

Facies A distinctive group of characteristics within part of a rock body (such as defined by composition, grain size, or fossil assemblages) that differ as a group from those found elsewhere in the same rock unit

Fault-block mountains A mountain or range formed when the crust is broken into blocks of different elevations by normal faulting

Faunal succession, principle of The principle that fossils in a stratigraphic sequence succeed one another in a definite, recognizable order or pattern (Fig. 6.4)

Feldspar Family of silicate minerals containing varying amounts of potassium, sodium and calcium along with aluminum, silicon and oxygen such as potassium feldspar, calcium plagioclase, and sodium plagioclase

Felsic A light-colored igneous or metamorphic rock that is poor in iron and magnesium and contains abundant quartz and feldspars (Fig. 16.7)

Firn Sometimes it is referred as Neve. It is a granular ice formed by the recrystallization of snow. It is an old, dense, compacted snow and is intermediate between snow and glacial ice

Firn limit See equilibrium line (Fig. 11.5)

Fissure An open, elongate and narrow fracture within a rock

Fissure eruption A volcanic eruption (the extrusion of lava) coming out from an elongate fissure rather than from a central vent (Fig. 16.21)

Fjord A former glacial valley with steep walls and a U-shaped profile, now occupied by sea forming a long, narrow, steep-walled inlet (Fig. 11.9)

Flood basalt A basaltic plateau extending many kilometers in flat, layered flows originating from fissure eruptions. It is synonymous with Plateau Basalt (Fig. 16.23)

Floodplain A level plain of stratified, unconsolidated sediment on either side of a stream, submerged during floods and built up by silt and sand carried out of the main channel (Fig. 8.22)

Fluvial Pertaining to a river or rivers

Fluvial environment The sedimentary environment of river systems

Focus It is sometimes also called the Hypocenter. It is an area within earth where an earthquake originates (Fig. 15.2)

Folded mountain belt A long, linear zone of earth's crust where rocks have been intensely deformed by horizontal stresses and generally intruded by igneous rocks. The great folded mountains of the world have formed at convergent plate margins (such as the Appalachians, the Himalayas, the Rockies, and the Alps) (Fig. 14.20)

Foraminifer A group of single-celled organisms whose secretions and calcite shells account for most of the oceans carbonate sediments. They are an important source of biochemical sediment in the oceans

Foraminiferal ooze Calcareous pelagic sediment composed of shells of dead foraminifera (Fig. 12.2)

Forearc At a convergent plate margin, it is the region between the trench and the volcanic arc. Forearc is underlain by a long sedimentary basin and accretionary prism (Fig. 14.11)

Foreset bed One of the inclined beds found in cross-bedding; also an inclined bed deposited on the outer front of a delta

Foreshore The marine zone between the upper limit of wave wash at high tide and the low-tide mark and is the seaward part of the shore or beach (Fig. 12.6)

Formation A body of rock of considerable extent with distinctive characteristics (such as lithology) that can be mapped, described, and named. It is the basic unit for the naming of rocks in stratigraphy

Fractional crystallization The separation of a cooling magma into components by the successive formation and removal of crystals at progressively lower temperatures. The early crystallized mafic minerals commonly are separated by gravitational settling, so that the residual magma is left enriched in silica, sodium, and potassium

Fracture The irregular breaking of a crystal along a surface not parallel to a crystal face when it does not have cleavage. Fracture serves to identify minerals

Fracture zone A zone of long, linear fractures on the ocean floor, expressed topographically by ridges and troughs. Fracture zones are the topographic expression of transform faults

Fringing reef A coral reef that is directly attached to a landmass not composed of coral and that lies alongside the shore of a landmass (Fig. 12.13)

Frost heaving The lifting of unconsolidated material by the freezing of subsurface water

Frost wedging Forcing apart of rocks by the expansion of water as it freezes in fractures and pore spaces (Fig. 7.4)

Fumarole Small volcanic vent that emits gas and steam from which minerals precipitate onto surrounding surfaces

G

Gabbro A dark-colored, coarse-grained, intrusive igneous rock, composed of calcic feldspars (Ca-plagioclase) and Pyroxene, possibly Olivine, but no Quartz. It is the intrusive equivalent of Basalt (Figs. 16.5, 16.7)

Gaining stream A stream that receives water from the zone of saturationzone of saturation

Geothermal energy Energy generated by using the heat energy (extracted from steam and hot water found within) of the earth's crust, especially in volcanic regions

Geothermal gradient The rate at which temperature increases with depth (it is ~25°C per km)

Geyser A thermal spring that intermittently erupts steam and boiling water (Fig. 16.28)

Glacial environment The sedimentary environment of glaciers and their melt waters

Glacier A permanent mass of ice formed from compacted, recrystallized snow that is thick enough to flow plastically. It shows evidence of downslope or outward movement due to the stress of its own weight. The size range is from 100–10,000 km (Fig. 11.3)

Glass A rock formed when magma or molten rock is cooled too rapidly to allow time for crystal growth

Glassy texture The texture of igneous rocks in which the material is in the form of natural glass rather than crystals. It resembles the properties of glass, as in smoothness, brittleness, or transparency

Glossopteris flora Assemblage of late Paleozoic fossil plants named for the seed fern *Glossopteris*, one of the plants in the assemblage. These flora are widespread in South America, Africa, Australia, India, and Antarctica and provide important evidence for the theory of continental drift (Fig. 14.2)

Gneiss A coarse-grained, foliated regional metamorphic rock that commonly has alternating bands of light and dark-colored minerals (i.e. banding and parallel)

Gondwanaland The ancient continental landmass that is thought to have split apart during the Mesozoic period to form the present-day continents of South America, Africa, India, Australia, and Antarctica (Fig. 14.2)

Graded bedding A bed in which the coarsest particles are concentrated at the bottom and grade gradually upward into fine silt; the whole bed having been deposited by a waning current

Graded stream A stream that has attained a state of equilibrium, or balance, between erosion and deposition, so that the velocity of the water is just great enough to transport the sediment load supplied from the drainage basin; neither erosion nor deposition occurs (Fig. 8.2)

Gradient (stream) The slope of a stream channel measured along the course of the stream (Fig. 8.2)

Granite A felsic, coarse-grained, intrusive igneous rock composed of quartz, orthoclase feldspar (K-feldspar), sodium-rich plagioclase feldspar, and micas with small amounts of mafic minerals (Figs. 16.5, 16.7)

Granodiorite A plutonic rock similar to granitic composition, except that plagioclase feldspar is present in greater abundance rather than orthoclase feldspar. It is the intrusive equivalent of Dacite (Figs. 16.5, 16.7)

Gravel The coarsest (>2 cm diameter) clastic sediment (sedimentary rocks). They include cobbles and boulders

Groundmass Matrix of relatively fine-grained material between the phenocrysts in a porphyritic rock

Ground moraine A blanket of till deposited by a glacier or released as glacier ice melted (Fig. 11.14)

Guyot A flat-topped submarine mountain or seamount

Gypsum An evaporite mineral composed of calcium sulfate with water ($CaSO_4.2H_2O$) (Fig. 7.8)

H

Hadean Eon The oldest eon

Half-life The time required for half of a sample of a given radioactive isotope to decay to its daughter isotope (Fig. 6.13)

Hanging valley The valley left by a melted glacial tributary that enters a larger glacial valley above its base, high up on the valley wall. Such valleys are commonly created by the deepening of the main valley by glaciation, but they can also be produced by faulting or rapid retreat of a sea cliff (Fig. 11.12)

Headland An extension of land seaward from the general trend of the coast; a promontory, cape, or peninsula (Fig. 12.7)

Headward erosion Extension of a stream headward, up the regional slope of erosion (Figs. 12.6, 12.7)

Hogback A formation similar to a cuesta in that it is a ridge formed by slowly eroded hard strata (resistant rock), but having two steep, equally inclined slopes (Fig. 8.26)

Holocene Epoch The youngest epoch began around 10,000 years ago and is continuing presently

Horizon A layer of soil distinguished by characteristic physical properties(Fig. 7.2)

Horn A sharp peak formed at the intersection of the headwalls of three or more cirques (Figs. 11.8–11.12)

Hornblende Amphibole family of silicate minerals forming prism or needle like crystals; generally containing iron, magnesium, calcium and aluminum in varying amounts, along with water. Hornblende is the most common dark green to black variety of amphibole

Hotspot The volcanic surface expression of a mantle plume. It is a column of hot, buoyant rock rising in the mantle beneath a lithospheric plate (Figs. 16.11, 16.12)

Hummock A small, rounded or cone shaped, low hill or a surface of other small, irregular shapes. A surface that is not equidimensional or ridge-like

Hydration The absorption (chemical combination) of water by a mineral, usually in weathering (Fig. 7.8)

Hydraulic head The pressure exerted by a fluid at a given depth beneath its surface. It is proportional to the height of the fluid's surface above the area where the pressure is measured

Hydrologic system The system of moving water at earth's surface (Fig. 9.1)

Hydrolysis Chemical reaction wherein hydrogen ions replace other ions in a mineral. This commonly results in the production of hydrous minerals such as clay or the complete dissolution of Calcite

Hydrostatic pressure The pressure within a fluid (such as water) at rest, exerted on a given point within the body of the fluid

Hydrosphere The water on or near earth's surface (Fig. 4.1)

Hypothesis A tentative theory

I

Ice sheet A thick, extensive body of glacial ice that is not confined to valleys. Localized ice sheets are sometimes called Ice caps (Fig. 11.10)

Ice wedging A type of mechanical weathering in which rocks are broken by the expansion of water as it freezes in joints, pores, or bedding planes. This expansion is as much as 9 % of the total volume. Ice wedging is synonymous with Frost Wedging (Fig. 7.4)

Iceberg calving The breaking off of blocks of ice from a glacier when it moves to a shoreline, forming IcebergsIcebergs

Igneous rock A rock formed by cooling and solidification of magma (molten silicate minerals). Igneous rocks include volcanic and plutonic rocks (Fig. 16.7)

Incised meander A meander that retains its sinuous curves as it cuts vertically downward below the level at which it originally formed (Fig. 8.9)

Inclusion, principle of Fragments included in a host rock are older than the host rock (Fig. 6.4)

Index fossil A fossil from a very short-lived species known to have existed (Figs. 6.7, 6.8)

Influent stream A stream or portion of a stream that recharges groundwater through the stream bottom because its elevation is above the groundwater tablewater table

Intermediate focus earthquake An earthquake with a focus located at a depth between 70 and 300 km (Fig. 15.12)

Intermittent stream A stream through which water flows only part of the time

Internal drainage A drainage system that does not extend to the ocean

Interstitial Pertaining to material in the pore spaces of a rock. Both petroleum and groundwater are interstitial fluids

Intrusive igneous rock Rock that, while it was fluid, penetrated into or between other rocks (country rock) and solidified. It may later be exposed at the earth's surface after erosion of the overlying rock. This is also called an "Intrusion." Igneous rock that cools and solidifies beneath the earth's surface. (= plutonic rock) (Fig. 16.7)

Ionic substitution The replacement of one kind of ion in a crystalline lattice by another kind that is of similar size and electrical charge. Ions are charged atomic particles, produced when an atom gains or loses one or more electrons

Iron formation A sedimentary rock containing iron, usually more than 15 percent, as sulfide, oxide, hydroxide, or carbonate; a low-grade ore of iron

Iron meteorite A meteorite composed principally of iron-nickel alloy

Island arc A linear or arc-shaped chain of volcanic islands formed at a convergent plate boundary (generally convex toward the open ocean; like the Aleutian Islands). The island arc is formed in the overriding plate from rising melt derived from the subducted plate and from the asthenosphere above that plate (Fig. 14.11)

Isotope One of several forms of chemical element, all having the same number of protons in the nucleus, but differing in their number of neutrons and thus, in their atomic weight

Isotopic dating Determining the age of a rock or mineral through its radioactive elements and decay products (previously and somewhat inaccurately called radiometric or radioactive dating) (Fig. 6.12)

J

Joint A large and relatively planar fracture in a rock across which there is no relative displacement of the two sides

K

Kame An irregular ridge-like or hilly local glacial stratified deposit of coarse clastic sediments formed as an alluvial fan or a delta at the glacier front by meltwater streams (Fig. 11.14)

Karst topography An irregular landscape (a topography) characterized by sinkholes, caverns, solution valleys, and lack of surface streams produced by groundwater activity. These are formed in humid regions because an underlying carbonate formation has been riddled with underground drainage channels that capture the surface streams (Fig. 10.1)

Kettle A closed hollow or depression formed in glacial deposits when outwash was deposited around a residual block of ice that later melted (Fig. 11.14)

Kimberlite An ultramafic rock that contains olivine along with mica, garnet, or both. Diamonds are found in some kimberlite bodies (Fig. 16.18)

Kimberlite pipes Depending on the magma supply and depth of exposure, these pipes typically range from 50 to 500 m in diameter. The shape of Kimberlite pipes is steeply conical, resembling a carrot shape. They often contain diamonds (Fig. 16.18)

L

Lagoon A shallow body of seawater separated from the open ocean by a barrier island or reef (Fig. 12.10)

Lahar Also called a volcanic debris flow. A type of mudflow that originates on the slopes of volcanoes when volcanic ash and debris becomes saturated with water and flows rapidly downslope

Laminar flow A flow in which the fluid flow is streamlined. However, in realty, often the flow of the fluid is gently curved

Landform A characteristic landscape feature on the earth's surface that attained its shape through the processes of erosion and sedimentation. Landforms include major features (such as continents, ocean basins, plains, plateaus, and mountain ranges) and minor feature (such as hills, valleys, slopes, drumlins, and dunes)

Landslide A general term for relatively rapid types of mass movement, such as debris flows, debris slides, rockslides, and slumps (Fig. 7.14)

Lapilli (plural) Pyroclasts in the 2–64 mm size range (singular, lapillus) (Fig. 16.9)

Lateral continuity Principle that states that an original sedimentary layer extends laterally until it tapers or thins at its edges (Fig. 6.4)

Lateral erosion Erosion and undercutting of stream banks caused by a stream swinging from side to side across its valley floor

Lateral moraine An accumulation of till deposited along the side margins of a valley glacier. It accumulates as a result of mass movement of debris on the sides of the glacier (Fig. 11.15)

Lava Magma that reaches the earth's surface through a volcanic eruption. When cooled and solidified, forms extrusive (volcanic) igneous rock (Fig. 16.6)

Lava dome Bulbous lava flow or viscous plug of lava piled near its vent. Most are made of Rhyolite (Fig. 16.6)

Lava flow Flow of lava from a crater or fissure (Figs. 16.4, 16.6, 16.8)

Lava tube Tunnel-like cave within a lava flow. It forms during the late stages of solidification of a mafic lava flow

Leach To dissolve and remove the soluble constituents of a rock or soil (Fig. 7.2)

Lee slope It is synonymous with Slip face. The part of a dune that is sheltered or turned away from the wind (Fig. 13.13)

Levee, natural A ridge along a stream bank, formed by deposits left when floodwater slowed on leaving the channel; also, an artificial barrier (a broad, low embankment) to floods built in the same form (Fig. 8.22)

Limestone A sedimentary rock composed mainly of $CaCO_3$, usually as mineral Calcite

Linear dune A long elongate, narrow eolian sand dune aligned parallel to the direction of the prevailing wind (Fig. 13.14)

Liquefaction A type of ground failure in which water-saturated sediment turns from a solid to a liquid as a result of shaking, often caused by an earthquake (Fig. 15.15)

Lithification The chemical and physical diagenetic processes that bind and harden a sediment into a sedimentary rock. These processes include cementation and compaction

Lithosphere The outer layer of solid rock that includes the crust and uppermost mantle, situated above the weak Asthenosphere. This layer is $\sim$100 km thick, and forms the earth's tectonic plates (Fig. 5.1)

Load The total amount of sediment carried at a given time by a stream, glacier, or wind (Fig. 8.20)

Loess An un-stratified, unconsolidated, wind-deposited, dusty sediment rich in clay minerals (Figs. 13.11, 13.12)

Longitudinal dune (seif) Large, symmetrical ridge of sand parallel to the wind direction (Fig. 13.14)

Longitudinal profile A cross section profile of a stream or valley drawn from its head to its mouth, showing elevation versus distance to the mouth (Fig. 8.2)

Longitudinal wave It is synonymous with P-wave. A seismic body wave in which particles oscillate along lines in the direction in which the wave travels (Figs. 15.10, 15.11)

Longshore current A current in the surf zone that flows parallel to the shoreline. Longshore currents occur where waves strike (Fig. 12.8)

Longshore drift The zigzag movement of sediment along a beach by swash and backwash of waves that approach the shore obliquely (Fig. 12.8)

Losing stream Stream that loses water to the zone of saturation

Love waves A type of surface seismic wave that causes the ground to move side to side in a horizontal plane perpendicular to the direction the wave is traveling (Fig. 15.9)

Low velocity zone Mantle zone at a depth of about 100 km where seismic waves travel more slowly than in shallower layers of rock (Fig. 5.4)

M

Mafic A dark-colored mineral rich in iron and magnesium silicates and relatively poor in silica (such as pyroxene, amphibole, or olivine) (Fig. 16.7)

Magma Molten rock material (generally a silicate melt with suspended crystals and dissolved gases) that forms igneous rocks upon cooling. The molten rock that flows out onto the earth's surface is called Lava (Fig. 16.1)

Magma chambers A body of molten rock and solid crystal mush beneath the earth's surface. When this chamber cools and solidifies, it is called a Pluton. It is a magma-filled cavity within the lithosphere (Fig. 16.1)

Magmatic arc A line of batholiths or volcanoes. Generally the line, as seen from above, is curved (Fig. 14.11)

Magmatic differentiation A general term for the processes by which magmas differentiate. It includes fractional crystallization, magma mixing, and assimilation (Fig. 16.1)

Magnitude A measure of earthquake size, determined by taking the common logarithm (base 10) of the largest ground motion observed during the arrival of a P-wave or seismic surface wave and applying a standard correction for distance to the epicenter

Mantle The layer of the earth below the crust and above the core. Specifically, between the base of the crust (the Moho discontinuity) and the core, ranging from depths of about 40–2,890 km. Mantle is composed of dense, mafic silicate minerals and divided into concentric layers by phase changes that are caused by the increase in pressure with depth. The uppermost part of the mantle is rigid and, along with the crust, forms the 'plates' of plate tectonics. The mantle is made up of dense, iron and magnesium rich (ultramafic) rock such as dunite and peridotite (Fig. 5.1)

Mantle plume Rising jet of partially molten buoyant mass of hot material, thought to emanate from the deep mantle (or the base of the lithosphere) and responsible for intra-plate volcanism. Mantle plumes commonly produce volcanic activity as noted in Europe, Asia, North America, and Greenland (Fig. 14.21)

Mare (plural-maria) Any of the relatively smooth, low, dark areas of the Moon. The lunar maria were formed by the extrusion of lava

Mass wasting (mass movement) It is synonymous with Mass Wasting. A downhill movement of soil or fractured rock under the force of gravity without a flowing medium (such as a river or glacial ice) (Fig. 7.14)

Matrix The relatively fine-grained rock material occupying the space between larger grains in a sedimentary rock

Meander A broad, semicircular curve (a looping bend) in a stream that develops as the stream erodes the outer bank of the bend and deposits sediment (as point bars) against the inner bank (Figs. 8.5–8.7)

Meander cutoff A new, shorter channel across the narrow neck of a meander (Figs. 8.6, 8.7)

Meander scar An abandoned meander filled with sediment and vegetation (Fig. 8.6)

Medial moraine A ridge of till formed in the middle of a valley glacier by the junction of two lateral moraines where two valley glaciers converge (Fig. 11.15)

Mercalli scale A measure of earthquake intensity determined from the effects on people and buildings. This scale ranges from I (low) to XII (near total destruction)

Mesa A flat-topped, steep-sided upland capped by a resistant rock formation. A mesa is smaller than a Plateau but larger than a Butte (Fig. 8.27)

Mesozoic Era The era that followed the Paleozoic Era and preceded the Cenozoic EraCenozoic Era

Metamorphism A process whereby rocks undergo physical or chemical changes or both to achieve equilibrium with changing conditions. Agents of metamorphism are heat, pressure, and chemically active fluids

Meteor Fragment that passes through the earth's atmosphere, heated to incandescence by friction; sometimes incorrectly called "shooting" or "falling" stars

Meteorite Any particle of solid matter that has fallen to Earth, the Moon, or another planet from space

Microplate terrane A block within an orogenic belt containing rock assemblages that contrast sharply with those in the surrounding areas, interpreted as small continents, seamounts, or island arcs that were accreted onto the larger continent at a convergent plate boundary (Fig. 14.18)

Mid-ocean ridge It is synonymous with Oceanic Ridge. A major elevated linear feature of the seafloor in the ocean basins consisting of many small, slightly offset segments, with a total length of 200–20,000 km. A mid-ocean ridge occurs at a divergent plate boundary, a site where two plates are being pulled apart and new oceanic lithosphere is being created (Fig. 12.1)

Migmatite A rock with both igneous and metamorphic characteristics that shows large crystal and laminar flow structures and in which thin dikes and stringers of granitic material inter-finger with metamorphic rocks

Modified Mercalli scale Scale expressing intensities of earthquakes (judged on amount of damage done) in Roman numerals ranging from I to XII

Mohorovičić discontinuity (moho) First global seismic discontinuity below the surface of earth. It lies at a depth varying from about 5 to 10 km beneath the ocean floor to about 35 km beneath the continents. It is commonly referred to as the Moho and is marked by rapid increase in seismic wave velocity to more than 8 km per second (Fig. 5.4)

Moment magnitude An earthquake magnitude calculated from the strength of the rock, surface area of the fault rupture, and the amount of rock displacement along the fault

Moraine A glacial deposit of till left at the margins of an ice sheet. Subdivided into ground moraine, lateral moraine, medial moraine, and end moraine (Fig. 11.15)

Mud crack A crack in a deposit of mud or silt resulting from the contraction that accompanies drying

Mudflow A mass movement of material mostly finer than sand, along with some rock debris, lubricated with large amounts of water which makes the mudflow move faster than earth flow or debris flow (Fig. 7.14)

Mudrock A fine-grained sedimentary rock made of clay and silt-size particles (Shale is a finely laminated mudrock)

Mudstone The lithified equivalent of mud; a fine-grained sedimentary rock similar to shale but less finely laminated. Mudstone is a mixture of clay and silt-sized particles

N

Neap tide A tide cycle of unusually small amplitude that occurs twice monthly when the lunar and solar tides are opposite i.e. when the gravitational pull of the Sun is at right angles to that of the Moon (Fig. 12.5)

Neve It is synonymous with Firn. Granular ice formed by the recrystallization of snow

Natural levee Low ridges of flood-deposited sediment formed on either side of a stream channel, which thin away from the channel (Fig. 8.22)

Nebula A large volume of interstellar gas and dust

Nebular hypotheses The hypothesis that the Solar System formed from a rotating cloud of gas and dust, the solar nebula (Fig. 4.4)

Nonconformity An unconformity in which stratified rocks rest on eroded granitic or metamorphic rocks (Fig. 6.5)

Numerical age It is synonym with Absolute age. Geologic time measured in a specific duration of years (in contrast to relative time, which involves only the chronologic order of events) (Fig. 6.3)

O

Obsidian Dark volcanic glass, usually of felsic composition, equivalent to that of a Granite (Fig. 16.5)

Ocean basin A low part of the lithosphere lying between continental masses. The rocks of an ocean basin are mostly basalt with a veneer of oceanic sediment (Fig. 12.1)

Oceanic crust The type of crust that underlies the ocean basins. It is generally less than 8 km thick, composed predominantly of basalt and gabbro with a density of about 3.0 g/cm^3.The velocities of compressional seismic waves traveling through it exceed 6.2 km/sec (Fig. 5.1)

Oceanic ridge The continuous ridge, or broad, fractured topographic swell that extends through the central part of the Arctic, Atlantic, Indian, and South Pacific oceans. It is several hundred kilometers wide, and its elevation above the ocean floor is 600 m or more. The ridge marks a divergent plate boundary where new oceanic lithosphere is being formed (Fig. 14.9)

Oceanic trench A narrow, deep trough parallel to the edge of a continent or an island arc (Fig. 14.10)

Ooze Marine sediment consisting of more than 30 percent shell fragments of microscopic organisms such as Foraminifers, etc (Fig. 12.2)

Ophiolite A sequence of rocks characterized by ultramafic rocks at the base and (in ascending order) gabbro, sheeted dikes, pillow lavas, and deep-sea sediments (Fig. 5.3)

Ophiolite suite An assemblage of mafic and ultramafic igneous rocks with deep-sea sediments found on land, believed to be associated with divergent plate boundaries and the seafloor environment (Fig. 5.3)

Original horizontality, principle of The proposition that all sedimentary bedding is horizontal at the time of deposition (Fig. 6.4)

Orogenic belt A linear mountain belt that has been subjected to folding and other deformation in a mountain-building episode (Fig. 14.20)

Orogeny The tectonic process in which large areas are folded, thrust-faulted, metamorphosed, and subjected to plutonism. The cycle ends with uplift and formation of mountains

Orthoclase (potassium) feldspar A feldspar with the formula $KAlSi_3O_8$

Outcrop A segment of bedrock exposed to the atmosphere

Outwash A stratified sediment deposited (in front of the End moraine) by melt-water streams coming from a glacier (Fig. 11.14)

Outwash plain Area beyond the margins of a glacier where meltwater deposits sand, gravel, and mud washed out from the glacier (Fig. 11.14)

Oxbow lake A long, broad, crescent-shaped lake formed when a stream abandons a meander and takes a new course (Fig. 8.7)

Oxidation A chemical reaction in which electrons are lost from an atom and its charge becomes positive. It is the chemical combination of an element with Oxygen

Oxide mineral A mineral lacking Silicon, but containing Oxygen bound to a metal such as Hematite and Magnetite

Oxides Combinations of metal ions with Oxygen, comprises the major ores extracted in mining operations—Hematite (iron ore)—Magnetite (iron ore, magnetic mineral)—Corundum (gemstone, abrasive)

Ozone layer A zone within the stratosphere where Ozone (O_3) is abundant and forms a protection from some of the Sun's harmful ultraviolet radiation (Fig. 4.2)

P

P-wave (primary seismic wave) The primary or fastest seismic wave traveling away from a seismic event through solid rock and consisting of a train of compressions and dilations of the material (Fig. 15.10)

Pahoehoe A basaltic lava flow with a glassy, smooth, and ropy surface

Paleocurrent An ancient current, which existed in the geologic past, with a direction of flow that can be inferred from interpreting sedimentary structures such as cross-bedding, ripple marks, and others

Paleozoic Era The era that followed the Precambrian and began with the appearance of complex life, as indicated by fossils

Pangea Supercontinent that coalesced in the latest Paleozoic era and comprised of all the present continents. The breakup of Pangea began in the Mesozoic

Parabolic dune It is also called a Blowout dune. A dune shaped like a parabola with the concave side toward the wind (Fig. 13.14)

Partial melting A process in which heating melts some of the minerals in a mass of rock while the rest remains solid. Partial melting occurs because the minerals that constitute a rock, melt at different temperatures. Partial melting is believed to be important in the generation of basaltic magma from peridotite at ocean ridges and in the generation of granitic magma from the basaltic crust (Fig. 16.1)

Pebble A rock fragment with a diameter between 2 mm (size of a match head) and 64 mm (size of a tennis ball)

Pediment In an arid and semiarid region, a planar, gently sloping rock surface forming a ramp up to the front of a retreating mountain range. Locally, it may be covered by a thin alluvial deposit (Fig. 8.28)

Pelagic sediment A deep-sea sediment composed of fine-grained detritus that slowly settles from surface waters. Common constituents are clay, radiolarian ooze, foraminiferal ooze, and silica ooze (Fig. 12.2)

Period Each era of the standard geologic time scale is subdivided into periods (e.g., the Cretaceous Period)

Permafrost A permanently frozen aggregate of ice and soil in very cold regions

Permanent stream It is synonymous with the perennial stream. A stream that flows continuously throughout the year

Permeability The ability of a formation or material to transmit groundwater or other fluids through pores and cracks (Fig. 9.5)

Phaneritic texture The texture of igneous rocks in which the interlocking crystals are large enough to be seen without magnification. The crystals are approximately of equal size

Phanerozoic Eon Eon of geologic time. Includes all time following the Precambrian

Phenocryst A large crystal surrounded by a finer matrix in an igneous rock. Phenocrysts form during an early phase in the cooling of a magma when the magma cools relatively slowly. Those containing abundant phenocrysts are called Porphyry

Phreatic explosion A volcanic eruption of steam, mud, and debris caused by the expansion of steam formed when magma comes in contact with groundwater or seawater

Physical weathering It is synonymous with mechanical weathering. It is the disintegration of materials in which no new mineral or substance formed

Physical geology A large division of geology concerned with earth materials, changes of the surface and interior of earth, and the forces that cause those changes

Pillow structure Rocks, generally basalt, formed in pillow-shaped masses fitting closely together; caused by underwater lava flows

Pillow lava Basaltic lava that forms under water when many small tongues (ellipsoidal mass of igneous rocks) of lava break through the chilled ocean floor and quickly solidify into a rock formation resembling a pile of sandbags

Plagioclase A group of feldspar minerals with a composition range from $NaAlSi_3O_8$ to $CaAl_2Si_2O_8$

Planetary differentiation The process by which heating, cooling, and gravitation sorts the material according to the density of our planet so that it evolves into concentric layers that differ chemically and physically possessing a core (dense), mantle and crust

Planetesimal Small, planet-like body

Plankton A collective term for very small plants and animals that drift near the surface of water. Phytoplankton include bacteria, algae (including diatoms), and fungi. The small animals are called zoo-planktons

Plastic flow Deformation of the shape or volume of a substance without fracturing (Fig. 11.7)

Plate A broad segment of the lithosphere (including the rigid upper mantle, plus oceanic and continental crust) that floats on the underlying asthenosphere and moves independently. (Fig. 15.5)

Plate tectonics A theory that earth's surface is divided into a few large, thick plates that are slowly moving and changing in size. Intense geologic activity occurs at the plate boundaries. It is the study of plate formation, movement, interactions, and destruction; the attempt to explain seismicity; volcanism, mountain building, and evidence of paleomagnetism in terms of plate motions (Fig. 14.8)

Plateau An extensive upland region at high elevation with respect to its surroundings

Plateau basalt It is synonymous with Flood Basalt. Basalt extruded in extensive, nearly horizontal layers, which, after uplift, tend to erode into great plateaus (Fig. 16.23)

Platform A sediment-covered, tectonically stable, almost level region of a continent

Playa A depression in the center of a desert basin, the site of occasional temporary lakes (Fig. 8.28)

Playa lake The flat floor of a closed basin in an arid region, usually rich in evaporate minerals. It may be occupied by a shallow temporary lake after heavy rain (Fig. 8.28)

Pleistocene Epoch An epoch of the Quaternary Period characterized by several glacial ages

Plucking The process of glacial erosion by which large rock fragments are loosened by ice wedging, become frozen to the bottom surface of the glacier, and are torn out of the bedrock and transported by the glacier as it moves. The process involves the freezing of subglacial meltwater that seeps into fractures and bedding planes in the rock (Fig. 11.13)

Pluton A large igneous intrusion, at least 1 km^3, formed at depth in the crust and solidifies within the crust. Batholiths and stocks are types of plutons

Plutonic Rock formed by slow crystallization, which yields coarse texture. Once believed to be typical of crystallization at great depth, but not a necessary condition

Point bar A deposit of sediment on the inner bank of a meander that forms because the stream velocity is lower against the inner bank. A crescent-shaped accumulation of sand and gravel deposited on the inside of a meander bend (Fig. 8.6)

Pore fluid A fluid, such as groundwater or liquid rock material resulting from partial melting that occupies pore spaces of a rock

Pore space The spaces within a rock body that are unoccupied by solid material. Pore spaces include spaces between grains, fractures, vesicles, and voids formed by dissolution

Pores An interstice between the constituent particles or molecules of a body

Porosity The percentage of the total volume of a rock or sediment that consists of pore space i.e. not occupied by mineral grains (Fig. 9.4)

Potassium feldspar A feldspar with the formula $KAlSi_3O_8$

Pothole A hemispherical hole in a stream bed formed by the abrasion of small pebbles and cobbles in a strong current swirled around in one spot by eddies

Precambrian The vast amount of time that preceded the Paleozoic Era

Precambrian shield A complex of old Precambrian metamorphic and plutonic rocks exposed over a large area (Fig. 14.19)

Precipitate A mineral deposited from a water solution in pores or other openings in rocks. Chemical reaction with the surrounding rock, changes in pressure or temperature, or just drying up (evaporation) can cause a mineral to precipitate out of solution. Quartz veins are common products of mineral precipitation; The process that separates solids from a solution

Pressure-release A significant type of mechanical weathering that causes rocks to crack when overburden is removed (Fig. 7.3)

Proterozoic Eon Eon of Precambrian time

Precipitation Discharge of water, in rain, snow, hail, sleet, fog, or dew, on land or water surface. Also, process of separating mineral constituents from solution by evaporation (halite, anhydrite) or from magma to form igneous rocks

Pressure ridge An elongate uplift of the congealing crust of a lava flow, resulting from the pressure of underlying and still fluid lava

Primary sedimentary structure A structure of sedimentary rocks (such as cross bedding, ripple marks, or mud cracks) that originates contemporaneously with the deposition of the sediment (in contrast to a secondary structure, such as a joint or fault, which originates after the rock has been formed)

Proton A subatomic particle that contributes mass and a single positive electrical charge to an atom (Fig. 11.5)

Pumice A form of volcanic glass, usually of felsic composition (hence, light-colored), so filled with holes (vesicles) from the escape of gas during quenching that it resembles a sponge and has very low density

P-wave A compressional wave (seismic wave) in which rock vibrates parallel to the direction of wave propagation (Fig. 15.10)

P-wave shadow zone The region on earth's surface, 103–142° away from an earthquake epicenter, in which P waves from the earthquake are absent. The P-waves spread out from an earthquake until, at 103° of arc (11,500 km) from the epicenter, they suddenly disappear from seismograms. At more than 142° (15,500 km) from the epicenter, P waves reappear on seismograms. The region between 103–142°, which lacks P waves, is called the P-wave shadow zone (Fig. 15.11)

Pyroclastic flow A glowing cloud of volcanic ash, fragments of volcanic rock, and gases that moves rapidly downhill away from the eruptive center during a volcanic eruption

Pyroxene A group of rock-forming silicate minerals composed of single chains of silicon-oxygen tetrahedra. Amphibole have double chains

Q

Quartz One of the most common minerals in earth's crust. It is an important rock-forming silicate mineral made up of silicon dioxide (SiO_2). Its crystals are clear and glassy 6-sided prisms. It is distinguished by its hardness, glassy luster, and conchoidal fracture

Quaternary Period The youngest geologic period; includes the present time

R

Radial drainage A system of streams running in a radial pattern away from the center of a circular elevation, such as a volcano or dome (Fig. 8.18)

Radioactive decay The spontaneous nuclear disintegration of certain isotopes (Fig. 6.10)

Radioactivity The spontaneous emission of energetic particles and/or radiation during radioactive decay (Fig. 6.12)

Radiocarbon A radioactive isotope of Carbon, ^{14}C, which is formed in the atmosphere and is absorbed by living organisms (Figs. 6.14, 6.15)

Radiometric dating The method of obtaining ages of geological materials by measuring the relative abundances of radioactive parent and daughter isotopes in them (Fig. 6.13)

Rayleigh waves A type of surface seismic wave that behaves like a rolling ocean wave and causes the ground to move in an elliptical path (Fig. 15.9)

Reaction series A series of chemical reactions occurring in a cooling magma by which a mineral formed at high temperature becomes unstable in the melt and reacts to form another mineral (Fig. 7.12)

Recent (Holocene) Epoch The present epoch of the Quaternary Period

Recharge The replenishment of groundwater reservoir, usually by infiltration of meteoric water through the soil

Recrystallization Growth of new crystals with no changes in the overall chemistry. It is the reorganization of elements of the original minerals in a rock resulting from changes in temperature and pressure and from the activity of pore fluids

Rectangular pattern A drainage pattern in which tributaries of a river change direction and join one another at right angles (Fig. 8.18)

Reef A mound or ridge-shaped organic structure that is built by calcareous organisms, particularly coral, is wave resistant, and stands in relief above the surrounding seafloor (Fig. 12.13)

Refraction Deflection from a straight path undergone by a light ray or energy wave in passing obliquely from one medium (as air) into another (as glass) in which its velocity is different (Fig. 5.6)

Regolith The layer of loose, unconsolidated, heterogeneous material lying on top of bedrock. It includes soil, unweathered fragments of parent rock, and rock fragments weathered from the bedrock (Fig. 7.2)

Relative age The age of a geologic event relative to another expressed in terms of the geologic time scale

Relative dating Determination of the chronologic order of a sequence of events in relation to one another without reference to their ages measured in years. Relative geologic dating is based primarily on superposition, faunal succession, and crosscutting relations (Fig. 6.3)

Relief Maximum regional difference (high and low parts of an area) in elevation (altitude)

Reservoir A source or place of residence for elements in a chemical cycle or hydrologic cyclehydrologic cycle

Rhyolite The fine-grained volcanic or extrusive equivalent of a Granite. It is light-brown to gray and compact. It is composed of quartz, K-feldspar, and plagioclase

Richter magnitude A measure of earthquake size, determined by taking the common logarithm (base 10) of the largest ground motion observed during the arrival of a P-wave or seismic surface wave and applying a standard correction for distance to the epicenter

Richter scale A logarithmic scale for expressing the magnitude of an earthquake in terms of the energy dissipated in it. A modified version of this scale is commonly used

Ridge push The concept that oceanic plates diverge as a result of sliding down the sloping lithosphere-asthenosphere boundary (Fig. 14.21)

Rift system A system of faults resulting from extension (Fig. 15.13)

Rift valley A fault trough (the down-dropped block) bounded by normal faults formed at a divergent plate boundary on continents and along the crest of the mid-oceanic ridge (Fig. 15.13)

Rip current A current formed on the surface of a body of water by the convergence of currents flowing in opposite directions. Such currents are common along coasts where longshore currents move in opposite directions

Ripple marks Small waves produced on a surface of sand or mud by the drag of wind or water moving over it

River system An integrated system of tributaries and a trunk stream, which collect and funnel surface water to the sea, a lake, or some other body of water. A river with all of its tributaries

Roche moutonnée An abraded knob of bedrock formed by an overriding glacier. It typically is striated and has a gentle slope facing the upstream direction of ice movement (Fig. 11.13)

Rock avalanche The rapid, downhill-flowing mass movement of broken rock material, during which further breakage of the material occurs (Fig. 7.14)

Rock cycle Explains how the three major groups of rock are related to each other. Igneous, sedimentary and metamorphic (Fig. 7.1)

Rock glacier A mass of poorly sorted, angular boulders cemented with interstitial ice. It moves slowly by the action of gravity

Rockfall The relatively free rapid falling (mass movement) of a newly detached segment (ranging from large masses to small fragments) of bedrock from a cliff or other steep slope (Fig. 7.14)

Rockslide The mass movement (a landslide) of large blocks of detached bedrock sliding more or less as a unit over an inclined surface of weakness (such as a joint or bedding plane) (Fig. 7.14)

Runoff The amount of rainwater that does not infiltrate the ground but leaves an area in surface drainage or that flows over the land surface

S

Saltation The movement of sand or fine sediment in a current of wind or water by short jumps above the streambed under the influence of a current too weak to keep it permanently suspended (Fig. 8.20)

Sand dune Landform made from wind blowing on sand that makes ridges. Stronger winds make bigger dunes and lighter winds make smaller dunes (Figs. 13.13, 13.14)

Sandblasting A physical weathering process in which rock is eroded by the impact of sand grains carried by the wind, frequently leading to ventifact formation of pebbles and cobbles (Fig. 13.10)

Saturated zone The zone in the subsurface of soil and rock in which pores are completely filled with groundwater (Fig. 9.3)

Scarp A cliff produced by faulting or erosion

Scoria A dark colored volcanic rock containing abundant vesicles

Sea arch An arch cut by wave erosion through a headland (Fig. 12.7)

Sea cave A cave formed by wave erosion (Fig. 12.7)

Sea cliff A cliff produced by wave erosion (Fig. 12.7)

Sea stack A small, pillar-shaped, rocky island formed by wave erosion through a headland near a sea cliff (Fig. 12.7)

Seafloor spreading The mechanism by which new seafloor (new lithosphere) is created by igneous activity at oceanic ridges (divergent plate boundaries) as adjacent plates move apart. This process may continue at a few cm per year through many geologic periods (Fig. 14.10)

Seamount An isolated, tall mountain on the seafloor that may extend more than 1 km from the base to peak. Seamounts are submerged shield volcanoes (Fig. 12.1)

Sedimentary basin A region of considerable extent (>10,000 km^2) that is the site of accumulation of a large thickness of sediments (Fig. 12.1)

Sedimentary environment A sedimentary environment is an area of the earth's surface where sediment is deposited and is distinguished from other areas on the basis of its physical, chemical, and biological characteristics

Sedimentary rock A rock formed by the accumulation and cementation (consolidation) of mineral grains by wind, water, or ice transportation to the site of deposition or by chemical precipitation at the site. Sedimentary rocks often have distinctive layering or bedding (Fig. 7.1)

Seif dune A linear dune of great height and length (Fig. 13.14)

Seismic discontinuity A surface within the earth at which seismic wave velocities abruptly changes (Fig. 5.4)

Seismic reflection The returns of part of the energy of seismic waves to the earth's surface after the waves have bounce off a rock boundary (Fig. 5.6)

Seismic refraction The bending of seismic waves as they pass from one material to another (Fig. 5.6)

Seismic surface wave A seismic wave that follows the earth's surface only, with a speed less than that of the S-waves (Fig. 15.9)

Seismic wave An elastic wave or vibration produced within the earth by earthquakes or explosions (Fig. 5.5)

Seismicity The worldwide or local distribution of earthquakes in space and time; a general term for the number of earthquakes in a unit of time

Seismograph An instrument for magnifying and recording the motions of the earth's surface that are caused by seismic waves (Fig. 5.5)

Serpentine A magnesium silicate mineral. Most asbestos are a variety of serpentine

Settling velocity The rate at which a grain falls through water or air

Shadow zone A zone 103–142 ° from the epicenter of an earthquake in which there is no penetration of seismic waves through the earth because of wave refraction or because the waves are not transmitted upon entering the liquid core (Fig. 5.7)

Shale A very fine-grained clastic sedimentary rock composed of silt and clay that tends to part along bedding planes. Particles in shale are commonly clay minerals mixed with tiny grains of quartz eroded from pre-existing rocks

Shear wave It is synonym for S-waves. A type of seismic wave wherein the elastic vibrations of the particles are transverse to the direction the wave is moving. Shear waves do pass through liquids (Fig. 15.9)

Shield An extensive area of a continent where igneous and metamorphic rocks are exposed and have approached equilibrium with respect to erosion and isostasy

Shield volcano A large, broad volcanic cone with very gentle slopes built up by non-viscous basaltic lavas. The slopes rarely exceed 10°, so that in profile they resemble a shield or a broad dome (Fig. 16.21)

Shock metamorphism Metamorphism that occurs when minerals are subjected to high pressures and high temperatures of shock waves generated by an impact

Shore The zone between the waterline at high tide and the waterline at low tide. It is a narrow strip of land immediately bordering a body of water, especially a lake or an ocean

Shoreline The straight or sinuous, smooth or irregular interface between land and sea (Fig. 12.6)

Silica ooze Pelagic sediments consisting of the remains of tiny organisms that have shells made of amorphous silica (Fig. 12.2)

Silicate rock An igneous or metamorphic rock made largely of silicate minerals, such as feldspar, mica, or garnet

Siliceous sedimentary rock Rock containing abundant free silica of either organic or inorganic origin, formed by biochemical, chemical, or physical deposition of silica (Fig. 12.2)

Sill A sheet-like or tabular-shaped igneous intrusion that cuts across sedimentary layering, metamorphic foliation, or other texture of a pre-existing rock. It runs between parallel layers of bedded country rockcountry rock

Silt Sedimentary material composed of fragments ranging in diameter from 1/265 to 1/16 mm. Silt particles are larger than clay but smaller than sand particles

Sinkhole Characteristic of Karst topography and formed by the dissolution and collapse of subterranean caverns (the collapse of a cavern roof) in carbonate formations. It is a small, steep depression (Figs. 10.4, 10.5)

Slab pull The concept that subducting plates are pulled along by their dense leading edges (Fig. 14.21)

Slip face The steep downwind face of a dune on which sand is deposited in cross-beds at the angle of repose (Fig. 13.13)

Slump A slow mass movement of unconsolidated materials that slides as a unit. The material moves along a curved surface of rupture

Snowline The line on a glacier separating the area where snow remains from year to year from the area where snow from the previous season melts (Fig. 11.6)

Soil profile A vertical section of soil showing the soil horizons and the parent material (Fig. 7.2)

Solar nebula The rotating disk of gas and dust from which the Sun and planets formed

Solar system The Sun, planets, their moons, and other bodies that orbit the Sun (Fig. 2.2)

Solar wind The outflow of low-density, hot gas from the Sun's upper atmosphere. It is partially this wind that creates the tail of a comet by blowing dust and gas away from the comet's immediate surroundings

Solid solution The substitution of atoms of one element for those of another in a particular mineral (Fig. 7.12)

Solifluction A type of mass movement (creep) in which material moves slowly downslope in areas where the soil is saturated with water and/or ice caused by alternate freezing and thawing; most common in polar regions (permafrost areas) (Fig. 7.14)

Solubility (mineral) The extent to which a mineral can dissolve in water; the amount of the mineral dissolved in water when the solution reaches the saturation point

Sorting A measure of the homogeneity of size, shape, or weight of particles in a sediment or sedimentary rock. It occurs during transportation by running water or wind

Spatter cone A low-steep-sided volcanic cone built by the accumulation of splashes and spatters of lava (usually basaltic) around a fissure or ventvent

Specific gravity The ratio of the density of a given substance to the density of water

Spheroidal weathering It is synonymous with Exfoliation. A physical and chemical weathering process in which curved layers split off from a rounded boulder, leaving a spherical inner core. Thus, the rock acquires a spheroidal or ellipsoidal shape (Fig. 7.3)

Spit A sandy bar projecting from the mainland into open water deposited by longshore currents and longshore drift where the coast takes an abrupt inward turn. It is attached to land at the upstream end (Fig. 12.10)

Spreading axis (or spreading center) The crest of the mid-oceanic ridge, where sea floor is moving away in opposite directions on either side (Fig. 14.10)

Spring A place where groundwater flows or seeps naturally to the surface (Fig. 9.8)

Spring tide A tide cycle of unusually large amplitude that occurs twice monthly when the lunar and solar tides are in phase (Fig. 12.5)

Stable platform The part of a continent that is covered with flat lying or gently tilted sedimentary strata and underlain by a basement complex of igneous and metamorphic rocks. The stable platform has not been extensively affected by crustal deformation

Stalactite An icicle-like or tooth-like deposit of calcite or aragonite hanging from the roof of a cave. It is deposited by evaporation and precipitation from solutions seeping through limestone (Fig. 9.11)

Stalagmite A conical or inverted icicle-shaped deposit that builds up on a cave floor beneath a stalactite and is formed by the same process as a stalactite (Fig. 9.11)

Star A massive, gaseous body held together by gravity and generally emitting light. Normal stars generate energy by nuclear reactions in their interiors

Star dune A mound of sand with a high central point and arms radiating in various directions (Fig. 13.14)

Stony meteorite A meteorite made mostly of plagioclase and iron-magnesium silicates

Stony-iron meteorite A meteorite composed of silicate minerals and iron-nickel alloy in approximately equal amounts

Strata (plural of stratum) Layers of rock, usually sedimentary

Stratification Thc characteristic layering or bedding of sedimentary rocks

Stratigraphic sequence A set of deposited beds that reflects the changing conditions and sedimentary environments that define the geologic history of a region (Fig. 6.6)

Stratosphere The upper atmosphere, 11–50 km above the surface, where a protective Ozone layer forms and in which temperature increases gradually to about 0° C and clouds rarely form (Fig. 4.2)

Stratovolcano It is synonymous with Composite volcano. A steep-sided volcano built by ash, lava flows, and shallow intrusions (Fig. 16.21)

Stream A general term for any body of water, large or small, that moves under the force of gravity in a channe

Stream capacity The amount of sediment and detritus a stream can transport/carry past any point in a given amount of time

Stream capture See stream piracy (Fig. 8.14)

Stream channel A long and narrow depression more or less filled by a stream

Stream discharge Volume of water that flows past a given point in a unit of time (Fig. 8.20)

Stream gradient Downhill slope of a stream's bed or the water surface, if the stream is very large (Fig. 8.2)

Stream load The total amount of sediment carried by a stream at a given time

Stream order The hierarchical number of a stream segment. The smallest tributary has the order number of 1, and successively larger tributaries have progressively higher numbers (Fig. 8.17)

Stream piracy The erosion of a divide between two streams by the more competent stream, leading to the capture of all or part of the drainage of the slower stream by the faster. This results in the diversion of headwaters of one stream into another stream due to the headward erosion of a stream having greater erosive power than the stream it captures (Fig. 8.14)

Stream terrace One of a series of level surfaces in a stream valley representing the dissected remnants of an abandoned floodplain, stream bed, or valley floor produced in a previous stage of erosion or deposition (Fig. 8.12)

Stream-dominated delta A delta with finger-like distributaries formed by the dominance of stream sedimentation; also called a bird-foot delta (Fig. 8.24)

Striation Scratches or grooves left on bedrock and boulders by the overriding ice, showing the direction of glacial motion (Fig. 11.15)

Subduction The sinking of an oceanic plate beneath an overriding plate; occurs at convergent plate boundaries (Fig. 14.9)

Subduction zone An elongate zone between a sinking oceanic plate and an overriding plate, descending away from a trench and characterized by high seismicity. A subduction zone is typically marked by an oceanic trench, lines of volcanoes, and crustal deformation associated with mountain building (Fig. 14.9)

Submarine canyon An underwater canyon (a V-shaped trench or valley with steep sides) in the h or slope (Fig. 12.1)

Submarine fan A terrigenous cone- or fan-shaped deposit located at the foot of a continental slope, usually seaward of large rivers and submarine canyons

Subsidence A gentle epeirogenic movement in which a broad area of the crust sinks with respect to the surrounding parts without appreciable deformation

Superposed stream A stream that flows through resistant formations because its course was established at a higher level on uniform rocks before downcutting began

Superposition principle of, A bed that overlies another bed is always younger (applicable to a series of sedimentary strata and not valid for extremely deformed strata) (Fig. 6.4)

Surf The foamy, bubbly surface of water waves as they break close to shore

Surf zone An offshore belt along which waves collapse into breakers as they approach the shore

Suspended load Fine sediment kept suspended in a stream as the settling velocity of the sediment is lower than the upward velocity of eddies (Fig. 8.20)

Swash The landward rush of water from a breaking wave up the slope of the beach

S-wave A secondary (or shear) seismic wave that consists of elastic vibrations transverse to the direction of travel (i.e. at right angles to the direction in which the wave travels). S-waves travel slowly than the P-waves. They also cannot penetrate liquids. They cause the rock to vibrate perpendicular to the direction of wave propagation (Fig. 15.10)

S-wave shadow zone The region on earth's surface (at any distance more than 103° from an earthquake's epicenter) in which S-waves from the earthquake are absent. S-waves can travel only through solids. The S-wave shadow zone seems to indicate that S waves do not travel through the core at all. If this is true, it implies that the core of earth is a liquid, or at least acts like a liquid. The S-wave shadow zone (where S-waves are absent) is larger than the P-wave shadow zone (Fig. 15.11)

T

Talus A deposit of large angular fragments of physically weathered bedrock. These usually accumulate at the base of a cliff or on a steep slope (Fig. 7.14)

Tectonic forces Forces generated from within the earth that result in uplift, movement, or deformation of the earth's crust (Fig. 15.1)

Tensional forces Forces that stretch a body and pull it apart. Such forces dominate at divergent plate boundaries (Fig. 15.1)

Tephra A general term for all sizes of particles ejected into the air during a volcanic eruption. It includes ash, dust, bombs and blocks, and other types of fragments

Terminal moraine A ridge of material deposited by a glacier at the line of maximum glacier advance (Fig. 11.14)

Terrace (stream valley) A flat, step like surface above the floodplain in a stream valley, marking a former floodplain that existed at the higher level before regional uplift or an increase in discharge caused the stream to erode into the former floodplain (Fig. 8.12)

Terrigenous sediment Sediments eroded from the land surface

Theory An explanation for observed phenomena that has a high probability of being true

Tidal delta A submerged body of sediment formed by tidal currents passing through gaps in barrier islands (Fig. 8.24)

Tide The rise and fall of the water level of the ocean. It occurs twice a day and is caused by the gravitational attraction of the Moon and, to a lesser extent by the Sun (Fig. 12.5)

Tide-dominated delta A delta formed by the reworking of sand by strong tides (Fig. 8.24)

Till An un-stratified and poorly sorted glacial deposit containing all sizes of fragments from clay to boulders. (Fig. 11.10)

Tillite The lithified equivalent of a till (unsorted, un-stratified glacial sediment (Fig. 11.10)

Tombolo A beach or bar connecting an island to the mainland (Fig. 12.11)

Transform fault plate boundary A boundary at which earth's plates slide horizontally past each other, approximately at right angles to their divergent plate boundaries (Fig. 14.17)

Transverse dune A dune that has its axis perpendicular (transverse) to the prevailing winds or to a current. The upwind or upcurrent side has a gentle slope, and the downwind or downcurrent side lies at the angle of repose (Fig. 13.14)

Travertine terrace A terrace formed from calcium carbonate ($CaCO_3$) deposited by water on a cave floor (Fig. 9.11)

Trellis drainage pattern A pattern of streams in which the tributaries tend to lie in parallel valleys formed in steeply dipping beds in folded belts (Fig. 8.18)

Trench A long, narrow, deep trough in the seafloor oriented parallel to the trend of a continent or an island arc. It marks the line along which a plate bends down into a subduction zone (Fig. 12.1)

Tributary A stream that discharges water into a larger stream (Fig. 8.2)

Troposphere The lowermost zone of the atmosphere, where most of the weather occurs (Fig. 4.2)

Tsunami A seismic sea wave caused by an earthquake, faulting, or a landslide on the sea floor. Its velocity can reach 800 km per hour. Tsunamis are commonly and incorrectly called Tidal waves (Fig. 12.4)

Tuff Volcanic rock made up of rock and mineral fragments in a volcanic ash matrix. Tuffs commonly are composed of shattered volcanic rock glass-chilled magma blown into the air and then deposited. If volcanic particles fall to the ground at a very high temperature, they fuse together, forming a welded tuff (Fig. 16.9)

Turbidite A sedimentary rock deposited by turbidity currents. The unit typically shows graded bedding (a series of graded sediments of sand, silt and mud) (Fig. 12.2)

Turbulent flow A high-velocity flow in which streamlines are neither parallel nor straight but curled into small tight eddies and swirls (a fluid with very irregular flow)

U

Ultramafic rock An igneous rock consisting mainly of mafic minerals and containing <10 percent feldspar. It includes peridotite, amphibolite, dunite, and pyroxenite (Fig. 16.7)

Unconfined aquifer An aquifer that is not overlain by an aquiclude, thereby causing the level of water in a well that penetrates the aquifer to be at the level of the surrounding groundwater table (Figs. 9.8, 9.9)

Unconformity A surface that separates two strata. It represents an interval of time in which deposition did not occur (Fig. 6.5)

Uniformitarianism, principle of The theory that the processes which have shaped earth through geologic time are the same as those observed today (Fig. 6.4)

Universe The largest astronomical structure we know of which contains all matter and radiation and encompasses all space

Unloading The removal of a great weight of rock (Fig. 7.3)

Unsaturated zone The zone below the surface and above the water table in which pores are not completely filled with water (Fig. 9.3)

U-shaped valley A deep valley with steep upper walls that grade into a flat floor (caused by glacial erosion) (Fig. 11.12)

V

Valley (stream) The entire area between the top of the slopes on either side of a stream

Valley glacier A glacier that is smaller than a continental glacier or an Ice cap and that flows mainly along well-defined valleys in mountainous regions. It is confined to a stream valley and is synonymous with alpine and mountain glaciers (Fig. 11.8)

Varve A thin pair of sedimentary layers grading upward from coarse to fine and from light to dark, found in a glacial lake and representing one year's deposition. The coarse-grained layer is formed during spring runoff, and the fine-grained layer is formed during the winter when the surface of the lake is frozen (Fig. 6.17)

Vent The opening at the earth's surface through which volcanic materials come out (Fig. 16.4)

Ventifact A rock that exhibits the effects of sandblasting or snowblasting on its surface that have been abraded, pitted, etched, grooved, or polished by wind-driven sand or ice crystals. These with time become flat with sharp edges in-between (Fig. 13.10)

Vesicle A small hole formed in a volcanic rock by a gas bubble that became trapped as the lava solidified

Vesicular Vesicles are holes, usually spherical, to tubular, to oval in shape

Viscosity A measure of a liquid's resistance to flow. Increased viscosity implies decreased fluidity, the ability to flow (Fig. 16.6)

Volcanic ash Dust-sized particles ejected from a volcano (usually glass, <2 mm in diameter), that is formed when escaping gases force out a ine spray of magma. (Fig. 16.9)

volcanic bomb A hard fragment of lava that was liquid or plastic at the time of ejection and acquired its form and surface markings during flight through the air. Hence, they are often tear-drop shaped with a long tail. Volcanic bombs range from a few mm to more than a m in diameter; often >64 mm (Fig. 16.8)

Volcanic dome A rounded accumulation around a volcanic vent of congealed lava that is too viscous (like the rhyolite lava) to flow away quickly. Diameters of domes can range from a few m to several km. They can as high as a 1,000 m (Fig. 16.21)

Volcanic Explosivity Index (VEI) An index used to measure, on a scale of 1-10, the power of a volcanic eruption (Figs. 16.19, 16.20)

Volcanic neck The solidified magma that originally filled the vent or neck of an ancient volcano and has subsequently been exposed by erosion (Fig. 16.27)

Volcanic tuff A consolidated rock composed of pyroclastic rock fragments and fine volcanic ash welded together by their own heat (Fig. 16.9)

Volcanism The processes that form volcanoes; the progress of magma and gases as they rise up through the earth's interior onto the surface as lavalava, and solidifies into volcanic rocks and landforms (Fig. 16.11)

Volcano A hill or mountain that forms from the accumulation of molten matter that erupts at the surface (Fig. 16.1)

W

Wave base The lower limit of wave transportation and erosion, equal to half the wavelength

Wave crest The highest part of a wave (Fig. 12.3)

Wave height The vertical distance from trough to crest of a wave (Fig. 12.3)

Wave-built terrace A terrace built up from the wave-washed sediments. These terraces usually lie seaward of a wave-cut terrace (Fig. 12.10)

Wave-cut cliff A cliff formed along a coast by the undercutting action of waves and currents (Fig. 12.6)

Wave-cut platform A terrace cut across bedrock by the erosion of waves. It is synonymous with the wave-cut terrace (Fig. 12.7)

Wave-cut terrace A level surface formed by wave erosion of coastal bedrock to the bottom of the turbulent breaker zone. It may also appear above the sea level due to uplift or if the sea level drops (Fig. 12.7)

Wavelength The distance between two successive peaks, or between troughs, of a wave (Fig. 12.3)

Weathering The destruction of rocks at the earth's surface through chemical alteration, biological mediation and by mechanical breakdown of rock materials during exposure to air, moisture, and organic matter. Hence, three mechanisms operate for weathering: chemical, physical and biological

X

Xenolith Fragment of rock distinct from the enclosing igneous rock

Y

Yardangs A streamlined, sharp-crested ridge aligned parallel to the direction of the prevailing wind in arid regions. The silt and dust carried by the wind erode and abrade to give them their typical form (Fig. 13.7)

Z

Zone of ablation That portion of a glacier in which ice is lost (Fig. 11.5)

Zone of accumulation (1) It is the portion of a glacier that has a perennial snow cover; (2) see B horizon (a soil layer) (Fig. 11.5 and 7.2)

Zone of leaching See A horizon (a soil layer) (Fig. 7.2)

Zone of saturation See saturated zone (Fig. 9.3)

Index

A

Aa lava flows, 380, 381, 396
Abandoned channels, 205, 209
Abrasion, 131, 133, 162, 191, 199, 209, 251, 254–256, 261, 285, 286, 292, 308, 429, 461
Absolute ages. *See* Numerical ages
Absolute dating, 95, 98, 99, 103, 113, 124
Abyssal hills, 430
Abyssal plains, 263, 265, 266, 282, 295, 430, 439
Accreted terrane margins, 328, 336
Accretionary wedge, 324, 430
Accurate radiometric dates, 109
Achondrites, 43–45, 47, 52
Acidic solutions, 143
Acidification, 137, 142, 162
Active continental margins, 349
Active volcanoes, 384, 388, 390–393, 415, 418, 424
African Rift Valley, 325, 378
Aftershocks, 160, 338, 430
Aggradation, 181, 204, 209
Alaska earthquake, 155, 347, 355, 359
Albedo, 22, 31, 40, 41, 50, 65
Aleutian Islands, 324, 384, 389, 451
Aleutian Trench Alaska, 387
Algae, 143, 232, 266–268, 460
Alluvial fans, 196, 197, 202, 205, 207, 209, 289, 303, 304, 309, 430, 433, 442
Alluvium, 181, 183, 202, 204, 205, 208, 209, 365
Alluvium cones, 292, 430
Alpha decay, 107
Alpine glacier, 242, 247, 253–256, 261
Alpine-Himalayan orogenic belt, 82
Amphibolite, 332, 430, 431, 473
Amphibolite facies, 431
Ancient rifting, 70, 75, 98
Andes Mountains (South America), 386
Andesite, 89, 378, 379, 414, 415, 431, 442
Andesitic magma, 324, 376–377, 401
Andesitic volcanoes, 324, 399, 401
Andromeda, 14
Angle of repose, 152, 159, 160, 297, 414, 431, 433, 467, 472
Angle of slope, 160, 180, 201, 265, 276
Angular unconformity, 102, 124, 431
Annular, 189, 190, 209, 281, 283
Antarctic ice shelves, 249, 250, 261
Antecedent stream, 177, 431
Anthracite, 431
Aphanitic texture, 431
Aphelion and perihelion, 30, 41
Appalachian Mountains, 316, 317
Aquicludes, 219–221, 225
Aquifers, 10, 212, 219–221, 225, 230, 432
Arabian plate, 357
Archean Eon, 123
Arêtes, 253, 254, 261
Argon, 23, 58, 59, 70, 72, 108
Arid regions, 132, 165, 183, 215, 289, 299, 303, 305, 309, 430, 433, 475
Artesian well, 219, 225, 432
Ash, 266, 296, 308, 374, 383, 399, 401, 403, 405, 412, 414, 415, 417–420, 424, 432, 438, 452, 462, 469, 471, 472, 474
Ash falls, 296, 308, 383, 393, 412, 415, 417, 424
Ash flow, 417, 432
Asteroid belt, 37, 39, 44, 71, 73
Asteroids, 14, 15, 21, 32, 33, 37–40, 42, 45, 48, 50, 69, 71, 73, 269, 271, 282, 419, 432
Asthenosphere, 60, 74, 78, 85, 88–91, 93, 321, 324, 333–336, 358, 432, 451, 453, 460, 464
Atens, 41
Atlantic Ocean, 313, 316, 317, 335, 389, 404
Atmosphere, 16, 20, 21, 23, 24, 26, 29, 31, 33, 34, 38, 42, 46, 52, 57, 59, 61, 63, 65, 69, 73, 74, 109, 111, 145, 150, 212, 219, 222, 232, 269, 287, 296, 297, 307, 308, 354,

S. Jain, *Fundamentals of Physical Geology*, Springer Geology
DOI: 10.1007/978-81-322-1539-4, © Springer India 2014

371, 403, 432, 435, 437, 442, 455, 458, 463, 468, 469, 472
Atmospheric pressure, 22, 23, 59, 302, 415
Atolls, 266, 281
Attenuation, 84
Attrition (Aeolian), 285, 287, 302, 308, 399
Attrition (Stream), 152, 156, 165, 168, 179, 184, 186, 191, 192, 208, 209, 255, 260, 284, 285, 303
Avalanches, 154, 155, 162, 362, 394, 401, 405

B

Back-arc basin, 433
Backshore, 277, 433
Bacteria, 143–144, 162, 432, 460
Badlands, 191, 202, 209, 303, 309, 433
Badland topography, 202, 433
Bajada, 202, 203, 209, 304, 309
Barchan dunes, 297, 298
Barchanoid dune, 297
Barrier islands, 281, 283, 472
Barrier reefs, 280, 434
Bars, 22, 169, 181–183, 202–203, 208, 209, 278, 281, 436, 455
Basal slip, 243, 245, 246, 261
Basalt, 78, 80, 91, 218, 325, 349, 375, 376, 378, 396, 398, 404, 410–412, 415, 420, 431, 434, 440, 445, 447, 457, 460
Basalt lava flow, 381, 383, 424, 458
Basalt plateaus, 396
Basaltic lava, 20, 33, 64, 81, 380, 381, 424, 458, 460, 467
Basaltic magma, 78, 88, 93, 325, 375–376, 410, 411, 420, 459
Base level (Streams), 165, 175, 179, 181
Basins, 20, 33, 90, 129, 165, 175, 176, 184, 185, 190, 191, 208, 214, 259, 263, 282, 303, 313, 321, 335, 430, 433, 443, 444, 456, 457
Batholiths, 434, 454, 461
Bay barriers, 277, 278, 281, 283
Beaches, 151, 277, 283, 294, 352
Bed load, 151, 182, 193, 194, 208, 209, 435
Bedding, 102, 148, 155, 199, 216, 217, 227–229, 431, 435, 440, 447, 448, 450, 458, 461, 462, 465, 466, 469, 473
Bedrock, 101, 132, 133, 153, 189, 191, 198, 199, 205, 209, 227, 232, 238, 243, 251, 255, 256, 286, 289, 292, 293, 301, 303, 304, 309, 366, 368, 435, 436, 444, 458, 461, 463–465, 470, 475
Beds, 8, 100, 102, 103, 105, 112, 114, 156, 190, 203, 211, 219, 225, 228, 294, 440, 443, 447, 469, 472
Beheaded river, 183
Benioff zones, 354, 359, 435
Berm, 277
Beta decay, 107
Biochemical sedimentary rocks, 129, 141, 435
Biogenous sediment, 266, 268
Biological weathering, 143, 144, 162
Biosphere, 57, 58, 60, 63, 64, 73, 74, 212, 435
Bi-parental hypothesis, 65–69, 74
Bird's-foot deltas, 435, 470
Blind valley, 237
Block lava flows, 381
Blowouts, 298, 435, 459
Body waves, 87, 343, 344, 367, 454
Bombs, 282, 394, 398, 399, 405, 414, 420, 424, 471, 474
Bornhardt, 291, 292
Boundary, 59, 73, 74, 80, 84–94, 102, 115, 134, 230, 264, 275, 283, 285, 322–325, 327–328, 332, 335, 338, 345, 358–360, 368, 384, 385, 387, 439, 443, 451, 456, 457, 464, 466, 472
Braided channels, 180–183, 208
Braided stream, 202, 203, 260, 303, 436
Breaker, 271, 274, 281, 283, 434, 436, 471, 475
Breaker zone, 281, 475
Brittle fracture, 229
Butte, 201, 207, 209, 290, 308, 336, 436

C

Calcareous ooze, 267, 268, 282
Calcite ($CaCO_3$), 137, 147, 150, 151, 223, 231, 268, 282, 436, 446, 450, 453, 469
Calcite compensation depth (CCD), 268, 282, 436
Calcium carbonate ($CaCO_3$), 70, 223, 232, 267, 268, 281, 282, 436, 472
Calderas, 23, 392, 416–418, 425, 436
Calving, 252, 261, 364, 429, 436, 450
Cambrian period, 115
Canyons, 173, 429, 470
Capillary zone, 213, 214, 225
Carbon (C), 23, 29, 30, 32, 42–44, 48, 51, 58, 77, 108–110, 124, 137, 431, 436, 438, 463
Carbon dioxide (CO_2), 23, 51, 58, 110, 137, 223
Carbon isotopes, 70, 109, 111, 124

Carbon monoxide, 23, 30, 48
Carbon-14 (C-14) dating technique, 69, 429
Carbonate (CO_3), 70, 142, 194, 216, 223, 224, 228, 231, 232, 267, 268, 281, 378, 435–437, 451, 452, 467, 472
Carbonatite, 378, 379
Carbonic acid, 137, 139, 143, 192, 223, 231, 234
Carboniferous age, 114, 315, 431
Cascade range, 405, 413–415
Caverns, 137, 216, 217, 222, 224, 226, 229, 230, 354, 452, 467
Caves, 137, 216, 217, 222, 224, 226, 227, 229, 236, 238, 275, 283, 381, 424, 442
Cenozoic Era, 455
Centaurs, 41, 42
Channel bars, 202, 203, 209
Channel flow, 179, 195
Channel morphology, 180
Chemical weathering, 129, 130, 136, 137, 139, 142–144, 146–148, 161, 162, 165, 194, 224, 303, 437, 468
Chondrites, 40, 43–45, 47, 52, 71, 73
Chromosphere, 16, 17
Cinder cones, 325, 393, 398, 412, 413
Cinder volcano, 412, 414, 425
Circum-Pacific belt, 415, 437
Cirques, 253, 254, 261, 449
Clay, 43, 112, 124, 129, 133, 134, 137, 139, 141, 144, 145, 149–151, 156, 193–195, 205, 218, 220, 256, 266, 287, 288, 294, 308, 348, 371, 433, 438, 441–443, 450, 453, 456, 459, 466, 472
Clay minerals, 43, 149, 150, 438, 453, 466
Clints, 234, 237, 238
Collapsed sinkholes, 238
Collision boundaries, 360
Colluvial fan, 202, 209
Colluvium, 196, 205
Columbia river basalt, 398
Coma, 38, 48, 50, 51
Comet, 14, 15, 28, 33, 37, 38, 41, 46, 48–52, 65, 74, 269, 420, 438, 468
Composite volcanoes, 373, 376, 413–415, 420, 425, 437
Compound sinkholes, 234, 238
Conduit, 227, 372, 373, 401, 405, 414, 423
Cone of depression, 438
Confined aquifers, 219, 220
Confluent glacier, 249
Conglomerate, 29, 217, 218, 438
Consequent streams, 176, 190, 208
Continental accretion, 439
Continental crust, 77, 80, 82–84, 89, 90, 93, 325, 330–333, 378, 386, 420, 439, 440, 460
Continental deserts, 299, 300, 309
Continental drift, 313, 316, 319–321, 335, 439, 448
Continental fit, 74, 313
Continental glaciers, 242, 247, 249, 250, 255, 257, 259–261
Continental lithosphere, 324–326, 359, 360, 385
Continental margins, 268, 282, 333, 349, 431
Continental plates, 271, 325
Continental rise, 263–266, 430, 439
Continental shelf, 263, 264, 270, 282, 439
Continental slope, 263–265, 439, 470
Continental-continental plate boundaries, 271
Convergent plate boundaries, 83, 322, 324, 358, 359, 368, 433, 470
Core, 17, 20, 21, 23, 24, 27, 33, 44, 57, 60, 64, 69, 74, 77, 79, 80, 84, 88, 89, 91–93, 330, 345, 346, 365, 367, 440, 454, 460, 466, 468, 471
Core-mantle boundary, 60, 93
Coriolis effect, 274, 283
Corona, 16, 17
Correlation, 3, 8, 102, 103, 105, 124
Corrosion, 191–193, 233
Country rock, 404, 434, 440, 451, 467
Crater Lake Oregon, 413
Craters, 24, 44, 46, 47, 302, 352, 393, 403, 410, 412, 416, 417, 420, 425
Cratons, 330, 332, 336
Creep, 152, 157, 159, 162, 187, 288, 294, 308, 440, 468
Crest, 184, 260, 270, 273, 276, 282, 436, 464, 467, 475
Cretaceous, 114, 115, 317, 410, 459
Crevasse, 204, 209, 243, 246, 261, 440
Crust-mantle boundary, 80, 84, 94
Cryosphere, 241
Crystallization, 130–132, 148, 149, 162, 436, 440, 447, 454, 457, 461
Cuestas, 199–201, 207, 209
Current ripple marks, 98
Cut bank, 199, 209
Cutoff meander bends, 181
Cynognathus, 316

D

D layer, 93
Dacite, 377, 378, 415, 423, 441, 449
Darcy's law, 222, 226
Dating techniques, 69

Daughter element, 108, 124
Daughter isotopes, 107–109, 124, 463
Debris, 38, 48, 129, 132, 150, 152, 155–157, 161, 162, 202, 205, 241, 251, 252, 259, 261, 269, 294, 362, 391, 394, 396, 418, 441, 452, 453, 456, 459
Debris avalanche, 152, 155, 162, 362, 394, 441
Debris flows, 129, 157, 441, 452
Debris slides, 156, 162, 452
Decomposition, 129, 143, 161, 429
Deep focus earthquakes, 360
Deep seafloor, 268, 282, 328
Deflation, 285, 308, 429, 441, 442
Deltas, 196, 198, 205, 209, 277, 278, 283, 293, 441
Dendritic drainage, 187, 441
Density, 13, 18–22, 28, 31–33, 60, 65, 73, 74, 77, 80, 81, 85–87, 89, 90, 92, 93, 184, 190–192, 202, 209, 274, 283, 323, 324, 343, 346, 348, 367, 375, 431, 439–441, 457, 460, 462, 468
Denudation earthquakes, 349
Depth of focus, 349, 368, 442
Deranged drainage, 209, 442
Desert, 45, 147, 285, 286, 289, 290, 295, 299–310, 320, 335, 429, 433, 442, 445, 460
Desert pavement, 285, 286, 289, 308, 442
Devonian, 289, 290, 299–310, 429, 433, 445
Differential erosion, 275, 283, 304, 442
Differential weathering, 207
Dikes, 314, 352, 356, 434, 457
Dilution, 268, 422
Diorite, 379, 431, 434, 442
Disappearing streams, 224
Discharge, 165, 166, 169, 172, 175, 178, 179, 181–183, 194, 196, 197, 202, 204, 208, 209, 227, 238, 383, 391, 399, 422, 424, 436, 443, 444, 462, 470–472
Disconformity, 101, 124, 443
Disintegration, 90, 129, 132, 133, 161, 459, 463
Dissolution, 137, 143, 146, 151, 162, 223, 224, 226, 228, 231, 250, 268, 443, 461, 467
Dissolved load, 151, 194, 443
Distributary, 443
Divergent plate boundaries, 358, 368, 457, 465, 471, 472
Divides, 173, 184, 186, 190, 242, 436, 443
Dolines, 227, 232, 233, 238
Dolomite, 150, 151, 218, 227–229, 238, 436, 443
Dolostone, 227, 436
Domes, 133, 134, 190, 377, 401, 415, 416, 418, 431, 434, 474
Dormant volcanoes, 357
Downcutting, 470
Drainage basins, 184, 185, 191, 443
Drainage density, 184, 190–192, 202, 209
Drainage divide, 184, 185, 209
Drainage patterns, 187, 189, 201, 209, 259
Drainage texture, 187, 190, 209
Drapery, 223, 443
Drift, 252, 256, 259, 261, 262, 274, 277, 278, 280, 281, 283, 313, 315–317, 319–321, 325, 335, 419, 439, 443, 445, 448, 454, 460, 468
Dropstones, 252, 261
Drumlins, 256, 452
Dune migration, 294, 296
Dunes, 203, 209, 277, 285, 293, 296–299, 301, 302, 308, 432, 444, 445, 452, 465
Dustbowl, 285, 308
Dust storms, 287, 294, 308

E

Earth creep, 157
Earthflows, 156
Earthquakes, 2, 5, 60, 84, 86, 89, 91, 92, 154–157, 160, 265, 271, 273, 282, 323, 325, 327–329, 337–368
East African Rift, 357, 378
East Pacific Rise, 359, 383
Elastic rebound theory, 337, 339, 366, 444
Elbow of capture, 183
Emperor Seamount, 387
End moraines, 259
Ephemeral streams, 199
Epicenter, 92, 339, 341, 342, 348, 349, 365–368, 442, 445, 454, 462, 466, 471
Epochs, 102, 124, 296, 445
Equilibrium line, 242, 252, 261, 445, 446
Eras, 102, 124
Ergs, 22, 293, 294, 304, 308, 347
Erosion, 6, 9, 96, 97, 101, 113, 124, 129, 130, 133, 134, 161, 166, 168, 173–179, 183, 194, 196, 198, 199, 201, 207, 208, 227, 232, 238, 251, 253, 270, 274, 275, 283, 285, 287, 289, 290, 304, 308, 349, 416, 419, 441, 445, 474, 475
Erosional coasts, 9, 274, 276
Erratics, 256, 257, 262
Escarpment, 199, 200, 207, 209, 237, 445
Eskers, 260, 262
Estuaries, 271, 434
Evapotranspiration, 2
Exfoliation, 133–135, 145, 162, 445, 468
Exfoliation domes, 133, 134

Exfoliation sheets, 133
Extinct volcanoes, 391, 424

F
Feldspars, 77, 89, 93, 133, 148, 150, 151, 376, 446, 447
Felsic igneous rocks, 80, 90, 434, 446, 449
Firn, 242, 261, 446, 457
Firn limit, 446
Fission-track dating, 109
Fissure eruptions, 328, 446
Fjords, 255, 261
Flood basalt, 410, 411, 425, 446, 460
Floodplain, 16, 95, 173, 180–182, 196, 197, 204, 205, 209, 446, 470, 471
Fluvial landforms, 191, 196, 209, 303, 309
Focal depth, 342, 348, 358, 359, 367
Focus, 86, 229, 273, 328, 342, 343, 349, 358, 360, 367, 368, 442, 445, 446, 451
Foreshore, 277, 447
Freezing and thawing, 135, 147, 162, 468
Frost action, 147
Frost wedging, 135, 136, 143, 254, 437, 447, 450
Fumaroles, 421–423, 425
Fungi, 131, 460

G
Gabbro, 80, 376, 379, 434, 442, 447, 457
Gaining stream, 175, 447
Gamma Ray, 106
Gas planets, 18, 21, 24, 33
Gaseous tidal theory, 68, 74
Gelifluction, 153, 161, 162
Geological column, 102, 113
Geysers, 421, 422, 425
Glacial deposits, 318–320, 452
Glacial drift, 252, 256, 261, 445
Glacial erratics, 257
Glacial evidence for continental drift, 313, 318, 335
Glacial marine sediments, 252
Glacial polish, 256
Glacial rock sequences, 315
Glacial striations, 255, 261
Glaciation, 248, 319, 320, 443, 449
Glaciers, 5, 133, 191, 241–261
Glauconite, 109
Glossopteris flora, 315, 335, 448
Gobi desert, 295, 300, 305, 307, 309
Gondwana continents, 315, 316, 318, 319
Gondwanaland, 448
Gorges, 24, 33, 198, 209
Graded bedding, 448, 473
Graded stream, 175, 448
Gradient, 59, 73, 83, 90, 165–169, 171, 172, 175, 178–180, 189, 196, 204, 205, 208, 222, 226, 231, 380, 448
Grain flows, 159, 162
Grand Canyon, 198
Granite, 80, 133, 134, 141, 142, 150, 151, 159, 216, 218, 349, 379, 434, 440, 442, 449, 464
Granitic continental crust, 80, 90, 440
Granitic magma, 459
Granodiorite, 80, 141, 142, 150, 151, 440, 441, 449
Granular flows, 158, 162
Gravel, 169, 191, 193, 202, 205, 215, 252, 260, 279, 289, 294, 301, 351, 382, 430, 433, 434, 437, 442, 449, 458, 461
Great Red Spot, 24, 29, 33, 34
Grikes, 234, 235, 237, 238
Grinding or rubbing, 133, 162
Ground moraine, 259, 449, 456
Ground shaking, 354, 364–366, 368
Groundwater, 10, 62, 74, 172, 175, 179, 208–225, 229, 309, 412, 417, 420, 422, 432, 437, 444, 451, 459, 461, 465
Guide fossils, 103, 105
Gully, 6, 196, 197, 209
Gutenberg discontinuity, 80, 88, 91, 93
Guyots, 263, 266, 282

H
Half-lives, 108, 109
Hanging valleys, 253, 255, 261
Hawaii volcano observatory, 374
Hawaiian Island chain, 387
Hawaiian volcanoes, 372, 381, 396–397, 403, 406, 418
Hawaiian-type eruptions, 380, 381, 387, 396, 398
Headward erosion, 176, 177
Hiatus, 101, 124
Himalayan mountains, 327
Hogbacks, 199–201, 207, 209
Honeycomb, 286, 287
Horizon A, 132
Horizon B, 132
Horizon C, 132
Horn, 254, 261, 297
Hot spots, 265, 287
Hydration, 137, 139, 141, 162
Hydraulic action (Stream), 191
Hydrogenous sediment, 269, 282

Hydrologic cycle, 211, 212
Hydrolysis, 133, 137, 139–141, 146, 151, 162
Hydrosphere, 57, 58, 60–64, 73, 74
Hydrovolcanic eruptions (Surtseyan eruption), 403–404, 424

I
Ice age, 244, 249
Ice cap, 63, 243, 247, 249–250, 252, 256, 261, 302
Ice sheets, 241, 249, 250, 252, 256, 260, 261
Ice shelves, 250, 261
Icebergs, 28, 250, 252, 261
Igneous activity, 324
Ilulissat (Jakobshavn), 241, 243
Incised meanders, 174
Index fossils, 103, 124
Initial (Youthful stage; upper course), 166–168
Inner core, 79–81, 92, 93, 345
Inselbergs, 291–292, 304, 308, 309
Insequent streams, 177, 208
Intensity of earthquakes, 347
Intermediate (Mature stage; middle course), 168–171
Intermediate focus earthquakes, 349, 360, 368
Intermediate igneous rocks, 376, 379
Intermediate magma, 378, 391
Intermediate zone (Groundwater), 214, 225
Intermittent streams, 173, 303
Internal drainage, 303
Interstellar matter, 13, 14, 32, 33
Intraplate earthquake, 358
Intraplate volcanism, 384, 424
Intrusive igneous rocks, 189, 376, 379
Iron meteorites, 1, 43, 44, 52, 70
Isotopes, 7, 70, 71, 90, 103, 107–109, 111, 124

J
Jovian planets, 20
Juan de Fuca ridge, 383
Jupiter, 14, 15, 18, 20–29, 32–34, 37, 41, 50, 371
Jurassic age, 315, 327

K
Kame terraces, 260, 262
Kamenitzas, 238
Kames, 260, 262
Kaolinite, 133, 134, 141, 145, 151
Karren, 234
Karren features, 234, 238
Karst lakes, 237, 238
Karst landforms, 230, 232–237
Karst topography, 137, 192, 224–228, 230, 238
Karst valleys, 237, 238
Karstification, 227, 228, 230–232, 238
Kettle lakes, 259–260, 262
Kettles, 259, 260
Kilauea volcano, 389, 392
Kimberlite pipes, 420
Kirkwood gaps, 41, 42
Kluftkarren, 234, 238
Krumbein phi (φ) scale, 288

L
Lahars, 157, 405
Laminar flow, 179, 193
Landslides, 9, 155, 271, 273, 282, 348, 352, 362, 364
Lapilli, 382, 398, 424
Lassen peak, 413
Lateral erosion, 166, 169, 181, 191
Lateral moraines, 259
Lava, 81, 130, 371–374, 377–383, 391–393, 396, 398, 399, 401, 403–405, 410–412, 414, 415, 417–419, 423, 424
Lava caves, 381–382, 424
Lava domes, 393, 413, 415, 416, 425
Lava flows, 23, 70, 75, 380–381, 383, 396, 411, 414, 415, 424
Lava fountains, 403
Lava plateau, 411
Lava tubes, 381–382, 412, 424
Levees, 173, 196, 197, 204, 209
Linear, 179, 181, 189, 208, 209, 304, 329, 360, 387, 396
Liquefaction, 352, 364
Lithogenous sediments, 266, 282
Lithosphere, 6, 57, 58, 60, 64, 73, 74, 77, 81, 82, 88, 90, 321, 322, 324–328, 330, 334–336, 358–360, 385
Loess, 252, 260, 293–295, 308
Longitudinal dunes, 298
Longshore currents, 276, 277
Longshore drift, 277, 278, 280, 281, 283
Loosing stream, 175
Love waves (L-waves), 84–86, 94, 343, 367
Lower mantle, 79, 80, 85, 90, 91, 93
Lystrosaurus, 315–316, 335

M

Maars, 403, 420, 425
Mafic igneous rocks, 457
Mafic lava, 379, 453
Mafic magma, 148, 379
Mafic plutons, 134, 461
Mafic rocks, 80, 457
Magma, 4, 78, 79, 89, 91, 93, 130, 148, 149, 324, 325, 328, 334, 336, 371–379, 381, 391, 398, 399, 402, 403, 405, 409, 410, 414–417, 420, 422, 423, 425, 434, 436, 447, 448, 451–454, 459, 462, 463, 473, 474
Magma chamber, 372, 373, 391, 401, 402, 416–418, 436, 454
Magnitude, 6, 31, 65, 271, 273, 282, 341, 346, 347–349, 355–360, 364, 366–368, 454, 456, 464
Manganese nodules, 269
Mantle, 21, 23, 33, 44, 45, 60, 64, 69, 70, 74, 77–85, 88–94, 265, 321–325, 328, 335, 345, 376, 379, 384, 387, 422, 440, 442, 450, 453–455, 460
Marine terrace, 275, 283
Mars, 14, 18, 19, 21–25, 32, 33, 37, 41, 43–45, 50–52, 64, 74, 91, 294, 297, 302, 371, 376, 380, 405, 432
Mass Wasting, 3, 5, 129, 152–154, 159–162, 303, 362, 455
Mauna Loa Hawaii, 396, 412
Meander cutoff, 455
Meandering channels, 180, 181, 204, 208, 436
Meandering streams, 181, 183, 204
Meanders, 165, 168, 171–174, 181
Mechanical weathering, 134, 303, 450, 459, 462
Medial moraine, 259, 262, 455, 456
Mercury, 14, 18–23, 32, 33, 44, 45, 52
Mesa, 201, 209, 290, 291, 455
Mesopause, 59, 60, 73, 74
Mesosaurus, 316
Mesosphere, 58–61, 73, 78, 88, 89, 93
Mesozoic Era, 103, 316, 445, 455
Metallic core, 20
Meteor, 38, 42, 44, 46, 48, 49, 52, 269, 420, 455
Meteorite hypothesis, 67
Meteorites, 14, 37, 38, 40, 42–45, 47, 48, 50, 52, 67, 69, 71, 73, 84, 94, 109, 269, 271, 282, 420, 430
Meteoroids, 14, 27, 28, 33, 37, 38, 44, 50
Mid-Atlantic Ridge, 328, 358, 359, 383, 386, 430
Mid-Oceanic ridge, 359, 379, 415, 464, 468
Milky Way Galaxy, 13, 14, 33
Mississippi River, 295, 441
Modified Mercalli intensity scale, 346, 348–350, 368, 456
Moho, 80, 84, 85, 88, 94, 454, 456
Mohorovičić discontinuity, 80, 84, 85, 88, 90, 94, 440, 456
Moment magnitude, 456
Monadnocks, 172
Monsoon desert, 302, 309
Moon, 13, 15, 17, 20, 21, 25, 28–30, 33, 40, 64–65, 74, 271, 282, 294, 431, 455, 456, 472
Moons of Mars, 24
Moons of Uranus, 29
Moraines, 248, 256, 258, 259, 262, 455
Mount Baker, 415
Mount Etna, 392, 398, 413, 415
Mount Hood, 413
Mount Pele, 383, 386, 401
Mount Rainier, 413, 414
Mount Shasta, 155, 413
Mount Saint Helens, 407
Mount Vesuvius, 371–373, 399, 401, 415
Mountain glaciers, 254, 259, 473
M-type asteroids, 40
Mud cracks, 456, 462
Mud flows, 152, 156, 162, 405, 452, 456
Mudslide, 351

N

Nebular hypothesis, 66
Natural levee, 173, 457
Near-Earth asteroids, 41, 42
Nearshore currents, 279
Neptune, 15, 15, 21, 28–30, 32–34
Niagara falls, 177
Nickpoint, 177, 178, 199, 208
Nonconformity, 102, 124, 457
Non-graded stream, 208
Non-volcanic craters, 420–421, 425
Nucleus, 17, 48, 50, 51, 66, 103, 104, 106, 107, 109, 332, 451
Nuées Ardentes, 383, 396, 401, 424
Numerical ages, 98, 124, 429, 457
Nunataks, 256

O

Obsequent streams, 176, 208
Ocean currents, 274, 283, 306, 307, 309, 310, 418, 419
Oceanic crust, 78, 80–83, 89, 93, 264, 265, 323, 325–327, 333, 354, 430, 440, 457
Oceanic lithosphere, 81, 82, 322, 324–326, 359, 385, 456, 457
Oceanic plates, 271, 324, 328, 404, 464, 470
Oceanic ridges, 89, 263, 282, 325, 334–336, 359, 379, 386, 415, 465
Oceanic trenches, 265, 322, 323, 335, 384
Oceanic-continental plate boundaries, 83
Ooze, 266–268, 282, 447, 457, 459, 467
Ophiolites, 81–84, 93
Ordovician, 115
Original horizontality, 99, 113, 458
Orogeny, 82, 332, 458
Orthoclase, 133, 134, 151, 449, 458
Outer core, 79, 88, 91–93, 345, 346, 440
Outwash plain, 260, 262, 285, 294, 308, 458
Outwash terraces, 260, 262
Oxbow lakes, 172
Oxidation, 139, 141, 146, 151, 162, 458

P

Pacific Ocean, 263, 265, 266, 270, 324, 384, 385, 387, 389, 393, 404, 424, 431, 437, 457
Pacific plate, 330, 359, 387
Pahoehoe, 380, 381, 383, 396, 412, 424, 458
Paleontology, 3, 8, 313, 316, 335
Paleozoic Era, 317, 318, 320, 455, 458, 459, 461
Pallasites, 44, 52
Pangaea, 313, 316, 335, 390, 459
Pans (Karst)
Parabolic dune, 298, 435, 459
Parent element, 108, 124
Parent-daughter isotope pairs, 70
Parent-daughter ratio, 70, 108, 109
Pedestal rock (Mushroom rock), 289
Pediments, 199, 206, 209, 303, 309
Peléan, 396, 401, 424
Peneplain, 172, 201, 209
Pennsylvanian age, 315, 318
Perched water table, 221
Perennial, 161, 173, 302, 459, 475
Peridotite, 80, 81, 84, 455, 459, 473
Periods, 50, 52, 82, 102, 109, 124, 147, 155–157, 182, 211, 271, 303, 346, 398, 459, 465
Permafrost, 157, 161, 459, 468
Permeability, 187, 191, 209, 216–222, 225, 226, 229, 459
Permian, 114, 115, 315, 316, 410
Phanerozoic Eon, 123, 459
Phosphatic material, 42, 43
Photosphere, 13, 16, 17
Phreatic eruptions (Steam-blast eruptions), 405, 424
Phreatomagmatic eruptions, 403, 405, 424
Physical geology, 1–4, 10, 460
Physical weathering, 129, 131, 136, 139, 143, 144, 146, 147, 161, 162, 303, 309, 445, 459, 465
Piedmont glacier, 249
Pillow lava, 383, 424, 457, 460
Pinnacles, 201, 291
Pits (Karst), 216, 234, 237, 238, 418
Plagioclase feldspars, 148
Planetesimal hypothesis, 67, 74
Planetesimals, 68, 69
Plastic flow, 243, 245, 246, 261, 460
Plate boundaries, 83, 271, 321, 322, 324, 328, 335, 337, 338, 340, 349, 358–360, 364, 366, 368, 433, 457, 460, 465, 470–472
Plate divergence, 386
Plate tectonics, 69, 75, 313, 321, 330, 333–336, 454, 460
Plateau, 60, 173, 174, 199–201, 207, 224, 236, 237, 289, 291, 294, 295, 300, 303, 309, 325, 396, 404, 410, 411, 425, 446, 455, 460
Plateau basalts, 410, 425
Playa lakes, 303, 304
Pleistocene Epoch, 445, 461
Plinian volcanic eruptions, 399
Pliny the Younger, 371, 373, 401
Plucking, 251, 254, 261, 437, 461
Pluto, 14–16, 18, 21, 22, 30–32, 34, 50, 130, 133, 134, 434, 442, 449–451, 454, 461
Point bars, 181, 202, 203, 209, 455
Polar deserts, 299, 301, 305, 309
Polar glaciers, 245, 247, 261
Poljes, 230, 237
Pompeii Italy, 371, 373, 399, 401
Ponors, 233, 238
Porosity, 215–218, 220, 221, 225, 228, 231, 294, 438, 461
Potholes, 191, 199, 209
Precambrian Era, 332
Pressure ridges, 382, 424
Pressure-release, 131, 133, 134, 162, 462
Primary waves, 343, 367
Principle of cross-cutting relationships, 101
Principle of fossil succession, 100

Principle of inclusions, 100
Principle of lateral continuity, 100, 113
Principle of original horizontality, 99, 133
Principle of relative dating, 98, 113
Principle of superposition, 100, 101
Principle of uniformitarianism, 98
Prominences, 16
Proterozoic Eon, 462
Protoplanetary hypothesis, 67, 68, 74
Proxima Centauri, 14
Pumice deposits, 412, 416, 418, 462
Pyroclastic cones, 412, 425
Pyroclastic flow, 383, 401, 415, 417, 418, 462
Pyroclastic layers, 414
Pyroclastic materials, 412
Pyroclastic rock, 438, 474
Pyroclastic sheet deposits, 416, 417
Pyroxene, 43, 45, 78, 80, 90, 137, 139, 146, 150, 151, 376, 377, 431, 434, 438, 442, 447, 454, 462

Q

Quartz (SiO_2), 137
Quaternary, 114, 115, 156, 445, 461, 462

R

Radial drainage, 463
Radioactive decay, 90, 103, 106–108, 463
Radioactive isotopes, 108, 109, 124
Radioactive parent and stable daughter atoms, 108, 124
Radiometric dating, 44, 52, 69, 103, 108–109, 111, 429, 463
Rainshadow deserts, 301, 309
Rayleigh wave, 343, 367, 463
Recessional moraine, 259, 262
Rectangular drainage, 189
Reef, 266, 280, 281, 283, 320, 335, 432, 434, 447, 452, 463
Reflection (Seismic waves), 81, 85–87, 333, 466
Refraction, 85- 87, 92, 273, 274, 279, 364, 463, 466
Regolith, 64, 132, 152, 153, 155, 157–160, 162, 232, 435, 440, 463
Rejuvenated rivers, 172
Relative dating, 98, 113, 464
Relative geologic time scale, 113, 124, 445, 459, 463
Resequent streams, 176, 208
Resurgent calderas, 417, 418, 425
Rhyolytic magma, 376, 378
Richter magnitude scale, 341, 347, 356, 367, 464
Rift valley, 325, 329, 358, 359, 378, 464
Rift zones, 263, 282
Rifting, 325, 329, 333, 335, 433
Rills, 187, 197
Rings of Jupiter, 26, 27
Rings of Neptune, 30
Rings of Saturn, 27
Rings of Uranus, 28, 29
Rinnenkarren, 234, 238
Rip current, 464
Ripples, 98, 203, 209, 287, 294, 297, 308, 348
River terraces, 201, 209, 259
Roche moutonnée, 254, 255, 261, 464
Rock avalanche, 152, 154–155, 162, 464
Rock cycle, 130, 465
Rock slide, 153
Rockfalls, 153
Rotational slide (Slump), 152, 156, 157
Runnels, 234, 237, 238

S

Sahara Desert, 45
Salt crystal growth, 131, 132, 162
Saltation, 193, 194, 287, 288, 308, 465
San Andreas fault, 327, 328, 330, 337, 360
Sand seas, 293, 294, 302, 308
Sand sheets, 294, 308
Satellites, 14, 15, 21, 28, 29, 33, 34, 37, 50, 66
Saturn, 14, 15, 18–23, 25, 27–29, 32, 33, 37, 38, 294
Sea arches, 275, 283, 358, 465
Sea caves, 275, 283
Sea stacks, 275, 276, 283, 465
Seafloor spreading, 321, 335, 433, 465
Seamounts, 263, 265, 266, 282, 387, 404, 456, 466
Secondary waves, 343, 344, 367
Sediment flows, 157, 158, 162
Seif dunes, 298
Seismic discontinuity, 456, 466
Seismology, 3, 84, 85, 94, 341, 367
Semiarid regions, 305, 309, 430
Shadow zone, 92, 345, 346, 367, 462, 466, 471
Shallow-focus earthquakes, 325, 443
Sheet flow, 380, 416, 417
Sheet lava flows, 380
Sheetwash, 199
Shield volcanoes, 375, 396, 411, 412, 414, 418, 466
Shields, 331–333, 411

Shoreline, 24, 33, 273, 274, 276, 277, 279, 281–283, 434, 435, 439, 450, 454, 467
Siliceous oozes, 268, 282
Sills, 377, 467
Silurian, 115
Sinkholes, 137, 143, 216, 217, 223, 224, 226, 227, 229, 230, 232–234, 238, 452
Slaking, 134
Slip face, 433, 435, 453, 467
Slope angle, 160, 265
Slump, 129, 152, 153, 156, 157, 162, 362, 418, 452, 467
Slurry flows, 158
Snowline, 242, 244, 245, 261, 468
Soil water (moisture) zone, 213, 225
Solar nebula theory, 69, 457, 468
Solar system, 7, 13–16, 20, 21, 24, 29–31, 33, 34, 37, 38, 40–42, 44, 45, 48, 50, 52, 57, 64–71, 74, 75, 84, 444, 457, 468
Solfataras, 421–423, 425
Solifluction, 153, 157, 159, 161, 162, 468
Solution, 137, 139, 141, 143, 144, 151, 162, 180, 191–193, 208, 215–218, 225, 227, 229, 231–234, 237, 238, 375, 376, 436, 440, 443, 445, 452, 461, 462, 468, 469
Solution sinkholes, 232, 233, 238
Spalling, 133
Spectra, 40, 50
Speleothems, 223
Spheroidal weathering, 145, 146, 162, 468
Spit, 277–281, 283, 468
Spreading center, 357, 386, 468
Stability of silicate minerals, 149
Stalactites, 101, 222, 223
Stalagmites, 101, 222, 223
Star dunes, 299, 308
Stony meteorite, 43–45, 52, 84, 469
Stony-Iron meteorite, 44, 52, 469
Straight channel, 179–181, 208
Stratified drift, 252, 256, 259, 261, 262
Stratified drift deposit, 256, 259, 262
Stratopause, 59, 73
Stratosphere, 58, 59, 61, 73, 109, 399, 458, 469
Stratovolcanoes. *See* Composite volcanoes
Stream capacity, 194, 209, 469
Stream capture, 183, 184, 208, 469
Stream channels, 6, 178, 180, 181, 197, 203, 204, 208, 259
Stream discharge, 179, 202, 470
Stream erosion, 191, 433
Stream gradient, 165, 167, 179, 208, 470
Stream order, 186, 209, 470
Stream piracy, 183, 208, 469, 470
Stream terraces, 470
Stream transport, 152, 162, 193
Stream velocity, 178, 179, 193- 196, 202, 203, 209, 433, 461
Streamlines, 473
Streams and rivers, 63, 329
Striations, 243, 255, 256, 261, 319
Stromboli, 390, 393, 397, 406
Strombolian volcanic eruptions, 393, 396
S-type asteroids, 40
Subduction, 265, 271, 273, 322–324, 326, 333, 359, 385, 393, 404, 439, 470
Subduction zones, 265, 322, 323, 335, 337, 358–360, 366, 368, 431, 433
Subglacial eruption, 405, 424
Submarine canyons, 429, 470
Submarine eruption, 391, 404, 418, 424
Submarine fans, 470
Submarine ridges, 266
Submarine volcanism, 418
Subsequent streams, 176, 208
Subsidence sinkholes, 233, 238
Subtropical desert, 299, 300, 309
Sun, 13–23, 30, 31, 33, 34, 37–41, 45, 48, 50–52, 57, 65–68, 74, 271, 273, 282, 432, 438, 444, 456, 458, 468, 472
Supercontinent, 316, 459
Superposition, 100–102, 111, 113, 124, 464, 471
Supraglacial sediment, 252
Surf zone, 274, 283, 454, 471
Surface waves (L-waves), 86, 343, 367
Suspended load, 151, 180, 193, 194, 208, 471
Suspension, 179, 193, 194, 196, 197, 209, 287, 293, 294, 308, 441
Swallow hole, 233, 238
S-Wave, 84- 86, 88, 91–94, 344–346, 367, 466, 467, 471

T

Tail, 38, 48, 50, 51, 438, 468, 474
Talus, 153, 292, 415, 436, 471
Tambora eruption, 406
Tarn, 253, 255, 261
Tellus, 57, 73
Temperate glaciers, 246, 247
Tephra, 296, 382, 383, 398, 399, 403, 407, 412, 414, 424, 425, 471
Terminal moraine, 444, 471
Terminal (Old stage; lower course), 171
Terra, 57, 73
Terranes, 329, 339
Terrestrial planets, 18, 33, 44, 45, 380

Terrigenous sediments, 266, 268, 430
Tertiary, 114, 115, 396, 410
Thawing, 135, 147, 162, 468
Thermal contraction, 405
Thermal expansion, 135, 162, 325, 405
Thermohaline circulation, 274, 283
Thermopause, 60, 74
Thermosphere, 58–61, 73, 74
Tidal flats, 273, 282
Tidal wave, 472
Tide-dominated deltas, 472
Tides, 65, 271, 273, 278, 282, 433, 456, 469, 472
Till, 218, 252, 256, 258, 259, 262, 443, 444, 449, 453, 455, 456, 472
Tillite, 256, 472
Tombolos, 277, 278, 281, 283
Topographic relief, 228, 231, 238
Topsoil, 349
Traction, 193, 194
Transform fault boundary, 358, 360, 368
Transform faults, 83, 263, 282, 330, 337, 358, 366, 368, 447
Transition zone, 91, 93
Transverse dunes, 297, 298
Travertine terraces, 472
Tree-ring dating, 111, 112
Trellis drainage, 187, 472
Trench, 197, 263, 265, 282, 321, 323, 324, 335, 354, 359, 384, 387, 430, 433, 435, 439, 447, 457, 470, 472
Triassic, 114, 116
Tributary, 176, 179, 186, 209, 255, 441, 443, 449, 470, 472
Trittkarren, 234, 238
Trojans, 42
Tropopause, 59, 73
Troposphere, 58, 59, 61, 73, 109, 472
Trough, 270, 273, 433, 437, 447, 457, 464, 472, 475
Tsunami, 46, 265, 271, 273, 282, 343, 364–366, 368, 472
Tuff, 403, 412, 415, 420, 425, 432, 472- 474
Turbulent flow, 179, 193, 473

U

Udden-Wentworth scale, 288
Ultramafic crust, 90
Ultramafic rocks, 452, 455, 457, 473
Unconfined aquifers, 219
Unconformities, 101, 124
Undercutting, 160, 162, 177, 199, 275, 283, 452, 475
Uniformitarianism, 98, 473
Uniparental hypothesis, 65, 66, 74
Unloading, 131, 133, 134, 162, 473
Unstratified drift, 252
Upper mantle, 60, 74, 78, 80–83, 85, 88, 90, 91, 93, 321, 335, 460
Uranium, 69, 70, 71, 90, 103, 107–109, 124, 429
Uranium-lead pair, 70
Uranus, 14, 15, 18, 21, 22, 28–30, 32–34
U-shaped valleys, 253, 254, 256
Uvaluas, 234, 238

V

Vadose zone, 211, 214, 225
Valley glaciers, 247–249, 252, 259, 455
Varve, 112, 113, 124, 473
VEI. *See* Volcanic explosivity index
Velocity, 22, 31, 50, 52, 65, 80, 84, 85–87, 90, 93, 98, 152, 155, 156, 166, 169, 171, 172, 175, 178, 179, 181, 183, 191, 194, 203–205, 208, 209, 251, 270, 273, 282, 285, 286, 296, 304, 308, 343–345, 367, 387, 433, 440, 448, 454, 456, 461, 463, 466, 471
Vent, 372, 373, 380, 382, 396, 398, 399, 403, 411, 412, 414- 416, 418, 420, 423, 436, 437, 440, 442, 446, 453, 474
Ventifacts, 292, 293, 308
Venus, 14, 18, 21, 23, 32, 33, 44, 52, 294, 376, 380
Vesuvian, 396, 399, 401, 424
Viscous magmas, 399, 415
Volatiles, 40, 48
Volcanic, 23, 78, 81, 89, 93, 100, 130, 149, 150, 155, 156, 160, 162, 189, 190, 216, 265, 266, 271, 280, 281, 283, 296, 323, 325, 328, 329, 354, 368, 372, 373, 377, 382–384, 386, 389, 393, 396, 399, 402–411, 415, 417, 419, 420, 423–425, 432, 436, 437, 440, 442, 452, 459, 462, 472, 432
Volcanic ash, 266, 296, 308, 414, 415, 432, 452, 462, 472, 474
Volcanic block, 405
Volcanic bomb, 394, 398, 405, 474
Volcanic breccias, 436, 442
Volcanic deposit, 189
Volcanic domes. *See* Lava domes
Volcanic explosivity index (VEI), 401, 406–408, 474
Volcanic gases, 375, 422
Volcanic igneous rocks, 130

Volcanic island arc, 323, 384
Volcanic islands, 281, 387, 451
Volcanic landforms, 409, 411, 413, 415, 417
Volcanic mudflows, 415
Volcanic necks, 419, 425
Volcanic pipes, 420
Volcanic plugs, 419
Volcanic rocks, 23, 81, 216, 383, 445
Volcanism and volcanoes, 1, 3, 4, 24, 70, 75, 78, 83, 94, 190, 211, 265, 281, 283, 321, 324, 325, 328, 357, 358, 364, 371, 373, 374, 378, 380, 381, 384–394, 396, 399–401, 404, 405, 408, 409, 411–416, 418, 420, 424, 434, 438, 439, 443, 455, 460, 469, 474. *See also* Lava flows
V-shaped valleys, 198, 209
Vulcan God of fire, 392
Vulcanian, 394, 396, 398, 399, 403, 406, 424

W
Water table, 172, 175, 189, 208, 214, 217, 219, 221, 222, 226, 229, 305, 309, 438, 443, 451, 473
Waterfalls, 165, 166, 173, 198, 199, 208, 209, 255
Wave action, 434
Wave base, 270, 273, 282, 475
Wave height, 269–271, 365, 368, 475
Wave period, 270, 346
Wave refraction, 466
Wave-built platform, 475
Wave-cut platforms, 475
Wave-dominated deltas, 198
Wavelength, 270, 271, 275, 475
Waves. *See also* Seismic waves, 81, 84–94, 269, 282, 334–346, 348, 351, 354, 360, 362, 365–368, 371, 373–379, 433, 444, 454, 457, 462–464, 466, 467, 471, 472, 475
Weathering, 3, 5, 129–137, 139–152, 155, 159, 161, 162, 165, 194, 198, 207, 224, 303, 309, 349, 437, 438, 442, 445, 450, 459, 462, 465, 468, 475
Weathering rinds, 145, 146, 162, 442
Wind abrasion, 286
Wind deposits, 296, 453
Wind erosion, 285, 289, 290, 294, 308, 429, 435
Wind gap, 183

X
Xenoliths, 81, 84, 93

Y
Yardangs, 291, 475
Yellow river (China), 186, 294
Yellowstone caldera (Wyoming), 408, 418

Z
Zeugen, 289–291, 308
Zircons ($ZrSiO_4$), 70, 75
Zone of ablation, 242, 261, 475
Zone of accumulation, 242, 261, 433, 435, 475
Zone of aeration, 211, 213, 214, 217, 219, 225, 229
Zone of leaching, 132, 475
Zone of saturation, 211, 214, 221, 225, 447, 454, 475